# CELLULAR IMMUNOLOGY

# CELLULAR IMMUNOLOGY

## SIR MACFARLANE BURNET
O.M., F.R.S., Nobel Laureate

BOOKS ONE & TWO

MELBOURNE UNIVERSITY PRESS
CAMBRIDGE UNIVERSITY PRESS
1971

PUBLISHED BY

MELBOURNE UNIVERSITY PRESS

Carlton, Victoria 3052, Australia

AND

THE SYNDICS OF THE CAMBRIDGE UNIVERSITY PRESS

Bentley House, 200 Euston Road, London NW1 2DB

American Branch: 32 East 57th Street, New York, N.Y. 10022

Library of Congress Catalogue Card Number: 69–12162

Dewey Decimal Classification Number: 576.23

International Standard Book Numbers:

MUP 0 522 83895 2

CUP 0 521 07217 4

First published 1969

Reprinted 1970, 1971

Printed in Great Britain

at the University Printing House, Cambridge

(Brooke Crutchley, University Printer)

# CONTENTS

*Contents*

# PREFACE

This book has an unusual form and an unusual purpose. For the last ten years I have been interested, perhaps obsessed, by the potentialities latent in what I called the clonal selection theory of immunity. It was a point of view that seemed to be in accord with the way biology was developing, and even in a rather crude early form provided a better frame for the facts of immunity than any variant of the dominant instructive theory. It was not enthusiastically received and has been 'disproved' on several occasions. But it is the only theoretical approach to immunology which has been able to adapt itself successfully to the accelerating flood of new techniques and new facts coming from experimental immunology. At the 1967 symposium, 'Antibodies', at Cold Spring Harbor the central postulate of the clonal selection theory—that the immune pattern of antibody is determined by genetic processes in the somatic cell line which produces it—was implicitly and explicitly accepted by the great majority.

There were in fact signs of a rapidly increasing interest in these ideas at the Prague Conference on Antibody in 1964 and about that time I decided to attempt a comprehensive discussion of the whole immunological field from a consistently 'selective' angle. In an attempt to make such a treatment reasonably short and readable but at the same time to provide as much factual justification as was needed to establish points regarded as vital, the present form was adopted.

Book I is an attempt to present the current picture of immunology in terms essentially of a Darwinian process of inheritable change and selection by proliferation and death amongst the mobile cells of the mammalian body. It is written as an essay for readers interested, as biologists of one sort or another, in immunology but not necessarily concerned directly with immunological research. Only occasional references are made to the workers on whose findings the discussion is based and little or no mention is made of work in fields which seem to have no special relevance to the central theme.

The primary object is to present a modernized version of clonal selection theory which could form a satisfactory background for reading or investigation in immunology.

Experience in lecturing and discussion over the past few years has made me very much aware of the difficulty many students and even some investigators find in grasping the nature of the selective approach. I have therefore deliberately presented the general concept from different points of view and at different levels of elaboration. This may make Book I appear unduly repetitious to some. However, I have tried to apply the central theme to all those aspects of immunology which are currently under investigation, and I hope that the treatment will also interest professional immunologists.

It is inevitable that a number of interpretations have emerged which involve significant modification of current opinion. Book II is designed to offer a more technical and documented justification for such interpretations as well as to provide a fairly critical summary of the overall selective approach. In the course of working on the book I found that there were a surprising number of areas which had not been seriously discussed from the selective standpoint. For obvious reasons the whole body of immunology must have a concordant theoretical basis and I have found a good deal of pleasure in trying to bring some of the more peripheral fields into relation with the central theme. It may be of interest to some readers to indicate the main topics where the treatment adopted has been significantly changed by a thoroughgoing application of clonal principles. In the following list the chapters (which are similarly numbered in both Books) in which the topics are discussed are also shown.

A. The thymus as the principal centre for the differentiation of stem cells to immunocytes and for the elimination of immunocytes reactive with self-components (3).

B. Cellular phenomena associated with local depletion of the thymic cortex (3).

C. The significance of thymic pallor (3).

D. The existence of a hormone derived from gut-associated lymphoid tissue concerned with plasma cell development (4).

E. High local oxygen tension as a prerequisite for plasma cell development and accumulation (4).

F. A duplex structure for the combining site on antibody (5).

G. The special importance for antibody production of the dendritic phagocytic cells of lymph follicles (4, 7).

H. The drug-like character of the action of antigenic determinant on receptors of immunocytes (8).

I. The liberation, after antigen–immunocyte reaction, of agents activating adjacent cells to immunological activity: its bearing on nonspecific immunoglobulin production (9).

J. Myelomatosis as failure of maturation of an antibody-producing plasmablast (9).

K. Tolerance as resulting from destruction of *all* immunocytes carrying receptors with more than minor avidity for the corresponding antigenic determinant (10).

L. The special conditions necessary for the development of delayed hypersensitivity: antigen in mobile cell surface (11).

M. The active role of the thymus in the pathogenesis of systemic lupus erythematosus (12).

N. α-Methyldopa action as a phenocopy of somatic mutation (12).

O. The origin of histocompatibility antigens in relation to adaptive immunity (13).

In summary, the primary objective of the book is to provide an outline and justification of the clonal selection theory of immunity and to suggest a number of areas where a whole-hearted application of the theory offers new scope for experiment and interpretation.

The work was commenced some years ago, but the book has been written essentially as a post-retirement activity since I left the Hall Institute in 1965. I am indebted to the University of Melbourne for a Rowden White Fellowship over the period, to Professor Rubbo for providing office space in the School of Microbiology, to the Wellcome Trustees who have made it possible for me to employ a secretary, and to a grant from Merck Sharp and Dohme Research Laboratories, New Jersey, to cover stationery and other office expenses. I owe a special debt of gratitude to Mrs L. Nillson who has carried out many retypings and assisted in all other aspects of its production with intelligence, accuracy and cheerfulness.

# BOOK ONE

# 1 The history of immunological ideas

The history of immunological ideas must have begun when people began to realize that a man or a child with a pock-marked face did not take smallpox again. No doubt it was also old wives' knowledge that other recognizable childhood infections struck once only, but the virulence of smallpox in the eighteenth century in Europe, when it was one of the main causes of death in childhood, gave it pre-eminence. The history of variolization as a deliberate attempt to provoke immunity and of its replacement at the end of the century by Jenner's 'vaccination' with cowpox is known to all. For all its practical importance there was no possible basis at that time for any significant theoretical ideas about the way in which protection was achieved.

The first approach to a general method of immunization against an infectious disease came with Pasteur's work (1880) on the protection of fowls against chicken cholera by inoculation with 'attenuated virus'; in this case, old cultures of the micro-organism called *Pasteurella aviseptica*. To explain his results, Pasteur suggested that the immunizing infection 'exhausted' something necessary for the proliferation *in vivo* of the virulent culture.

With the recognition around this time of the universality of bacteria and the existence of specifically pathogenic types, a new problem began to be recognized—the normal resistance of animals against most bacteria. Apparently related to this was the capacity of blood held outside the body to resist putrefaction much longer than most organic materials. Freshly drawn blood was able to kill at least some types of bacteria and gradually the idea of special agents in the blood adapted to defend the body against bacterial invasion gave birth to the concept of specific antibodies. Concurrently, however, there developed an alternative way of looking at the defence against bacteria as a function of the white cells of the blood. The classical controversy between cellular and humoral theories of natural immunity flourished from about 1884, when Metchnikoff described his researches on phagocytic cells in the crustacean *Daphnia*, until 1903, when Almroth Wright's ideas on opsonins began gradually to lead to a recognition of the complexity of the processes concerned and the importance of more than one type of cell and many humoral factors.

The future course of immunological theory, as something concerned with a much wider field than immunity against infection, was laid down in 1898 by Bordet's recognition of immune lysis of foreign red cells followed in 1904 by Landsteiner's discovery of the ABO blood groups. From this time onward there was a steady increase in interest in the immunological behaviour of cells and body fluids and soon a realization of the fact that only *foreign* material was antigenic added an important new problem for understanding.

With von Behring's discovery of potential therapeutic agents against diphtheria and tetanus in 1895 and Ehrlich's subsequent studies on the nature of antitoxin, it was inevitable that the main stream of immunological thought for the first forty years of this century should be concerned with the problem of antibody. How was it possible for the body to produce something which would neutralize specifically any one of a large group of poisonous substances against which it had been immunized? Ehrlich's side-chain theory was the first serious attempt to explain the origin of antibody. It was based on a primitive picture of a living protoplasmic molecule which carried a variety of side-chains by which food molecules could be taken in and which could equally serve as receptors for the attack of damaging substances like bacterial toxins. If the molecule was not 'lethally' damaged it responded by an overproduction of the receptors involved with their liberation into the blood as antitoxin. The theory was obviously designed to deal with toxins and antitoxins and depended, of course, on there being available in the body pre-formed molecular groupings which could unite specifically with what we should now call antigenic determinants.

When Landsteiner developed methods of studying artificial antigens made by the chemical union of small molecules—haptens— to carrier proteins, it soon became evident that there were far too many possible types of antibody to allow each to be accepted as representing a pre-formed receptor on molecules or cells liable

4

to attack by the antigen. By 1930 it was clear that antibodies were associated with serum globulins and, independently, Breinl and Haurowitz, Alexander and Mudd all suggested that antibody might represent globulin which had been synthesized in contact with antigen and, in so doing, had taken on a complementary steric configuration which would ensure that, on renewed contact with the antigen, a firm union could occur. This was the first form of what Lederberg subsequently called the 'instructive' theory of antibody formation.

The whole initial concept of immunity was in relation to infectious disease in man or his domestic animals. Pasteur was trained as a chemist, but once he had shown the potentiality of immunization with attenuated pathogens the central objectives of immunological research were defined for the next sixty or seventy years. Immunology was one of the practically important aspects of medical bacteriology and almost all those concerned in its advance were medically trained. There were, of course, other interests in immune processes that went much wider than their applicability to the cure or prevention of disease. The specificity of immunity called for laboratory study of its basis and led to serological methods both for the classification of pathogenic micro-organisms and for retrospective diagnosis of the infecting organism. The triumphs of immunology were practical ones: the production of antitoxins against tetanus and diphtheria and, later, toxoids for active immunization; a variety of bacterial vaccines, most of which have suffered a progressive diminution in reputation with the years, and eventually immunization against poliomyelitis and other virus diseases.

### THE BIOCHEMICAL APPROACH

The theoretical approach over most of the period was at a rather superficial level and it is probably correct to say that not until Pauling became interested in immunology around 1940 was there any effective association of immunology with the developing principles of biochemistry. The concepts of Ehrlich, for instance, were *ad hoc* constructions with only a minimal relevance to the chemical form in which they were cast. Landsteiner's work in the 1930s was of immense importance in establishing a chemical basis for the

5

specificity of immune pattern and almost automatically suggested that antibodies were produced in the body by some impression of complementary pattern on normal serum globulin during the process of its synthesis.

The development of quantitative methods by Heidelberger about this time largely substituted the techniques of the biochemist for the medical immunologists' methods of titration to a limit dilution. When mixtures of a soluble purified antigen with a corresponding antiserum reacted to produce a precipitate, Heidelberger was concerned to know the amounts of the two reagents in the precipitate in terms of milligrams or micrograms. This provided immunological information in what a chemist could regard as a meaningful fashion. Antibodies were clearly proteins and with the rapid development of an understanding of the polypeptide structure of proteins and, in particular, of the potential ways in which the basic polypeptide chain could be folded to give secondary and tertiary structure, Pauling was able to give a relatively precise formulation to what has since become known as the classical instructive theory of antibody formation. All theories must be produced within the limitations of contemporary knowledge. In 1940 there was hardly a hint of the part played by nucleic acid in coding the amino acid sequence of polypeptide chains and very little conception of the complexity of $\gamma$-globulin and other proteins. The essence of Pauling's concept was that, as synthesized, a polypeptide chain had no inbuilt compulsion to adopt any specific type of folding and intramolecular bonding. A vast variety of arrangements were thermodynamically all equally admissible, normal $\gamma$-globulin representing a random mixture of the various possible three-dimensional configurations. When, however, the newly synthesized polypeptide chain was brought into contact with the steric patterns of parts of the antigenic molecule or particle, it would develop a stable configuration in a pattern complementary to that of the antigen. Secondary hydrogen bonding between the coils in their new position then stabilized the structure. When the antigen was in one way or another separated and made available to mould another polypeptide chain, a cavity appropriate to fit firmly with the antigenic determinant on a subsequent occasion was left imprinted on the antibody molecule.

6

Much of this picture of the nature of antibody has been retained. Most writers conceive the two combining sites of a single antibody molecule to be cavities formed by amino acid residues so distributed as to provide a complementary steric pattern to the corresponding antibody. It is more than possible that this is too naïve a concept but from the point of view of providing an acceptable picture of how antigen and antibody react it has been valuable. Where the instructive approach has proved insufficient is its failure (*a*) to conform to the new understanding of the nature of protein synthesis, (*b*) to provide an interpretation of immune tolerance and (*c*) to account for the changing character of a given antibody during the course of immunization. Another aspect of antibody which has become very prominent recently is its heterogeneity. With every refinement for the physical separation of related protein molecules, it has become more and more evident that each antiserum is made up of a heterogeneous population of molecules with the one common feature of a recognizably specific capacity for union with the antigen in question.

## THE APPEARANCE OF SELECTION THEORIES

In the period between 1940 and 1955 there was rapid technical advance in immunology as in other biological sciences. The Second World War had a major influence in two directions, due to the immense practical importance of blood transfusion and plastic surgery in dealing with war casualties. The ABO blood groups had been discovered many years before but now there was a much greater volume of laboratory work going on in this field. As a result, the Rh groups and their clinical significance were discovered in 1940. The various immunological disorders associated with pregnancy and transfusion now became matters of major interest and, in particular, underlined what Medawar called the 'uniqueness of the individual'. In the field of transplantation it was discovered and rediscovered that while plastic surgery could deal almost without limitation with the patient's own tissues, most attempts to use any other person's skin or other tissue to replace deficiencies were wholly unsuccessful. The time was ripe to emphasize the importance of 'self' and 'not-self' for immunology and to look for the

7

ways in which recognition of the difference could be mediated. In due course, the concept of tolerance and the experimental demonstration of acquired immunological tolerance became known to all immunologists and immediately raised grave difficulties for 'instructive' theories.

New experimental work on antibodies themselves greatly complicated the picture. In human beings it became clear that there were three main types of antibody-carrying immunoglobulins (Ig G, Ig M, Ig A), each with their own physical characteristics and with at least one distinguishing antigenic quality. In all mammals examined there are at least three such types and wherever intensive search has been made, as in mice or human beings, a complex of antigenic groups and subgroups has been found. The elaboration of this field has depended first on the application of simple electrophoresis (Tiselius), then on the refinement of antigen–antibody precipitin reactions that becomes possible when the reactions take place in agar (Ouchterlony, 1948) and finally on the combination of both techniques in immunoelectrophoresis (Williams and Grabar, 1955).

This work has no direct bearing on the nature of the combining site—which in fact has only been directly studied in Ig G antibodies—but there is a good deal to suggest that the combining site may be the same for any type of immunoglobulin. Clearly the situation was far more complex than Pauling had envisaged in 1940, and over the last decade the main theoretical discussion on immunity has been concerned with the possibility of replacing instructive theories of the Pauling type by selective theories of antibody formation. The two types of approach are similar in principle to what has occurred in another contemporary field, the acquisition by a bacterial culture of 'adaptive' capacity to resist an antibiotic or to ferment a sugar against which it is normally inactive. In the one field we ask whether the antibiotic impresses some change on a small proportion of bacteria so that they produce descendants genetically resistant (instructive approach) or whether in any large population there will be so many mutants of all sorts that some will be resistant to the antibiotic and in its presence will be able to proliferate selectively. In the field of antibody production, selective theories endow a unique quality on the genetic mechanism of anti-

8

body-producing cells and the forms ancestral to them. Within that population by some process, which may be accelerated somatic mutation or some operationally similar random re-assortment of pre-existent patterns, cells develop each of which can produce one of a vast number of different patterns of antibody. It is a further necessary postulate of selective (or genetic) theories that each cell, each *immunocyte*, must also possess receptors (or 'fixed antibody') which on contact with antigen of one particular type will cause the cell to proliferate selectively.

The first suggestion that a selective theory of antibody production was possible came from Jerne in 1955. In the course of studies of the production of antibody against bacteriophage, he found an unusual type of antibody present in small amounts in normal serum. It is well known to all serologists that for almost any sort of antigen very refined tests will show 'traces' of antibody in most normal sera and sometimes surprisingly large amounts. Jerne felt that these traces were real pre-existent antibody and speculated that the first step in the production of antibody was the union of antigen with appropriate molecules of natural antibody. The complex was then taken up by macrophages of the reticulo-endothelial system and stimulated them to produce more antibody of the same pattern as had been drawn into them by union with the antigen. It was left uncertain how the original normal antibodies had arisen and no reasonable analogy was available to indicate how or why an accidentally chosen cell should produce replicas of a protein taken into its cytoplasm. There are still supporters of a modified version of Jerne's theory but most of them would now assume that the uptake of the antigen–antibody complex was merely the first step in processing the *antigen* into a suitable form for its basic function of stimulating lymphoid cells of the same clone that had produced the natural antibody.

One of the principal virtues of Jerne's natural selection theory was that it offered an interpretation of the failure of the body to make antibody against its own constituents—its natural tolerance. It was regarded as self-evident that any natural globulins capable of reacting with cell components accessible to the circulation would be eliminated from the plasma. This did not, however, deal particularly adequately with the phenomenon of neonatal tolerance

9

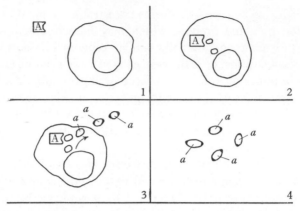

Fig. 1. The instructive theory of Pauling (1940) and his predecessors. Non-specific polypeptide chains are moulded after synthesis to take on specific immune pattern by physical contact with antigen.

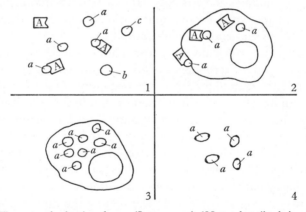

Fig. 2. The natural selection theory (Jerne, 1955). 'Natural antibody' unites with antigen; the complex is ingested by a phagocytic cell and more antibody is produced to the pattern of that ingested.

Figs 1–4. A series of diagrams to show in oversimplified form the essential features of four concepts of the process of antibody formation.

which had recently been demonstrated by Billingham, Brent and Medawar (1953) and had provoked widespread interest. It seemed that if a foreign antigen were implanted early enough, an animal would fail to become immunized and would subsequently be incapable of producing antibody or its equivalent against the antigen.

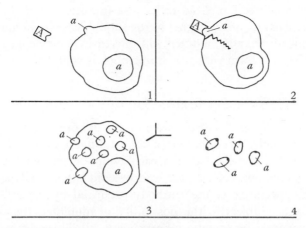

Fig. 3. The clonal selection theory (Burnet, 1957). Stem cells by a process of randomized somatic genetic change develop different immune patterns, one only being expressed for each cell and clone. Contact with the corresponding antigen provokes antibody formation or other specific response.

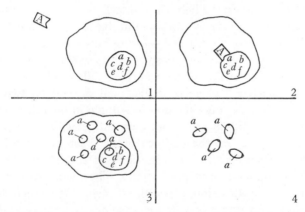

Fig. 4. Subcellular selection theory (Lederberg, 1959). Each cell has pre-existent capacity to produce large numbers of antibody patterns. Entry of antigen selects the appropriate pattern for expression.

The situation subsequently became more complex, but it was very largely from consideration of the implications of immune tolerance that concepts of cellular (or clonal) selection approaches began to develop. In 1957 Talmage and Burnet suggested independently that it was more satisfactory to consider the process primarily in terms of clones of lymphoid cells. The implications of this view were stated *in extenso* by Burnet in 1959 and since then the approach has been widened and modified with increasing knowledge until it has reached the form with which this book is concerned.

### THE CLONAL SELECTION THEORY OF IMMUNITY

In its original form this clonal selection theory of immunity postulated a randomization of pattern amongst differentiating lymphoid cells in embryonic life, so that each lymphoid cell in the body carried one immunological pattern expressed either as a 'receptor' or as the specificity of the antibody produced by the cell or its descendants. Subject to the possibility of mutation, the pattern was transmitted by somatic inheritance to all descendants giving rise to large numbers of clones of cells, each with a distinct immunological specificity.

The novel features of the approach were all related to a concentration of attention on the behaviour of cells as units rather than on the process of antibody synthesis. The basic concept was of the immunologically competent cell—a term now replaced by 'immunocyte'—that is, a cell which is susceptible to specific stimulation by contact with the appropriate antigenic determinant. On this hypothesis the result of such contact will depend on two major factors, the effective concentration of antigen in the cell's environment and the avidity of union between antigenic determinant and the combining site on the cell receptors. Factors dependent on the local environment of the immunocyte and its physiological state will presumably also play a part. The main possibilities of reaction are (a) destruction, especially if physiologically immature, (b) proliferation without essential change of character—to what we would now call memory cells and (c) proliferation and antibody production as a clone of plasma cells.

It was immediately evident that such an approach provided an alternative mechanism by which the amount and type of antibody and immunocytes could be adjusted to the current needs of the body. It also provided the simplest possible interpretation of how the body's own constituents are shielded from immunological attack.

At the time of writing it is probably correct to say that most experimental immunologists are averse from supporting any comprehensive statement of immunological theory. There is a general feeling that current activity in the elucidation of the molecular structure of the immunoglobulins will within a few years provide a solid background not now in existence against which theoretical concepts with some hope of being definitive may be developed.

This must be accepted but it seems that already the major decision between instructive and selective theories has been made. It is now mainly a matter of working out what mechanism is concerned in generating the diverse genetic patterns which govern the production of the vast repertoire of specific antibodies which can be produced more or less on demand.

My own preference has been to look to somatic mutation for the origin of this diversity, but there can be no doubt that genetic influences of the normal sort are important. With the development of ways by which the individualities of immunoglobulin molecules could be recognized irrespective of their specificity as antibodies, it became clear that complex inheritable genetic information from more than one source in the genome was converging to determine the structure of antibody. When the first complete sequences of amino acids in one set of the polypeptide chains that make up Ig G were established in 1965–6, a much more direct approach to the problem of genetic origins lay open. Already it seems virtually certain that the evolution of the immunoglobulin molecule has involved several successive duplications of a primitive cistron coding for 100–110 amino acids. Many immunochemists have been impressed with the probable resemblance between the processes which have produced the haemoglobins and the emerging evolutionary history of the immunoglobulins.

There can be no doubt that in part any antibody, any immunoglobulin, is produced by the interaction of a number of genetic units. There is wide sympathy for the view not yet elaborated in detail that there is a store, as it were, of genetic information which can be drawn on to provide a very large number of combinations each of which can code for a functioning antibody molecule. Somatic mutation is still a curiously unpopular concept amongst biologists and many would look to genetic processes (in the normal sense) to provide all the necessary variety of antibody pattern.

13

There is no sort of unanimity on the matter. None can be possible till much more structural information is accumulated. It is, however, not quite pure guesswork to suggest that both (germinal) genetic and somatic genetic processes play an important role and that in broad terms the former can be compared to the coarse adjustment and the latter to the fine adjustment of a microscope. Again relying on an intuitive sense of what is biologically likely in the absence of relevant data, one would expect genetic effects to be more important in small short-lived animals like mice, and somatic mutation to play a much greater part in large long-lived animals like men.

Whatever the eventual assignment of responsibility for antibody diversity, there is a quality about immunological phenomena which calls for a random process by which antibody patterns are distributed, as it were, to immunocytes and established for the descendant clone by strict phenotypic restriction.

The other important feature that has been revealed by the structural studies of immunoglobulin light chains is the much greater diversity of the N terminal 107 residues than of the other half of the chain. Circumstantial evidence points strongly toward 'labile' sectors of this type in both chains being concerned with the formation of the combining sites and therefore with the specificity of the antibody. For the present it is probably true to say that technical achievement is not yet adequate to support the speculations on the nature of specificity.

There have also been a number of attempts to devise a quasi-instructive theory in which the choice amongst some tens of thousands of potentialities in each cell is actually made by the antigenic determinant. Theories of this general character are easy to endow with *ad hoc* qualifications which make them almost impossible to disprove. The main virtue of the clonal selection theory is that it can be disproved immediately it can be shown that from a single cell (or its operational equivalent) three or more unrelated primary antibodies can be produced to nominated antigens.

In parallel with the ideas on the nature of the relationship between antigen and antibody and the gradual elaboration of the complexity and heterogeneity of the immunoglobulins, the cells

The clonal selection theory of immunity

responsible for antibody production have also been the subject of much discussion. One of the most fruitful lines of experimental advance since 1957 has been the development of technical methods by which single cells or pure clones derived from a single cell can be studied immunologically, and these techniques have now brought a much clearer understanding of the interrelationships of the cells concerned. In parallel with these single-cell studies there has been a steady growth in the understanding of the cytological character of the various cell populations available for experimental transfer and manipulation.

Any attempt to disprove or modify a clonal selection theory would necessarily have to make much use of these new techniques, and it is in fact the ability of the theory to absorb in natural fashion the various concepts as they arose from the new experimental approaches that has been its main claim to validity. It is worth while therefore in this historical introduction to devote a few pages to the significance of the new techniques that have been developed in recent years for handling immunologically significant cells.

### THE CELLULAR ASPECTS OF IMMUNITY

The first set of cells to be considered as producers of antibody were the macrophages of Metchnikoff or the reticulo-endothelial system of Aschoff, both defined by their capacity to take in foreign particulate matter or macromolecular dyes. With the prevalent instructive view of antibody formation it seemed reasonable that the cells taking up the antigen should be those making the corresponding antibody. It was evident that two important sites for antibody production were spleen and lymph nodes, in both of which cells of the reticulo-endothelial system were numerous; so also were lymphocytes and plasma cells, while there were also smaller numbers of mast cells and a variety of supporting cells. In each site, activity associated with artificial immunization was evidenced by the appearance of germinal centres (reaction centres, secondary follicles) in the primary lymph follicles of lymph nodes and spleen, and of plasma cells at the periphery of the lymph follicles and in the red pulp of the spleen and the medullary cords of lymph nodes. On the necessary but unprovable and improbable assumption that the three cell types were completely distinct it was felt that either the macrophage, the lymphocyte or the plasma cell must be the producer of antibody.

In 1948 Fagraeus showed conclusively that antibody production was correlated with the presence of plasma cells, and since then there has been no serious alternative to the view that antibodies and immunoglobulins are predominantly secreted by plasma cells and the intermediate cells of the lymphoblast to plasma cell series. The application of all the modern methods of cytological study leads to the same conclusion. Electron micrographs show abundant rough endoplasmic reticulum in plasma cells and special methods have identified globulin and antibody between the lamellae. The numerous polyribosomes and the active nucleolus also indicate a rapid synthesis of protein and are responsible for the characteristic staining of the plasma cell by the Unna–Pappenheim method.

The first of the direct histochemical methods to be adapted to immunological experimentation was Coons's 'fluorescent antibody' approach (Coons *et al.* 1942). There are relatively simple ways of attaching fluorescent dyes to soluble proteins using the same principles that Landsteiner applied in binding antigenic determinants of known character to carrier proteins. If a semi-purified antibody is labelled in this fashion with fluorescein and used to stain a thin section of fresh tissue, there will be a precipitation of antibody at any site where the corresponding antigen is present in adequate concentration. The most important antigen for such experiments is paradoxically enough antibody itself or, more correctly, immunoglobulin. If human tissues are to be studied, purified antibodies are prepared from rabbits immunized each with a pure solution of one of the antigenically distinguishable human immunoglobulins and linked to a suitable fluorescent dye. Any cell in a tissue section which is secreting significant amounts of the appropriate immunoglobulin will be 'stained' and the fluorescent areas can be recognized microscopically with appropriate illumination. All studies are in agreement that plasma cells are the predominant producers of each of the defined immunoglobulins. By suitable refinement using treatment first with antigen then with specific fluorescent antiserum (the sandwich technique) it can also be shown that in an immunized

16

animal many plasma cells are producing antibody of corresponding specificity.

A few years later, autoradiographic techniques were introduced and soon concentrated on the use of tritiated thymidine as a marker by which the lineage of a given cell could be traced. If a cell in the process of duplicating its DNA before mitosis is exposed to labelled thymidine, this is rapidly taken up and incorporated into the DNA of the chromosomes. Owing to the conservative quality of DNA replication, the thymidine taken up is retained by the nucleus and distributed approximately equally to any descendant nuclei. The location and amount of label is determined by overlaying thin sections or smears with photographic film. After processing and staining, the number of silver grains associated with a cell nucleus is a measure of the amount of label present. Subject to a variety of technical safeguards, cells showing nuclear label are lineal descendants of cells which at the time the tritiated thymidine was administered were actively synthesizing DNA. The application of this technique to immunological problems is only limited by the ingenuity and dexterity of the experimenter, and many results will be discussed in later chapters.

Other radioisotopic techniques have also been applied, one of the most important being to provide labelled amino acids to a cell population in order to show by subsequent measurement of activity in specific antibody that antibody was being actually synthesized by the cells during the time the labelled amino acids were present in the system.

From these and other types of experiment it became clear that Fagraeus was correct in ascribing antibody production predominantly to immature and mature plasma cells.

It is still uncertain to what extent other morphological cell types are associated with antibody. When appropriate antisera against the different immunoglobulins are used to detect cells carrying or producing the antigens in question, the strongest reactions to each type are given by plasma cells, but there is frequently a general glow over germinal centres and on scattered reticular cells. There are some small cells, not morphologically different from small lymphocytes, which may be present in thoracic duct fluid, that show strong staining for $\gamma$-globulin in their thin rim of cytoplasm;

the significance of these lymphokinocytes is unknown. Quite recently it has been found that many lymphocytes can produce at least small amounts of immunoglobulins and, in suitable material, actual production of antibody can be demonstrated.

Perhaps the most important available generalization on which all interpretations must be based is that antigen is taken up by cells that do not produce antibody—macrophages of various types—and that antibody is produced by plasma cells that contain no detectable antigen. There are two tenable interpretations that will have to be discussed. The first is that within some of the macrophages the antigen is broken down to give rise to a certain number of antigenic determinant–RNA complexes which have a special capacity to stimulate competent lymphoid cells to which they are transferred. The second (which does not necessarily exclude the first) is that the main immunological function of the reticulo-endothelial system is to reduce the concentration of circulating antigen to a level low enough for an effective response by the antibody-producing mechanism.

With the statement of the clonal selection theory it seemed immediately evident that the theory would be disproved if it were shown that a single cell could regularly produce more than two antibodies. With the increasing information about immunoglobulin types and the as yet unexcluded possibilities of interchange of immunological information between cells, we should nowadays be sceptical about so simple an approach. Nevertheless, for one reason or another, the last decade has seen great activity in devising ways of demonstrating and measuring antibody production from single cells and in seeking to demonstrate cells producing more than one type of antibody after immunization with multiple antigens.

For obvious reasons, very sensitive assay methods must be available if the antibody produced by a single cell isolated in a droplet of fluid is to be measured. Neutralization of bacterial viruses and immobilization of bacterial flagella have both been successfully used. As is usually the case with biological experiments the results are not in themselves decisive but they have contributed valuable new information.

A less direct but more fruitful approach was devised by Jerne and Nordin, who showed that single cells could produce sufficient haemolytic antibody against foreign red cells to allow the development of a plaque-counting technique. If a mouse is making antibody after an injection of sheep red cells four days previously, most of the antibody-producing cells are in the spleen. Their number can be obtained by emulsifying known numbers of spleen cells with a suspension of sheep red cells in melted agar. The agar is poured and, after setting, incubated for 30 minutes during which any antibody diffuses from the cell and attaches to adjacent red cells. By flooding the plate with guinea-pig serum (complement) the areas where antibody is attached to red cells become visible as clear areas (plaques) of haemolysis. By appropriate artifices, similar plaques can be used to measure the number of cells producing other antibodies and from 1964 onwards the technique became standard in most immunological laboratories.

Two other methods of handling single cells for quite different purposes have also been significant for immunological theory. The first was developed in my laboratory by Boyer (1960), who found that when adult fowl leucocytes were placed on the chorioallantois of chick embryos, rather large opaque foci developed whose number was proportional to the numbers of leucocytes added. The evidence subsequently obtained is interpreted as showing that an adult immunocyte reactive against an antigenic determinant of the recipient membrane will initiate a focal lesion in which both host and donor cells are involved. It is in fact a typical graft-versus-host reaction.

The second is a by-product of studies on the recovery of mice lethally irradiated and subsequently given bone-marrow cells from normal animals. Such mice take up to thirty days to die in the absence of treatment. Till and McCulloch found that if such mice were given a small dose of bone-marrow cells intravenously and killed five to ten days later, their spleens showed discrete white masses of cells a millimetre or two in diameter. Irradiated mice receiving no further treatment showed only shrunken spleens without nodules. Again, the number of these foci was proportional to the number of bone-marrow cells administered and each was predominantly a clone derived from a single initiating cell. Just as in the foci of the chorioallantoic membrane, these nodules are open to the entry of circulating cells and the idea that they are 'clones' must be used very circumspectly.

19

## THE CONTRIBUTION OF IMMUNOPATHOLOGY

Running in parallel with experimental immunology has been a growing interest in immunopathology and, as has been the case in many fields of medicine, pathological findings have thrown much light on the normal functioning of the immune mechanism.

For fairly obvious reasons, acquired haemolytic anaemia was the first autoimmune disease to be clearly recognized. Many other human diseases in which there is at least a major autoimmune component are now known and similar conditions have been recognized in mice and dogs. One's first reaction to autoimmune disease is that it must represent a disturbance in the development of natural tolerance and almost all who have written on the nature of immune tolerance have been influenced by ideas drawn from pathology.

Of even greater influence on the development of immunology has been the progressive understanding of the nature of the myeloma proteins, Bence Jones protein and of the conditions with which they are associated. As soon as electrophoretic methods were applied to these pathological sera it became clear that there was a great excess of a homogeneous population of immunoglobulin molecules. In Waldenström's words these were 'monoclonal gammopathies' in which the myeloma protein is operationally and perhaps absolutely equivalent to antibody produced by a single clone of immunocytes. As will be shown in later pages, this realization not only gave a great impetus to chemical studies on immunoglobulins but was also probably the most potent factor in bringing about a more tolerant attitude to the clonal selection theory.

We can omit in this outline other interesting leads from pathology, particularly in relation to the thymus, but mention must be made of congenital agammaglobulinaemia. This was a by-product of the antibiotic treatment of ailing infants. It was of special importance in showing that a normal complement of lymphocytes could be present without plasma cells and that while antibody production failed to occur, a variety of skin hypersensitivities could develop, measles ran a normal course with subsequent immunity and skin homografts could be rejected. No theoretical approach to immunology could subsequently be complete unless this dissociation of immune functions could be covered.

# 2    The general character of immune phenomena

*Evolutionary considerations*

Until the early 1950s it was implicit in medical and microbiological thought that immunity was concerned wholly with defence against micro-organismal infection. Various clinical and laboratory pheno-mena such as anaphylaxis which, though obviously immunological in character had no bearing on protection against infection, were well known but could be interpreted as artefacts arising from un-biological contrivance or circumstance. In recent years this attitude has changed, perhaps mainly because of the upsurge of interest in tissue transplantation in surgery. Tissue grafting is something un-known in nature, so at first sight it is hard to understand how an impressively large and self-consistent body of scientific knowledge could have been accumulated around something that from the evolutionary standpoint was meaningless. There are, however, two very important natural mammalian situations which in signi-ficant ways are analogous to an experimental graft of foreign tissue, pregnancy and cancer. A pregnancy represents the implantation of foreign tissue in the uterus—it is in most respects formally equivalent to grafting skin from an $F_1$ mouse (AB) on to an individual of one or other of the pure line parent strains AA and BB from which it was derived. The AB skin graft on an AA recipient is foreign in respect of its B component and will be rejected. The AB foetus in an AA mother thrives. Clearly, some special mechan-ism must have been evolved to allow placental reproduction.

A cancer results from the multiplication of cells within the body which are alien in the sense that they are not adequately subject to the controls that ensure the morphological integrity of the body. It is now realized that most malignant cells are antigenically dis-tinguishable from related normal cells of the animal in which the tumour has arisen, but the difference in histocompatibility antigens is always slight. There is a steadily increasing opinion that the immunological difference between normal and malignant cells is biologically and even clinically very important. In addition there is a still more important point which depends on the fact that a spontaneous tumour initially at least retains the essential antigenic character of the host. If it arises in a mouse of pure line strain A, it will behave on transplantation very like a piece of A's normal tissue. It will 'take' in almost 100 per cent of mice of strain A or in $F_1$ hybrids obtained by mating A mice with any other pure line strain. It will be rejected like a foreign skin graft in any other type of host.

There are no populations of genetically uniform mammals in nature. Given the general properties of the evolutionary population, every naturally occurring individual will differ in its inheritance from every other individual. There is, however, no *a priori* reason why two animals of the same species should not accept transplants from one another. This can, in fact, be done in insects and in many types of embryonic or immature vertebrates. The characteristic incompatibility of mammalian skin grafts is of relatively late evolutionary origin. One of the possible reasons for the existence of this immunological individuality within a vertebrate species can be put succinctly by saying that, in its absence, cancer could be a contagious disease.

Evolution, therefore, may have moulded immune mechanisms for other reasons than defence against micro-organisms. Invertebrates require just as effective means of countering infections as vertebrates but they produce no antibodies. Amongst the lower vertebrates, the hagfish shows no evidence of an immune response to any of the standard tests but a more advanced cyclostome, the lamprey, shows for the first time the attributes of the immune mechanism which is present in all higher forms. It appears, therefore, that at an early stage in vertebrate evolution a need arose for some mechanism beyond what had been developed against bacterial infection in the invertebrates. Discussion of the vertebrate, and specifically the mammalian, immune mechanism leads almost inevitably to the conclusion that it is more basically concerned with the control of tissue integrity and reaction against recognized anomaly in tissues than in defence against micro-organisms and the production of antibody. One can imagine that with the emergence of the new tissue-controlling mechanism new possibilities of effective defence arose and thereafter the two interrelated functions evolved together.

A third type of immunological phenomenon distinct from either of these categories, though in curious ways related to both, is currently described as 'delayed hypersensitivity'. The tuberculin reaction is the classical example but there are, in addition, various types of skin hypersensitivity to natural or synthetic chemicals and clear indications that many of the manifestations of chronic infectious or autoimmune disease are based on the same type of reaction.

There are, then, three main biological functions of the immune system to be considered: tissue control as exemplified by the rejection of foreign skin grafts, tissue reactions that are not mediated by antibody such as the tuberculin reaction, and antimicrobial functions such as protection by antiviral antibody or the production of antitoxin in diphtheria and similar bacterial infections.

*Transplantation immunity*

Transplantation immunity is usually demonstrated by skin grafting. It is well known that a plastic surgeon can graft a piece of skin from any part of his patient's body to any other part. If, however, he attempts to borrow skin from a donor—for example, to make good skin lost in very extensive burns—the best he can hope for is that the borrowed skin will be retained for a little over a week. By fourteen to twenty days it will have been completely rejected. There are three exceptions to this rule which, in themselves, almost define the processes involved:

(*a*) When two people are identical twins they will accept and retain skin grafts from each other. Identical twins arise by the splitting of the fertilized ovum after the first division, with the two cells both going on to normal development. The twins have exactly the same genetic endowment.

(*b*) Very rarely twins are found who are clearly not identical but differ from ordinary two-egg twins by sharing *in utero* a blood supply from fused placentas. Despite the fact that the twins are genetically distinct, arising from two separately fertilized ova, blood

23

cells of all types move freely between the two embryos and they are born with this mixed blood condition. If one is genetically blood group O and the other group A, each will have the same mixture of O and A cells and retain the mixture for life. Such twins will also accept and retain reciprocal skin grafts.

(*c*) There is an extremely rare condition in which a baby is born with a thymus which fails to develop. Such infants have virtually no lymphocytes in their blood, are extremely prone to infection and, even with the best of medical care, usually die within two years. If a child with this condition is grafted with a piece of un-related skin the graft heals satisfactorily and is retained as long as the child survives.

Clinical observations of this sort would never have been made without the background of years of work on experimental skin grafting with pure line strains of mice, rats and hamsters. For the present, however, we can use the clinical findings to introduce the essential features of tissue immunity.

In the first place we have the demonstration that for a tissue to be rejected it must be recognizably *different* and that the differences involved are genetic in origin. By rigid and prolonged inbreeding, pure lines of mice can be obtained, each member of which is genetically equivalent to any other. The standard test for such genetic purity is to interchange skin grafts. All should be retained.

Genetic identity is not, however, a *necessary* condition for skin graft acceptance. The intermingling of placental blood of two dissimilar twins is a natural experiment which shows that tolerance of another individual's tissues is possible if the body has experienced the presence of the foreign cells from a period early in embryonic life. From this deduction the whole topic of *immunological tolerance* has developed and in a sense the present hope that organ transplantation will one day be regularly possible.

When the thymus fails to develop, lymphocytes are few or absent and standard antigens fail to provoke antibody formation. If one removes the thymus from baby mice on the first day of life, broadly similar changes result. In particular, such mice will accept skin from foreign strains of mice and even from rats. The implication is clear that an immunological process is concerned in skin graft rejection and that the presence of lymphocytes is essential.

In summary, studies of transplantation of skin and other tissues in experimental animals supported by such findings in man as have been mentioned have provided the following conclusions:

(*a*) Skin rejection requires the recognition by the recipient that the graft contains antigenic determinants not present in the recipient's cells. All antigenic determinants are genetically controlled. They may change by mutation in the germ cells or by somatic mutation.

(*b*) Recognition that an antigenic determinant is foreign requires that it shall not have been present in the body during embryonic life. Conversely, any foreign cells introduced early enough in life will be accepted as if they were the body's own cells for as long as they persist. There is more than one explanation of tolerance but the existence of the phenomenon was the stimulus that led to the conception of the clonal selection approach to immunity.

(*c*) Skin rejection requires the action of lymphoid cells. It can be accelerated by the presence of specific antibodies but these are probably not necessary for the reaction.

(*d*) If a mouse of strain A is grafted with skin B and rejects it in 10 days, a second graft of B on the same animal will be rejected more rapidly. The effect is specific, not being shown against skin from unrelated strains, and represents a typical secondary immune response.

(*e*) The reverse of skin rejection is the so-called graft-versus-host reaction which occurs when mature lymphoid cells are introduced into a recipient of another strain which, for one or other reason, is unable to destroy the cells. They can react in various ways against the host, particularly when introduced into very young recipients. These are liable to die with the expressively named 'runt disease'.

(*f*) There are several phenomena which seem to establish that there are in normal lymphocyte populations a proportion of cells which are immediately capable of recognizing and reacting against foreign tissue antigens.

## Delayed hypersensitivity reactions

In 1891 Koch described the tuberculin reaction by which a guinea-pig actively infected with tuberculosis responded to a subcutaneous injection of a sterile filtrate from a tubercle bacillus culture. There was a sharp rise of temperature and the development of severe inflammation and necrosis at the site of the injection. In normal guinea-pigs there was virtually no reaction of any sort. With purified reagents and small intradermal injections the same reaction in the form of the Mantoux test is widely used in human medicine. In experimental animals it has become the prototype of *delayed hypersensitivity* reactions.

Such reactions, which in various more or less similar forms may be given by a wide variety of antigens, reach their peak around 24 hours after injection in most of the commonly studied examples. There are, however, some much slower reactions of otherwise similar quality which are seen in the lepromin and Kveim tests for leprosy and sarcoidosis respectively. All such delayed reactions are clearly differentiated both from the acute Arthus reactions seen in experimental animals immunized with standard soluble antigens and with acute allergic responses such as are seen when hay fever patients are tested with the incriminated pollen extract. The latter may for the present be dismissed as resulting from the fixation of a special type of antibody in the tissues of the skin and its reaction with the antigen. Delayed hypersensitivity has proved more difficult to interpret, and since the rejection of foreign tissues is basically due to very similar processes, there has been a big surge of new interest in the phenomenon.

It has not been possible to show that circulating antibody is in any way involved in delayed hypersensitivity though the possible activity of antibody adsorbed to the surface of leucocytes or macrophages has not been finally excluded. Most immunologists regard as the main key to the nature of delayed hypersensitivity the fact that if a very large mass of lymphoid cells is transferred from a sensitized guinea-pig to a normal one, the recipient also becomes reactive within a few hours. When an injection of antigen is made into the skin the response is essentially the same in the passively sensitized recipient as in the actively sensitized animal. What is observed is redness and swelling of the reaction site represented histologically by dilatation and stasis in small blood vessels and a vigorous migration of mononuclear cells which, as they pass through the walls of small venules and capillaries, appear to be mainly small lymphocytes. As they pass away from the capillary into the tissues they become larger and most would be called histiocytes.

There are many unsolved aspects of this reaction. What stimulates the movement of cells through the vessel wall? What proportion of the migrating cells have a specific immune relationship to the antigen deposited in the tissues? How does contact of antigen with cell give rise to the vascular responses we observe? In addition, delayed hypersensitivity has served as a central point from which many related experimental phenomena have been developed, only one or two of which need be mentioned in this orienting approach.

Transplantation immunity as demonstrated in experiments with skin grafts may well be a slightly modified delayed hypersensitivity response. Histologically the early stage of rejection is shown by a local accumulation of mononuclear cells including lymphocytes and plasma cells. Homograft immunity can be transferred to another animal by lymphoid tissue cells but not by serum. Finally, if we have an animal *A* which has developed homograft immunity against *B*, an intradermal inoculation of *A*'s lymphocytes into *B*'s skin will give a reaction of the same quality as a delayed hypersensitivity reaction.

In the field of clinical immunology there has been much recent interest in the reactions of human blood lymphocytes in the presence of kidney bean phytohaemagglutinin and certain antigens. It has been known for many years that a variety of plant extracts, particularly from leguminous seeds, could agglutinate red cells, and sometimes showed useful specificity for one or other of the ABO human blood groups. As a logical development from this work, such phytohaemagglutinins were used to separate leucocytes from red cells in blood. The addition of the extract to citrated or heparinized blood caused rapid clumping and sedimentation of red cells, leaving leucocytes in suspension in the supernatant fluid.

It had been well known that blood leucocytes held under ordinary tissue culture conditions did not proliferate. Cultures made from leucocytes obtained by the use of phytohaemagglutinin, however, rather surprisingly showed large numbers of mitoses that resulted from the enlargement and then division of small lymphocytes. The next step was the observation of Pearmain and his colleagues in New Zealand that tuberculin or PPD added to blood lymphocytes from tuberculin-positive individuals produced enlargement and mitoses but not cells from non-reactors. The present position which could readily be changed by a few intelligently planned experiments suggests that the phenomenon may be a measure of the range of delayed hypersensitivity reactions shown by the donor of cells.

## Immunity against micro-organismal infection

The third broad area of immune function is in many ways the most obvious and, by far, of the greatest practical importance. We can prevent diphtheria, tetanus, poliomyelitis and whooping cough, smallpox and yellow fever with almost complete certainty by a simple application of immunological principles. Yet, looked at from the evolutionary angle, there are puzzling features. In underdeveloped regions, which included Europe until a little over a hundred years ago, by far the greatest concentration of mortality was in infants around the time that their maternally transmitted immunity had disappeared. Once the period from 6 months to 5 years had been survived, children had usually sufficient immunity to the common pathogens to be immune to them for life. It appears that the best evolutionary solution to the problem of survival in a heavily contaminated environment is to accept the unprotected state for the first impact of any particular infection and provide the survivors with a solid immunity that will render an otherwise dangerous pathogen harmless. The classical example is yellow fever. European children brought up in a country where the disease was prevalent might suffer from other diseases but never, as adolescents or adults, from yellow fever. Unprotected adults such as the European soldiers brought to the West Indies during the Napoleonic wars had an appalling death rate from yellow fever.

The mechanism of protection against second or subsequent attacks of infectious disease is not quite as clear now as it seemed twenty years ago. In virus diseases it was regarded as self-evident that antibody persisting in circulation plus the capacity to produce further antibody at short notice were the protective agents. The fact that children with congenital agammaglobulinaemia who can produce no detectable antibody show a normal course during an attack of measles and have a solid subsequent immunity was disconcerting. There are indications that other virus diseases do not behave so normally as measles in these children but the nature of measles immunity remains unexplained. The main disability in agammaglobulinaemic children is the occurrence of repeated bacterial infection of the respiratory tract, which suggests that long-held ideas about pneumococcal infections and the part played by antibody in their control are probably correct. Nothing discovered since the great days of research on pneumonia in the 1930s has diminished the importance of the opsonizing function of antibody in allowing control of bacterial infections.

THE THEORETICAL APPROACH

In the immediately preceding sections, I have outlined the range of phenomena with which this book is concerned and in the rest of the chapter I shall summarize the theoretical approach to be adopted. That approach represents an attempt to restate in terms of recent work the essential features of the clonal selection approach which I developed in 1957–8.

*The premises of clonal selection theory*

Clonal selection theory in a modern guise must be expressed a little differently from the form used in 1958 but basically the premises of the theory are the same.

(*a*) There exist in the body populations of cells differentiated for immune function by their ability to produce antibody and to react in other ways to contact with specific antigen. These cells are referred to as 'immunocytes'. In the adult mammal they comprise a substantial proportion of the cells with the morphology of small lymphocytes as well as all plasma cells and a variety of immature and intermediate forms.

(*b*) Like the patterns which endow enzymic activity or antigenic individuality on other proteins, the specific pattern of antibody is determined by the genetic endowment of the cell which produces it.

(*c*) The range of immune reactivity of any immunocyte or clone of immunocytes is sharply limited by a process of phenotypic restriction. It may be subject to limited extension by subsequent somatic mutation.

(*d*) The origin of the diversity of immune pattern is to be sought at the genetic level. The important processes are probably (i) random choice for phenotypic expression of alternative gene combinations, and one or both of two types of somatic genetic change, (ii) 'scrambler' mechanisms of somatic recombination of the type recently suggested by Smithies and (iii) somatic mutation occurring within the relevant cistrons during the individual's life.

(*e*) Antigen acts essentially only as a signal or stimulus to such cells as are competent by the possession of antibody-like receptors to react to it.

Such a formulation is highly flexible in the sense that in any field of immunological investigation the precise form of theoretical interpretation can be modified to conform to the convenience of the situation. At the present time, for instance, there are a number of contentions opposed to aspects of my own interpretation of what is on record in the immunological literature. Some workers hold that it will soon be unequivocally demonstrated that transfer of capacity to produce immune pattern can occur from one cell to another. Others would base what most writers regard as the direct action of immunocytes on the adsorption of cytophilic antibody to immunologically neutral cells. Many immunologists are unconvinced that tolerance is wholly a matter of the *absence* of the corresponding immunocyte and would look for a more active process.

If any of these contentions are established, an extensive reorientation will become necessary but the central problem of accounting for the immune pattern in genetic terms will remain. At the present time the chief virtue of the clonal selection approach is that it provides a reasonable framework within which to interpret findings which involve both the population dynamics of lymphoid cells and specific immunological responses such as the production of antibody. This, in fact, is a concise statement of my present objective.

In this chapter I am concerned essentially with an orientating discussion of the biological principles germane to the development of selective theories of immunology. In writing this book, the greatest difficulty I have found is the utter impossibility of providing a logical sequence that will cover the subject matter of immunology in a single one-dimensional movement. Tolerance, for instance, can only be discussed in terms of interference with antibody production but the phenomena of tolerance are highly relevant to the nature of the differentiation of stem cell to immunocyte, which seems the logical point at which to start discussion of the whole immune process.

The solution I have adopted is, first, to assume some background of biological and immunological knowledge in the reader and to provide in this chapter a generalized summary of the approach that will be adopted. It is divided into two main topics which in one form or another cover most of the themes that need consideration:

1. Aspects of differentiation and somatic mutation which may be relevant to antibody diversity.

2. The origin and population dynamics of the lymphoid cells and the concept of the immunocyte.

*The nature of antibody*

There is no way of defining an antibody in any simple fashion. At the present time we regard as antibody any population of immunoglobulin molecules which has a specific capacity to unite preferentially with a definable chemical substance or configuration. The preferential capacity to combine with a given antigenic determinant is ascribed to the 'combining sites' of the antibody and its specificity is a manifestation of the 'immune pattern' of the antibody.

Before entering on any discussion of immune pattern, the concept of antigenic determinant needs to be clarified. It comes originally from Landsteiner's experiments with artificial antigens in which many molecules of some simple pattern, an arsanilic acid for instance, were chemically combined to a simple protein such as egg albumin. The antisera produced were first treated with the carrier protein to remove antibodies reacting with components other than the artificial hapten. Then, by appropriate precipitation tests with proteins carrying a series of related haptens plus inhibition tests in which the simple chemical blocked the capacity of the antibody to precipitate the full antigen, it became evident that the hapten was the only structure reacting with this population of

31

antibodies. This conclusion has been confirmed and elaborated by refined experimentation in recent years. On current views any antigen (that is, a macromolecule or particle capable of reacting with antibody) can be regarded as a mosaic of configurations, of molecular twists and outlines, which are potentially antigenic determinants. Many of them, because they are identical or very similar to configurations in the animal injected, will be inert immunologically.

With this preliminary we can use the term immune pattern for the specific aspect by which antibodies of all types react with a given antigenic determinant. In man and most other mammals there are three major types of immunoglobulin which can function as antibody. The one present in highest concentration both in the total immunoglobulins of serum and in most standard antibodies is immunoglobulin G (Ig G or $\gamma$ G). This form has been sufficiently studied at the chemical level to allow us to picture immune pattern on that molecule as the disposition of amino acid residues in two equivalent combining sites, one at each end of an elongate molecule. The molecule of a typical antibody is a highly complex structure. Strictly speaking, only myeloma proteins are available for the detailed study needed to define antibody protein structure. There is, however, no evidence whatever against the tacit assumption made by almost all immunochemists that an Ig G myeloma protein is a homogeneous population of a single type of antibody molecule produced by a clone of plasmacytes. There are indications that under special conditions large amounts of monoclonal antibody can be produced, but all that have so far been studied are heterogeneous. It cannot be wholly excluded that an antibody has some quality not expressed in a myeloma protein of equivalent type, but in this discussion it will be assumed that there is no such difference.

The Ig G antibody is a symmetrical molecule made up of two light chains each approximately of 22,000 mol. wt and two heavy chains of 50,000 mol. wt united by a relatively small number of disulphide bonds. Both chains are complex. The light chain of 214 residues is made up of two parts, each of precisely 107 residues. That including the N terminal group is rather highly variable from one Ig G myeloma protein to another, while the C terminal half is essentially identical except for the single amino acid change associated with the Inv a or Inv b antigenic character. The larger heavy chain is not so well investigated. The fact that a relatively large stable portion, now known as Fc, could be split off with pepsin without modifying antibody activity has been known for many years. The remaining Fd section has not yet been closely studied at the chemical level. It does, however, include a portion that varies in peptide structure from one myeloma to another, and it is the one section of the antibody molecule which all writers agree carries all or part of the combining site.

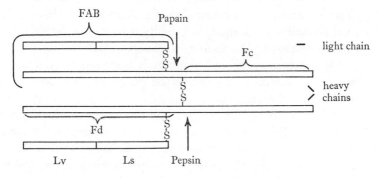

Fig. 5. The Porter diagram (modified) of Ig G structure, to indicate the various fractions and the site of action of papain and pepsin.

The highly variable character of the N terminal end of the light chain which is at least closely related to the combining site, plus the probability that the portion of the heavy chain which carries antibody specificity is also of variable constitution, points strongly but not absolutely to the currently popular view that these variations in amino acid sequence are in fact the basis of antibody specificity, of what we have called immune pattern. For reasons which may or may not stand up to future research, I have suggested that the variable portion of the heavy chain may be identical with the variable half of the light chain and therefore governed by the same cistron. A majority would probably regard the combining site as being derived from the mutual interaction of dissimilar segments on light and heavy chains. Irrespective of this diversity of opinion about the nature of the chains involved in the combining site, it seems probable that in general only a few residues (perhaps, as

Karush has suggested, about fifteen) are specially relevant to its specific character. Having regard to the very much greater extent of the 'variable' segment of the light chain (107 residues) than the length presumably involved in a combining site (7–10 residues), it must be kept in mind that the term 'combining site' has relevance only to the antigenic determinant being considered. If, as is suggested in a later diagram, the light and heavy chains may be closely related over a region of 20–30 residues, there could be possibilities of the same combining *region* carrying combining *sites* for two or more quite different antigenic determinants. If randomness of origin with determinate transmission of pattern is, as I believe, the essential character of antibody specificity, this would be bound to happen on occasion. Discussion of experimental results on the basis that any combining site is uniquely appropriate to a single antigenic determinant is as naïve as the belief that it was made to order to fit that determinant.

Irrespective of its relationship to the two chains, everything is consistent with the view originally due to Pauling that the combining site has a steric pattern and electronic configuration which make it unite more readily with the complementary configuration that we call an antigenic determinant than with any other molecular configuration. To be a little more precise, if we have a population of antibody molecules with identical combining sites so that we can speak of the population as a chemical species—antibody $x$—there will be a very limited class of chemical configurations which we can call collectively antigenic determinant $X$ which will unite specifically with antibody $x$ through the same combining site. Specific union can only be defined as firmer union than that shown by other substances tested. There is no absolute chemical equivalence between $X$ and $x$. In one well-studied example, antibody to the glucose polymer dextran was shown to be able to react to some extent to the disaccharide maltose and successively more firmly up to the 5- or 7-hexose molecules.

We can also look at the relationship between antigenic determinant and combining site from the opposite side. In practice this will usually involve using a chemical hapten of known structure united to a natural protein or a synthetic polypeptide. Experiment has consistently shown that all available antibody populations—

those produced, for example, by immunizing rabbits with the artificial antigen—are highly heterogeneous.* In particular, there is still a wide variation in the avidity of union of different sections of the antibody population with the fully defined antigenic determinant. One must bear in mind that any soluble protein, notably

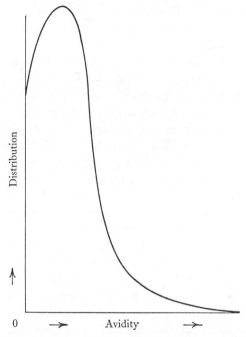

Fig. 6. To indicate the expected distribution of affinity of mixed immunoglobulins for any antigenic determinant. Only the small proportion of high avidity patterns to the right will be recognized as antibody.

serum albumin, has a capacity to adsorb a wide variety of simple chemical substances whose configurations are more or less equivalent to those groupings which can function as antigenic determinants. If it is correct to regard the emergence of new immune patterns as a random process, then for any given antigenic deter-

---

* The exceptions to this rule are those myeloma proteins which have a definable status as specific antibodies. The description by Eisen's group (1967) of a human myeloma protein active against antigens carrying dinitrophenyl as hapten is of special relevance.

minant there will be a fairly standard type of distribution, according to avidity, of the $\gamma$-globulin molecules present in a representative pool of serum from many members of the same species. The vast majority of immune patterns will have a negligible avidity of the same order as could be expected with any unspecialized protein. The distribution to be expected for higher degrees of avidity will be as shown in fig. 6. The highest possible degree of avidity will depend both on what arrangement of amino acid sequence and mutual disposition provides the strongest union and on how close to that optimal arrangement the initial pattern provided genetically can be brought by mutation or whatever other process is concerned in the 'generation of diversity'.

This character of the relationship between antigenic determinant and antibody is basic to any understanding of the concept of immunological pattern. The main objective of the present work is to seek an understanding of the nature of specificity and the 'soft edges' of specificity are an essential aspect of the problem.

The evidence is now overwhelming that the classical interpretation of protein synthesis as the transcription of information (coded pattern) in DNA to messenger RNA (m-RNA) and ribosomes where it is translated into an equivalent pattern of amino acids in a polypeptide chain is correct within its limitations. It is implicit in this that any protein that maintains its quality through successive cell generations is endowed with that quality by, in the last analysis, information held in the genome of the cell line concerned. There is nowadays no place for the 'instructive' theory that polypeptide chains are mechanically moulded into shape by contact with antigenic determinant. In one way or another there must be a genetic origin for the information that allows an antibody to be produced against a virus that may never previously have been met in the whole evolutionary history of the species.

## The basis of diversity

The most evident fact about antibody is its heterogeneity. There are at least three different types of immunoglobulin which can carry immune reactivity in all mammalian species that have been studied. The range of immune pattern appears to be virtually unlimited. Antibody is produced predominantly by plasma cells

36

and if we agree that instructive theories are now untenable there are only two sources for that diversity of cells which is needed to account for the heterogeneity of antibody. One of these is differentiation, by which in some way a choice is made amongst the various potentialities developed over the course of evolution and stored in the genome. The second is by inheritable change within a somatic cell line, either by somatic mutation in the ordinary sense or by some type of intrachromosomal rearrangement or recombination. Whichever of these somatic mechanisms is concerned the effect will be the emergence of patterns which differ in random ways from the pattern of the zygote from which they derive.

Orthodox theory holds that every diploid cell in the body contains potentially all the information contained in the genome of the fertilized ovum. The process of differentiation consists of a programmed series of changes in the regions of information-containing DNA which are allowed to function. The essence of differentiation is that the changes are programmed under the control of the genome. The best criterion of differentiation is that it occurs uniformly in all individuals of a homozygous population in which environmental factors are held constant. Since all rabbits, homozygous or not, produce the three physically defined immunoglobulins A, G and M, we must accept the differences as resulting from differentiation in the cells that produce them.

Similarly, if it should be demonstrated that a stem cell from the bone-marrow is multipotent and, depending on the particular internal environmental niche in which it lodges, will develop into erythropoietic, granulopoietic or lymphopoietic cell lines, this would clearly also be the manifestation of a process of differentiation.

When, however, we come to look at specific immune pattern the picture is quite different. If we inject a series of animals of the same species with the same antigen and maintain all the conditions as uniform as possible, it is usual to find a proportion which fail to respond and a wide range of titre in those that do. This holds even in homozygous animals especially when a 'poor' antigen is used.

When an antiserum against a well-defined antigen like a hapten–protein complex is produced in a rabbit, detailed study of the antibody population will show a gross heterogeneity of the combining sites. The picture is not of something made to the pattern of the antigen but of a random population of diverse patterns whose only common feature is ability to react with the hapten.

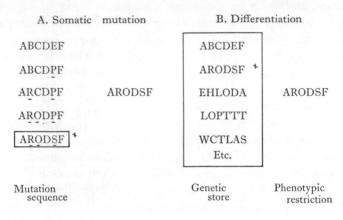

Fig. 7. Three origins for diversity: Schematic alternatives as to how the pattern ARODSF might arise in relation to a 'standard' pattern ABCDEF. A. Somatic mutation; B. differentiation; C. intergene recombination. Phenotypic restriction is operative in all.

From a quite different point of view the phenomena of tolerance necessarily imply that some random process is responsible for the production of cells carrying antibody patterns which, unless they are eliminated by the presence of circulating antigen, would be directed against antigens present in the body.

38

The essence of the situation is the *random* quality of the emergence of diversity and the evolutionary necessity that it should be random and that the random populations must be generated during the lifetime of the individual. If we use the conventional analogue of a protein pattern, a meaningful word in which letters represent amino acids, there are two basic ways by which a computer could be programmed to produce a random series of 10-letter words. One could provide in the memory all the 10,000 words that could conceivably be wanted and pick at random from these. Alternatively, starting with any 10-letter word, single-letter changes could be made at random, the result duplicated, other random single-letter changes made in each, and so on indefinitely. It is also obvious that the two processes could be combined in various ways. The problem at the immunological level is to discuss the pros and cons of 'randomized differentiation' or of 'accelerated somatic mutation' as the generators of diversity.

*Diversification as a process of differentiation.* Differentiation during embryonic and later development is basically a process of expression or inhibition of genetic potentialities in the fertilized ovum and the subsequent generations of descendant somatic cells. The complex diversity of function which we must postulate for antibody-producing cells is not unique in the vertebrate body. Primitive neurons seem almost to be programmed individually, each to move and direct its axon along an elaborately predetermined course. The control of pigmentation patterns in some organisms is almost equally elaborate. What is now known of the migration of pigment-controlling cells from the neural crest of the embryo does not make it any easier now than in Darwin's time to detail the process that gives the peacock its tail.

It must therefore be well within the capacity of evolution to devise a process by which genetic potentialities to cover every possible type of antigenic pattern could be present in the genome of the species. What is probably the most popular current interpretation of the heterogeneity of immunoglobulins and antibodies in one individual takes this point of view. Put perhaps oversimply, it is suggested that over the period of vertebrate evolution the primitive cistron responsible for the structure of a blood protein underwent repeated duplication, each new cistron then being susceptible to point mutation and, of course, selection for survival. Eventually, very large numbers of these mutated replicas of the original gene accumulated. Along with this process the rule was established that only a limited number could achieve expression in any one cell, the current mammalian rule being that two are used for the light chain, and four for the heavy chain of Ig G antibodies. There is no clue as to how many alternatives are maintained in the memory store, what the rules are for choice of this or that alternative and whether or not special rules apply to the cistrons whose product we recognize as variable segments in the myeloma proteins.

Many aspects of this approach must be accepted. It is almost inconceivable that the halves of the light chain do not represent the product of duplicated genes. The extensive range of antigenic types amongst immunoglobulins with determinants located on the heavy chain and in many cases precise genetic determination, demands a complex genetic situation that could well be based on duplicated cistrons.

*Somatic mutation.* It is an inevitable consequence of the evolution of a genetic system based on the replication of linear pattern in DNA that a proportion of errors should occur. In a very small proportion of the replicated chains there is a disturbance of the nucleotide sequence, and if the change has taken place within a structural gene the corresponding protein gene product will show an equivalent disturbance of amino acid sequence. At the level of germinal mutations the process is familiar in the changes observed in amino acid sequence in the variants of human haemoglobin, of insulin in different species, and of fibrinopeptides in the ungulates. The changes have in fact been used as one of the several independent methods of determining the triplet nucleotide code for amino acids.

There seems to be no reason for believing that error in replication is less likely to occur in somatic cells than in the germ cell line. On *a priori* grounds, one might believe that when a given segment of the genome in a somatic cell is regularly concerned with directing the synthesis of protein through the intermediary of messenger RNA, it would be *more* liable to accident than the corresponding region in a cell line concerned only with replication.

We will therefore assume that Orgel's figure of $10^{-8}$ per nucleotide per cell generation is of the right order of magnitude for point mutation, that is, for the simplest type of error in replication in somatic as well as for germinal cells. This does not rule out the likelihood that 'accidents' of one sort or another may also involve the genome of 'resting' cells which replicate only under emergency conditions.

There are in round figures $3 \times 10^9$ base pairs in the human diploid content of chromosomes or the likelihood of sixty errors at each replication of each cell. With perhaps $10^{12}$ cells replicating once a day or more frequently, there are almost hourly possibilities of every conceivable mutation occurring somewhere in the body. It becomes immediately evident that if large long-lived animals are to maintain functional and structural integrity, means of dealing with this flood of mutations must have evolved. Later on a case will be made for believing that the whole immunological process in vertebrates may have evolved to deal with this inherent weakness of living structure. For the present, all that need be emphasized is that if for any explanation of bodily function or disorder we have to call on somatic mutation there is a virtually unlimited reservoir of genetic accident available.

Any discussion of the significance of somatic mutation within the body must in principle take a similar form to the standard considerations of Darwinian evolution. Mutation gives rise to phenotypic change, any change will favour or prejudice chances of survival and any favourable mutation will eventually displace the former standard form of the species—or at the somatic level, of the cell clone concerned. If we confine ourselves to structural genes, the phenotypic result of point mutation or of grosser changes such as inversion or somatic recombination will be a modification of one or more points in the polypeptide sequence of a certain protein, enzyme, antibody or something else. Most changes will involve non-critical parts of the sequence, the gene product functions normally and the mutation in $10^{-6}$ or $10^{-8}$ of the cells of an organ or tissue will be unrecognizable. In another, probably large, proportion of mutant cells a vital region of a functional protein will be distorted and within a few cell generations the cell will become ineffective. In one way or another it will be eliminated and except perhaps for the transitory appearance in a stained section of a pyknotic nucleus, there will be no way of recognizing the occurrence of the mutation. A somatic mutation will in general only be recognizable if in one way or another the initial phenotypic effect can be magnified.

The effects of mutation are by definition inheritable and in broad terms no mutation will be recognized unless the mutant cell gives rise to a large number of descendant cells. There are three important ways by which this can happen. The mutation can occur in a cell at an early stage of embryonic development; it can give rise to a mutant that can proliferate faster and/or survive longer than its congeners; or it is so changed that it is stimulated to selective proliferation by some agent available in the internal environment. One example of each may be given:

(*a*) The first is exemplified in fleece mosaicism as described in Australian sheep by Fraser and Short. On rare occasions sheep-breeders find a lamb conspicuously different from its fellows by having patches of long loosely crimped wool contrasting sharply with the compact texture of the rest of the fleece. Search for such exceptional lambs amongst a sheep population, probably effectively ten million animals, provided a group of 30. Amongst these, one animal had approximately 50 per cent of its skin surface covered with the abnormal long wool, two had 20–25 per cent, and as the proportion of skin area diminished so the number of animals showing that proportion increased. The interpretation by the investigators was that the mutation in question could occur at any stage of development. If it involved one of the two cells from the first division of the fertilized ovum, half the fleece would be involved, at the next division the likelihood of the mutation would be doubled but the proportion of fleece affected would be halved, and similarly for the further stages of segmentation.

(*b*) Proliferative advantage from mutation is probably responsible for every spontaneous appearance of a malignant tumour as well as for all the increase in invasiveness (progression) observable either in the original host or on transfer to suitable animal recipients. An example of non-malignant change may be taken from experiments using Till and McCulloch's method of clonal growth of haemopoietic cells in the spleen of lethally irradiated mice. If cells from bone-marrow or foetal liver are injected intravenously into such a mouse, white 'colonies' of multiplying donor cells are easily visible in the spleen 10 days later. A suspension of cells from such colonies can be similarly injected into another irradiated mouse and the passage sequence carried on. With adult cells the passage series eventually fails but, on two occasions, passage of foetal liver cells gave at second or third passage a few larger colonies. At the next passage the spleens were fully colonized by cells of uniform primitive type resembling lymphoma cells. Passage could be continued indefinitely. The two strains differed in significant properties and it was evident that they were two independent mutants which had a strong proliferative advantage compared to normal cells in the depleted splenic environment. Neither had any malignant characters and they differed completely from malignant lymphoma cells by being unable to colonize the spleen of normal mice.

(c) It is the essence of the clonal selection approach that what determines the immunological specificity of a lymphoid cell clone is the pattern of a particular segment of the genome and that differences between such patterns have arisen by a randomly acting genetic process. A significant part or almost the whole of this process seems likely to result from somatic mutation. This cannot be established directly, but it is possible to show experimentally that when such cells have arisen they are specifically stimulated to proliferate by the corresponding antigen. Jerne's technique of producing discrete plaques of haemolysis each centred on a cell producing and liberating antibody against sheep red cells provides a way of directly counting antibody-producing cells. It can readily be shown both *in vivo* and *in vitro* that secondary stimulation with sheep red cells will stimulate a sharp proliferation of effective cells. A dozen different types of less direct evidence have led to the same conclusion.

There is a strong body of opinion which holds that the normal process of ageing represents essentially the accumulation of somatic mutations throughout the body and secondary results that may arise from the increasing loss of efficiency of various organs. In the opinion of Curtis, the most important aspect is genetic change in cells such as those in the liver which are normally replaced only at very long intervals or following emergency damage. In thinking of mutation in the ordinary sense it is convenient to consider mutation rate in terms of error per replication but even in bacteria it is known that mutation may occur in resting organisms. If the quality of cells in the mouse liver is tested by provoking mitosis during regeneration after damage by carbon tetrachloride, it is found that there is a steady accumulation of abnormalities of mitosis with age and there is other indirect evidence from functional anomalies in liver tumours that mutations are frequent. In view of the intense and varied biochemical functions of the liver we can be certain that there is a constant process of transcription from DNA to m-RNA. Although the precise physical nature of the transcription process is unknown it undoubtedly involves partial uncoiling of DNA and separation of the two strands, possibly with breaking and repair of the transcribing strand. Error and accident could well be frequent enough in these processes of activation, transcription and repression to account for the observed frequency of mutation.

Wherever there is a potentially harmful bodily function, there will usually accompany it some form of homeostatic mechanism to minimize the likelihood of damage. One of the most interesting recent developments in bacterial genetics is the description of a mechanism, apparently involving three enzymes, by which it is possible to excise and repair short segments of DNA within which incompatible chemical changes have occurred. It is not known whether this can also occur in vertebrate somatic cells but the possibility is worth bearing in mind that if any region of the genome should for evolutionary reasons benefit by having a high rate of mutation, such a mechanism for excision and repair could be intimately concerned. An abnormally high local concentration of such enzymes could conceivably be adapted to make small excisions and inversions much more frequent in certain sections of the genome than elsewhere. If somatic mutation is the essential origin of differences in combining site pattern, this local mutation rate must necessarily be higher than in other regions of the genome. There is a rather distant analogy to this in the highly mutable 'hot spots' of Benzer in the phage genome.

*The balance of probability.* At the cytological and serological level there seems to be no means by which one can differentiate between the two hypotheses: (*a*) that diversity of immune pattern depends wholly on a strictly random selection from a polycistronic mechanism within which germinal mutations accumulated over the period of vertebrate evolution are stored and (*b*) that superimposed on a complex genetic control of the various segments of the immunoglobulins there is a local process of accelerated somatic mutation or some more extensive but still random chromosomal rearrangement which involves the cistron(s) responsible for the segments that give rise to the combining site of antibody. On either alternative it is axiomatic that there is a strict process of phenotypic restriction by which the segments selected remain constant throughout the clone of descendant immunocytes. This would leave open the possibility of further somatic mutation within the immunocyte clone.

It is obvious that both alternatives may be broadly correct, and that either accelerated somatic mutation or the currently popular suggestion of intrachromosomal interaction of genes with recombination could greatly reduce the number of duplicated cistrons required to give the observed diversity of pattern.

If any decision is possible with available techniques it will have to be on a basis of analyses of amino acid sequences and, unless some other material than myeloma proteins becomes accessible, success seems to be extremely unlikely. Even the study of the variable chains in, say, 100 myeloma proteins produced in a strictly homozygous subline of BALB/C mice might not provide enough material for a definite decision.

In all subsequent discussion when the term 'somatic mutation' is used in relation to diversity in specific immune pattern it can be read as equivalent to any type of random inheritable change differentiating one line of immunocytes from another. Perhaps the next major development in experimental immunology will be the provision of a technical approach by which the present dilemma of interpretation can be resolved.

### The role of the lymphocyte

The immunological significance of the lymphocyte was probably the most active topic of immunological research in the period 1960–5. Before 1960 it is fair to say that the function of the lymphocyte was unknown, although many felt that it must be primarily immunological, and at one period there had been strong claims that it was the antibody-producing cell. In 1958 I wrote that the only cellular basis for a clonal selection theory must be the lymphoid cells comprising in a single series both lymphocytes and plasma cells. This, however, was based on general considerations, mainly that there was no other available system of cells, rather than on direct experimental evidence.

Three converging lines of experimental approach have led to the recognition of the lymphocyte as of central importance.

The thoracic duct lymph in the rat contains an almost pure suspension of lymphocytes, most of them small lymphocytes. From a study of the effects of depletion of the lymphocyte population by prolonged drainage of the thoracic duct Gowans was led to a broad series of investigations in which various modifications of this technique were used. These showed that depletion of lymphocytes could prevent a primary antibody response, that this incapacity could be remedied by infusion of small lymphocytes, and that under appropriate conditions of antigenic stimulation a proportion of labelled small lymphocytes developed into large pyroninophilic cells capable of mitosis.

The thymus in the young animal is a very active site of lymphocytic multiplication and for a long time it has been known that lymphatic leukaemia in certain strains of mice was initiated in the thymus and could largely be prevented by thymectomy. In extending work of this sort J. F. A. P. Miller examined the results of removing the thymus surgically on the first day of life. With appropriate strains, an important series of observations could be made. First the mice developed normally for some weeks and then suffered a chronic fatal disease with wasting and loss of lymphoid tissue. During the period of apparent health, homografts and even heterografts of skin were not rejected and there was failure to produce antibody in response to several common antigens. Lymphocyte levels were low but there was often a normal level of $\gamma$-globulin in the serum, and plasma cells were common in lymphoid tissue. None of these findings were completely regular, even in Miller's hands, and with their extension to other strains and species a rather confused situation developed that has not yet been fully

46

explained. At the present time it is clear that the thymus in the young animal is a very active centre for lymphopoiesis and that it produces humoral agents intimately concerned with the differentiation of lymphocytes and with the functioning of the immune system. There is strong evidence that a large proportion of lymphocytes produced in the thymus do not reach peripheral lymphoid tissues and almost equally valid evidence that the lines of immunocytes which develop are derived from progenitors differentiated in the thymus. Irregularities in the experimental results have been ascribed to racial variations in the time and rate of liberation of lymphocytes in the prenatal period and to the type of microorganismal infestation that develops after birth.

The third approach derives from the use of phytohaemagglutinin to induce mitosis in lymphocytes, as mentioned earlier in relation to delayed hypersensitivity. Under appropriate conditions, the small lymphocytes enlarge to form pyroninophilic DNA-synthesizing cells which subsequently divide. This behaviour fell so neatly into line with Gowans's findings *in vivo* that there was a violent 'gold rush' of clinical investigators to the new field.

The small lymphocyte can best be regarded as a highly mobile carrier of genetic information with no more executive capacity than is needed to stimulate it to take the form of a functioning cell. There is a growing consensus of opinion that the circulating lymphocytes represent a morphologically uniform population of cells with a wide range of functional potentialities. In bone-marrow there are numerous cells morphologically resembling small lymphocytes which have no immunological function. On rather slender positive evidence these are widely regarded as stem cells from which any of the mobile mesenchymal cell types, monocyte, granulocyte and immunocyte, as well as the red cell series, can develop. The line to be taken is ascribed to the action of appropriate hormones present in high concentration in particular areas of differentiation. If this is true, it becomes illegitimate to equate the lymphocyte morphology with any particular functional capacity. There may be excellent evidence that macrophages are derived through circulating monocytes from small lymphocytes in the bone-marrow. This does not mean that small lymphocytes of established immune function can give rise to peritoneal macrophages. Again, everything suggests that immune lymphocytes can pass through a pyroninophil blast intermediary to initiate a clone of plasma cells—a bone-marrow lymphocyte is almost certainly incapable of this. If we omit the bone-marrow and the other haematopoietic regions for the time being, lymphocytes arise by mitosis and proliferation in the thymus, and in peripheral lymphoid tissue in spleen, lymph nodes, Peyer's patches and elsewhere. In the young mammal the thymus is the most active centre of lymphocyte production and for reasons which will need later discussion it can be regarded as the standard site for the differentiation of stem cells to lymphoid cells of immune function which henceforward will be referred to as *immunocytes*. In addition there is a growing opinion that similar primary differentiation to immunocytes can also take place in other lymphoid tissue sites along the gastrointestinal tract. It is becoming common to differentiate between thymus-dependent cells, differentiated in the thymus, and those not so dependent. With some residual uncertainty the functions of delayed hypersensitivity and homograft immunity are ascribed to thymus-dependent cells, while antibody production in general is not thymus dependent. From what is known of the bursa of Fabricius in fowls it can be taken that the process of differentiation in the still largely hypothetical 'gut-associated lymphoid tissue' has the same general character as in the thymus.

Based largely on evidence from grafting infant thymus from mice whose cells are acceptable by, but karyotypically distinguishable from those of the host, it is now accepted that there is a rapid turnover of cells in the thymus. The view that differentiation to immunocyte takes place in the thymus is becoming increasingly popular and, if this is accepted, the existence of what I have called the 'censorship function' of the thymus seems to be unavoidable.

As a prelude to the discussion of this function of the thymus in relation to tolerance, the following outline of cellular changes in the normal thymus of a young mammal may be offered.

Stem cells from the bone-marrow reach the thymic cortex via the blood. Here, they settle in close relation to the reticulo-epithelial cells and under the influence of the local hormonal environment become actively mitosing large lymphocytes. The descendants are medium and small lymphocytes and at a certain point the proliferative potential of the clone is exhausted. The small lymphocytes produced remain for 3-4 days in the thymus and then either die *in situ* or pass to the circulation, probably via the local lymphatics. There is much histological evidence of cell destruction in the form of pyknotic nuclear fragments, and calculations based on the intensity of mitosis in the thymus suggest that 90-99 per cent of the cells produced may be destroyed.

A small proportion of the cells leaving the thymus can be experimentally identified in peripheral lymphoid tissue. If the view is correct that the function of the thymus is to differentiate stem cells to immunocytes, these cells go to swell the reservoirs of immunocytes available to be called on when any antigen with which they are capable of reacting reaches the tissue. Peripheral production of lymphocytes takes place in germinal centres and the rather inconclusive evidence is consistent with the view that a germinal centre is a clone of cells developing from a single stimulated immunocyte. Germinal centres always arise in lymphoid tissue, but opinion at present is against the view that most of the adjacent small lymphocytes are derivatives of the germinal centre they surround. Details are still lacking but there is convincing evidence that newly produced small lymphocytes are rapidly distributed throughout the lymphoid tissues of the body with the notable exception of the thymus. In every primary lymph follicle there must be thousands of different immune patterns represented by the cells which compose it. This becomes of great importance when we are concerned with following the path by which antigen stimulates the cells which will produce antibody.

Always with the reservation that it is not possible to equate small lymphocyte morphology with immunocyte function, it still remains highly probable that the peripheral lymphocytes observed in spleen lymph nodes and aggregations along the gut are in fact immunocytes of various types. There are three features of these cells which are of great immunological importance.

The first is their characteristic mode of entry and exit into lymphoid tissue. Entry is from the bloodstream by passage through the cytoplasm of swollen endothelial cells of capillaries or venules. These swollen cells are characteristic of lymphoid tissue and appear to have a specific stickiness or other form of attraction for circulating lymphocytes. The traffic is apparently strictly one-way, from capillary lumen into the substance of the lymphoid tissue. Movement *from* a lymph node is solely via the efferent lymphatics.

The second point of importance in this preliminary discussion is the relationship of the plasma cell to the lymphocyte. This is still controversial and some immunologists would keep the two cell types quite distinct, deriving the plasma cell from perithelial cells around small blood vessels. On the other hand, a majority now take the view that a small lymphocyte antigenically stimulated to a pyroninophil blast can be directed either to multiply as lymphocytes or, in an appropriate environment, to initiate a clone of plasma cells. If, as seems likely, the non-thymic origin of plasma cells becomes firmly established, it will depend on the site of differentiation how the activated cell will develop.

The third concerns the fate of lymphocytes and plasma cells. Both may be very long lived. After a single exposure of a rat in the early stage of antibody production to a large dose of tritiated thymidine, labelled small lymphocytes and mature plasma cells can be seen for at least six months. The great majority of both types are much shorter lived and only a very small proportion of small lymphocytes undergo activation to blast cells. Small lymphocytes are extremely susceptible to cytotoxic agents including at least one physiological substance, cortisol, and reasons will be given later for regarding this vulnerability as a way of making available the amino acids and nucleotides of their substance to local or general metabolic pools as needed. Plasma cells are more resistant to such agents but tend to disappear rapidly after an antigenic stimulus ceases to act. There is no evidence that mature plasma cells ever develop capacity to de-differentiate to blast form and renew mitotic activity.

*Tolerance in relation to immunocyte origins*

Without being aware of its significance, experimental embryologists who found that a piece of skin or a limb bud could be transferred from one species of amphibian embryo to another were demonstrating immunological tolerance many years ago. Quite extensive grafting of chick embryos was done in 1937–40 by Willier's group but the importance of the difference between embryo and adult was not recognized. Within the conventional field of immunology the first important findings were Owen's discovery that bovine dizygotic twins regularly showed two blood groups and that the presence of two distinct types of red blood cell usually persisted for life in both twins. In 1949 Burnet and Fenner discussed Owen's results along with those of Traub, who found that congenital infection of mice with the virus of lymphocytic choriomeningitis gave rise to a long-lasting asymptomatic infection. They regarded both as evidence of toleration by the embryo of what would in the adult be rejected as foreign, and introduced the concept that the differentiation of self from not-self was the central problem of immunology. It was predicted that appropriate injections of antigens in the embryo would give rise to subsequent tolerance of that antigen. Eventually the prediction was abundantly fulfilled but, as is the way of simple biological concepts, new experimental and clinical phenomena make it impossible now to present any unitary picture of immune tolerance.

Although it is relatively simple to produce tolerance by neonatal injection of a variety of soluble proteins, serum albumin for example, there are many antigens such as fibrinogen, bacterial flagella and bacteriophage where only partial degrees of tolerance can be produced in some species, and none at all in others. In most cases homograft tolerance can be produced but it is much rarer to be able to induce acceptance of a heterograft. There are now many ways by which degrees of tolerance or, better, unresponsiveness, can be produced in adult animals by, as it were, flooding the animal with a soluble foreign protein, by X-irradiation or by the use of immunosuppressive drugs and even (with a properly chosen antigen and a suitable mouse strain) by the injection of a very small dose of antigen.

The original concept of tolerance was that an essential part of the process of embryonic development was the establishment of a taboo on immune response to any of the normally circulating components of the body. In most respects this seems to be a self-evident necessity but it is a tradition of immunological experimentation to press far beyond the bounds of the normal and there has been no serious difficulty in showing that, with intense immunological stimulation, antibodies against almost any natural component of the body can be produced. This, however, no more proves that there is no normal homeostatic process preventing such responses than the appearance of hyperpyrexia in malaria indicates that there is no physiological control of body temperature.

The early work on tolerance was concerned with the situation in which a foreign antigen persists in the body either in the form of an alien cell chimera or a tolerated virus infection. Such conditions could readily be tested by grafting appropriate skin or examining for the continuing presence of the virus. Even in these fields, however, there were indications of partial tolerance, a condition that is much more frequently observed when attempts are made to establish tolerance to soluble antigens.

If a neonatal animal is given a series of relatively large injections of a soluble antigen it is usual for no antibody to be produced, and for this unresponsiveness to a challenge by the same antigen to persist for a variable period. The results of such experiments vary greatly from one individual to another, and if several different methods are used for measuring antibodies reactive with the antigen it is rare to find an example of tolerance that will embrace them all.

It is now clear that there are a great many conditions when an animal will fail to respond, or respond poorly, to an antigenic stimulus which in a majority of 'normal' control animals will produce easily demonstrable immune responses. No single interpretation can be valid and there are several ways of looking at even 'classical' neonatal tolerance.

(*a*) In the thymus or in any other site of primary differentiation, it is believed that a stem cell becomes an immunocyte because it has been stimulated to produce a receptor equivalent, at least, to the combining site of antibody in the sense that it can react with an appropriate antigenic determinant. This immune pattern of the receptor is of random origin and as likely to react with a body component as with a foreign antigen. It is a necessary part of any clonal selection theory that whenever a cell differentiates as an immunocyte carrying a pattern complementary to any antigenic determinant present in the thymic environment, contact with that antigen will initiate a destructive or inhibitory reaction which will prevent the development of the cell as an immunocyte. In the absence of cells which can react to produce antibody there can be no production of antibody.

Within the thymus there is very active cell proliferation and destruction so that one can confidently assume that in addition to potential antigens in plasma and intercellular fluids, all the vast number of potential antigenic determinants within any unspecialized cell will be available in the thymic environment. If the basic hypothesis of destruction by antigen–receptor contact in the thymus is correct, there will be a sufficiency of 'self' antigens there to ensure that a very large proportion of newly differentiated immunocytes should be destroyed by this process.

(b) For some substances to be antigenic, particularly in young animals, it may be necessary for them to be coated with $\gamma$-globulin (natural antibody). Otherwise they will not be taken up by phagocytic cells in such a way that they can be presented to immunocytes in an effectively antigenic form. Since normal antibody is presumably produced by immunocytes which have passed the thymic 'censorship' and have been unspecifically stimulated, this is essentially only an operational variant of (a).

(c) The immune response to any antigen is always complex and will normally include production of antibody in the form of all three types of immunoglobulin and the appearance of a variety of immunocytes. The relative intensity of each component will vary greatly from one situation to another and there are circumstances where one type of antibody can mask some other type of immune response against the same antigen. Antibody itself has a well-marked feedback capacity to prevent the initiation of any more production of antibody by the same antigen, and this phenomenon can be responsible for some examples of partial unresponsiveness. Experimental allergic encephalomyelitis in rats can be prevented by immune serum; enhancement of transmissible mouse cancer by antiserum is well known. When tolerance to a certain soluble antigen is measured by one method, for example the nonimmune curve of removal of labelled antigen from the circulation, it is quite common to find discrepancies in two directions. An apparently tolerant mouse may show antibody detectable by some other method, and a rabbit which has apparently 'broken tolerance' may fail to produce any precipitating antibody. The possibility almost diametrically opposite to (*b*) that, on occasion, tolerance may be mediated by antibody, must always be kept in mind.

(*d*) If unresponsiveness is due basically to the absence of cells which can respond to antigen, it could result either from destruction of such cells by antigenic contact or from the descendants of a stimulated cell all being forced into an irreversible phase of differentiation. Most immunologists regard the mature plasma cell as an end cell no longer capable of replication or of differentiation into a cell which can replicate. If all the available immunocytes capable of reacting with antigen $X$ are irreversibly committed to plasma cell formation, a state of unresponsiveness must develop until new cells of the required pattern emerge as a result of mutation and differentiation.

Some writers have suggested that there is no such thing as tolerance on the grounds that by sufficiently strenuous treatment an animal can be made to produce antibody against any of its components and even to die of autoimmune disease. This is an illegitimate argument unless we assume that tolerance must be all or nothing. Partial tolerance is easily demonstrated and the effect of large doses of antigen given with Freund's complete adjuvant is merely the end of a continuous spectrum.

My own view would be that (*a*) is the predominant mechanism by which tolerance, including natural tolerance to body components, is mediated and that in experimental studies it may often function by way of the mechanism described in (*b*). The other two approaches are particularly important in relation to partial tolerance.

It must be emphasized here as elsewhere in immunology that tolerance is a soft-edged phenomenon and always partial. When we say that a given antigenic determinant reacts specifically with a combining site on a receptor, this can mean only that the union is a reversible one of such affinity that it can be manifested in some experimentally demonstrable fashion. The affinity level necessary will vary with the reaction being used. This is of special relevance in studying the phenomenon of loss of tolerance when the antigen in question is withheld for long periods.

If tolerance or unresponsiveness is due to the temporary absence for any cause of cells with which the antigenic determinant can react significantly and if there is a constant flow of new mutant patterns finding expression in the thymus, it is self-evident that in the absence of the antigen there will be an opportunity for new reactive immunocytes to emerge. The likelihood of their doing so will diminish with thymectomy and with age. For a time, at least, the clones concerned will probably be directed against only one antigenic determinant of the antigen or represent low avidity patterns with only partial reactivity for the antigen.

## SUMMARY

This outline should give a broad indication of the theoretical approach to be adopted. It is essentially Darwinian natural selection applied at the cellular level to clones instead of at the level of the organism and the species. This general concept can be summarized as follows.

The specific immune pattern presented by the combining site of an antibody molecule or the receptor of an immunologically competent cell is genetically derived. It is still uncertain whether the pattern of the combining site is the responsibility of a single cistron or whether, like the immunoglobulin molecule as a whole, it arises from the convergence of several streams of genetic information. Whatever the source of genetic control, the pattern arises by a random process operationally equivalent to somatic mutation. Associated with the random determination of the immune pattern a strict phenotypic restriction must become operative which limits the cell to the production of one type of immunoglobulin and one pattern of combining site.

It is a function of the thymus and any other sites of primary differentiation to eliminate any newly differentiated immunocytes reacting with sufficient avidity with any antigens that are accessible in the organ.

Immunocytes that pass this censorship colonize lymphoid tissue and are ancestral to the general lymphocyte–plasma cell population of the body.

Any such immunocyte carrying an immune pattern comple- ˙ mentary to an antigenic determinant present in the body, for example as part of a bacterial invader, will be liable to proliferate and so give rise to a clone of cells with similar potentiality. Once any newborn mammal has had an opportunity to experience the impact of antigens from the environment, the lymphoid system can be pictured as an intermingled population of cell lines or clones, those present in largest number being reactive against commonly experienced foreign antigens. The number of distinct clones probably increases to a plateau at adult life but the relative populations of each clone will be highly variable, a reflection at all times of the recent antigenic experience of the individual. The element of a Darwinian situation are all present. The numbers of essentially autonomous genetic elements, the lymphoid cells, are large, perhaps around $10^{12}$ in man, giving adequate opportunity for mutation, and the selective advantage of an accidentally appropriate pattern is enormous and extremely rapidly expressed.

# 3　The thymus in relation to the origin and differentiation of lymphoid cells

The small lymphocyte is probably the commonest mobile nucleated cell of the body. Its origin and functional differentiation must always remain the central problem of immunology viewed from the cytological aspect. An immediate difficulty arises in the rapidly developing view that under the name 'lymphocyte' or 'small lymphocyte' we may be lumping together many different types of cell. In particular, the lymphocytes commonly seen in the bone-marrow may have different potentialities from those in the circulation, lymph nodes and spleen, and these again from those in the thymus. Other workers, however, while accepting these observable differences, see no reason why genetically similar cells should not show different behaviour as a result of their sojourn in different internal environments.

All body cells, apart from haploid germ cells, are assumed to contain the full complement of genetic information present in the fertilized ovum from which they developed. The problems of defining the limits of a certain morphological type of cell and tracing its line of origin by progressive differentiation are probably insoluble by any single experimental approach. The lymphocyte is particularly difficult to define as its morphology is undistinctive. It is known that some small lymphocytes may become large pyroninophil cells capable of mitosis, and there is strong evidence that either directly or by way of these active cells, even macrophages, plasma cells, mast cells and fibroblasts may be derived from cells with the morphology of small lymphocytes. If this is true it suggests the need for adopting an attitude to the small lymphocyte that marks it off from all other cell types. We shall adopt the hypothesis that a lymphocyte is a mobile repository of genetic information without any executive organization of the cytoplasm beyond

57

that necessary for its free mobility through tissues, plus receptors of some sort which can mediate the capacity of environmental stimuli to mobilize the particular store of information required. If we include all populations of small lymphocytes defined simply on a morphological basis, they presumably include cells whose genome is at various stages of differentiation not necessarily always directed toward immunological function.

In any emergency situation one can see the value of having always available large numbers of cells, morphologically small lymphocytes but representing perhaps many thousand clones each of which has a differentiated potentiality, which can be called into action by appropriate stimulation. If only 1 in 1,000 cells are of the needed potentiality the number can be rapidly raised to 10 per cent by a series of six or seven binary divisions of the stimulated cell within a few days. It would be reasonable, too, that to provide the special building blocks needed for the rapid expansion of a badly needed clone, lymphocytes not needed should be broken down at random. This would present the trophic function of lymphocytes referred to or hypothecated by numerous writers and provide an evolutionary rationale for the abnormal vulnerability of lymphocytes to irradiation, corticosteroids and cytotoxic drugs.

## THE ROLE OF THE THYMUS

From this operational outlook by which the lymphocyte is regarded simply as a mobile store of genetic information with no immediate executive responsibility, it is not necessary to assume that all lymphocytes arise from a common source at some stage in embryonic development. In recent years, however, it has become clear that the thymus plays an important part in the development and functioning of lymphocytic populations in both mammals and birds. There is strong evidence that in the placental mammals that have been studied in any detail the embryonic thymus is the source of the first lymphocytes, and through late embryonic life, and at least until adolescence, is an important centre for the control of lymphocytic activity. There is no good evidence against the speculation that all lymphocytes could be traced back to origin in the epithelial cells of the thymus primordium but, equally, there is nothing to counter the assumption that primitive blood-forming cells at any stage of development could also take on the appearance of lymphocytes. The evolutionary approach suggests that thymus, lymphocytes, $\gamma$-globulin and capacity for immune response appeared together in the higher cyclostomes.

Embryologically the epithelial cells of the early thymus seem to take on lymphocytic character under the stimulus of adjacent mesenchymal cells at a time when there are no other lymphocytes in the body. Failure of the thymus to develop may be associated with almost complete absence of lymphoid tissue and virtually no circulating lymphocytes, but these cases seem to vary considerably from one to another and some show only moderate reduction of blood lymphocytes. Removal of the thymus on the first day of life in certain strains of mice gives rise to a situation of generally impaired immunological effectiveness. J. F. A. P. Miller's first accounts suggested that virtually total loss of immune responses was the rule if thymectomy was complete. The results with other strains of mice and other species have been in the same general direction but less striking. It appears that in mice and placental mammals generally there is active production of lymphocytes from the thymus before birth and neonatal thymectomy has no special virtue in itself. It merely cuts off a process at a variable time after its initiation.

Opinion is still fluid in regard to the relative importance of the thymus as a source of lymphocytes or as an endocrine organ producing hormone(s) controlling immunological function. Any general statement must therefore involve an arbitrary choice between alternatives at certain points and will eventually be found to be unduly simplified. With these reservations we can use experimental, comparative and clinical sources of information to summarize the function of the thymus as follows:

(*a*) The thymus is the primary centre for differentiation of immunocytes and provides the progenitors of the various clones of lymphoid cells which eventually colonize the peripheral lymphoid tissues. The actual proportion of the cells produced in the thymus which find it possible to do this is probably very small in adult life but considerably higher in prenatal and neonatal periods.

(*b*) For the first few days of life the colonizing lymphocytes and the clones to which they give rise need the persistence of a humoral agent produced by the non-lymphocytic tissues of the thymus if they are to remain indefinitely viable.

(c) In the neonatally thymectomized mouse there is a gross deficiency in the types of lymphocyte available and there are patchy deficiencies in the types of immune responses that can be initiated. In general the capacity to reject homografts or to display delayed hypersensitivity reactions is depressed more than the ability to produce conventional antibodies.

### THE STRUCTURE OF THE THYMUS

In a young animal, a 4- to 8-week-old mouse or a child of ten for example, the thymus is still large and very actively producing lymphocytes. Yet its removal at this stage appears to have practically no effect in impairing immunological responses. The only way to demonstrate impaired function is to expose the animal to some severe stress such as near-lethal X-irradiation. Under such circumstances the full recovery of immune function is more rapid in the presence of a thymus than in its absence.

A great deal of attention has been paid to the dynamics of the lymphoid cells which make up the main mass of the thymus in the young animal. The basic finding is that there is a rapid turnover of cells which, apart from proliferation, destruction and liberation into the general circulation, involves a relatively slow but continuous entry of new stem cells from extra-thymic sources. When two mutually compatible mouse strains, one of which carries a recognizable marker chromosome, are available, parabiosis experiments can be used to provide an almost physiological process of lymphocytic interchange. Under these circumstances it was possible to show that after five weeks with a common blood circulation the two partners had almost reached an equilibrium in lymph nodes and spleen, and even in the thymus 12 per cent of the mitoses were of the partner's type. Grafts of neonatal thymus undergo atrophy and depletion before redeveloping normal histological structure. The conditions are therefore more or less unphysiological but the findings are consistent that within three weeks the mitoses seen in the graft are wholly of host type.

To provide a structural basis for envisaging the cellular inter-change we may adopt the interpretation that has been derived from electron microscope studies by Clark (1963). In fig. 8 it is modified slightly to include the macrophages ('PAS cells') which are be-lieved to have an important functional role. In Clark's view, the thymus is an epithelial organ densely honeycombed with packets of proliferating lymphocytes in the cortex, and with the medulla irregularly infiltrated with lymphocytes and other cells mostly associated with blood vessels. An individual packet is shown in the

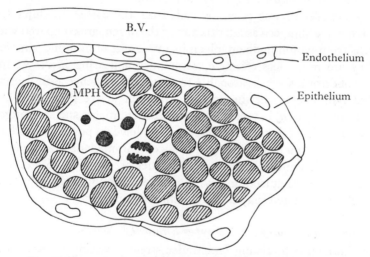

Fig. 8. The unit of cortical cells in the thymus (Clark packet).
B.V. = blood vessel; MPH = macrophage.

diagram. Essentially it is a nest of actively multiplying lymphocytes plus one (or occasionally more) macrophage usually recognizable by its content of phagocytosed nuclear debris. The packet is com-pletely enclosed in a tenuous capsule of epithelial cells with a minimal amount of supporting tissue, including collagen and elastic fibres.

The great majority of the thymic cells are lymphocytes; large, medium and small. In the cortex their close-packed distribution in Clark packets provides a different texture from the randomly dis-tributed lymphocytes in primary follicles of lymph node or spleen.

61

In smears or impression preparations the cells are typical lymphocytes. If collections of thymic lymphocytes are compared with those of the thoracic duct or with circulating lymphocytes there are certain differences, usually rather subtle ones that do not necessarily hold for other strains or species. In mice there are at least two antigens which can differentiate thymic lymphocytes from the great majority of peripheral lymphocytes. These are shown by cytotoxic tests and may, in part, depend on the susceptibility of thymic cells to damage. Absorption experiments, however, indicate that a real, if labile, antigen is concerned.

Whenever thymic cells, i.e. cells from the cortex, medulla and connective and perivascular tissues of the organ, are compared with cells from lymph node or spleen in any aspect of immune function they are much less effective, sometimes appearing inert. At the risk of error it seems reasonable to accept the evidence as indicating that lymphocytes in the cortex with their active but controlled multiplication (*a*) have a serologically differentiable surface antigen not found in peripheral lymphocytes, (*b*) have no power to produce antibody *in situ* or *in vitro*, (*c*) produce no immediate graft-versus-host reaction and (*d*) when opportunities are provided for thymic cells to pass to spleen and lymph nodes, immune competence to react with antigen may become demonstrable.

*Changes in cell numbers and types within the thymus*
In the course of ageing and under stress a variety of structural changes can be observed in the thymus. It is well known in human pathology that at autopsy after any illness lasting more than a day or two there is thymic atrophy, most marked in the cortex. The picture can be simulated in experimental mice by administration of a milligram of hydrocortisone, moderate local or general irradiation, a large but sublethal dose of almost any cytotoxic agent and any severe, acute or subacute infection. Any of these will, in the mouse, result first in the necrosis of many cortical cells with pyknosis and active phagocytosis of nuclear debris, followed by the progressive depletion of the lymphocytes in the cortex. Within a day or two the cortex may be almost empty of lymphocytes. It appears greatly shrunken and composed of epithelial cells, macrophages, occasional plasma cells, and structural elements associated

with the blood vessels. After removal of the stress the cortex refills with lymphocytes.

In ageing there is a general diminution of thymic weight with a relative expansion of medullary area in relation to cortex. In man both cortex and medulla are also progressively replaced by fat; and in old individuals, although an atrophic thymus can be found at autopsy, section shows only shreds of thymic medullary tissue.

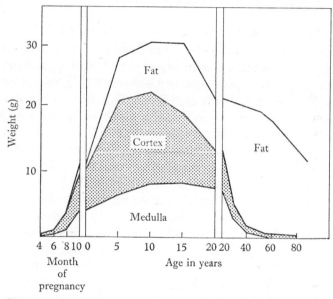

Fig. 9. Changes with age in the weight and composition of the human thymus. Redrawn from J. A. Hammar (1936), *Die normalmorphologische Thymusforschung im letzten Vierteljahre.* Leipzig: Barth.

In mice one finds a variety of appearances in older mice including cellular accumulations in the medulla. In the autoimmune strain NZB and its hybrids with which I have had much experience, there is an almost regular development of germinal centres in the medulla and a variety of secondary changes. In these strains both mast cells and plasma cells are common and I have seen a number of sections in which large areas of cortex have been converted *en masse* to mast cells and in the hybrid (NZB × NZW) F$_1$ areas of cortex almost wholly replaced by pyroninophil cells. The impres-

63

sion is overwhelming in both conditions that lymphocytes in the cortical packets have been transformed directly either to mast cells or pyroninophil cells closely resembling immature plasma cells.

These are essentially pathological conditions but they point toward an unsuspected lability of differentiation of thymocytes that may be relevant to the theme that their differentiation to immunocytes is the main function of the thymus.

*The significance of the thymus in the perinatal period*

Congenital failure of the thymus to develop in man or its removal surgically in the immediate neonatal period in mice, both produce striking effects which are not seen as a result of thymectomy later, even if this is only a week after birth. The discovery of the effect of neonatal thymectomy in mice by Miller in 1960–1 was almost wholly responsible for current interest in thymic function. As shown in the previous section, the fact that later removal of the thymus has little effect on the apparent well-being of a mouse does not mean that the thymus functions only in the perinatal period. The main virtue of neonatal thymectomy is that the effects produced are gross and can be readily demonstrated within a few weeks or months. Few investigators like to wait for a year or more before the result of an experimental manipulation becomes apparent.

The effect of neonatal thymectomy varies greatly with the species of animal used and in the mouse (the species most commonly studied) with the strain. There is no visible effect of neonatal thymectomy in the dog and very little in the rabbit. The effect in the rat is clearly demonstrable but there is only a partial and quantitative depression of immune function. Only in certain strains of mice and in that immunological curiosity, the golden hamster, is the effect of neonatal thymectomy of the classical type described by Miller in his first papers. Initially then, we may consider the nature and significance of the classical result.

In the strains C3H and (AK × T6) $F_1$ neonatal thymectomy has no effect on the growth of the infant mouse for several weeks. Until then it gains weight normally and looks healthy and, if it is raised in germ-free conditions, may remain healthy indefinitely. Reared conventionally, a wasting disease whose nature is still controversial, sets in, and the mice die at around three months of age. This, however, allows a reasonable period over which tests of immunological capacity can be carried out. Restricting ourselves to this period, it is found that the mice show a level of circulating lymphocytes about half the normal value and lymphoid tissue reduced to about the same extent. Plasma cells are present in the intestinal tissues and in lymph nodes and spleen. There are variable changes in the serum proteins but usually all the immunoglobulins can be detected.

Of immunological deficiences the most outstanding is the capacity to accept homografts of skin and even heterografts of rat skin. Antibody-producing capacity is reduced but when a number of antigens are used there is an almost random-appearing mixture of ability and inability to produce individual antibodies.

The capacity of various manipulations to restore immune competence to neonatally thymectomized animals has been extensively studied. Amongst the effective procedures are early grafting of neonatal syngeneic or allogeneic thymic tissue, and repeated injections of syngeneic thymic cells or of relatively large amounts of syngeneic spleen cells. Thymic tissue implanted in Millipore chambers intraperitoneally is also effective. Injection of syngeneic bone-marrow cells or foetal liver does not protect nor does a 'successful' graft of rat thymus. Most workers have failed to show significant effects of cell-free thymic extracts but this may merely reflect inadequate dosage.

Perhaps the most significant findings are (*a*) the failure of bone-marrow cells to return immune capacity either in the neonatally thymectomized mouse or in the older mouse subjected successively to thymectomy and lethal X-irradiation (although in the latter case bone-marrow can 'rescue' the animal and, with thymic tissue in Millipore chambers, reconfer immune capacity) and (*b*) the ability of adult syngeneic spleen cells in adequate dose to return immune capacity. In terms of our general hypothesis these are compatible with the views: (i) That bone-marrow contains cells which can give rise to immunocytes but only in the presence of a thymus or an adequate concentration of thymic hormone. In their absence bone-marrow cells are immunologically inert. (ii) If an adequate number of compatible adult lymphoid cells are provided, the neonatally thymectomized animal can remain immunologically competent at least for a considerable period.

*The comparative anatomy of the thymus*
More is known about the function of the thymus in the mouse than in any other mammal and there is enough to suggest that apart from some differences in the timing of its development the situation in other placental mammals, including man, is functionally similar. There are, however, anatomical and embryological differences. In most mammals the thymus develops from the third branchial pouch and in the course of development migrates into the thorax. In the guinea-pig the organ remains in the neck. Many marsupials have two sets of thymuses, a pair of cervical thymuses derived from the third branchial pouch and a thoracic thymus from the fourth and fifth pouches. All have a microscopic structure similar to that of placental mammals, and there is evidence from the American opossum that their function is broadly similar.

*Entry of cells into the thymus in the postnatal period*
It is now certain that among the circulating cells of the blood are some from which may be derived lymphocytes, plasma cells, erythrocytes, macrophages and granulocytes and, in all probability, vascular endothelium and fibroblasts. It is the simplest of a number of as yet unverifiable alternative hypotheses to believe that the 'stem cells' concerned may initially be capable of differentiation in any of these directions and that the actual route along which they are differentiated is determined by their sojourn in an appropriate internal environment for an adequate period. There is relatively firm evidence that such stem cells arise from the bone-marrow (and perhaps any other sites of haemopoietic activity).

Evidence from parabiosis indicates a continuous changeover of the proliferating cortical lymphocytes in the thymus. Our current hypothesis is that any stem cell, that is, a relatively undifferentiated cell, which can enter the structural units of the thymic cortex is there subject to differentiation to an immunocyte. This is a term which it is convenient to use for any cell, lymphocyte, plasma cell or other, which has specific immune reactivity of any sort. In our usage it will imply a good deal of interchangeability between

66

different morphological or functional characters within a single clone of immunocytes (fig. 10).

Autoradiographic studies using tritiated thymidine, plus more old-fashioned methods using counts of mitotic rates under a variety of conditions, have given a reasonably consistent picture of the cell dynamics of the mouse thymus. There is nothing substantial to suggest that the behaviour of the thymus in the mouse is in any way exceptional amongst mammals. There is a very high level of proliferative activity amongst the large and medium lymphocytes; 70–80 per cent being labelled by a single intravenous injection when fixed an hour later. A proportion of all mitoses involve differentiation toward the standard form of the small lymphocyte,

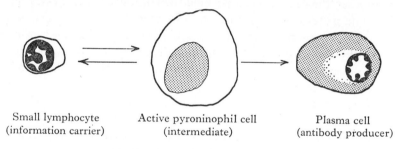

| Small lymphocyte (information carrier) | Active pyroninophil cell (intermediate) | Plasma cell (antibody producer) |

Fig. 10. The immunocyte concept.

and at a certain point two small lymphocytes are produced by division of a medium lymphocyte and do not multiply further in the thymus. On the average, such small lymphocytes spend 3–4 days in the thymus from their time of birth. Then they disappear. It is still an unsolved problem as to what happens to the great majority of cells produced by mitosis in the thymus. A proportion undoubtedly die *in situ* as judged by the constant presence of pyknotic debris in macrophages. A common belief is that large numbers leave the thymus as small lymphocytes but, for some reason, the great majority are autolysed and disappear as individual entities soon after entering the circulation. However, even if the output of cells which develop into viable clones is something less than 1 per cent, this might well be adequate for any immunological function, such as the provision of new immune patterns, which may be dependent on the thymus.

In discussing the basis of immunity it is impossible to start at one point and progress logically to a complete synthesis. In addition to the incompleteness and sometimes the contradictions of the experimental evidence, each set of phenomena is influenced by a variety of factors that it would be more convenient to discuss in some other setting. The significance of the entry of perhaps 5 per cent of new clones per day to the thymic cortex is one of these difficulties. It can only be discussed adequately in terms of immune pattern and the origin of diversity within immune pattern (see chapter 6). Here then, in the context of thymic function, we can simply accept the existence of specific immune pattern as something based on genetic information and phenotypically expressed only when the cell has been enabled to synthesize protein carrying immune pattern. The first function of that protein is to serve as one or more cell receptors, the essential feature of clonal selection theory being that the receptor has broadly the same pattern as that of the combining site of antibody produced by other cells of the clone.

The picture of thymic function which has been rapidly developing, viz., that it is the most important and perhaps the only source of new immune pattern, seems to require two distinct capacities of the thymus—that it should be the sole site at which stem cells can become immunocytes and that it should release into the general circulation only immunocytes which are not reactive against body components. This requires some preliminary discussion of the process of differentiation in general and of the thesis that in immunocytes new patterns arise by somatic mutation or some operationally equivalent process.

On modern genetic theory, every cell of the body carries the whole of the genetic information present in the diploid complement of chromosomes as soon as fertilization is complete. The cistron or cistrons concerned in determining immune pattern are present in every cell and, more particularly, in every stem cell capable of entering the circulation. If the region concerned is subject to normal mutability or has developed an increased mutability as an evolutionary requirement, the stem cells, by the time of birth, will have accumulated a wide range of potential patterns. In the absence of appropriate differentiation, however, the patterns will never be expressed. It matters not in the least that a skin cell or an epithelial cell in the small intestine is potentially an ancestor of an auto-immune forbidden clone. If it is not differentiated to an immuno-cyte it can do no harm.

The nature of differentiation is still unknown but it is reasonably clear that *adjacent cells* play a very important part. In the thymus, entering cells are intimately associated with the reticulo-epithelial cells that seem to endow the thymus with its individual character as an organ. They almost certainly provide the differentiating environment. The stem cell enters a Clark packet, proliferates and, by hypothesis, the 'immune cistron' is de-repressed. Synthetic

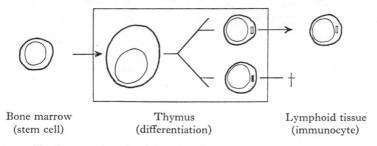

| Bone marrow | Thymus | Lymphoid tissue |
| (stem cell) | (differentiation) | (immunocyte) |

Fig. 11. To illustrate the role of the thymus in differentiation and 'censorship'. With differentiation, immune pattern emerges and the cell becomes reactive to contact with antigen. Open rectangle = 'non-self' pattern; black rectangle = 'self' pattern.

activity for the product, in the first instance as receptor rather than as antibody, is initiated. By the time the terminal small lympho-cytes are produced, the receptors must be functional.

At this point the second postulated function of the thymus comes into play. When a newly emergent receptor can react specifically with any antigenic determinants present in the thymic environment, the cell carrying it must be destroyed. Depending on a number of possible factors, destruction may take place as soon as the receptor is present at all, perhaps in the large or medium lymphocyte phase, or it may be delayed until the 'sensitized' cell is liberated into the circulation. Whatever the details, the essential feature claimed for this hypothesis is that the only newly differ-entiated cells which can give rise to normal clones of immunocytes are those which fail to react with accessible body components.

This postulated double function of the thymus is virtually the only way in which the facts can be interpreted within the framework of clonal selection theory. A point which has not so far been mentioned in this context is that in the thymus a cell being made ready for one defined type of differentiation is likely to be temporarily highly labile and easily influenced by abnormal stimuli to develop along an aberrant line of differentiation. This is the most likely interpretation of the very wide range of morphological cell types to be seen when thymic lesions develop in the 'autoimmune' mouse strain NZB and its hybrids. In addition to occasional massive change to pyroninophilic cells approaching plasma cells in character, or to mast cells, a variety of other cell types may be present in affected areas. Some may have come via the circulation but, even so, the form they take in the thymus must be determined largely by the nature of the environment provided.

The existence of a thymic agent active on lymphocytic function has been known since Metcalf's work in 1956 and it is probable that this plays an important part in the various actions that have been ascribed to the internal environment of the thymus. In appropriate experimental situations it can be shown that the presence of thymic tissues in Millipore chambers implanted in the peritoneal cavity can restore capacity to reject homografts both to neonatally thymectomized animals and to animals thymectomized at a later stage and then subjected to a lethal irradiation dose with 'rescue' by bone-marrow cells. This is perhaps a little at variance with the special importance ascribed to the actual thymic environment, since it indicates a capacity of thymic agents to allow manifestation of immunocytic capacity in other environments than the thymus. It does, however, strengthen the claim that the internal thymic environment may have a potent effect on cells differentiating within it.

In summary then, the function of the thymus in postnatal life is to serve as a source of new immune patterns to ensure an adequate supply of 'information' to deal with any significantly likely emergency. This is done by making use of all the genetic information relevant to immune pattern that has been developed in the cell genome either during the course of vertebrate evolution or as a result of somatic mutation or equivalent processes during the life of the individual. This information will be accumulated in the first instance within the stem cells of the haematopoietic system. Each cell must carry a number of alternative potentialities as to the type of immunoglobulin it will produce and each, depending on its line of somatic descent, will have accumulated changes due to somatic mutation. The nature and extent of the potentialities in each stem cell must be deferred for later discussion but the materialization of one potentiality only is the essence of differentiation. Those stem cells which enter the thymus are constrained to differentiate to immunocytes, that is, by expressing one potentiality to produce receptor protein or antibody of immune quality. Once such receptors are in existence the cell becomes susceptible to positive or negative stimulation by the antigenic determinants which are sterically complementary to the immune pattern. In the thymic environment such contact results in death of the cell either within the thymus or soon after its liberation to the circulation. As will be discussed later, this provides the essential basis for natural immune tolerance of body components.

*The function of thymus and bursa in birds*

The avian thymus is unlike any seen in mammals, being composed of a string of ovoid lobules running along each side of the neck. In the chicken there are usually seven lobules, each in a young bird of 10–12 weeks being about 6–10 mm in longest diameter. Embryologically the epithelial thymus first shows the appearance of lymphocytes at ten days' incubation. It persists through life but with a seasonal cycle of partial atrophy with relative medullary enlargement followed, after the breeding season, by a redevelopment of cortical tissue and increase in weight.

Another organ whose function is in many ways parallel to that of the thymus is the bursa of Fabricius, which develops as an epithelial sac budding from the dorsal region of the cloaca. From and after the fifteenth day of hatching, lymphocytes appear; probably by the entry of stem cells from the blood, although direct conversion of cells in the epithelial nodules to lymphocytes is perhaps not wholly excluded. In one way or another a multilobular lymphoid organ develops. It reaches a maximum size of 1·5 to 2 g and atrophies at sexual maturity. This atrophy is presumably a response to hormonal stimulation and, following this clue, an important means of inhibiting the development of the bursa has been developed in the form of treatment of the embryo with testosterone or related androgens.

Either by hormonal treatment in the embryo or by surgical bursectomy immediately after hatching, chickens lose completely or almost completely their capacity to produce antibody against any of the standard types of antigen. This can occur without any significant reduction in the immunoglobulin content of serum.

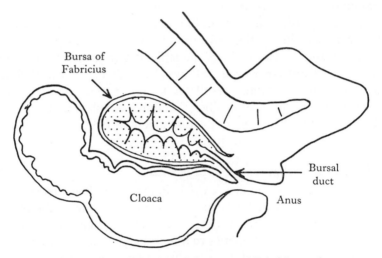

Fig. 12. The relationship of the bursa of Fabricius to the cloaca in the chicken.

Bursectomized animals with intact thymuses show no difference in the time of rejection of homografts or in the capacity to produce graft-versus-host reactions in appropriate recipients. There is some difference of opinion in regard to sensitization by tuberculin, but there is acceptable evidence that bursectomized chickens incapable of producing antibody can respond to treatment with CNS tissue plus adjuvant by the development of 'allergic' encephalomyelitis.

Neonatal thymectomy is difficult to carry out successfully in chickens but all reports consistently show a prolongation of the time of skin homograft rejection. Other features reported are a diminution in the number of circulating lymphocytes and *no* diminution in the capacity to produce graft-versus-host reactions on the chorioallantois.

There is an unfortunate lack of information on many aspects of bursal and thymic function in chickens and one would feel that an intensive study of this system is urgently needed. If, as appears superficially to be the case, we have the mammalian thymic function divided between two avian organs, there must be great scope for the further analysis of that function. The bursa is *not* an antibody-producing organ and, in a bird immunized with sheep red cells, contains no cells capable of producing antibody plaques by Jerne and Nordin's technique. The fate is not known of the lymphocytes produced so abundantly in the bursa. There is, however, evidence that (as in the mouse thymus) multiplying clones are constantly reinforced by new stem cells from perhaps thymus or bone-marrow.

Two groups have provided evidence that soluble products from bursal cells in diffusion chambers can regenerate immune capacity to some extent in bursectomized chickens. There is evidence that plasma cells are absent or very rare in the spleens of bursectomized chickens but there appears to be no consistent diminution in the concentration of immunoglobulins in serum.

Without a great deal more information no satisfactory interpretation of the avian thymus–bursa relationship is possible. I have suggested as a hypothesis with heuristic potentialities—and under the influence of a related hypothesis due to Good—that the thymus is responsible, as in mammals, for the differentiation and 'censorship' of immunocytes but that if they are to give rise to proliferating cells of antibody-producing type they must be exposed to a hormone produced by 'gut-associated lymphoid tissue'. This exposure could occur by the entry into the bursa of cells partially differentiated in the thymus or by contact with bursal hormone at a distance. Present opinion, however, is swinging rather rapidly to the view that the bursa and its mammalian analogues receive stem cells direct from the bone-marrow and differentiate them to potential antibody-producing immunocytes. Any further elaboration of this hypothesis can be left until p. 92, where it can be brought into relation with some of the features of human immuno-pathology.

# 4   The immunocyte: its definition, recognition and distribution

At the cellular level the central problems of immunology concern the differentiation, genealogy and interconvertibility of the various types of mobile nucleated cells present in the circulation, wandering through tissues or accumulated in bone-marrow, thymus or the peripheral lymphoid organs. These include the granulocytes (polymorphonuclear leucocytes, eosinophils and basophils) and the rather enigmatic tissue mast cell. The cells of the lymphoid series from blast through medium to small lymphocyte and the related plasmablast to mature plasma cell series, contain what, following Dameshek, we shall call 'immunocytes', but it should be emphasized at once that it is not possible to equate morphology with immune function. Finally, there are the cells of the reticulo-endothelial system primarily defined on their phagocytic capacity but differing very considerably from one site to another.

No immunologist has any doubt that the process of antibody formation and all the other basic activities of the immune response take place within cells of these groups. The problem is to co-ordinate what is known at the morphological level, including the sequence of progenitors and the range of differentiation, with the immunological behaviour of tissues, discrete cell populations or single cells.

## THE EXPERIMENTAL DEMONSTRATION OF IMMUNE FUNCTION IN CELL POPULATIONS

There are a number of possible approaches, most of which will be mentioned only briefly as now having little more than historical interest.

Surgical removal of spleen or thymus has only minor effects on the immune reactivity of an animal. Rather more has been learned by using the technique of shielding a particular part of the body or a single organ when an animal is heavily X-irradiated using a dose sufficient to annul all immunological response to an antigen given in the next day or two. The most interesting finding was that, in the rabbit, shielding the appendix which has much lymphoid tissue and many plasma cells allows almost full retention of antibody-producing capacity. In a sense this is taken to indicate not that antibody is mostly synthesized in the appendix but that there are mobile cells in the appendix which within a few hours can effectively restock the various lymphoid organs depleted of their lymphocytes by the irradiation. An operationally equivalent procedure would be to obtain from a syngeneic animal an adequate number of living lymphocytes and use them to restore immune capacity to an animal rendered immunologically inert by whole-body irradiation. There are many variants of this approach. Nearly all require that the recipient must not actively reject the transferred cells and that it should be incapable of producing the antibody under study. Both ends can be attained by recent X-irradiation or the use of embryonic or neonatal animals.

The principal difficulty in all such experiments is that any suspension of cells from an animal, whether from spleen, bone-marrow, blood leucocytes or thoracic duct, is a heterogeneous population. Usually the most that can be said is that from a mixture of cells of which the differential count in a stained smear is such and such, a certain immune response is obtained with doses of cells equal to or greater than $x$, $x$ being usually of the order of some millions of cells.

For many years now, it has been known that suspensions of cells from any organ of a primed animal that contained relatively large numbers of lymphocytes, plasma cells and macrophages could give rise to antibody production under such experimental conditions. The role of the plasma cell as a major producer of antibody has been universally accepted since 1948 and modern work has been largely directed toward establishing the immunological role of the lymphocyte.

From all points of view the most suitable source of lymphocytes is the thoracic duct, and recent opinion in this field has been greatly influenced by Gowans's exploitation of the opportunities so provided. Most thoracic duct cells are small lymphocytes but there is always a proportion of large and medium lymphocytes some of

75

which may be indistinguishable from plasmablasts. By suitable manipulation almost all the large cells can be eliminated and by administration of such material Gowans and collaborators have shown that immune capacity may be lost and regained under the following circumstances:

(*a*) Chronic drainage of the thoracic duct in rats produces severe depletion of lymphocytes, and such animals can neither produce antibody against a standard antigen such as sheep red cells, nor reject a skin homograft which differs by only minor histocompatibility factors.

(*b*) Such depleted animals can be restored to normal immunological capacity by the administration of thoracic duct cells from a compatible donor.

(*c*) Rats rendered tolerant to skin homografts lose their tolerance when given syngeneic normal thoracic duct cells.

(*d*) Thoracic duct cells from a donor immunized with tetanus toxoid conferred on a syngeneic recipient the ability to give a secondary-type response to challenge with the same antigen.

(*e*) Using suitable pure strains of rats differing by major histocompatibility factors, injection of thoracic duct cells into an allogeneic recipient produces a severe graft-versus-host reaction. In the course of this a proportion of small lymphocytes in the inoculum are converted to large pyroninophil cells capable of mitosis.

These findings are consistent with what has been obtained by the use of less suitable material in the form of cell suspensions from spleen and lymph nodes. No one now has any serious doubt that within any large population of lymphocytes taken from an animal of defined immunological status there are subpopulations of lymphocytes capable of transferring capacity to react in fashions corresponding to the immunological ability of the donor. This does not of course mean any more than that small lymphocytes can carry specific immunological information. It does not allow identification of individual capacity in any given cell or provide a measure of the proportion of the lymphocyte population with specific activity. For this, methods applicable to single cells are necessary.

## THE FUNCTIONAL RECOGNITION OF INDIVIDUAL
## IMMUNOCYTES

An immunocyte is a cell that can react with a specific antigen in some recognizable fashion, and a number of ways have been devised by which the specific immunological reactivity of single cells can be demonstrated. The most direct is to use a sensitive test to measure the production of antibody by an isolated cell. Another method is to observe under the microscope whether visible antigenic particles adhere to the surface of a given cell. Immunofluorescent methods are more generally applicable to detect either immunoglobulin or, by the 'sandwich' method, specific antibody. Finally the recognition of immunoglobulin by using ferritin-labelled antiglobulin can be used to identify immunocytes in electron micrographs of tissue or cell sections.

### The role of the plasma cell

Since Astrid Fagraeus's work in 1948 there has been little doubt that the plasma cell is the predominant producer of antibody and in this section it will be convenient to sketch the confirmation of this opinion by the use of modern methods.

In ordinary histological sections or smears, plasma cells are easily recognized by their round nuclei with lumps of chromatin giving a clock-face appearance and fairly abundant basophil cytoplasm. The basophilia is due to large amounts of RNA and sections made for electron microscopy show that this RNA is part of the characteristic mechanism of protein-synthesizing and -secreting cells, the endoplasmic reticulum. In plasma cells, as in the cells of the pancreas or any other protein-secreting gland, there is a dense accumulation of close-packed flattened vesicles, smooth on the inside, roughened on the outside, with enormous numbers of ribosomes. The ribosomes represent the site of protein synthesis and it has now been clearly demonstrated by the use of an antigen (ferritin) which can be seen in electron micrographs, that antibody accumulates within the vesicles. It is reasonable to believe that this accumulated immunoglobulin is liberated either continuously or intermittently into the surrounding fluid. This has not yet been experimentally established and the possibility exists that what is

77

seen within the vesicles is a stored reserve rather than antibody at an intermediate stage of a continuous process of synthesis and secretion.

A more general approach is by the use of fluorescent antibody. By appropriate chemical manipulation it is possible to attach a highly fluorescent dye (e.g. fluorescein) to an antibody molecule without damaging its power to combine with antigen. The combination can thus be used as a specific stain for antigen if we wish to find, for instance, where a certain antigen is located in a section of lymph node in an immunized animal. If the animal was inoculated 24 hours previously it will be found that the reticulo-endothelial cells (macrophages) will be stained and fluoresce brilliantly under appropriate illumination. Neither lymphocytes nor plasma cells are stained by this method.

To use the method to detect immunoglobulin in or on a cell and, by implication, so to identify it as an immunocyte, the immunoglobulin is regarded as an antigen. The best technique if human cells are to be studied is to produce in rabbits specific antisera against human A, G and M immunoglobulins and prepare specific fluorescein-coupled antibodies. Further refinements can be introduced by preparing specific sera against one or other type of light chain (see chapter 5). In this way it has been found that all three immunoglobulins may be produced by plasma cells, mature and immature.

The sandwich technique makes it possible to show that plasma cells in an immunized animal contain specific antibody. A section usually made from frozen tissue is first treated with *antigen* and washed. Antigen is thereby attached to antibody if this is present in the cell and a considerable proportion of antigenic determinant groups will be left on the surface available for binding to fluorescent antibody and specific for the antigen being studied.

Finally, there are two direct methods by which single cells can be shown to produce functional antibody. Both will be discussed at more length in subsequent chapters.

The first is to separate a single cell from the local lymph node of an immunized animal, wash it and allow it to secrete antibody into a tiny volume of fluid. The antibody can then be detected if a sufficiently sensitive test is available. One method, developed by

Nossal and Lederberg, is to immunize rats with the flagellar antigen of a *Salmonella* strain. Flagellar antibody attaches to and entangles the flagella responsible for the motility of bacteria. If a washed lymphoid cell from a rat immunized with flagellar antigen of *Salmonella adelaide* is left in a microdroplet of saline for an hour and then five or ten motile bacteria of the right strain introduced, we have a sensitive system for the detection of antibody. If the cell is producing antibody, the bacteria will be immobilized within a minute or two. If it is not, the bacteria will swim around actively for an indefinite period. With this method, virtually all the cells which produce antibody are plasma cells or early members of the plasma cell line.

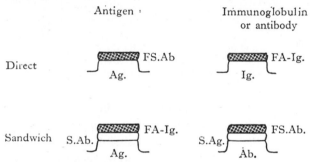

Fig. 13. Detection of antigen, immunoglobulin and antibody by fluorescent antibody. FA-Ig. fluorescent antiglobulin; FS. Ab. fluorescent specific antibody; S. Ab. specific antibody; S. Ag. specific antigen.

Antibody against certain bacteriophages can be detected in very small amount and a basically similar method has been to take the fluid secreted into a microdrop by a single cell and test this for phage-neutralizing power.

Another method involving individual cells uses a different principle. If the antibody being studied is one that can damage red cells and produce haemolysis, it is possible to incorporate cells being tested for their production of this antibody and an excess of 'target' red cells in a thin agar layer. A cell-secreting antibody will then be surrounded by red cells to which the antibody becomes attached. Such cells are vulnerable to haemolysis by fresh guinea-pig serum (complement) and after such treatment one sees scattered

79

over the uniform layer of red cells small circular holes (plaques) which are colourless and almost transparent. Under the microscope a single cell will be found in the centre of each plaque; obviously the source from which the antibody was derived. Again it is found that some of the cells are plasma cells, but most of them are large and medium lymphocytes or immature plasma cells, presumably capable of further proliferation.

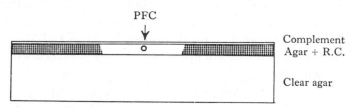

Fig. 14. The principle of antibody plaques by the Jerne–Nordin method. The upper layer contains a mixture of splenic cells and red cells (antigen). After incubation the plaques are 'developed' by complement.

By all these methods, plasma cells can be identified as antibody or immunoglobulin producers but it is recorded in almost all the relevant papers that occasionally or frequently positive results were obtained from cells morphologically not plasma cells. Some had the appearance of small lymphocytes, and some have been shown to contain only weakly developed endoplasmic reticulum.

*Immune function in individual lymphocytes*

Basically similar methods have been used and in fact many of the positive findings have been in the course of experiments in which plasma cells gave the most frequent and conspicuous positive appearances.

It is perhaps significant that positive findings are more common in conditions associated with typical delayed hypersensitivity reactions or with conditions such as homograft immunity and graft-versus-host reactions which are more analogous to delayed hypersensitivity than to antibody production, i.e. in 'thymus-dependent' conditions.

The only impressive results by the use of the sandwich technique are those of Raffel's group in guinea-pigs sensitized respectively

Plates 1, 2, 3. Phases of the immunocyte. Electron micrographs at a magnification around 15,000 × of cells from rat lymph nodes. (Provided by Mr A. Abbot of the Walter and Eliza Hall Institute.)

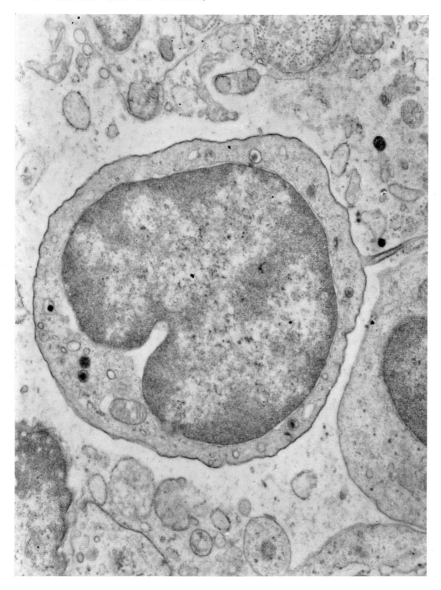

1. *Small lymphocyte.* The standard carrier of cellular information. Note the high nuclear–cytoplasmic ratio, the relative emptiness of the cytoplasm showing only one mitochondrion in this section, and the density of the nucleus without a clearly-visible nucleolus.

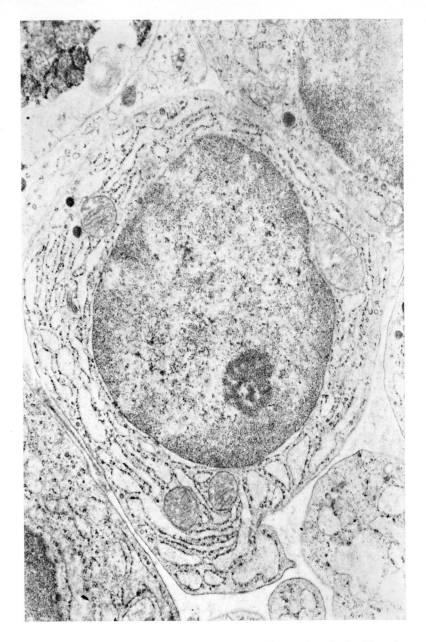

2. *Immature plasma cell.* The most significant producer of antibody. Note the large nucleus with conspicuous nucleolus, well-marked endoplasmic reticulum with ribosomes, and three mitochondria.

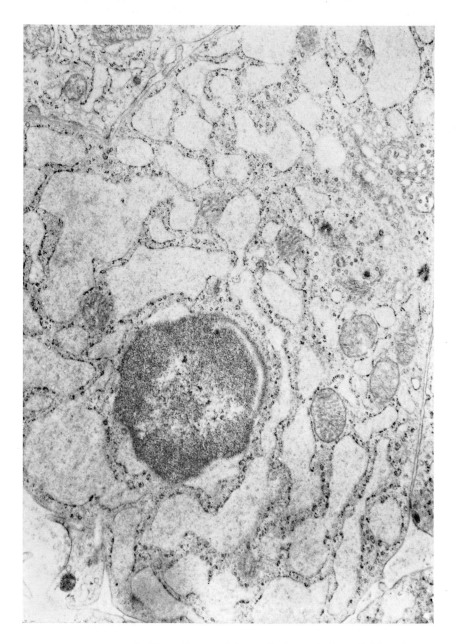

3. *Mature plasma cell.* The end stage of the antibody-producing cell. Note that the nucleus is small and beginning to show pyknotic degeneration. The endoplasmic reticulum is extensive and vacuolated and there are numerous mitochondria.

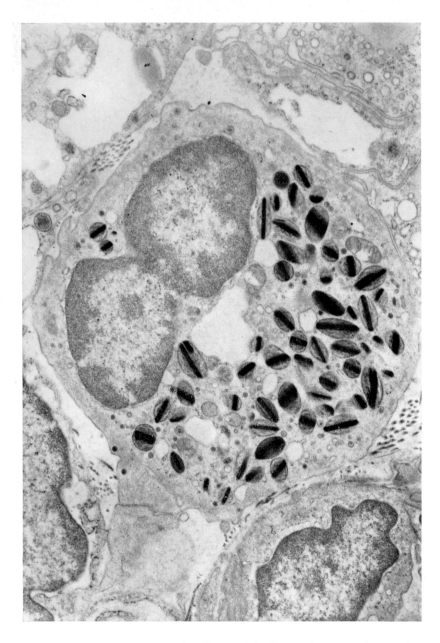

4. Eosinophil granulocyte (rat), characteristically reactive to antigen–antibody complexes. Note the bilobed nucleus and the characteristic granules with crystalloid inclusions. (Provided by Mr A. Abbot.)

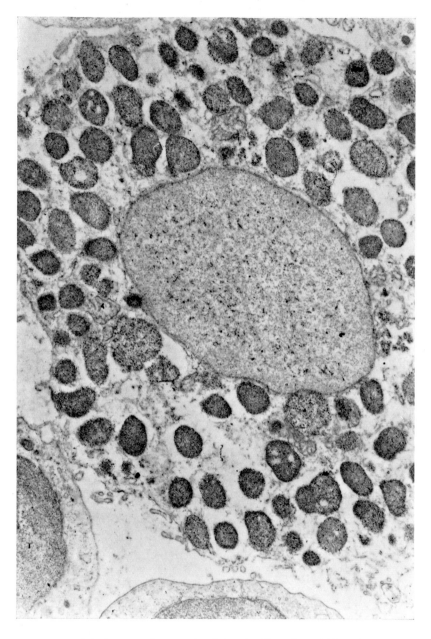

5. Mast cell (rat) responsive to stimulation by liberation of histamine, serotonin and heparin. Note the large rather variable granules responsible for the characteristic metachromatic staining and which contain the pharmacologically active agents. (Provided by Mr A. Abbot.)

Plates 6 and 7. *The relationship of macrophages to the uptake of antigen.* Electron-micrographs of lymph node sections from rats given heavily labelled flagellar antigen. By a combination of autoradiographic and electron microscopic techniques the location of antigen is shown by the dense black deposits of metallic silver. (Electron micrographs provided by Professor G. J. V. Nossal, Dr Judith Mitchell and Mr A. Abbot.)

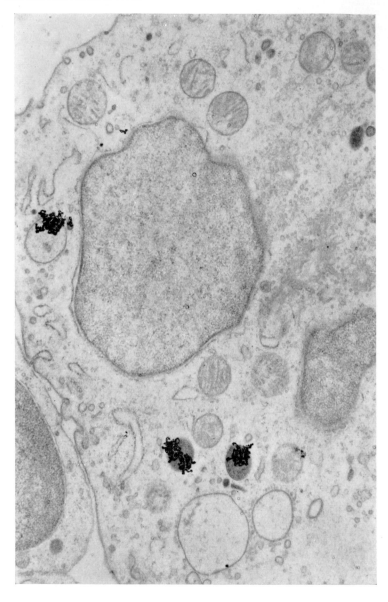

6. Medullary macrophage showing concentration of antigen in lysosomes (phagosomes). None is present on the surface membrane of the cell.

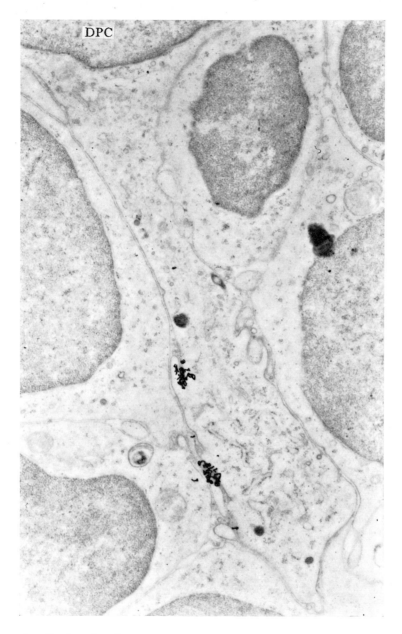

7. Localization of antigen on the cell membrane of the dendritic phagocytic cell. DPC = the nucleus of the phagocytic cell involved; the other nuclei are those of small lymphocytes.

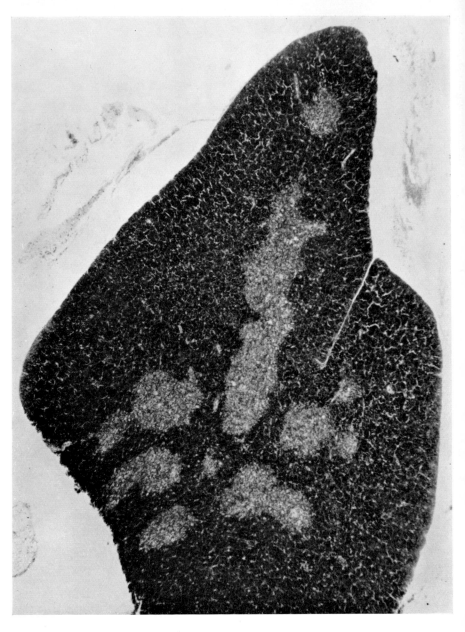

8. Section of thymus from a young mouse to show the sharp demarcation of cortex (dark with close-packed nuclei) and medulla.

by tuberculous infection and by the injection of encephalitogenic protein. Animals showing positive tuberculin reactions or typical experimental allergic encephalomyelitis gave lymphocyte populations of which '6–20 per cent' or '5–10 per cent' of small lymphocytes showed specific fluorescence when treated with antigen and fluorescent antibody. There are other less fully established reports that tuberculoproteins are more readily taken up by lymphocytes from sensitive individuals than from normals.

Basically similar studies have been made in which the tuberculoprotein (PPD) was conjugated with fluorescein and used to determine the percentage of reactive lymphocytes. In positive tuberculin reactions the percentage reacting ranged from 1·5 to 9 per cent; in negative reactors ten of eleven showed less than 1 per cent.

Evidence for the presence of immunoglobulins in or on lymphocytes is more extensive. Using immunofluorescence techniques to study human cells, van Furth found that cells, mostly medium lymphocytes, could be found in thoracic duct lymph, giving active fluorescence for M, G or A immunoglobulins; small lymphocytes both in the thoracic duct lymph and in the blood showed weak staining for M only. In all situations, medium lymphocytes showed more marked fluorescence and included some positive for A and G. In line with much other evidence indicating that the 'small lymphocytes' of bone-marrow may be a distinct cell type, these showed no evidence of any immunoglobulin. Equally interesting was the finding that cells in foetal thymus showed no fluorescence at a time when spleen cells showed M or G immunoglobulins. The general impression from these results is that many large and medium lymphocytes are, in fact, cells in the process of becoming plasma cells rather than small lymphocytes. The latter cells produce either M immunoglobulin or none at all.

The other field of interest for the individual characterization of lymphocytes is stimulation of human small lymphocytes to blast formation and mitosis *in vitro*. It is too early to generalize, but what has been published seems to establish that a proportion of small lymphocytes can be stimulated by specific agents as well as by a variety of nonspecific ones such as phytohaemagglutinin. Among confirmed examples are the regular occurrence of blast transformation in lymphocyte populations from tuberculin-positive individuals when the cells are treated with tuberculoprotein; transformation is also seen regularly when lymphocytes from two different individuals are mixed. There are also accounts of individual cases of severe drug reaction to 'Dilantin' and of infantile eczema with specific response to the appropriate antigen. In experimental animals the only well-documented account is of stimulation by allotypic antiglobulin serum which is not relevant here (see p. 168).

TABLE I. *Presence of immunoglobulins G, A and M in cells*

|  | Plasma cells | Small lymphocytes |
| --- | --- | --- |
| Thymus | G A | — |
| Spleen | G A M | M |
| Lymph node | G A M | M |
| Thoracic duct | G A M | M |
| Bone-marrow | G A M | — |

— no immunoglobulin detected.
Based on R. van Furth (1964). Doctoral thesis, Leiden.

It is rather striking that so far there are no reports of equivalent reactions in immunized animals and no recent confirmation of claims that any type of immunization not associated with the production of delayed hypersensitivity in human beings will produce appropriately reactive lymphocytes. As a tentative conclusion we may state:

(*a*) that many large and medium lymphocytes are cells of the plasmacyte series and can produce one or other of M, G or A immunoglobulins;

(*b*) that the only immunoglobulin produced by small lymphocytes is M—most produce none;

(*c*) that small lymphocytes associated with delayed-hypersensitivity-type reactions may have an exceptional type of surface reactivity for the sensitizing antigen;

(*d*) that the bone-marrow population of cells with the morphology of small lymphocytes is not part of the immunocyte group.

COMMITMENT OF IMMUNOCYTES

In the second half of this chapter the aim is to describe the population dynamics of lymphoid cells, their movements, sites of proliferation, morphological and functional modification and final disposal. At this stage we assume that each immunocyte, that is, every descendant of a lymphoid cell differentiated to immune function in the thymus or elsewhere, carries receptor sites of specific character capable of reacting with a restricted range of antigenic pattern and often only with a single group of closely related determinant patterns. This is more fully discussed in chapter 8. It will also be accepted that, in mammals, cells carrying new immune patterns first emerge from the thymus, retaining, however, the qualification that future research may require the recognition of other primary sources of new cells possibly analogous to the bursa of Fabricius in chickens.

In the previous chapter it was concluded that immunocytes are differentiated in the thymus and there subjected to a scrutiny for immune patterns capable of reacting with antigenic determinants of accessible body components. Only a minor proportion of the lymphocytes produced in the thymus give rise to continuing clones after leaving it. Many are destroyed in the organ itself, others may enter the circulation and for one reason or another prove non-viable. Nevertheless, the mass of direct and circumstantial evidence points to the conclusion that all the immunocytes in the body are descended from cells which have been differentiated in the thymus and have passed thence to the circulation.

Most deductions as to what happens when differentiated cells leave the thymus must be made from indirect evidence. Such deductions give rise to a fairly well-established picture, many details of which are subject to change with the accumulation of future information. With this qualification, what happens seems to be as follows.

About three days, on the average, after their last mitosis, small lymphocytes leave the thymus and enter the blood circulation either directly or indirectly. A majority of these are lost, perhaps as a result of stimulation received in the thymus. Of those which can be traced, most lodge in lymphoid tissue. Indirect but persuasive evidence indicates that most of those that lodge in lymph nodes are to be found in the paracortical areas rather than in lymph follicles. If we slightly anticipate later discussion, these small (and medium) lymphocytes of immediate thymic origin bring immune patterns whose origin is unrelated to the immunological experience of the body. They are uncommitted or progenitor immunocytes, each with its characteristic immune pattern awaiting stimulation, and they represent the only way in which fresh supplies of 'new' immune patterns can arise. Everything suggests that these newly produced cells are vulnerable. In all probability any high concentration of antigen will be lethal, perhaps not so much in virtue of the amount but of the sequence of specific stimuli. There is good evidence for the existence of a thymic hormone necessary for the viability of cells recently released from the thymus. In the absence of continued production of some humoral product by thymic epithelial tissue the new cells cannot mature to the level at which they can manifest any of the demonstrable functions of immunocytes. Once cells have passed this vulnerable stage there is no evidence that thymic hormone is necessary. Economy of hypothesis is to assume that thymic epithelium secretes a single protein hormone which, in addition to the well-documented effect of mediating the survival of newly liberated cells, is also concerned in stimulating proliferation and differentiation of cells entering the thymus and in facilitating the elimination of 'self-reactive' cells. The additions, however, are at the present time wholly speculative.

Contact of an uncommitted cell with an antigenic determinant capable of union with its specific receptors can have, of course, other effects than the destruction of the cell. There is rather uncoordinated evidence that (a) the cell may be induced directly to synthesize and liberate Ig M antibody and (b) it is de-differentiated to a blast-type cell from which a relatively small descendant clone of 'committed immunocytes'—morphologically lymphocytes—is derived. The conditions that determine whether destruction or some type of positive immune reaction follows contact with specific determinant are largely unknown, and what is known is better left to chapter 8.

Here, our concern is with the fact that after preliminary contact with a given antigen an animal gives a much more active immune response. In cellular terms it now has a population of committed immunocytes. In a sense the uncommitted immunocyte is a hypothetical entity needed to account indirectly for many observational and experimental facts. Committed immunocytes are the lymphocytes which can be demonstrated to respond specifically to antigen.

The change from the uncommitted to the committed immunocyte can be most readily and probably correctly pictured as involving an increase in the number of specific immune receptors (which may now be in the form of Ig M antibody molecules) and in their accessibility to antigen in the environment.

THE STRUCTURE OF PERIPHERAL LYMPHOID TISSUE

For many years it has been clear that spleen and lymph nodes are the tissues most directly concerned with immune processes. For technical reasons, most recent experimental work on cellular immunology has been concerned with the cells of accessible lymph nodes, either popliteal or auricular, and it will be convenient to base general discussion of the function of lymphoid tissue on this work. In all probability the lymphoid tissue of the spleen and the extensive accumulations of lymphoid tissues along the alimentary tract, from tonsils downward, play a basically similar role.

*Lymph node structure*

The essential features of a peripheral lymph node are shown in fig. 15. The afferent lymph carrying any antigenic material from its catchment area enters at the periphery and is distributed over the lateral sinus, from which it percolates through more or less definite channels into the cortex and eventually is collected into the lymph sinuses associated with the medullary cords and brought to the efferent lymph vessels at the hilus. Lying immediately beneath the lateral sinus is the cortical tissue with a number of areas of more closely packed small lymphocytes, the primary lymph follicles. Within these, secondary follicles or germinal centres develop under antigenic stimulation. Toward the hilum of the node the medullary cords, more or less closely packed, represent essentially small blood vessels surrounded by a sheath of

85

tissue infiltrated by lymphoid cells and delimited by the lymphatic endothelium lining the channels which come together to form the efferent lymph vessels. Between the cortex and medulla is the paracortical lymphoid tissue which is not clearly demarcated into primary follicles and is labile in amount and distribution according to the degree of stimulation of the node. In general, small lymphocytes predominate in the primary follicles and the paracortical areas; the germinal centres characteristically contain large lymphocytes and variable numbers of medium lymphocytes and lightly

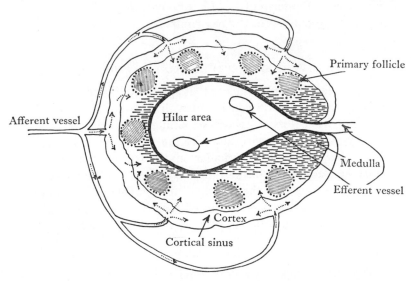

Fig. 15. Schematic diagram of a popliteal lymph node from a young rat. From G. J. V. Nossal, G. L. Ada and C. M. Austin (1964). Antigens in immunity. IV. *Aust. J. exp. Biol. med. Sci.* **42**, 311.

pyroninophilic cells. In active lymph nodes the main concentration of plasma cells is in the medullary cords but there are also many pyroninophilic cells in the paracortical area. Phagocytic cells of the reticulo-endothelial system are numerous and are most conspicuous in the wall of the peripheral sinus and of the lymph channels in the medulla. In addition there are the dendritic cells which form an inconspicuous reticulum through the cortex and which appear to play a major role in presenting antigen to cells of the lymphoid series.

The distribution and character of the blood vessels have an important bearing on the immunological function of the lymph node. In a typical rat lymph node the arterial supply enters at the hilum, and primary arteriolar branches, each within a medullary cord, run directly to the cortex where they divide into capillaries. These are relatively inconspicuous compared with the smallest venules, which show many areas of cuboidal endothelium and are tortuous and numerous. They leave the cortex via the medullary cords and at the hilum join to form the draining venous trunk. The medullary cords are most clearly seen in an oedematous lymph node. They are surrounded by lymphatic endothelium, on the lymph sinus side of which are numerous macrophages (littoral cells). There is a single unbranching central arteriole or venule with venules about ten times as frequent as arterioles. Patches of cuboidal endothelium may be present at any part of the venule but it is more frequent near the cortex.

The generally accepted view of the embryology of lymph nodes is that they arise by the accumulation of lymphocytes around appropriate regions of the developing network of lymph channels. Lymphocytes and lymph nodes do not appear until the time that the thymus contains lymphocytic cells, and the view that all or most primitive lymphocytes arise from the thymus, and by colonization and proliferation give rise to the peripheral lymphoid accumulations, is widely but by no means universally held.

### The immune function of the lymph node

In the developed lymph node the most important functional feature is the extensive entry of small lymphocytes from the blood and equivalent discharge of cells by the efferent lymphatic. In the popliteal lymph node of the sheep this may correspond to a complete turnover of cells once every seventy hours. Entry from the blood is across the relatively thick-walled postcapillary venules which are characteristic of all lymphoid tissue. Lymphocytes are highly mobile cells and they have been noted to have a special tendency to enter the cytoplasm of certain types of epithelial cell in tissue culture and of intestinal epithelium *in situ*. On the other hand those who have studied the movement of leucocytes microscopically in rabbit ear chambers or preparations of mesentery

87

have never recorded the passage of a small lymphocyte through the capillary wall. Polymorphonuclear leucocytes, eosinophils and monocytes pass readily at the junctions between two endothelial cells.

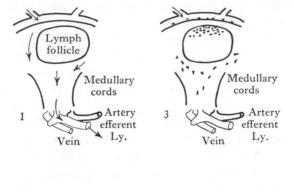

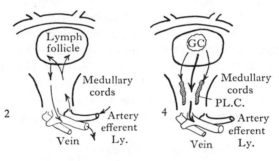

Fig. 16. Movement in a peripheral lymph node. A series of simplified diagrams to indicate: 1. The flow of lymph; 2. The movement of lymphocytes entering by the blood, leaving by the efferent lymph trunks; 3. Distribution of antigen reaching the node by afferent lymphatics; 4. Probable movement of proliferating immunocytes. Pl. C. = plasma cells; GC = germinal centre; Ly = lymphatic.

The capacity of small lymphocytes to leave the bloodstream via the postcapillary venules of lymphoid tissue in large numbers has been adequately established. It must follow that the cuboidal endothelium of the venules concerned must differ significantly and adaptively from normal capillary and venular endothelium. The patchy distribution of the cuboidal character suggests that a local, rather than a genetically determined process may be concerned.

There are interesting implications of the simplest formulation of such a process—that in any accumulation of lymphocytes and antigen-retaining reticulum there will be a continuing liberation of an agent with the effect of making venular endothelium 'attractive' or 'sticky' for lymphocytes passing through in the blood. 'Permeability factors' can be extracted from lymph node cells and it is possible that such extracts contain the hypothetical agent responsible for maintaining the cuboidal state of the venular endothelium in lymphoid tissue.

Such a hypothesis might be of value in interpreting the process by which lymph nodes and other lymphoid accumulations develop. If, whenever for any reason a group of lymphocytes come together, the adjacent postcapillary venule develops this specific permeability for small lymphocytes, a 'gravitational' process will give rise to an increasing aggregation around which presumably the characteristic architecture and vascular supply of the lymph node will develop.

The afferent lymphatics are so disposed as to bring to the node any antigenic material appearing in their field of drainage as well as any lymphocytes or other mobile cells present in the tissue fluids. They embouch into the peripheral sinus and so bring antigen into close relationship to the cortical aspect of the primary lymph follicles. The functional anatomy of the lymph follicle can best be elaborated by following the fate of a labelled antigen, such as the *Salmonella* flagellin of Nossal's group injected into a footpad of the rat. There are many phagocytic cells lining the various lymph paths and sinuses in the draining popliteal node and a large amount of antigen is taken up by these. From the immunological angle these seem to be unimportant compared with the dendritic phagocytes that are inconspicuously interspersed amongst the lymphocytes of the primary follicles. When lymph nodes are examined at daily and weekly intervals, after such an injection the dense accumulations of labelled antigen seen at 24 hours soon vanish from most of the sites. A proportion of macrophages in the medullary and paracortical areas remain heavily loaded for weeks and with even a very small dose of antigen there is prolonged persistence in the dendritic reticulum cells of the primary follicles. With the appearance of germinal centres the labelled area is distorted to form a long-lasting cap on the cortical side of the germinal centre.

89

Our hypothesis is that these reticulum cells are important repositories for antigen—perhaps sites where its physical character is modified. In the living follicle we are almost compelled to think of the lymphocytes as individually mobile so that a continual sequence of cells can come into contact with extensions of the cytoplasm of the dendritic antigen-containing cells. Quite rapidly, any lymphocyte in the follicle capable of reacting with the antigen present will have an opportunity to do so, and it is probably legitimate to regard any such population as a random sample of the whole complement of lymphocytes (or immunocytes) in the body.

Functionally speaking, the most important feature of a lymph node is the mobility of the cell population it contains. In the lymph leaving the popliteal node of a large animal, such as the sheep, are large numbers of lymphocytes which can be shown by appropriate labelling experiments not to be produced in the node. Approximate calculation suggests that about 10 per cent of the lymphocytes entering the node in the blood pass through vessel walls into the node, while an equivalent number leave by the efferent lymphatics. Similarly, if such a node is subjected to heavy (2,000 r) local irradiation to destroy all its contained lymphocytes and the situation examined 24-48 hours later, it is found that the cell content of efferent lymph is virtually unaltered, as is the capacity of the node to produce antibody on appropriate stimulation.

Evidence from a wide variety of experimental procedures shows that the lymphoid cell population of every primary lymph follicle in the body is constantly changing. Cells enter from the blood and for the most part leave the node by the efferent lymphatics, eventually to return to the blood circulation by one of the main lymphatic trunks. It follows that every accumulation of lymphocytes in the body that is not under immediate antigenic stimulation will contain representatives of virtually every clone of immunocytes that has been produced. This high degree of mobility is of great importance for anyone interpreting results of immunological experiments on individual lymphoid regions *in vivo*, whether they are lymph nodes, Peyer's patches or focal accumulations produced experimentally.

Antigenic stimulation from, say, a foot-pad injection, has two main morphological effects on the lymph node: (*a*) an increase in the numbers of plasma cells largely in the medullary cords and (*b*) the appearance of germinal centres (secondary follicles). At an appropriate time, antibody is detectable in the efferent lymph or in extracts of the node. There is still uncertainty about the cellular interpretation of these changes. One feels, however, that there is significant evidence in favour of one interpretation and none yet available to disprove it. According to this interpretation, an immunocyte activated by contact with the corresponding antigen lodges in lymphoid tissue in one of two situations: (*a*) in the substance of a lymph follicle or (*b*) in a medullary cord. In the lymph follicle, which may be in the same or in a lymph node distant from that in which the cell was stimulated by antigen, the activated cell proliferates to give rise to a clone of large and medium lymphocytes which remain in contact until the stage of small lymphocytes is reached, so giving rise to a germinal centre. All the progeny will, apart from incidental mutation, be of the same immune pattern as the initiating immunocyte. Everything indicates that a high proportion of newly formed small lymphocytes pass rapidly into the circulation. Certainly few of those present around the germinal centre are derivatives of it.

A proportion of lymphocytes undergoing mitosis can be seen in regions of the lymph node other than the germinal centres, and the relative contribution made by germinal centres to new cell production in the node is not clear. If one accepts the demonstration that medium lymphocytes synthesizing DNA are present in the circulation, it seems likely that some of these have been produced in germinal centres and also that on lodgment in a lymph follicle they may continue to divide for a generation or two. It is not impossible, therefore, that outside the thymus all production of small lymphocytes takes place in germinal centres. One senses, however, a growing opinion that lymphocytes associated with delayed hypersensitivity are *not* produced in germinal centres.

The origin of the numerous plasma cells in the medullary cords is controversial. It may be still too early to discard the old view that they arise from perivascular reticulum cells, but the whole present trend is to regard them as derivatives by direct descent from activated immunocytes. Our own experience in examining lymph nodes of mice and rats seems to be consistent with the view that at least a substantial proportion of plasmablasts arise in germinal centres and move through the paracortical region to the medullary cords. During this process, proliferation and maturation occur. The possibility that stimulated immunocytes in the circulation may pass through the endothelium of the capillaries or venules in medullary cords and there develop as plasma cells is by no means excluded.

There are some important aspects of plasma cell physiology still to be elucidated. One of the most important medical problems in the immunological field is the nature of agammaglobulinaemia where lymphocytes are normal and appear capable of a number of immunological functions. Plasma cells, however, do not appear in response to antigenic stimulation, antibodies are not demonstrable and only very small amounts of immunoglobulins. A rather similar picture is seen in bursectomized chickens and the best current hypothesis is probably that in both instances a hormonal factor is missing that is needed to allow the initiation by an activated lymphocyte of a plasma cell clone. What evidence there is points to the hormone being derived from associations of lymphoid cells with epithelium of entodermal origin.

The other important problem concerns the regions in which plasmablasts multiply and mature plasma cells accumulate. In the normal animal, plasma cells are conspicuous in the medulla of lymph nodes and in the red pulp of the spleen where most are found immediately adjacent to Malpighian bodies. In pathological situations (autoimmune disease) they are characteristically found at the periphery of lymphoid cell accumulations. In lupoid hepatitis in man they may outline the periphery of the liver lobules. All these findings suggest that plasma cells develop and function preferentially where there is an adequate supply of oxygen and nutrients via the blood. The requirements for lymphocytic multiplication in germinal centres, thymic cortex or fortuitous accumulations seem to be less critical.

MOVEMENT OF LYMPHOCYTES

Every cubic millimetre of normal mammalian blood contains a few thousand lymphocytes, mainly small lymphocytes, and this number remains approximately constant. Appropriate studies with labelled cells, however, indicate an average stay of only a few hours in the circulation. The population is being constantly changed and one of the major problems in cell dynamics for the future is the means by which homeostatic control of the numbers of circulating lymphocytes and other leucocytes is mediated. A closely related problem may well be the homeostasis of immunoglobulin levels in the plasma. Both must be postulated as background conditions in considering other aspects of the population dynamics of immunocytes.

In discussing the population dynamics of lymphocytes it is first necessary to repeat what has already been said about the heterogeneity of small lymphocytes. Everything points to the small lymphocyte as the common morphological form of a cell that is functionally inactive but has potentialities of function that may be needed elsewhere in the body. From bone-marrow one can obtain populations of small lymphocytes capable of 'rescuing' lethally irradiated mice. Small lymphocytes from thoracic duct lymph are inert in this respect. In any discussion of the immunological function of small lymphocytes it must be understood that all statements made should be construed as referring to 'a significant proportion of the cell population under study', not to morphological small lymphocytes as a whole, or even to what may often be a majority of the cells in the experimental population.

It is now certain that many small lymphocytes are immunologically competent in the sense that they can be shown to react specifically with antigens and/or contain demonstrable immunoglobulin in or associated with cytoplasm. On the now almost axiomatic assumption that many different functional or potentially functional types are included under the morphological guise of the small lymphocyte, it becomes necessary to stress that any discussion of any particular category of these cells must consist rather largely of hypotheses designed to offer indirect experimental approaches for their verification or disproof. With the general acceptance of a clonal selection approach as background one such hypothesis is as follows.

Small numbers of viable small lymphocytes leave the thymus and settle for longer or shorter periods in lymph follicles or elsewhere in lymphoid tissues. These are differentiated immunocytes in the sense that contact with a certain antigenic determinant can be recognized by its capacity to initiate functional changes of one sort or another in the cell. In view of the capacity for nonspecific stimuli like phytohaemagglutinin to induce mitosis in lymphocytes, we can probably assume with reasonable certainty that there are also generalized physiological stimuli other than antigen which can stimulate activity of more or less equivalent character. The characteristic feature of the response to an *antigenic* stimulus capable of inducing antibody production is the appearance of germinal centres and plasma cell accumulation in the lymph nodes draining the area involved.

The situation can be represented by the entry of antigen $A$ via the afferent lymphatics to a lymph node where the cortical areas contain, amongst thousands of other types of lymphocytes, a few representatives of clones carrying receptors (or fixed antibody) $a$, which can unite with antigenic determinant $A$. Contact of such a cell with $A$ may be either direct or by cell-to-cell contact with a dendritic macrophage which has taken up antigen $A$. The result will depend on many factors of which the concentration encountered of the antigen and the local internal environment where the stimulated cell lodges may be the most important. The *significant* response may well be one in which the primary immunocyte takes on the character of a large pyroninophilic cell, synthesizes small amounts of Ig M-type antibody and replicates to produce eventually a clone of small lymphocytes, perhaps in the process giving rise to a germinal centre. The result will be the development of a clone of 'committed' immunocytes which will be rapidly distributed to all lymphoid accumulations. These committed immunocytes are probably those which can be implicated in the following phenomena:

(*a*) They carry small amounts of Ig M in the cytoplasm.

(*b*) On stimulation by antigen they liberate M-type antibody and undergo mitosis.

(*c*) In suitable internal environments some of the descendants follow the plasma cell sequence, synthesizing and liberating first M-type and later G-type antibody.

(*d*) Other descendants take the form of small lymphocytes—memory cells—which persist as carriers of immunological memory.

94

In general the more intense the antigenic stimulus the larger the population of specific immunocytes that will develop.

In some such fashion the overall population of lymphocytes in the body will come to be made up of an immense variety of clones, with the number of representatives of each being a rough measure of the frequency and intensity of exposure to the antigen concerned.

Large numbers of lymphocytes are present in normal tissues other than lymphoid tissues, and there are accumulations in regions subject to any of a wide variety of subacute or chronic pathological processes. It is probable that lymphocytes pass more readily through the postcapillary venules of lymphoid tissues than through any other type of vascular endothelium, but the total of lymphocytes wandering through the general tissues of the body must be very large. This holds especially for the submucosal tissues of the intestinal tract.

On almost any conceivable formulation of the life history of the lymphocyte only a very small proportion of small lymphocytes will give rise to descendant clones. The vast majority are end-cells and their eventual fate requires discussion.

One of the corollaries of the concept that the body's lymphocyte population maintains a vast number of clones with differing immune patterns, from which a relatively small number of individual cells are 'chosen' by antigen for selective proliferation, is that large random populations could, if necessary, be sacrificed to provide building blocks for the synthesis of nucleic acid and protein by the proliferating clones. It has already been suggested that this may take place on a large scale in the thymus. The small lymphocyte is notoriously the most vulnerable cell of the body to the destructive action of X-irradiation, corticosteroids and cytotoxic drugs. Everything points to this vulnerability as having an important evolutionary role. In clonal selection theory it is important that a cell capable of reacting with an antigen should under certain circumstances be susceptible to destruction by contact with that antigen. This is the basis of tolerance and immune unresponsiveness. It is equally likely and reasonable that when active immunologically stimulated proliferation of immunocytes is taking place in the body, many other lymphocytes are being broken down to return their components to the metabolic pool. It is universal

95

to find pyknotic fragments of lymphocytic nuclei wherever lymphocytic proliferation is taking place. The evidence in regard to the disintegration of lymphocytes *in vivo* is still fragmentary and no quantitative data are available. Until methods to obtain the facts have been worked out it will be impossible to estimate the relative proportions of lymphocytes which (*a*) give rise to descendant cells, (*b*) are disintegrated *in vivo* or (*c*) are released into the external environment on mucous membranes or in secretions. There is no doubt that large numbers of lymphocytes are shed into the alimentary canal.

SUMMARY

The picture of the lymphocyte and lymphatic tissue that has emerged with the application of modern methods is very different from what it was only a decade ago. It is an altogether more dynamic concept and presents some lessons that are still incompletely learnt. It is evident that histologists have always been inclined to interpret appearances in tissues as being derived from other cells which are either present or have been present in the same tissue. This holds to a less extent but still significantly for autoradiographic and other methods of 'functional' histology.

It seems now that we must picture all accumulations of small lymphocytes as transitory populations always changing in two important respects. Within the tissue itself the constituent lymphocytes are presumably moving in random fashion and constantly changing their situation both in relation to other lymphocytes and to the cells more firmly attached to the collagen and reticulin fibrils that give form to the structure. Within that framework the lymphocytes can be pictured almost as the proverbial 'bag of worms'. The actual cells making up the local population are also changing at a rate which has been estimated for the popliteal lymph node of the sheep as corresponding to a complete turnover every seventy hours. Cells are constantly leaving by the afferent lymph vessels while others are entering from the blood. As far as lymphocytes are concerned, the whole lymphoid tissues of the body, with the important exception of the thymus, form a single compartment with essentially random, but also randomly restricted, movement throughout the whole system.

Although cells of the lymphocytic series are highly mobile, the cells which are associated with them are much less so. The individuality of a lymph node is preserved by the cells which are not lymphocytes, and everything points to the importance of some of these cells in taking up and retaining antigen. This is a point to be discussed in more detail in relation to the cellular aspects of antibody production but it is essential to an understanding of the circulation of the immunocytes.

If antigenic molecules are held by cells fixed to the sponge-like matrix of the peripheral lymphoid organs, the combination of short-range mobility and progressive replacement from elsewhere seems ideally suited to ensure that every lymphocyte that can react with an antigenic determinant which has entered the body will have an opportunity to do so.

# 5 The nature of antibody

As soon as methods for the demonstration and titration of antibodies had been established, efforts were begun to identify chemically the constituent of serum responsible. It was evident that antibody had the general characteristics of protein and that it was predominantly present in the globulin fractions as defined by contemporary methods of salt precipitation. By the 1930s it was established that antibody was globulin, resembling any other serum globulin in its chemical and physical characteristics except for a specific capacity to combine with the corresponding antigen.

The modern phase of immunochemistry can be dated from Tiselius's work on the electrophoresis of serum proteins by which he defined $\alpha$-, $\beta$- and $\gamma$-globulins. Separation of these fractions indicated that nearly all antibody was contained in the most slowly moving fraction, the $\gamma$-globulin. For many years $\gamma$-globulin was generally used as the appropriate chemical designation for antibody, although from an early stage it was well known that, while standard rabbit antibodies had a molecular weight around 160,000, the intensively studied equine antibodies to pneumococcal polysaccharide antigens were approximately $10^6$ mol. wt.

## THE IMMUNOGLOBULINS

With the progressive development of more refined chemical, physical and immunological methods for studying soluble proteins, it has become clear that there are several types of globulin which can carry the specific reactivity of antibodies. It is now conventional to refer to these as immunoglobulins, of which the three types commonly recognized in mammalian (including human) sera are Ig G, Ig M and Ig A.

It is certain that these are not the only physical types of immunoglobulin even in healthy animals. There are already indications that in mice and guinea-pigs there are two varieties of G, and a fourth type, D, has been described in man. Another type, Ig E, which may be pathological, has been recently demarcated. Immunoglobulins are usually studied in the form of antibodies obtained from normal individuals, but a great deal of information has also been obtained from immunoglobulins produced in pathological excess by patients with multiple myelomatosis and some related conditions. Rightly or wrongly, many deductions about the nature of antibody have been based on findings with such myeloma proteins and from this angle they will need to be discussed at some length in this chapter.

TABLE 2. *Human immunoglobulins and related abnormal proteins*

| Normal (Concentration, mg/ml) | Pathological | Molecular weight |
|---|---|---|
| Ig G 0·8–1·5 | Ig G (My. pr.) | 150,000 |
| Ig A 0·06–0·2 | Ig A (My. pr.) | 150,000 and multiples |
| Ig M 0·04–0·12 | Ig M (My. pr.) | 900,000 ± |
| Ig D trace | Ig D (My. pr.) | 150,000 ± |
| Light chain (K or L) | Bence Jones protein | 22,000 |
| Heavy chain (Ig G) | — | 50,000 |

My. pr. = Myeloma protein.

The primary differences between immunoglobulins by which the three types are defined are as follows:

(*a*) Molecular weight around 160,000 for G and A, and around 1 million for M, with corresponding sedimentation coefficients of 7 (6·6)S and 19S. Ig A is prone to aggregation and it is common to find a fraction with a sedimentation coefficient S9–11 which may represent a dimer.

(*b*) Specific *antigenic* characters differentiating A, G and M which allow their recognition either by gel-diffusion reactions in agar or by immunofluorescence methods.

(*c*) A characteristic distribution of electrophoretic mobility within each molecular type giving easily recognized lines in the patterns obtained by immunoelectrophoresis.

The biological characteristics of the three types of antibody will need to be discussed at length in relation to many facets of immunology, but for preliminary orientation the differences may be summarized as follows.

Antibodies of Ig M structure are the first to appear on immunization with almost all types of antigen and make up most 'normal' antibodies. In the later stages the titre of M antibodies is reduced and their place taken by antibodies of G type. When bacterial polysaccharides are used as antigen, M-type antibodies persist and their normal replacement by type G is not easily demonstrable. This is probably mainly due to the relative ineffectiveness of Ig G antibodies as compared with Ig M in producing some of the commonly used antigen–antibody reactions. Where direct comparison is possible, M antibodies attach more firmly to antigenic particles and against red cells as antigens are, molecule for molecule, many times more effective as haemolysins.

The classical antibody produced by deliberate immunization with protein antigens is of G type and almost all the refined chemical work to be mentioned later has been done with such antibodies. It is probable that biologically speaking they are less important than M and A types. G antibody is less readily attached to tissue cells than A and readily passes the human placenta into the foetus.

Ig A-type antibodies are found in all mammals and one subclass of these probably represents the antibody or reagin found in hay fever patients. The two important biological characters of Ig A antibodies are the ease with which they attach to body cells and, perhaps closely related to this, their special function of being concentrated in and secreted by glandular epithelium (in mammary, parotid and probably other glands).

Secondary differences at the chemical level include the failure of A antibodies to precipitate with antigen, perhaps, as suggested by Karush, because their two combining sites are too close together to allow lattice formation with antigen. M immunoglobulins contain a much higher content, 12·2 per cent of carbohydrate, than A with 10·5 per cent, or G with 2·8 per cent. It is an important point of similarity that all antibodies of M, G and A types may have in common the antigenic characters based on 'light' chains and the specific character of the combining site.

## THE HETEROGENEITY OF ANTIBODIES

Even within each of the immunoglobulins, antibodies are highly heterogeneous and with every refinement of physical, chemical and immunological technique for separating and characterizing soluble proteins the number of subpopulations of antibody molecules that can be separated increases. To anyone interested in a clonal approach to immunology this is the most important quality of antibody. Its significance has been greatly increased as a result of studies on myeloma proteins obtained from patients with multiple myelomatosis. Essentially this is a conditioned malignancy of relatively highly differentiated plasma cells. By what is perhaps in some sense a circular chain of reasoning, most of these myeloma proteins are believed to arise by the proliferation of a single mutant cell to form a uniform clone of thousands of millions of cells. Hundreds of myeloma proteins of human origin and considerable numbers of similar proteins from plasmacytomas of mice have now been studied in detail. The results can be simply, but I believe legitimately, summarized by saying that each myeloma protein provides an example of what a single antibody-producing cell can produce. Normal serum contains immunoglobulins in a heterogeneous mixture produced by thousands or millions of distinguishable clones of globulin-producing cells.

The great experimental advantage of myeloma proteins is the ease with which they can be isolated and purified from a patient's serum. Each patient provides an individual protein. A few are quite anomalous but 90 per cent of them can be identified as physically and antigenically typical A, G or M immunoglobulins. Within each Ig group, however, are individual differences and the results suggest strongly that *any* type of immature plasma cell can undergo this particular type of inheritable change. In addition there are anomalous conditions which produce proteins of the same general quality but not equivalent to any of the normal forms. These include the 'heavy chain' proteins and the Bence Jones proteins, both of which may be present as an abnormal component of the blood plasma. In addition, a large proportion of typical myeloma patients excrete large amounts of Bence Jones protein in the urine. This is now known to be composed of the light chains characteristic of the complete myeloma protein and, as such, having the same antigenic quality and amino acid composition. Urinary Bence Jones protein is therefore a convenient source of material for chemical studies of light chain.

The monoclonal character of a typical myeloma protein is evidenced by its uniformity. Instead of a broad zone on electrophoresis there is a single sharp spike; antigenically it reacts only with one of specific A, G, M or D antisera and, unlike any purified normal immunoglobulin, only one of the two alternative $\kappa$ or $\lambda$ antigens of the light chains is represented. While these chapters were being written it was reported that at least three myeloma proteins (Ig G) had a definite antibody specificity and were therefore true monoclonal antibodies with a specific combining site. This should lead to an important set of new advances. Already on the basis of the evidence from myeloma proteins it can be confidently assumed that any given clone of immunocytes will produce *physically* and *antigenically* uniform populations of antibody molecules. Most immunologists will probably soon be driven to agree that the immune pattern of all *antibody* produced by the clone will also be homogeneous.

Recognition of the heterogeneity of immune pattern in an antiserum produced against a single (but always complex) antigen goes back to the early days of bacterial serology. If a rabbit was immunized against bacterium $B$, an antiserum was produced which not only agglutinated $B$ but also a wide range of more or less closely related cultures, $A$ and $C$ for example. If, now, the antiserum was repeatedly treated with $A$ bacteria to remove all antibody capable of reacting with $A$, it would still react relatively strongly with $B$ and probably with $C$. After washing the $A$ bacteria, antibody might be eluted from them by a suitable technique. In such negative and positive manipulations it was possible to fractionate the antiserum into different subpopulations of antibody molecules to almost any degree of elaboration. Even when a single artificial hapten was used as antigenic determinant, cross-testing with haptens of related molecular structure revealed antibody molecules with different degrees of avidity for the homologous antigenic determinant and of cross-reactivity with related antigens.

A few myeloma proteins of Ig M type react as rheumatoid factors, i.e. as if they were antibodies against partially denatured γ-globulin. There are many difficulties in interpreting the meaning of this result and, at the experimental level, rheumatoid factor is an unsatisfactory type of antibody for detailed study. Nevertheless, the existence of one type of myeloma protein with antibody character resulted in a concerted effort to detect a myeloma protein that would react with a chemically definable antigenic determinant. Now that several of these have been found, it will be of immense interest to determine the range of cross-reactivity of a 'pure' antibody, something that has not yet been possible and which is vital for the understanding of many phenomena relevant to the nature and origin of immune pattern.

## THE CHEMICAL STRUCTURE OF IMMUNOGLOBULIN G

### The Porter diagram

In view of the heterogeneity of antibodies it has become necessary to concentrate chemical studies on examples of antibody which can be obtained in considerable amount and be readily purified. Most work has been done on G-type antibodies prepared in rabbits with artificial antigens carrying a suitable haptenic group as antigenic determinant. In man, chemical work has tended to be concentrated on G myeloma proteins or Ig G from normal individuals studied without regard to specific immune pattern, but there have been some important studies on specific antibodies.

In general, the approach has been to dissociate the immunoglobulin molecule into relatively large polypeptide chains and to ascertain the distribution on such fractions of the various functional qualities of antibody or immunoglobulin. It is now accepted that the 'Porter diagram', shown in slightly modified form in fig. 5, is a satisfactory schema for representing the structure of human or rabbit Ig G as deduced from such studies. The first methods to be used involved the use of proteolytic enzymes. In a commercial process developed in 1936 the fragment Fc was split off by pepsin leaving an active complex of two light chains and the Fd portions of the heavy chains. The modern approach was opened by Porter's use of papain with cysteine which produced essentially two fragments from each molecule, one containing a single antibody site and named Fab, and the other, the Fc fragment, which is part of the heavy chain. These procedures of limited proteolytic enzyme action have been supplemented by the development of methods to reduce S–S bonds by cysteine or mercaptoethanol and separate the chains by suitable manipulations while they were maintained under conditions preventing reunion.

The Porter diagram indicates that Ig G is built up of four polypeptide chains, two light of mol. wt 20,000 ± and two heavy of mol. wt 50,000 ±, united by disulphide bonds. There is a reasonable likelihood that the heavy chain may be composed of a large segment (Fc) and a smaller one (Fd) carrying the piece which plays the major part in forming the specific combining site. There is a real possibility that both the light and heavy chains are made up of more than one genetically distinct polypeptide chain. For the present, however, there is no decisive evidence that more than two genetic loci for light and heavy chains respectively are concerned. In addition, a relatively small oligosaccharide molecule is attached to an aspartic acid residue in the heavy chain (Fc segment). The carbohydrate contains 3 galactose, 5 mannose, 2 fucose, 10 acetylglucosamine and 1 sialic acid units.

*Antigenic qualities*

If we confine ourselves for the time being to human immunoglobulins, the first functional quality to be discussed is the distribution of antigenic determinants. These can be studied by producing antibodies against purified immunoglobulins or fractions thereof, either in a distant species such as the rabbit, or in the more closely related rhesus monkey. Even more illuminating results can be obtained by making use of the antibodies found under pathological conditions in humans which react against one or other determinant of their own immunoglobulins; the 'rheumatoid factors' represent the most important source.

Such serological studies of human immunoglobulins, in the first instance with rabbit antisera, nowadays almost always made against a single myeloma protein or its fractions, have shown that:

(*a*) The heavy chains characteristic of immunoglobulins A, G and M are serologically distinct.

(*b*) Light chains of Ig G can be divided into two serologically distinct types, K and L.

(c) A and M immunoglobulins also have the same K and L types of light chain.

(d) All human beings have both K and L light chains which are present in a ratio of $\pm 2$K:1L.

(e) A single plasma cell produces K or L but not both, while nearly all myeloma proteins are either wholly L-type or wholly K-type.

(f) The specificity of the combining site (that is, the immune pattern of the antibody) is uninfluenced by these antigenic characters. Any standard human antiserum will show antibody activity in immunoglobulins A, G and M of both K and L types.

(g) Using antisera made in rhesus monkeys, at least four subgroups can be recognized in the heavy chains from Ig G.

The findings from human antisera are complex and involve two systems. The Gm antigens (a+), (b+), (f+) are found in various combinations in individuals, (a+b+f+) being common and (a−b−) combinations being never seen. Again, myeloma proteins have only one antigen or none and studies of plasma cells by fluorescent technique show Gm(a+) or Gm(b+), but never both. The Gm determinants are located on the heavy chain, (a+) and (b+) on the Fc fragment, (f+) apparently close to the disulphide bond holding light and heavy chains together.

The Inv antigens are on the light chains. Inv a is found on K myeloma chains but never on L, while Inv b may be on both. Myelomas Inv(a−b−) are found, although normal globulin of this character is not known to occur. These points are important in relation to phenotypic restriction—an important topic in chapter 6.

*Electrophoretic qualities*

The immunoglobulins A, G and M are most clearly delineated by the use of Grabar's technique of immunoelectrophoresis. Ig G shows a particularly long arc which in itself is a clear indication of the heterogeneity of net charge within the population. More detailed information can be gained by using one of the modifications of starch-gel electrophoresis. When light chain molecules are separated from normal Ig G and examined by this method, a complex pattern of bands emerges which appears to indicate that ten levels of net positive or negative charge are possible for any mammalian species, the distribution in any given species being concentrated over three to five adjacent levels. Heavy chain shows a broader band with no sign that it could be resolved into a series of separate lines.

The regularity of the spacing of the light chains indicates that with the relatively small chains of a constant molecular weight around 20,000, each step represents one unit difference of net electric charge, the replacement of a neutral by a basic amino acid or vice versa. It indicates equally that the light chains present in an individual serum have a high degree of heterogeneity in addition to Inv a and b, and K or L characters.

This heterogeneity is not seen in myeloma proteins. Each shows a single heavy band at a position corresponding to one of the components of the light chain population in normal immunoglobulin. In most preparations there is a weak adjacent band which may represent an artefact arising in the course of preparation, or evidence of a secondary mutation within the plasmacytoma cell clone.

### Amino acid sequences of light chains

It has become almost an article of faith with immunologists that the differences between the myeloma proteins of two individuals are formally equivalent to the differences between the antibodies produced by distinct clones of immunocytes in the same individual. This may be wrong: myeloma proteins have not yet all been shown to carry a specific combining site. In several places in this book the question is raised of the possibility of the combining site being a distinct peptide unit with its own genetic determination. However unlikely this is it has not been excluded, and it could be that some myeloma proteins differ from antibodies in failing to possess any such unit. I believe, however, that more than one myeloma immunoglobulin has already been shown to have a typical combining site and it is on this assumption that the argument will be continued.

The light chains of a human immunoglobulin have a molecular weight of ± 22,000, corresponding to 214 amino acid residues. Perhaps the most interesting biological discovery of the last year

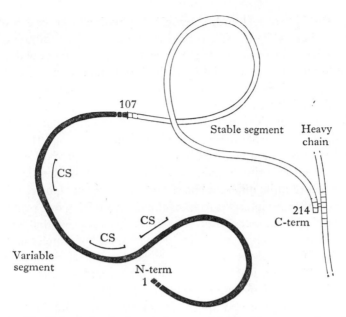

Fig. 17. The light chain of a human immunoglobulin drawn approximately to scale to show the 'variable' and 'stable' segments and the attachment to heavy chain. CS — approximate size of a combining site.

| 1 | | | | | | | | | | | | | | | | | | | | | | 25 |
|---|---|---|---|---|---|---|---|---|---|---|---|---|---|---|---|---|---|---|---|---|---|---|
| Asp | Ile | Glu | Met | Thr | Glu | Ser | Pro | Ser | Ser | Leu | Ser | Ala | Ser | Val | Gly | Asp | Arg | Val | Thr | Ile | Thr | Gys | Glu | Ala |
| Glu | | Val | Leu | | | | | Thr | | Leu | | Pro | | Glu | | Ala | Ser | Leu | Ser | | Arg | Ser |
| | | | | Thr | | | | Pro | Val | Leu | | | | Ile | Ala | | | | | | | |
| | | | | Asp | | | | | | | | | | | | | | | | | | |
| | | | | Leu | | | | | | | | | | | | | | | | | | |
| | | | | Gly | | | | | | | | | | | | | | | | | | |

Fig. 18. The first 25 amino acid residues in human K light chains to show variant and invariant positions. From Smithies's (1967) data.

White = invariant; hatched = showing one of two alternate amino acids; dotted = with one change in addition to any showing the alternate choice; black = hyper-variable. 1st row—full amino acid sequence; 2nd row—alternates; 3rd row—irregular changes.

or two has concerned the amino acid sequences of some selected K-type light chains from myeloma proteins. A comparison of three type K Bence Jones proteins shows that the chain is divided into two equal parts in which amino acid residues 1–107, numbered from the N terminal end, show great variation from one protein to the other, while the C terminal half, 108–214, is common to all three except for some single amino acid replacements of which the most interesting is associated with the change in no. 191 from leucine to valine. This corresponds to change in serological character from Inv a to Inv b.

In the N terminal end there are numerous differences but the length of the variable segment is 107 residues in all and there are a number of common features. The commonest type of difference is replacement of a single amino acid by another but there are short sequences of quite different character. To the present, no regularity in or convincing interpretation of the changes from one protein to another has been recognized. The differences suggest that, subject to the maintenance of certain invariant features of which one is the length of the segment, a completely random set of changes in the genome has been laid down at some phase and stabilized during the whole period of active proliferation. What bearing this may have on the nature and specificity of the combining site can be deferred to the next chapter.

Although the definitive work has been done on human material, workers with mouse myeloma and rabbit antibodies have obtained presumptive evidence that this division of the light chains into variable and constant halves holds also in these species. There is already a strong indication that the variable half may have an important role in relation to the combining site.

## The combining site

Physico-chemical studies have established the existence on the classical Ig G antibody of two identical and symmetrical combining sites to which the corresponding antigenic determinant can attach. From the ease with which a combining site can be blocked by relatively small hapten molecules it is deduced that the specific area involved is quite small, perhaps involving no more than ten to twenty amino acid residues. It is certain that the site is wholly a configuration of amino acid residues with a very strong likelihood that residues on both light and heavy chains are concerned.

The bivalence of Ig G antibody has been fully established by chemical methods (equilibrium dialysis and fluorescence quenching), by the immunological properties of fragments and reconstituted preparations, and directly by electron microscopy. The evidence that both combining sites on a given Ig G antibody molecule are identical is incomplete since all natural antibodies are heterogeneous. Most immunologists accept the likelihood that there is at least a very close approach to symmetry of the two sites. There is no evidence of the existence of hetero-ligating antibody with two different specificities although by appropriate recombination of half-molecules from different sources it can be shown that such hybrid molecules can be readily recognized. Fractionation of an anti-dextran by specific adsorptions gave results which on the whole were in accord with uniformity of the two combining sites but did not exclude a certain degree of disparity. It seems to be legitimate, until experimental evidence to the contrary is put on record, to accept the symmetrical character of the Ig G antibody and to expect no difference in the specificity and avidity of the two combining sites.

The size of the combining site has been studied in detail by Kabat and collaborators for human antibodies against the simple polysaccharide, dextran. It is possible to prepare a complete series of homologous oligosaccharides in the form of chains of glucose units from isomaltose (2) up to a 7-unit chain. These have a measurable blocking power in inhibiting precipitation of dextran which at first increases with increasing size. With most human antisera there was no improvement in inhibition beyond the hexasaccharide. This is therefore considered to have a configuration approximately complementary to that of the combining site and of about the same size. There is a widespread convention to think of the combining site as having the quality of a cavity into which the antigenic determinant fits, but I do not know of any documented discussion of the point. One can hardly doubt, however, that there is some special quality which allows only two small areas on a large protein molecule to serve as combining sites. The only hint available is that tyrosine appears to be present consistently in the combining site, perhaps indicating that two tyrosine residues in some special configuration is the important feature.

Opinion has varied from time to time as to the relative importance of heavy and light chains in their contribution to the specificity of the combining site. Separated chains are never as active as the original or the reconstituted molecule but in most experiments the heavy chain has been more active than the light one.

A more direct approach has been to use reagents which can modify the site by combining with certain amino acids. By using the reagent on antibody in which the combining site is protected by attachment of specific hapten and comparing this with the effect on unprotected antibody, an indication of the importance of a particular amino acid can be gained. Iodination, for instance, gives results indicating that tyrosine is significantly involved in most combining sites but to a much more important extent in some antibodies than in others. A modification of this experiment, in which the two isotopes $^{131}$I and $^{125}$I are used, allows a recognition of the peptides concerned in the combining site in the sense of being protected from iodination in the presence of hapten. The results obtained by Pressman's group suggest that the light chain carries the significant tyrosine residues and by implication plays the greater part in forming the combining site.

An elaboration of the same general approach due to Doolittle and Singer is to use the technique of affinity labelling. In this the hapten, dinitrophenyl, is presented to the antibody in the form of a compound, *m*-nitrobenzene-diazonium-fluoborate, which forms azo-links with tyrosine residues in the combining sites. If the reagent is suitably labelled with tritium, the tyrosine residue concerned can be recognized in peptide maps of the various fragments of the antibody. In this way it was shown that the tyrosine was present in both light and heavy chains. The peptide involved has an average size of $25 \pm$ units and a generally hydrophobic character. It is heterogeneous and in all these respects there is a striking similarity between the peptides from heavy and light chains. The heterogeneity points very strongly to the 'variable' segments of both heavy and light chains being involved.

One cannot resist trying to draw the modern findings on the fine structure of Ig G antibody into a speculative statement based on a suggestion I made at Canberra in December 1964. This is that there is a 'specific chain' involved directly in the structure of the combining site and that this chain appears four times in the immunoglobulin molecule as a portion of the Fd segment of the heavy chains and the variable segment of the light chains. On this view the combining site is made up of two adjacent identical sequences, one on the heavy chain, the other on the light, and the special quality of the combining site may well be due to this duplication of a potential adsorbing pattern. This hypothesis has the important advantage of requiring only one section of the genome to be subject to the 'randomization' process responsible for the heterogeneity needed to give an adequate number of immune patterns.

If it is established that the variable portions of light and heavy chains are quite different, as Putnam believes is the case, a more cumbersome picture must be accepted. One would have to assume that the interaction of two genetically distinct variable chains would result in certain regions being held together by appropriate short-range forces to produce a duplex in which the two sets of amino acid residues have only a general resemblance to each other. This is still needed to account for the Singer–Doolittle results. Provided the interchain forces are definite and regular enough to ensure that the product of all cells of the clone has the same combining site structure, no serious change in the argument is necessary. This modification does, however, make the emergence of the final pattern of the combining site an even more random process than we have envisaged.

## Versatility of the combining site

At this point it may be worth while to examine briefly the implications of the almost universally adopted picture of a small combining site made up of a few, perhaps twenty, amino acid residues. It is easy enough to say that a sequence of even ten amino acids can be arranged in $20^{10}$ or $10^{13}$ different ways, but this is meaningless without some indication of the evolutionary background by which adsorptive specificity arose. This involves a brief comment on the nature of enzyme action.

The extraordinary number and range of activity of the enzymes which can be extracted and isolated from tissues and micro-organisms must mean that there are as many molecular groupings which can serve as substrates for enzyme action or competitive inhibitors as there are potential antigenic determinants. All enzymes are wholly or predominantly protein and it is becoming clear that the active site is often (or always) a relatively short segment of polypeptide. In egg-white lysozyme, for instance, the site of attachment of competitive inhibitors has been accurately located. Generally, the active site seems likely to be considerably smaller than the combining site of an antibody.

It is clear that by suitable disposition of amino acids in sequence, active sites can be produced which are capable not only of allowing specific adsorption of almost any conceivable biological substrate but also, by joint action of the sequence, able to facilitate the molecular and electronic rearrangements that take place in the substrate. Enzymes are more fundamental than antibodies and no one would dispute that proteins have evolved primarily because of their unique suitability as biological catalysts. The same requirement has been presumably responsible for the 'choice' by evolutionary processes of the standard twenty amino acids. The first necessity for enzymic action is a capacity for specific adsorption, and protein has unique potentialities in this respect. Clearly, the same quality has been exploited in the evolution of specific immune pattern.

*Qualities of the Fc segment*

The Fc fragment of the Ig G heavy chain is of standard structure except for the changes corresponding to the Gm allotypes (in man). Its function seems to be to attach antibody to cells and to determine the passage of Ig G antibody through cellular membranes such as the foetal rabbit yolk-sac or the intestinal wall of newborn rats or mice. There is good evidence that the fixation of antibody which confers passive anaphylactic sensitization to reactive tissues is via some portion of the Fc fragment. Since in both these functions homologous antibody is significantly more effective than heterologous, it is necessary to postulate some specific relationship between an animal's immunoglobulins and something which can be regarded as a receptor on a wide variety of cells.

In seeking a general biological significance for this function one naturally looks at the basic function of immunoglobulins to serve as opsonins. It is probably significant that $\gamma$-globulin is the only common autologous protein that is readily taken up by macrophages, including the dendritic phagocytic cells of lymph follicles. The suggestion would be that the structure of some part of the Fc region is adapted to unite to some surface component common to a variety of body cells including macrophages and a variety of epithelial and connective tissue cells. It is highly probable that cytophilic antibody represents a fraction of immunoglobulin in which the Fc component is more accessible than normal. Since the carbohydrate unit of Ig G is associated with Fc, the possibility must be kept in mind that the carbohydrate may be wholly or partly responsible for this type of union.

There is also evidence that the catabolism of Ig G, including the three 7S immunoglobulins $\gamma1$, $\gamma2a$ and $\gamma2b$ in mice, is regulated by the concentration of these immunoglobulins in the plasma and that the responsible component is the Fc fragment. Brambell (1966) has offered a general hypothesis that all these functions depend on the uptake of Ig G molecules by pinocytosis with destruction of any proteins not capable of making specific union with receptors in the wall of the phagolysosome. The hypothesis has an *ad hoc* quality since it is difficult to see how union of a protein determinant to a vacuole wall receptor could protect the protein against a high concentration of protease in the vacuole or how it is subsequently released into the plasma or some other fluid.

*Recapitulation of the structure of G-type antibody*

Obviously the more we know about the chemical structure of antibody the better position we are in to understand the process of its production. It is the basic contention of this book that at every stage in the development of science there is need to choose the current interpretation which allows the clearest understanding and the best practical use of the facts as they have been determined. If that interpretation can be put in the form of a hypothesis which can, in principle, be disproved by experimental approaches, it will also help to speed further advance in understanding.

In the case of G-type antibody the picture that emerges in 1967 is as follows.

The antibody molecule is a complex structure of polypeptide chains based on the combination of two light and two heavy chains.

The genetic control of the various functionally distinct sections is still a matter for speculation and will remain so until the structure of the combining site and its relation to the light and heavy chains is clarified. We are almost compelled to accept as significant the resemblance in the length and location of cysteine residues between the variable and stable segments of the light chain. There is also the hypothesis, much less adequately based, that a segment of the heavy chain has a variable structure equivalent to that of the variable half of the light chain. It would simplify the situation, and be in general concordance with such processes as the evolution of the haemoglobins, to consider the immunoglobulin molecule as having evolved by the duplication and independent mutation of a single primitive cistron coding for 100–110 amino acids. The light chain would correspond to two and the heavy chain to four of the basic units and there is nothing to indicate how many of these duplicated and independently modified cistrons there are in the genome.

The structure of the oligosaccharide must have a complex set of genetic determinants but appears at present not to be directly relevant to the problem of immune specificity.

The molecule is symmetrical. Both combining sites have the same specificity and avidity, both light chains are of the same antigenic character, and what evidence there is suggests that the two heavy chains must also be identical. This symmetry has the implication that, having regard to the diploid character of somatic cells and the several loci concerned, *a high degree of phenotypic restriction must be operative*. If $n$ is the number of loci concerned and if there were no necessity for molecular symmetry, there should be $2^n$ patterns of globulin producible by any diploid cell instead of the one symmetrical pattern which is observed.

The combining site is produced by the juxtaposition of portions of the 'variable' segments attached to the light and heavy chains. The precise character of the site will depend mainly on the amino acid sequence of the variable chains but will also be influenced by factors such as the net charge on the whole light chains and the accuracy with which the two variable chains are mutually arranged. Here, a certain degree of accident may be involved, but otherwise it appears that in a given cell or clone all the details of the antibody globulin produced are genetically determined and the principle of phenotypic restriction operates. For any change to occur, an appropriate *genetic* change must first take place either by somatic mutation or by some alternative process with a similar overall result.

The classical instructive theory of antibody production has few points of contact with what interests us today, but it should not be dismissed without adequate reason. Broadly, it assumes that physically similar antibodies of different specificity differ only according to their secondary and tertiary folding and the subsequent formation of intramolecular disulphide and hydrogen bonds. It should follow, therefore, that when these bonds are broken and a wholly random secondary folding induced in strong urea or guanidine solutions, it ought not to be possible to restore immune specificity. Two well-substantiated claims to have successfully restored a large fraction of specific combining power by careful return to normal environments have been published, and there appears to be a growing consensus of opinion that the final configuration taken by any protein depends far more on the linear sequence of amino acid residues in its polypeptide chain(s) than on any other factors.

In some sense the picture we have adopted is derived from the preconception of a selective approach to antibody formation. It is still remotely possible that an antigenic determinant introduced into the nucleus of a globulin-synthesizing cell might in some at present inconceivable fashion lead to the replacement of an existent segment of DNA by one producing a pattern complementary to that of the antigenic determinant to which it was exposed. Any less cumbersome alternative way of accepting an 'instructive' approach to antibody formation hardly seems possible in the present state of knowledge.

THE STRUCTURE OF A- AND M-TYPE ANTIBODIES

Far less work has been done on the structure of the other types of immunoglobulin for quite simple technical reasons. Antibodies of type A do not precipitate with antigen so that it is virtually impossible to obtain specifically purified antibody for chemical studies. In one example of antibody of this type, where conditions for study were unusually favourable and defined chemical haptens were the antigenic determinants concerned, results were interpreted as corresponding to a molecule with two combining sites close together so that the molecule could not function as a bridge holding two antigen molecules together. This is an attractive picture which may well apply to all A antibodies for, in addition to accounting for failure of precipitation, it offers a reason for the readiness with which such antibody attaches to body cells. In G antibody, attachment to complement and to body components generally is by the Fc segment of the heavy chains, and the configuration of the A molecule may leave this region of the molecule more accessible for contact with a cell surface. The much larger carbohydrate content may also be significant in this respect.

The large M-type antibodies are approximately five times the size of A or G immunoglobulins and on reduction of S–S bonds give rise to 7S components. It is reasonable therefore to think of them essentially as pentamers of molecules resembling A with a considerable addition of polysaccharide. Where direct comparison is possible, as in haemolytic reactions, M antibody is much more active than G, perhaps to be related to the greater number of combining sites which can be brought to bear on a given surface. In most discussions, M-type antibodies are referred to as 19S and are recognized in practice by the fact that simple treatment with 2-mercaptoethanol destroys their specific reactivity. The correlation between 19S sedimentation and susceptibility to inactivation by 2-mercaptoethanol holds for adult mammalian sera but not necessarily for immunoglobulins from other classes of vertebrate.

Electron micrographs of G and M antibodies have been published but not of A-type antibody. The possibility has, however, been suggested that A antibody is bivalent but with the two combining sites very close together. It is also known that larger forms can be found of A antibody with a sedimentation constant around 11 S which may be dimers or trimers. I have found it useful to picture the physical form of the three immunoglobulins as shown in fig. 19. This provides a simple way of visualizing such phenomena as lattice formation by M and G but not A-type antibodies, the much greater complement-fixing power of M, and the special capacity of A to attach to tissues. The possibility that the pentamer of A shown as M may have five rather than ten accessible combining sites has been suggested.

The problem presented by the change from M to G antibodies in the early stages of the primary response will be discussed later, but here it should be noted that when a small hapten is used the avidity of the combining sites of G and M antibody present around the times of the change is similar. Although the character of the heavy chains differs in the three types, everything suggests that the portions which give rise to the combining sites are the same in each.

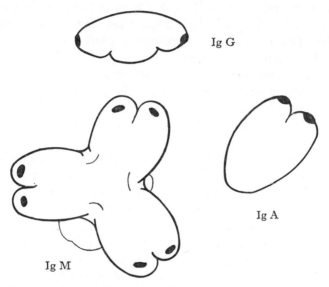

Fig. 19. Schematic illustration of the probable form of the three chief immunoglobulins.

## ANTIGENIC DETERMINANTS

The picture that has been built up of antibody as a large molecule with relatively small combining sites of precise molecular structure has important implications which have led to the concept of antigenic determinant. Like so much in immunology, the gradual development of concepts of the nature of antigens and antibodies has been a matter of reciprocal or even circular interactions. Anything new learnt about antibodies allows a greater precision in understanding the essential structure of antigens, and vice versa.

With the development of Landsteiner's methods of making artificial antigens by linking a relatively large number of small organic haptens to a single protein molecule, it became evident that a small molecular configuration could be wholly responsible for immunological specificity. Further, an antibody of this sort could be wholly inhibited from precipitating the complete antigen by an adequate concentration of the soluble hapten. This technique has subsequently been extensively used for studies of cross-reactivity and serological relationships. The interaction of antibodies with haptens of small molecular weight can also be studied by immunodialysis, particularly if the hapten is coloured, contains radioisotopes, or possesses some other character which allows convenient assay of its concentration. If, on one side of a semipermeable membrane, we have a solution of antibody reactive with hapten $H$ and, on the other, a similar solution of normal globulin, then if a solution of $H$ is added to both sides and time for equilibration allowed, there will be a higher concentration of hapten in the compartment containing the corresponding antibody.

By the use of such techniques it has been shown that a molecule as small as a monosaccharide can significantly block a combining site but that, in some instances, the combining site can 'differentiate between' two larger molecules, e.g. an anti-dextran human antibody is more effectively blocked by an oligosaccharide of six glucose units than by one of four. In a variety of bacterial antigens, di- or trisaccharides appear to be the most important antigenic determinants. The nature of the antigenic determinants in natural proteins is unknown. In some it is necessary for disulphide links to be intact. Many synthetic polypeptides are antigenic though some, such as poly-L-lysine, are not. In some synthetic polypeptides containing three or four component L amino acids (in random mutual sequences) gel-precipitation reactions show multiple lines indicating that several different antigenic determinants occur in the molecular population that makes up a preparation of such a polypeptide. We can reasonably conclude that relatively short sequences of a peptide chain can act as antigenic determinants, but that in some or all cases secondary configurations related to folding or intramolecular bonds may be important.

It is a postulate of clonal selection theory that in the healthy organism significantly active antibodies are not produced against any components (potential antigenic determinants) present in accessible form in the thymus or in any other site where stem cells are differentiated to immunocytes. This should ensure that the great majority of common amino acid configurations should be unable to function as antigenic determinants under any reasonably normal conditions. The same would hold for nucleic acid components and it is probably significant that the only type of DNA readily antigenic in rabbits is phage DNA containing the base hydroxymethylcytosine, and fucosyl in place of some of the ribosyl residues.

### THE FUNCTION OF A, G AND M ANTIBODIES

The functional relationships of A-, G- and M-type antibodies are still not fully understood, but it seems appropriate to conclude this chapter with a brief review of what is known in regard to function of the three types in relation to their structural differences. A certain regularity is beginning to emerge and it is worth making an attempt to define these functions.

The first antibodies to be detected when foetal or very young mammals are immunized are M immunoglobulins; so are all the common natural antibodies. When a grown animal is immunized with any standard antigen the first antibodies produced are M. With some antigens no other type develops. It is reasonable, therefore, to regard the M-type antibody (19 S) as the primary functional form. If we accept a modified Porter–Edelman structure for G antibody, in the form of an elongate flexible molecule with combining sites at each end, then the M antibody can be built up as a complex of five or six G molecules. To account for the various functional and physical characteristics of M antibody, we can picture first the folding of five G molecules into the configuration favoured by Karush's group for an equine A immunoglobulin, i.e. with the two ends carrying combining sites brought close together. This configuration may need the formation of additional intramolecular bonding and the incorporation of further oligo- or polysaccharide.

If these five primary molecules are bound into a three-dimensional configuration in which the binding forces are essentially disulphide links between the Fc regions, the product could well be identical with M antibody as we know it. Such a model is concordant with (a) the molecular weight of M antibody, (b) its susceptibility to mercaptoethanol with loss of precipitating ability, (c) the appearance of M antibody in electron micrographs, (d) the much greater effectiveness of M antibody against red cells in allowing fixation of complement and haemolysis, (e) the uniformity of the light chains from A, G and M antibodies and (f) the production of G and M antibody from the same cell. The only serious difficulty to be overcome is the clear antigenic individuality of M and G immunoglobulins and the lack of any evidence that, by dissociation of M immunoglobulin, fractions with the antigenic qualities of A and/or G can be obtained. I am not aware, however, that specific studies to test the latter point have ever been made.

Our picture of the function of M antibody is primarily to make a firm contact with antigenic molecule or particle. Most natural antigens are likely to have a surface carrying large numbers of uniform or nearly uniform antigenic determinants which will ensure that the six combining sites (of the total ten) on any side of the complex will make at least a proportion of effective contacts with antigenic determinants. Such union appears to be specially effective in initiating the very complex process we know as fixation of complement and the subsequent ingestion of the antigen–antibody–complement complex by phagocytic cells. For both the basic functions of the immune system, the recognition of foreign material in the body and its removal by phagocytosis, M antibody seems admirably adapted.

On any theory by which specific clones of immunocytes are stimulated to proliferate by antigen there must be an important danger associated with infection. In an infection there is a massive accumulation of one type of foreign antigen in the body. If such antigen could indefinitely continue to stimulate the corresponding clone of immunocytes to proliferation there would soon, in principle, be only insignificant numbers of other clones in the body and an enormous excess of specific antibody. Clearly there must be some effective and specific form of negative feedback to keep such a process within bounds. Two processes seem to be concerned

With any intense antigenic stimulation, a large number of the competent cells are pressed into plasma cell formation. This is generally taken as prohibiting any capacity for de-differentiation—once a mature plasma cell is produced it is an end-cell, actively producing immunoglobulin (A, G or M) but incapable of further proliferation under antigenic stimulus. In this way we can have exhaustion of all clones subject to stimulation by the antigenic determinants concerned. Although most M antibody is produced by plasma cells, there is direct evidence that some cells at a certain stage after immunization are producing both G and M, while at a later stage the great majority are producing only G. The obvious implication is that a substantial proportion of cells or clones initially producing M antibody subsequently switch to the production of G. Several workers have raised doubts about the reality of the M → G switch in an individual cell, but the overall evidence is in favour of the process. Experimentally, one of the most effective ways of reducing the proportion of M-antibody-producing cells is by exposing the animal to pre-formed G antibody of the same specificity.

The significance of this effect is unclear. One way of looking at the position is to assume that all cells on primary commitment are M producers and that a proportion change spontaneously to G producers. The introduction of pre-formed G antibody will impede any antigen in the circulation from effectively stimulating any progenitor cells. There will be no further recruitment of M producers and hence no new clones of G producers. Other interpretations are possible but this seems to be the most satisfactory way of covering the salient experimental findings. The essential result is a sharp diminution in the initiation of Ig M clones with an additional but less striking effect on Ig G production.

The primary biological function of G antibody production may therefore be to prevent too great a predominance of one type of immune pattern in the immunocyte populations of the body. Other incidental virtues of G antibodies may also have emerged. Only G antibody passes readily through the placenta and (in a completely different context) any virus or toxin molecule can be rendered inactive without need for the action of complement or phagocytosis.

There may be special importance in the conditions under which the normal transition from M to G antibody fails to occur. These include (*a*) with polysaccharide antigens, (*b*) in irradiated animals under some conditions, (*c*) in animals given mercaptopurine, (*d*) in allogeneic radiation chimeras and (*e*) in M-type myelomas. There is little that can be said in regard to these conditions. The fact that there is presumptive damage to the immunocytes in all instances except the first suggests that the change from M to G is intrinsic to the immunocyte and that any type of damage is likely to delay or prevent the switch. The reported failure of G antibody production toward polysaccharide antigens is apparently incorrect and based on the fact that M antibody is much more readily detected, molecule for molecule, than G. In all probability there is a considerable production of Ig G, according to the usual rules.

Antibodies of Ig A type have recently become of special interest. It has long been known that in mammals generally, but particularly in ungulates, large amounts of antibody are present in colostrum and that the antibody shows certain physical differences from that in the corresponding serum. It is only since 1965, however, that it has been generally recognized that the great mass of antibody in colostrum, saliva and some other secretions is of Ig A type. It has, however, an additional antigenic quality not found in serum Ig A, which is referred to as 'transport piece' by Good's group. There is some doubt as to whether circulating Ig A is taken up by secretory cells from the blood linked with transport piece and secreted as such. Intravenous infusion of normal serum into agammaglobulinaemic children results in the appearance of Ig A in saliva. On the other hand it is reported that the protein of Ig A myelomas is not found in the globulin of parotid saliva and that Ig A labelled with [131]I does not reach the saliva as such, all the label being present on small molecules. All are agreed that there is a close qualitative resemblance of antibodies (that is, corresponding immune patterns) in colostrum or saliva with those in the individual's serum. In the case of *Escherichia coli* agglutinins in man there is also a close quantitative relationship even though the serum antibody was almost wholly in the Ig M fraction. It was shown many years ago that plasma cells were numerous in the interstitial tissues of the bovine udder in the early stage of lactation. The possibility that much of the antibody in colostrum is of local origin must therefore be considered, particularly in view of a recent report that 80 per cent of the numerous plasma cells found in the rat duodenum are producing Ig A. I am not aware of similar studies on mammary or salivary glands but they are obviously called for.

As things stand, it seems likely that Ig A antibodies are secreted by the secretory epithelium of the gland or mucous surface concerned but reach the epithelial cells from two sources, the circulating blood and the antibody liberated into tissue fluids by adjacent plasma cells. The relative importance of the two sources may vary greatly with the glandular tissue concerned and its physiological activity. What happens within the epithelial cell is problematical. It seems most likely that the addition of 'transport piece' to the standard Ig A is all that is needed, but the possibility of a partial breakdown and reconstruction of antibody cannot be excluded.

An intriguing finding, in view of the special relation of Ig A to lactation, is that in one investigation women showed a much higher proportion of polio-antibody in that fraction than was ever observed in men.

For a number of years it has been regarded as likely that the reagins of patients with hay fever were of Ig A type. This was based largely on the separation of the immunoglobulins by physical means and there seems to be no doubt that the immunoglobulin concerned in man has the physical qualities of Ig A. The final diagnosis is, however, an immunological one and Ishizaka and his collaborators have provided convincing evidence that at least a large portion of the reagin demonstrated in ragweed-sensitive patients by the Prausnitz–Küstner test is not antigenically Ig A. It is now referred to as Ig E, and in normal human serum is present in only minute amount. As yet there is no indication of any physiological function of Ig E, just as none is known for the other rare type, Ig D.

SUMMARY

In this picture of the nature of antibody some of the essential features of clonal selection theory can be clearly seen. Basically all that is postulated is that by some process of randomization a great variety of genetic determinants of immune pattern can be generated. This is most simply thought of as arising by a high mutability (somatic mutation) of one or more cistrons concerned with defining the immune pattern of the combining region by the amino acid sequence of the 'variable' segments of light and heavy chains. There is still a preference amongst active workers in the field to seek the origin of diversity by a process which could legitimately be called 'random differentiation'. This would lead to the same operational result by making use of some as yet unparticularized process to establish one out of many thousand other possible configurations provided in the inherited genome. In whatever fashion the genetic mechanism emerges, its effect on the hypothesis we have favoured is to determine the production of a combining site where portions of the two chains are in a determinate duplex arrangement. Such regions are incorporated into each immunoglobulin molecule and the equivalent receptors in the immunocytes. The double loop of the 'variable' segments at each end of the Ig G antibody is specialized, perhaps simply by its duplex character, to serve as a highly specific adsorptive site, but the pattern is random. It may adsorb a common constituent of the body; it may be fitted to adsorb the characteristic antigenic determinant of a virus or it may be a meaningless combination which can adsorb nothing within the capacity of living organisms to produce. It is at this point that the biological quality of selection for survival enters the picture. Pure chance determines the pattern on the combining site, but what it reacts with will determine what happens to the cells carrying the pattern as an inheritable quality. If it reacts at once with accessible body components, the cell is destroyed; if it reacts later under appropriate circumstances with determinants of a foreign protein or micro-organism, the clone concerned will flourish; if the pattern finds no determinant to react with, the cell line can respond only to nonspecific stimuli if they exist or, by taking on myeloma character, provide its own intrinsic urge to proliferation and secretion.

# 6 The origin of immune pattern

The central problem that has worried theorists on immunology concerns the process by which from an animal stimulated naturally or experimentally by a certain antigen, there appears an antibody (or population of antibody molecules) with a specific complementary relation to the antigen used. Until the process of protein synthesis was understood it was possible to imagine a wide variety of processes with some analogy to various craftsmen's manipulations. The protein was moulded to fit the pattern of a molecular die or the antigenic determinant was intruded into the synthetic RNA, and there were many variants of these basic hypotheses.

With the clarification of the process of specific protein synthesis in bacterial viruses and bacteria during the last decade and the mounting evidence that, in all essentials, synthesis of protein by mammalian cells uses the same mechanisms, the scope for hypothesis has been greatly restricted.

PROTEIN SYNTHESIS

Protein, including antibody globulin, is synthesized on the basis of information held in the DNA of the genome of the cell concerned. Subject only to minor disagreement on detail, the process is as follows: a protein is composed of one or more polypeptide chains, often associated with other material such as oligo- or polysaccharide, and with inter- and intra-chain linkages by disulphide and hydrogen bonds. The synthesis of each component polypeptide chain is governed by a coded length of DNA in the genome. This transfers an equivalent pattern (transcription of information) to a similar length of single-stranded RNA (m-RNA) which, in association with one or a group of ribosomes and in the presence of transfer RNA and appropriate supplies of amino acids, induces the orderly synthesis of a peptide chain, each codon of three nucleotides defining the amino acid to be incorporated. The process by which the component chains of a complex molecule, like any of the

125

immunoglobulins, are put together is unknown. That it is genetically programmed is proven by the uniformity of the immunoglobulins produced by single clones of human myelomatosis or murine plasmacytomata. It has not, however, been proven directly that an identifiable antibody can be produced as a uniform molecular population, but at least one example has been reported where a sharp peak in the electrophoretic diagram of serum from a rabbit immunized with a streptococcal antigen corresponded to what was apparently a uniform population of antibody molecules. An accompanying statement that this result was obtained in several other rabbits of a particular breed suggests that a full analysis of these 'monoclonal antibodies' should soon be forthcoming. Indirect evidence that individual antibodies differ among themselves in the same way that myeloma proteins do is becoming stronger. When two antibodies of different specificity but the same general type are prepared in a single rabbit, there are definite differences in amino acid composition of the specifically purified antibodies. Antibody eluted from Coombs-reactive red cells from patients with acquired haemolytic anaemia nearly always shows only one antigenic type of light chain—presumptive evidence of molecular homogeneity.

Any standard antiserum contains a heterogeneous population of antibody molecules, but by appropriate manipulations it is relatively easy to obtain purified preparations of specific antibody molecules of type G immunoglobulin. The evidence that this antibody comprises only a very small number of distinguishable types suggests that these types are produced by about the same number of clones of immunocytes. This picture may be wrong but it seems to be the most satisfactory basis on which to consider currently tenable hypotheses of the origin of immune pattern.

### THE POSSIBILITY OF INSTRUCTIVE PROCESSES

Perhaps the most cogent approach to the problem (following Karush) is to consider the fact that the complementary steric relationship of antigen determinant to antibody combining site is three-dimensional. The information contained in the configuration of the combining site is not explicit in any one-dimensional arrangement of amino acid residues. It is inconceivable that any instructive transfer of information in regard to the three-dimensional structure of an antigenic determinant could modify the genetic process to allow the synthesis of an appropriately modified sequence of residues to allow the polypeptide eventually to form the corresponding combining site. There are, in fact, only two possibilities.

(*a*) The form of the final folding of the complex immunoglobulin molecule is determined by direct three-dimensional contact with an antigenic determinant followed by 'fixation' of the impressed pattern.

(*b*) There is no instructive relation between antigenic determinant and combining site, and the combining site configuration is of wholly genetic origin. If it 'happens to correspond' with an antigenic determinant in the body, their three-dimensional union sets in train the various processes that are postulated by selective theories of immunity.

The first requirement, then, is to consider the probability or even the possibility of an instructive tertiary folding of immunoglobulin against a physical template of antigenic determinant. There are some very difficult problems at the physico-chemical level here. The only serious suggestion as to how the modified tertiary configuration could be fixed is that of Karush, who assumes that in the lengths of polypeptide chain concerned there are enough cysteine residues to provide a sufficiently large number of ways of producing intramolecular disulphide bonds to hold any configuration stable. Then there are the problems as to how the antigenic determinant-combining site union is dissolved once fixation is complete and how the Ig G antibody comes to be symmetrically supplied with two identical combining sites.

Evidence against the importance of S–S linkages comes from two sources. General protein chemistry indicates strongly that the position of S–S linkages is strictly determined genetically and that the sequence of sulphur-containing and other amino acid residues in the polypeptide chain in itself determines the most frequent or the only way in which disulphide unions can form. Direct study of purified antibody has shown that by careful treatment all S–S links can be broken, allowing a wholly random configuration. When such reduced antibody is allowed to re-form the disulphide linkages, a high proportion of its specific reactivity returns. This represents perhaps the strongest argument against any direct template hypothesis of the Pauling–Karush type. The fact that there are differences in amino acid content of different Ig G antibodies in the same rabbit is incompatible with any simple instructive theory, but is not decisive for the selection approach.

The general indication from all the available information is wholly against the Landsteiner–Pauling–Karush theory and, unless some unexpected new experimental approach causes a re-evaluation, instructive hypotheses can be left out of further consideration. The problem remaining is to establish the way in which a multitude of patterns can arise in a single individual. At the cellular level it is well established that standard Ig G antibody is produced by cells of the plasma cell series, that except in very rare circumstances only one antibody is produced by a single cell and that an essential part of antibody production, as stimulated by the injection of a defined antigen, involves the formation by proliferation of a clone of plasma cells. The essence of the problem is to interpret the origin of the single pattern of antibody synthesis manifested by an active plasma cell clone.

## THE GENETIC BASIS OF ANTIBODY PATTERN

We have now reached the point where instructive theories of antibody production can be left out of consideration. With the evidence of the 'variable' half of the light chain in myeloma proteins and the virtual certainty that there is also a 'variable' segment which probably makes up a portion of the Fd segment of the heavy chain in both myeloma proteins (Ig G type) and antibodies, we have an important clue as to the type of genetic process that must be sought. It is perhaps still not quite fully established that the combining site is made up of portions of the variable segments of both light and heavy chains or that the physical basis of antibody diversity is, in fact, the variability of the sequence of amino acid residues in these chains. Both, however, seem to be so much more likely than any alternative that they will be accepted as a basis for all our subsequent discussion. As all who have followed the current literature on the stable and variable halves of the light chains will have realized, a genetic process of quite unusual quality must be concerned with coding for the variable segment of the chain.

There are two basically distinct possibilities in regard to the nature of this process. The first is that each pattern has arisen during the course of vertebrate evolution in analogous fashion to the way whereby the various patterns of the haemoglobin and myoglobin chains have come to vary in different species. An obvious corollary to this is that there must be a very complex set of cistrons in every cell and a mechanism for choice between them. This can be termed 'the hypothesis of polycistronic control' and will call for some introductory discussion of evolutionary changes in protein structure.

The second basic approach is to assume that the diversity of pattern arises by genetic processes taking place in somatic cells subsequent to fertilization, that is, by somatic mutation or some operationally equivalent process. At the present time there is much interest in the suggestion of Smithies that somatic recombination within an 'antibody–gene pair' offers the most economical model that would produce the required diversity. Both these 'somatic' approaches present the difficulty of (a) providing the very high mutation or diversification rates that are needed and yet (b) allowing the persistence of a standard immune pattern through an indefinite number of generations in a myeloma clone and, by implication, in an antibody-producing clone of cells. Whatever alternative or combination of these processes is responsible for the emergence of diversity of immune pattern, the actual expression of that pattern in the immunocyte will be subject to the principle of phenotypic restriction. This has been mentioned already on several occasions but it may be elaborated a little at this point, particularly in relation to the part that myeloma proteins have played in the development of immunological theory.

In any human individual there are at least twenty different immunoglobulins that can be produced, even when we confine ourselves to physical and antigenic qualities only. Every myeloma protein produces one class only. At the level of immune pattern, the individual has almost infinite potentiality to produce antibodies against foreign organic material. Any semi-purified antibody active against a single antigen is in fact a heterogeneous population of molecules of varying avidity for the antigen and of different antigenic types. On the other hand, when a myeloma protein is found with a recognizable antibody specificity, it is homogeneous in this respect. The myeloma protein described by Eisen which reacted with the artificial hapten, DNP, differed from every other 'DNP antibody' in having an extremely narrow range of affinity in combining with the hapten.

Subject only to rare and anomalous exceptions it appears to be a general rule that each immunocyte produces only one immune pattern—perhaps more correctly only one pattern of amino acid sequences in the variable chains making up the combining site— and only immunoglobulin of a single antigenic class. The nature of this process of phenotypic restriction is still uncertain. Although the three possible sources of diversity listed above are spoken of as alternatives they are not mutually exclusive and, in fact, probably the best provisional conclusion is that all three processes play a part. Before attempting to apportion their roles in producing a diversity of variable chains it is desirable to say something about the general quality of these processes.

*The evolution of change in protein structure*

Information on the theme of the changes in protein structure arising in the course of evolution will be drawn mainly from the report of a Symposium in 1965, 'Evolving Genes and Proteins'. With the development of satisfactory methods of obtaining the full amino acid sequence of small and medium-sized polypeptide chains, much attention has been turned on the species differences in small proteins and peptides such as insulin, the fibrinopeptides, pituitary hormones, cytochrome C and haemoglobin. Cytochrome C (104 residues) and the $\alpha$ and $\beta$ chains of globin (141 and 146 residues) are only slightly smaller than the light chain of immunoglobulin (214 residues). Both are functionally very important proteins with a long evolutionary history. Cytochrome C, which is functionally and to a large extent chemically similar in yeast and man, has presumably existed for $2 \times 10^9$ years, and haemoglobin for perhaps 400 million years. In the process both have undergone great changes in their amino acid composition.

Over the whole series of organisms from yeast to man, cytochrome C has 51 of its 104 residues present in the same positions.
The others are variable and the number of differences is roughly
proportional to the evolutionary distance between the forms concerned. In the haemoglobin chains, all of which, including the
single myoglobin chain, are presumed to have had a single evolutionary origin, only 8 per cent of the residues are in equivalent
positions in all chains. The changes must have arisen by mutation
plus the processes of recombination, selective survival and genetic
drift which allowed the replacement of one form by another. The
evidence suggests that in the vast majority of instances the mutation in question was a transition in which a single base pair in the
DNA was altered.

Evolutionarily effective mutations in globin chains seem to have
occurred about once in seven million years but, as Ernst Mayr
pointed out, many millions of mutations must have taken place in
the relevant region of the genome of individuals over the same
period but for one reason or another failed to replace the existent
pattern. It is reasonable to assume that the retention of nearly half
the residues in cytochrome C, virtually since the origin of aerobic
organisms, means that these are functionally necessary, or at least
that there is no biologically possible way by which they can be
replaced with an equally effective amino acid. In the globin chains
an examination of the number of alternative amino acids which are
known to be able to occupy the various positions gives something
very close to a Poisson distribution with a mean of 2·1. Random
processes clearly play an extremely important part.

In addition to these changes in the detailed sequence of amino
acid residues, there is a second evolutionary process of great
importance for understanding the history of the haemoglobins and
potentially of the immunoglobulins. This is the duplication of
genes in the course of evolution with subsequent independent
mutation. In normal human beings, haemoglobin contains $\alpha$ and
$\beta$ chains and, in the foetus, $\alpha$ and $\gamma$ chains. All three chains are
obviously of common origin but differ sharply in amino acid
sequence. Each must be represented separately in the genome and
in the case of the $\gamma$–$\beta$ relationship there must exist a mechanism by
which, in the course of differentiation, one cistron must be closed

down and another activated. The three cistrons clearly result from duplication of a primitive gene at least twice and in all probability at very different geological periods. Most geneticists believe that this process of gene duplication with subsequent independent mutational changes has played a major part in the development of functional proteins. In discussing the evolution of the immunoglobulins we can take it as almost axiomatic that similar processes have been involved. This does not, however, necessarily mean that the diversity of immune pattern has been so produced.

## Somatic mutation in vertebrates

In any discussion of somatic mutation as a source of variation in protein pattern, it is important to seek evidence on two points: (*a*) the legitimacy of comparing somatic and germinal mutations and (*b*) the frequency of somatic mutation of the type in which we are interested. Once again it is necessary to emphasize that only by refined genetic methods is it possible even in principle to differentiate between the various processes that can give rise to inheritable changes in somatic cell lines. Irrespective of whether they arise by point mutation, by deletions or inversions, or by any form of somatic recombination, provided they are qualitatively random in character the result can be accepted as 'somatic mutation' in the broad sense.

There is no doubt about the occurrence of somatic mutations in a wide variety of organisms and there is at least some evidence to suggest that they are of the same quality as germinal mutations. I have frequently quoted Fraser and Short's paper on fleece mosaics in sheep, which implies that in a mammalian species a single easily recognizable mutation occurs with approximately the same frequency ($10^{-7} \pm$) per cell in any stage of development of the fertilized ovum from the 2-cell stage onward. Mutant forms of fleece of basically similar character can also arise as a result of mutation in germ cells.

When somatic mutation occurs in a cell of a fully differentiated organism, it is always difficult to recognize its occurrence. If we accept a reasonable figure, say $10^{-6}$–$10^{-8}$ for the likelihood of mutation involving any particular locus per cell per generation, the most that can happen initially is that one cell amongst a million or more unchanged cells either dies or changes one of its functions. If it normally secretes a protein recognizable by its enzymic activity and, as a result of mutation, now secretes a protein antigenically similar but without enzymic activity, the modified enzyme molecules will represent 0·0001 per cent of the amount of active enzyme and will be quite undetectable. Only in quite exceptional circumstances can somatic mutational change in a single cell be detected. For a mutant to become demonstrable it must for one reason or another proliferate more extensively than its unchanged congeners so that a large number of descendants can manifest the change. The possibilities are:

(*a*) that the mutation directly involves some function controlling proliferation of the cell and allows rapid proliferation of neoplastic character;

(*b*) that, as a result of mutation, a stimulus present in the internal environment can stimulate the mutant cell to proliferate. This is the basic process postulated by clonal selection theory.

There is, however, a third situation in which a mutation (*c*) produces a functional change in a cell without influencing its proliferative potential. If subsequently an unrelated mutation of type (*a*) or (*b*) occurs allowing preferential proliferation of the cell, the function related to mutation (*c*) will become demonstrable. There are now many instances in which malignant tumours, notably hepatoma and bronchial carcinoma, have shown striking metabolic abnormalities, presumably as a result of this process.

*Somatic genetic processes in relation to the diversity of the 'variable' segments*

There are some obvious resemblances between the diversity of amino acid structure in proteins like cytochrome C or haemoglobin from a wide range of species, and the diversity found in the Lv (variable) segments of Bence Jones proteins and, by reasonable extension, in antibodies. In any attempt to equate the two processes, however, there are obvious difficulties particularly in regard to the time scale concerned. If a diversity of structure about as great as would be seen in the whole range of past and existent cytochrome C is needed to produce the required number of patterns, it must be produced in a few months instead of in one or two billion years.

It has become conventional to say that perhaps $10^4$ immune patterns would be needed to account for the range and flexibility of the immune response. On the line of thought we are following, mutations would be required in a particular cistron controlling a peptide chain of 107 residues, that is, of 321 nucleotide pairs. By the time of birth, stem cells, thymocytes and lymphocytes might make up about 1 per cent of the cells of the body and, in a newborn child, would number about $3 \times 10^{10}$ cells which would have required approximately twice that number of replications to produce them. If we adopt Orgel's figure of $10^{-8}$ 'errors' per nucleotide pair per replication, there would have been an average of 600 changes at each point in the cistron. For any given cell, however,

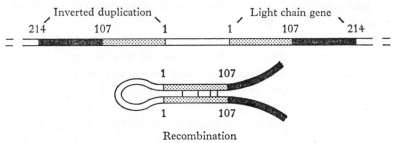

Recombination

Fig. 20. Smithies's 'scrambler hypothesis' for production of diversity by intra-chromosomal recombination of duplicated 'variable' genes.

with about thirty-two generations from the fertilized ovum there would be only a $10^{-4}$ chance of its differing by one single nucleotide pair replacement, $10^{-8}$ of two being present. All the rest should have the inherited quality—or more probably one of the two inherited qualities since the individual is sure to be heterozygous. If the calculation is even remotely correct, a much higher rate of mutation or some other type of change is called for. Mutation at the rate of $10^{-4}$ per nucleotide pair would be needed if virtually every differentiating cell were to show some change from the ancestral patterns. This does not seem to be impossible and if something of this degree of heightened mutability has been evolved for this particular cistron—and one can think of some local anomaly in the supply of DNA polymerases and repair enzymes at the site—none of the other difficulties is insurmountable.

If, in addition to accelerated point mutation, there is also available opportunity for somatic recombination—of the type suggested by Smithies in 1967, or in some other guise—a much wider range of patterns could be developed in the short period between fertilization and birth. Smithies makes no mention of somatic point mutation and considers that given a series of *ad hoc* but reasonable assumptions his 'scrambler' hypothesis of somatic recombination could allow the emergence in a single individual of a whole range of differences in Lv chains, equivalent to those seen in a comprehensive collection of Bence Jones proteins from many different patients.

Two comments may be relevant. (1) The recombination postulated must presumably take place at some period of embryonic development before the differentiation of immunocytes and with a relatively high frequency. After differentiation it must cease or occur at a very low rate. (2) As with the alternative of hyperactive point mutation concentrated on 'hot spots' within the variable segment, the changes produced by recombination will be wholly random in character from the point of view of the antigenic determinants with which the derivative immune pattern will react.

The feature in which any postulated process of randomization of pattern differs from every other type of mutational change is that there are no positive metabolic or other requirements for pattern in the 'variable' peptide chain. The pattern is required only in relation to factors external to the cell. The special type of selection for survival that it will undergo, according to our hypothesis, in the thymus or in any other organ of equivalent function is of quite a different quality from anything that involves a functioning component of the cell. Changes that in any other protein would necessarily lead to breakdown of cell or organism are in this context quite acceptable. Their action is nil unless they are brought into effective contact with something specifically adsorbable.

It is an interesting point that if mutational changes occur as frequently in multiplying somatic cells as in germ cells, there must be a constant wastage of cells through one or other functional lack. But, even if as many as 1 per cent of cells quietly died in every tissue, the effect would probably be recognizable only with very elaborate technique.

Whatever hypothesis is adopted to account for the generation of diversity in immune pattern and in the amino acid sequence of the variable chains, there seem to be two necessary conditions to provide the observed phenomena. The first is some form of phenotypic restriction by which only one of the alleles in a heterozygote is active. If the 'variable' cistron is on the X chromosome, the Lyon phenomenon would fit the circumstances admirably. At some stage in early embryonic life each cell in a female mammal elects at random to inactivate one X chromosome subsequently visible in interphase as a Barr body. Once the decision is made by the cell, its descendants are permanently controlled as to X characters by whichever X chromosome remains active. About half the cells have a paternal X, half the maternal one. Such restriction is not known for any autosomal chromosome pair but there seems no reason why specific types of phenotypic restriction should not occur.

The second condition is more questionable. At first sight it seems obvious that there must be a hypermutability (in the broad sense) of the relevant genes during embryonic development followed by stabilization once differentiation to immunocyte has occurred. If, however, one remembers that the effect of any inheritable somatic change is negligible unless it results in preferential proliferation of the phenotype, the necessity for postulating stabilization of the gene after differentiation becomes minimal. Even at the high rate of $10^{-4}$ mutations per necleotide pair per generation, in ten cell generations of a proliferating immunocyte only 3 per cent of the thousand descendants would have mutated. This assumes that all 320 nucleotide pairs were equally susceptible, while in actuality the percentage would probably be much smaller. There is no suggestion as to how frequently per cell generation Smithies's scrambler process occurs. It could hardly be more frequent than $10^{-4}$ per cell generation. Amongst the population of an established clone, mutants arising since its establishment would represent only a minute proportion unless they had some proliferative advantage. It would be impossible to detect in practice the fraction of 1 per cent which had undergone a significant change.

*The relative importance of the processes of diversification in the origin of immune pattern*

It has already been emphasized that the three processes that may be involved in producing the diversity of antibodies and immunoglobulins are not mutually exclusive. We have only to look at the multiplicity of genetic markers on human Ig G from a single individual to be certain that there must be polycistronic determinants for immunoglobulins. Most modern writers, too, are impressed with the evidence that each light chain is composed of *two* units of 107 amino acid residues and that in all probability the heavy chain of Ig G can be thought of as composed of four similar units plus the polysaccharide unit. Multiple cistrons arising by duplication is the most appealing implication.

The problem of accounting for the known diversity of structure in the Lv segments of human Bence Jones proteins is usually accepted as being equivalent to interpreting the diversity of immune pattern amongst antibodies. Provided there is a realization that other aspects of the immunoglobulin molecule may have some modulating influence on immune specificity, this is probably correct. Whatever the genetic mechanisms responsible, it is certain that their overall result is unique. No remotely similar phenomenon is known, and rather strained analogies therefore become permissible. Various types of intragenomic recombination can be imagined which would facilitate diversity. In microbial genetics, mutational processes of very high frequency have been observed and even within a limited region of a single cistron there are some loci vastly more mutable than others. Clearly, somatic recombination could involve a gene or portion of a gene which was also subject to abnormally high levels of point mutation at some regions.

From our present point of view the requirement is not to attempt any decision between alternatives but to assess the part different processes may play and to suggest the lines of work that can be expected progressively to clarify the position.

Most of those who have attempted a single interpretation have postulated processes which would allow a single cell to produce any one of a wide variety of immune patterns. One such hypothesis calls for a relatively small number of genetic loci being concerned in determining pattern, the diversity springing from the various combinations that can be produced. If, for instance, the number $n$ of loci involved in determining antibody specificity was 15, some 33,000 ($2^{15}$) patterns might be possible, provided a process of randomized phenotypic restriction were operative. The results of such a process would be operationally identical with a set-up in which $n$ is much smaller and mutational changes are freely operative. Most who favour such a view would think of the antigenic determinant as entering the cell and in some fashion instructing the cell as to the correct combination to be allowed phenotypic expression. To do so would need 33,000 ready-made molecules of each possible type to act as sensors of the antigenic determinant's three-dimensional configuration, and for each to have an individual 'circuit' to ensure de-repression of the proper allele of fifteen different loci. This is biologically inconceivable.

A slightly different approach would be to postulate a direct genetic specification of the combining site and to assume that a battery of $10^4$–$10^5$ small cistrons carried the information needed in the genome, one for each pattern. If the repressor for each cistron had the quality of the three-dimensional combining site, one might postulate that union of antigenic determinant with the repressor would activate the cistron. This hypothesis is operationally indistinguishable from a simple instructive one and equally unattractive.

These two hypotheses derive essentially from Lederberg's suggestion that subcellular selection was a legitimate alternative to clonal selection. Both require two extremely cumbersome postulates that sterically correct sensors must be present in each cell for all the antigenic patterns with which the animal can react and, to account for tolerance, that a very large number, perhaps half, of these sensors must be blocked by each one of the 'self' determinants to which the animal is tolerant.

It is highly improbable that these or any similar views have any relevance to the real problem. The relative importance of polycistronic information developed in the course of evolution and the processes of point mutation and recombination or other intragenomic processes in somatic cell lines, however, will remain subject to debate for a long time. The most significant evidence bearing on the part played by gene duplication is probably to be drawn from the antigenic character of heavy chains in human Ig G, while from the point of view of somatic cell processes the retention of a standard length and many stable sequences in the Lv chains is probably the most critical aspect. Both will be discussed, but it is perhaps of no great importance from the strictly immunological angle to try to reach any definite conclusion. It is inevitable that future developments will bring modifications of opinion at the genetic level but these may have little bearing on aspects of immunology susceptible to experimental study.

It is the essential theme of this book that immunology as a science has advanced far enough for us to be certain that any general approach will contain the central features of clonal selection theory:

(*a*) a genetic origin of pattern information;

(*b*) a randomized process determining phenotypic expression (somatic mutation or random differentiation);

(*c*) population dynamics of immunocytes determined by the results of specific antigenic contact.

Within that framework there is plenty of scope for modification to keep in line with new developments in cytogenetics and protein chemistry.

## *The significance of the stable components of the variable chains*

In ascribing immune pattern to diversity in the variable segments, no mention has yet been made of the probable significance of the fact that most of the 'variable' segment of the light chain is constant in character and has many sequences in common between species as far apart as mice and men.

In the first place this immediately rules out any suggestion that completely random base-changes in DNA are involved. If single changes in nucleotide pairs are mainly responsible they must be highly concentrated in certain positions in the cistron. The main point in favour of Smithies's 1967 hypothesis is the possibility of obtaining changes of this sort of somatic recombination between two 'half-genes' of similar general character.

The conservative elements of the variable segments could also have a bearing on a quite distinct set of phenomena. The disproportionately large size of the whole Lv segment, compared with the small extent of a typical combining site, suggests the possibility that more than one potential combining site might be present on a single immunoglobulin molecule. It would, for instance, not be too disconcerting to find a monoclonal myeloma protein or antibody showing two quite unrelated specificities. Quite small fragments of a haptenic complex, such as that of dextran, can block a combining site.

If our hypothesis is correct that the combining site is a small $10 \pm$ sector of duplex where equivalent portions of the variable sequences on light and heavy chains come together, one might well find that several quite distinct antigenic determinants could find appropriate regions with which to make effective contact. What is of special interest is the situation that would arise if a substantial sequence of constant character were associated with adjacent labile residues and the stable sequence had a significant adsorptive power for a certain range of related antigenic determinants. It could emerge that for some particular range of chemically related antigenic determinants of high evolutionary significance, a set of complementary combining sites might be developed by relatively few point mutations involving the adjacent labile positions of the variable chains. One has in mind the special problems of histocompatibility that are raised in chapter 13. On such a formulation, one could well believe that if a mouse histocompatibility determinant $A$ reacted with a combining site sequence $a$ of the type described, then if a point mutation produced a modification of the antigenic determinant to $A^1$, although it would fail to react actively with combining site $a$, yet a single change in one of the adjacent variable residues could allow it to do so. Some such interplay of stable and labile positions will almost certainly have a major role in determining the initial proportions of progenitor immunocytes that will react with this or that general type of antigen.

This is merely another implication of the contention that the sequence of amino acids in the combining site is wholly genetic in origin and has no instructive relationship whatsoever to the antigenic determinant with which it reacts. The only mandatory condition in regard to the range of reactivity of a given clone is that it shall *not* react significantly with any accessible determinant natural to the animal producing it. Within those limitations it may react significantly with any antigenic determinant that can find a sequence of appropriate complementarity on the combining site. Any argument against clonal selection theory based on the fact that 'too many' cells or molecules from a normal animal react with a single antigen must not ignore this situation.

## THE HEAVY CHAIN ANTIGENS OF IMMUNOGLOBULIN G

From the genetic point of view, the work largely centred on Kunkel's group at the Rockefeller University on the multiplicity of antigenic types in the Ig G of all normal human beings is of special importance. Omitting any reference to A, M or D immunoglobulins and confining ourselves to Ig G, we find four serologically different types of heavy chain, We, Vi, Ne, Ge. These are present in all normal Caucasians. In addition there are within each group 'Gm' factors. A man We and Gm(a f) will produce globulins which are (a), (f) or ( − ) in respect to these factors. There are other factors associated with Vi (b g) and Ne (n⁺ n⁻). There is evidence that antibody specificity may be associated with any of these but there are some restrictions which may but do not necessarily mean that capacity to produce a given antibody is always associated with one antigenic type. There is no evidence that involves the combining site of an antibody or the very poorly defined variable region of the heavy chain in these antigen-defined genetic qualities. So far as I am aware, no one has yet produced an authoritative interpretation of this antigenic diversity in the heavy chains of the same individual. The findings could be interpreted as depending on the existence of a series of, say, twenty complete cistrons each coding for one type of Ig G heavy chain.

In view of the close correspondence in physical and chemical features of the Fc segment of all G heavy chains, the postulated twenty cistrons must have arisen by duplication. This seems unduly cumbersome when one has the rather clear evidence that the halves of the light chain are so similar in general structure that they must represent a far distant duplication of the cistron concerned, with subsequent independent point mutation in each. It seems much more likely that, as several workers have already suggested, all parts of the immunoglobulin molecule are derived from a single primitive unit of 100–110 residues, equivalent to a primitive myo-globin molecule in the haemoglobin series. Two of these are concerned in the light chain, four in the heavy chain, with the possibility not yet finally excluded that in some immunoglobulin molecules one unit may be common to both chains. It would be completely premature to suggest what sort of a genetic and synthetic mechanism would be needed to account for chromosomal relationships, for the 'choice' of units for expression under rigid

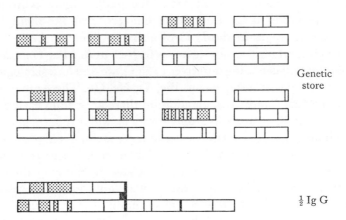

Genetic store

$\frac{1}{2}$ Ig G

Fig. 21. A way of visualizing the store of genetic information in an immunocyte and the way by which the information is expressed in light and heavy chain of Ig G under a process of phenotypic restriction. 'Variable' segments are shown with broad dotted bands.

phenotypic restriction and for the synthesis of light and heavy chains and their fabrication into immunoglobulin molecules. It is, however, a reasonable guess that perhaps ten to sixteen such cistrons present in each cell of the individual could provide the information needed for the construction of all the *antigenic* variants of heavy chain pattern.

Until it is replaced as a result of adequate experimental data, the hypothesis that the combining sites are solely composed of deter-minate portions of the variable segments in heavy and light chains still seems the most satisfying available. If it is accepted, the diversity of *antigenic* pattern of Ig G will have only a minor secon-dary influence on *antibody* pattern which is wholly a responsibility of the 'variable' chains.

THE GENETICS OF CAPACITY FOR ANTIBODY PRODUCTION

In summary, I have regarded antibody pattern as being derived (*a*) by random choice from the small number of cistrons (perhaps two to eight) that will define the variable segments of the light and heavy chains of the differentiating immunocyte, and (*b*) as a result of a high level of somatic mutation in the chosen cistron(s).

This still allows genetic factors in the conventional sense to play a significant part in defining the range of antigenic determinants for which the individual can produce effective immune patterns. Despite the random character of their emergence, immune patterns have an important determinate element in the presence of a number of stable sequences in the 'variable' segments responsible for the combining site. The existence of such stable groups is based not only on studies of the light chain sequences but also by analogy with the species differences that have evolved in standard proteins like haemoglobins and cytochromes, and with the results of Benzer's fine-structure work on phage mutation.

All immune patterns must be derived from the small number of relevant cistrons present in the fertilized ovum. It is obvious that some changes are more likely to occur than others and that some will be very unlikely to be derived from one starting pattern although readily obtainable from another. It would be expected, therefore, that one might find inbred strains of animals showing well-marked individuality in the type of their responses to a range of antigens and that even with outbred animals there should be many occasions on which failure of antibody response to one antigen should occur without any weakness being apparent against some others.

There are many such examples in the literature and anyone with experiences of experimental immunization will know that one will always obtain a wide range of titres, often with some below significant level, when a considerable group of animals is uniformly injected with a standard antigen. There has been less extensive work on pure line animals.

Recent work with pure line guinea-pigs and synthetic polypeptide antigens has added a new facet to the problem of genetic differences in capacity to produce antibody. A certain hapten–polylysine complex may be effectively antigenic in one strain of guinea-pig but completely inactive in a second (non-responder) strain. This can be shown not to be due to any inability to produce the necessary immune pattern, since complexing the hapten–polylysine to a protein carrier produces antibody against both carrier and the hapten. The genetic difference here is concerned with some phase of the preliminary process by which the antigen acts as a stimulus to the synthesis and liberation of significant amounts of antibody.

### SECONDARY MUTATION IN ESTABLISHED CLONES OF IMMUNOCYTES

By any acceptable hypothesis the cistron(s) coding for immune pattern must be abnormally labile at some stage of embryonic development. It would represent economy of hypothesis if that lability persisted through the proliferative phases of the immunocyte. In this section the pros and cons will be discussed.

Every classical antiserum against a conventional antigen is highly heterogeneous. The proportion of Ig M, Ig G and Ig A antibodies will vary characteristically from one stage of immunization to another but at almost every stage all three are present. In human beings, both K and L types of light chain are found and several electrophoresis bands for light chains from any subpopulation. Amongst all this heterogeneity it is hard to obtain evidence of secondary somatic mutation and it is best to look first at the myelomas.

When we are interested in mutant forms of protein there are two inbuilt difficulties: (a) we can only deal effectively with highly purified preparations and (b) the mutants we might expect to find will be in very small yield compared with the standard protein. Proteins are not easy to purify from associated proteins, and myeloma protein, for instance, needs strenuous treatment to eliminate more than traces of normal immunoglobulins or any other proteins not identical with the main mass of the myeloma proteins. The latter would include secondary mutants if they occurred. It may be significant that all the myeloma light chains tested by Cohen and Porter in starch electrophoresis showed in addition to a single very heavy band a single, much weaker, additional band one place on the negative side of the main band with some trailing between them. Subject to the exclusion of purely technical artefacts this would be consistent with the occurrence of a reasonably high incidence of single amino acid replacements by point mutations since the clone was initially established.

It is now well known that there are fairly numerous cases in elderly people of what has been called 'monoclonal gammopathy' without symptoms or radiological signs of myelomatosis bone lesions. These may persist for years but a proportion eventually develop malignant characteristics. Such patients might supply valuable material to determine whether the change to malignant character was associated with change in the amino acid sequence pattern.

Until homogeneous antibody populations can be obtained in reasonable quantity, direct examination of specific chain patterns is not likely to be productive. If our general point of view is correct, when an antigen is introduced into the body it has opportunity to contact a wide randomly established range of immune patterns carried by the immunocyte population. Depending on accidents of local concentrations and availability of patterned cells, it is likely that a wide range of separate clones will be stimulated with variable effectiveness. This will hold particularly for 'good' antigens which are 'good' presumably because they can react with a relatively wide range of patterns. These *a priori* deductions are in fact entirely in line with what has been observed in regard to the heterogeneity of specific antibodies.

Under the circumstances it becomes impossible to find functional evidence for the occurrence of secondary mutation in established clones of immunocytes. There are many instances in which secondary stimulation of a previously immunized animal with a related but not identical antigen gives rise to antibody different in quality from what would be produced by primary and secondary stimulation by either antigen alone. This may mean that secondary mutation has occurred, but it is much more likely that complex conditions for selective survival amongst the many clones of cells responding to the primary stimulus, and available for further selective response to the secondary one, could be responsible.

A superficially related problem arises from two types of experiment concerned with the repopulation of heavily irradiated animals or sequences of irradiated animals used in successive transfer experiments. The Harwell group showed by karyotype studies of heavily irradiated animals that, on occasion, decisive evidence could be obtained that almost the whole haematopoietic and lymphopoietic system had developed from a single surviving cell. The second line of approach was to pass stem cells through heavily irradiated hosts with the production of Till–McCulloch clonal nodules in the spleen, and to show that by repeated such passage commenced with a single nodule, reconstituted animals of relatively normal immune competence could be obtained. In both instances we have a situation in which one stem cell is provided with almost the same capacity to proliferate as the zygote or one of its early descendants. It is merely establishing in an unusual fashion the axiom that any relatively primitive cell possesses the whole repertoire of genetic information contained in the genome of the zygote.

# 7 The role of macrophages, eosinophils and other auxiliary agents in relation to adaptive immunity

There cannot be the slightest doubt that defence of the integrity of the body against parasitic invasion from the environment or anomaly from within requires much more than the functioning of the system of adaptive immunity with which this book is concerned. In all probability the polymorphonuclear leucocytes and the monocyte–macrophage system are in the direct line of the relatively nonspecific defence processes that have been an evolutionary necessity since the advent of the Metazoa. Their part in defence against bacterial infection and in the removal of damaged and effete cellular material is fundamental and in a sense the facilitation of some of these processes by opsonization is merely an evolutionary afterthought.

It is equally evident that throughout the vertebrates one finds eosinophil cells of characteristic morphology which in some only vaguely understood fashion are associated with immune reactions. So, too, are the basophil leucocytes and mast cells, but the significance of these 'pharmacologic time bombs' is even more obscure. Finally, we have the extraordinary system of laboratory phenomena which has been developed from Ehrlich's 'complement' or Bordet's 'alexine'. Just as has been observed in the study of blood coagulation, a dozen or more factors are concerned in complex adsorptive and enzymatic interactions. The development of these studies is something to chasten every laboratory biologist. It is more than probable that every biological phenomenon we study at some accessible level is at least as complex as either. From our present angle, however, all that we are concerned with is the relationship of complement to phenomena within the conventional field of immunology.

The aim of this chapter will be mainly to summarize the evidence *against* ascribing any major role to these components in the process by which immunocytes and antibody come into being and exert their specific immune functions. Their ancillary roles in the various processes that have a major immunological component have, however, provided fascinating subjects for investigation and must be covered even if only in summary fashion.

### THE MONOCYTE–MACROPHAGE SYSTEM

When Aschoff's formulation of the reticulo-endothelial system in 1922 became well known, it was regarded as almost self-evident that the cells of the system were responsible for antibody production. It was evident that all particulate antigens as well as a wide variety of non-antigenic foreign materials were taken up by macrophages and, if antigen was responsible for the production of antibody, the macrophage was the obvious site for its synthesis. With the recognition of the plasma cell series as the major and probably the only producer of antibody in more than trace amounts, the role of the macrophage has been reduced to that of an auxiliary whose exact function is problematical. Interpretations may still range from those which regard the macrophage as the cell in which the most important initial stages of antibody formation take place to those by which the macrophage's only function in immunity is to lower antigen concentration to a biologically tolerable level.

Perhaps the basic facts are: (*a*) Antibody production is a function of organs in which macrophages, lymphocytes and plasma cells are conspicuous and well-intermixed components.

(*b*) After appropriate injections of antigen, significant amounts are found to be present in all types of macrophage, but not in lymphocytes or plasma cells.

(*c*) Conversely, in a recently immunized animal, antibody (that is immunoglobulin) is present in large amount in plasma cells and is detectable in the germinal centres. None is found in typical macrophages although with immunofluorescent technique some littoral cells show evidence of immunoglobulin.

(*d*) Under carefully defined conditions, evidence has been obtained that exposure of cell suspensions containing macrophages to antigen can give rise to material with higher specific immuno-

148

genetic power than the original antigen. This modified antigen may contain low molecular weight RNA as an essential component.

(*e*) Within lymphoid tissue there is a system of dendritic phagocytic cells capable of concentrating both antigen and antibody on their cell surface. For a variety of reasons that will be discussed, I believe that these cells are of primary importance in the process of antibody production. Their origin is uncertain, and from the functional angle they are so distinct from the classical monocyte–macrophage series that they are being considered here under a separate major heading.

## The relevant types of phagocytic cell

Current usage is to refer to the mononuclear phagocytic cells of the blood as monocytes and to those of the tissues as macrophages. There are, however, very well marked differences amongst the phagocytic mononuclear cells in different tissues and it must remain an open question as to how legitimate it is to lump them into a single group. The following tabulation is concerned merely with the immunological significance of the different groups.

*Blood monocytes.* These have typically an indented reniform nucleus, numerous mitochondria and usually show lysosomes in the cytoplasm. The evidence is conclusive that the monocytes circulating in the blood are nearly all cells recently generated from stem cells in the bone-marrow. It is not excluded that the stem cells concerned have the morphology of small lymphocytes and are the same stem cells which, entering the thymic environment, would differentiate to immunocytes. Monocytes are mobile and leave the blood through intercellular junctions in the capillary endothelium. They accumulate in any area of subacute inflammatory change, whether produced by immunological activity or otherwise. In the tissue they take on the tissue macrophage form and may actively proliferate. There is still controversy as to whether the mononuclear cells infiltrating regions where a delayed hypersensitivity reaction is taking place are solely monocytes or include both lymphocytes and monocytes.

*Large macrophages of the spleen and lymph nodes.* The classical macrophages in lymphoid and other tissues are usually considerably larger than blood monocytes and much less uniform morphologically. In electron micrographs the most characteristic feature is the wide variety of cytoplasmic granules, some physiological (phagosomes, mitochondria, etc.), and some representing ingested material or its debris. In many, some rough-surfaced endoplasmic reticulum can be seen.

When an isotopically labelled antigen is injected locally and the draining lymph node is examined at appropriate times by auto-radiography, early uptake of antigen is predominantly in the littoral cells of the peripheral sinus and of the lymph channels in the medulla and in macrophages present in the diffuse areas of the cortex. There is relatively little in the lymphoid follicles. In the next day or two some of the label in these cells disappears, pre-sumably by digestion of the antigen. In the spleen such cells are numerous throughout the red pulp and are concentrated at the immediate periphery of the Malpighian areas of white pulp.

*Peritoneal macrophages.* When a suitable mild irritant such as liquid paraffin is introduced into the peritoneal cavity of any of the laboratory rodents the exudate produced contains large numbers of actively phagocytic mononuclear cells. Small and medium lymphocytes are also present and a number of authors have commented on the difficulty of clearly differentiating the two types morpho-logically. A fairly good functional separation can be obtained on the basis that cells that do not attach to glass are lymphocytes. Macrophages and polymorphonuclears rapidly develop a firm attachment to glass. Morphologically the peritoneal macrophages are mostly larger than blood monocytes.

There is evidence that a population of peritoneal cells from an immunized animal can transfer antibody-producing ability to a suitable test recipient but, where a differential study of cell types was possible, this facility was absent from the macrophage fraction.

A great deal of work has been done on various aspects of the functional activity of macrophages from the peritoneal cavity. The general conclusion has been that such cells have no intrinsic capacity of specific immune reactivity but that their behaviour is greatly influenced by the presence of 'natural' or 'immune' op-sonins in the experimental system. For example, washed peritoneal phagocytes from the rabbit will take up foreign red cells only if these are opsonized by treatment with normal rabbit serum. The opsonins are relatively highly specific and are removed by appropriate absorption with the antibody concerned. It is sometimes possible to confer specific capacity to phagocytose a particulate antigen, e.g., foreign erythrocytes, by treating the peritoneal macrophages with serum containing cytophilic antibody. Under certain conditions, macrophages can be stimulated to destroy ingested living bacteria but the same enhanced activity is also shown against quite unrelated bacterial species.

Even if the peritoneal macrophage is neither a producer of antibody nor has any specific capacity for antigen recognition, its reactions may well have important implications in regard to a more primitive type of recognition of foreignness and of defence against micro-organismal invasion.

At the technical level there has been much interest in work, largely due to Fishman, in which attempts have been made to establish the nature of the apparent association of macrophages and lymphoid cells in antibody-producing tissues. The basic experiment has been to expose a population of peritoneal exudate cells to an antigen (phage T2), extract RNA from the treated cells and apply this to a culture of lymph node cells. In about 20 per cent of such experiments, minute amounts of specific antibody were demonstrated and all controls were negative.

There seems to be little doubt that from mononuclear phagocytic cells that have taken up antigen, material can be extracted which is more immunogenic than the equivalent amount of original antigen and, in the techniques used, the modified antigen is associated with RNA. It must still remain doubtful whether it is justifiable to regard this as a model for all antibody production in the body. Discussion of its significance will be deferred until the next chapter.

*Kupffer's cells of the liver.* These are very numerous in the sinusoids of the liver and appear to possess cytoplasmic projections actually spanning the lumen of the sinusoid. They are highly active in removing particulate foreign matter and effete red cells from the circulating blood. There appears to be no direct evidence that Kupffer's cells have any immunological function.

*Alveolar phagocytes of the lung.* Since such cells can be readily obtained by washing out the bronchial tree of guinea-pigs and rabbits, they have been used in a variety of studies and show a number of differences from macrophages of the peritoneal cavity; for example, they contain higher concentrations of acid phosphatases and other hydrolytic enzymes found in lysosomes than any of the other types.

Such cells play an important part in dealing with micro-organisms and other foreign particulate matter entering the lung but, again, no special immunological function in our sense has been ascribed to them.

### The origin, movements and transformation of the mononuclear phagocytic cells

There is a current disinclination to make definite statements about the origin of a cell population specified by such a functional criterion as capacity to ingest foreign particles. By implication the population, here the mononuclear phagocytic cells, is a homogeneous system with a common developmental history. This is something that can never be established by the techniques that are nowadays obligatory. The most powerful is to find some way of labelling nuclei of a definable cell group with tritiated thymidine or some other DNA label and to examine the organism at successively later stages by autoradiography. There are always many difficulties in interpreting the results of such experiments but, with the macrophages, there is the special difficulty that these ingest nuclear fragments of damaged cells, particularly lymphocytes. The presence of label in a macrophage will therefore be almost as likely to have been derived from ingested material as to have been passed on from an ancestral cell.

Some limited evidence can be obtained in experiments in which distinctively labelled cells from one animal are transferred to another in which for one reason or another they are acceptable. The label may be a radioisotope, karyotypic marker or a histocompatibility antigen. By experiments of this general type, what appears to be good evidence has been obtained that (*a*) Kupffer's cells may be derived from thoracic duct lymphocytes, (*b*) when a lethally irradiated mouse is saved by the administration of bone-marrow cells, the phagocytic cells subsequently obtained from the peritoneal cavity are of donor origin and (*c*) the cells which migrate from capillaries to give macrophages in a lightly traumatized area are not derived from circulating small lymphocytes but come perhaps wholly from bone-marrow.

All these statements are applicable only to the special circumstances of the experiment concerned but, taken together, they suggest that the bone-marrow contains cells that have potentiality to produce any type of macrophage and that, while the great majority of small lymphocytes have no such potentiality, the possibility is not excluded that, under special circumstances, thoracic duct lymphocytes may give rise to descendant macrophages. Although the evidence is far from conclusive, there is a mounting opinion that the small lymphocyte-like cells in the bone-marrow may, for the most part, represent stem cells in the strict sense of being able to give rise to lymphocytes, macrophages and granulocytes. In addition there are hints from human pathology that in the absence of a functional thymus no 'lymphocytes' are seen in the bone-marrow.

It is at least conceivable that one of the reasons for the almost complete failure to obtain a clear picture of the interrelationships of wandering mesenchymal cells is that, in fact, morphological type is simply a reflection of functional activity and that, within the series, the normal sequence of specific differentiation along one or other of the paths from stem cells to specialized end-cell may, under appropriate circumstances, be interrupted by de-differentiation. Depending on the completeness of that de-differentiation and the local *re*-differentiating stimulus, a new line of differentiation may take place. This interpolation in a discussion of macrophage origins is admittedly based simply on the absence of definitive experimental evidence and can be refuted (or confirmed) if new experimental approaches are developed.

From the general immunological point of view, probably the most important outstanding problem in the cellular field is the origin of the mononuclear cells found in the tissues in the areas of delayed hypersensitivity and other subacute inflammatory reactions. The current disinclination to call such cells anything other than mononuclear cells is an indication of how difficult it has been to decide what type of blood cell, lymphocyte or monocyte has given rise to them. The pros and cons will be considered on p. 245.

In more chronic conditions the development of mononuclear phagocytic cells to epithelioid and giant cells is well known.

## Opsonization

An opsonin is a serum component which facilitates the ingestion of foreign particulate material by phagocytes; either polymorphonuclear leucocytes, or the various types of mononuclear phagocytic cells. In general, opsonin is synonymous with antibody or perhaps, more accurately, with adsorbed and partially denatured immunoglobulin. It is characteristic that when autologous proteins are administered to an animal the only one actively taken up by the phagocytic cells is immunoglobulin, necessarily partially denatured during the process of purification. The presence of homologous (or other) serum may be necessary even for the phagocytosis of synthetic particles such as polystyrene, and it is of much interest that the component of normal serum responsible for the opsonization of polystyrene particles is a specific one that can be removed by appropriate absorption.

This may be an appropriate place to attempt to assess the significance of opsonization in relation to macrophage function and antibody production. There is good evidence that antitoxin production in young animals is facilitated if very small amounts of antibody are administered artificially or received physiologically from the mother, before active immunization with toxoid is begun. This is reminiscent of Jerne's first theory of antibody production but most immunologists, probably now including Jerne himself, would interpret the findings differently.

In all mammalian and avian species, newborn animals receive antibody from the mother. The method of transfer and the type of antibody transferred will differ with the species. In many mammals, including man, transplacental passage is limited to Ig G-type antibody while, for those mammals in which transfer by colostrum is important, Ig A-type antibodies will predominate. In either case, a representative sample of the circulating antibodies of the mother will be received. The range of patterns will probably cover antigenic determinants of all the common micro-organisms the mother has encountered. This will ensure that when such micro-organisms begin to enter the tissues of the newborn animal

154

they will be opsonized and readily taken up by phagocytes, both polymorphonuclear and mononuclear. The primary function of the macrophage is to get rid of potentially dangerous foreign material by ingestion and enzymic breakdown to molecules which can either be contributed to the metabolic pool or prepared for excretion by one of the natural routes. As indicated earlier, another important function may be to reduce the concentration of all circulating antigens to a level that will prevent its undue activity at the immunological level. There is no evidence to convince me that the large macrophages of liver, spleen, lymph node and lung have any other significant immunological function than these. The results of Fishman and others can be best regarded as an indication that unspecialized phagocytic cells can carry out rather inexpertly a process characteristic of the specialized antigen-retaining reticular cells of primary lymphoid follicles.

## THE DENDRITIC PHAGOCYTIC CELLS OF LYMPHOID TISSUE

The primary lymph follicle of lymph node or other peripheral lymphoid tissue can be thought of as a spongy reticulum of supporting cells whose dendritic processes are in part directed along a light mesh of reticulin fibres. Within this framework lie the relatively closely packed small lymphocytes, in life probably a mobile mass of amoeboid individuals. These reticulum cells are actively phagocytic.

If a lymph node is examined three to seven days after the administration of a labelled antigen such as flagellin, label is conspicuously concentrated in the lymph follicles. It is held in the dendritic reticular cells fairly regularly dispersed throughout the follicle, and electron microscopic study points to most of the antigen being concentrated on the surface of the dendritic processes. When a germinal centre develops in the lymph follicle the usual effect is for the region showing label to be compressed into a cap on the outer side of the germinal centre. Basically similar appearances are to be seen in the lymph follicles of spleens in animals given antigen intravenously.

Studies of the presence of immunoglobulins in lymphoid tissue, using appropriate fluorescent antibodies for their detection, have consistently shown a reticular distribution of immunoglobulin within germinal centres. Unlike the situation in any isolated plasma cell, which shows only a single type of immunoglobulin, that present in the germinal centre is characteristically heterogeneous. Everything combines to suggest that antibody, both locally produced and circulating, is accumulated on the surface of these dendritic phagocytic cells and may in fact play a major part in retaining antigen in a situation where it is specially accessible to contact with committed immunocytes. It is in line with current opinion to believe that the retention of antigen in the surface layer of such dendritic cells may provide a particularly suitable means of stimulating any appropriately patterned cell receptors carried by small lymphocytes. The rapidity with which small lymphocytes move from one lymphoid cell collection to another throughout the body has been described earlier.

Basing our interpretation largely on the results of the group led by Nossal and Ada at the Hall Institute, special importance is attached to these reticulum cells of lymphoid follicles. They are actively phagocytic, are subject to the normal benefits of opsonization and are protected from too great an uptake of any individual antigen by a protective screen, in both spleen and lymph node, of standard macrophages. The special attributes of these cells, which are of significance for their role in antibody production, are:

(*a*) their capacity to retain antigenic label much longer than do standard phagocytes, most probably because of a less active lysosome mechanism for breakdown of ingested material;

(*b*) concentration of antigen at the cell surface and particularly at the surface of the dendritic processes. If the pictures published of the distribution of one labelled antigen, flagellin, can be given a general application, one can assume that each such cell in a lymph node carries on its surface a sample of most or all of the antigens that have been derived from the lymph node's area of drainage during the lifetime of the cell;

(*c*) if the accumulation of small lymphocytes in primary follicles is a mobile 'bag of worms', as seems likely from the *in vitro* behaviour of lymphocytes, large numbers of contacts between any given deposit of antigen on a dendritic process and a great many individual lymphocytes become inevitable;

(*d*) the frequently described picture of a rim of small lymphocytes forming a rosette around a macrophage could well represent an example or analogue of the process by which antigen carried on a reticulum cell surface can stimulate a lymphocyte with appropriate receptors.

Although we regard this as the standard intervention of the macrophage series in the process of antibody formation, the possibilities cannot be excluded, first, that antigen not held on a macrophage surface may be effectively immunogenic, and second, that other types of macrophage in special circumstances may play a similar role.

One final remark may be made in relation to the phenomena of tolerance to be discussed in a later chapter. If, by any means, cells capable of producing normal antibody reactive against antigen *A* are eliminated from the lymphoid cell population of the body and there is no opsonin for *A* in the circulation, this absence would in itself greatly reduce the chance that the antigen could establish itself on the reticular phagocytic cell surfaces.

## EOSINOPHIL LEUCOCYTES

Eosinophil leucocytes are well-defined cells present in all mammals and characterized by large cytoplasmic granules staining strongly with acid dyes. The granules are specialized lysosomes containing a variety of enzymes of which the most unusual is a very stable peroxidase of unknown function. In electron micrographs the most conspicuous feature is a large crystalloid of high electron density in each granule. Apart from the fact that it is protein in character, nothing is known of the chemical nature or functional activity of the crystalline material. Eosinophils develop in the bone-marrow and spend on the average only a few hours in the circulation. The cells are moderately mobile, leave the circulation through endothelial cell junctions, and are attracted to and phagocytic for antigen–antibody complexes. In the tissues, eosinophils are present in considerable numbers in the submucosa of the intestinal tract and other situations exposed to foreign antigens and, by implication, the site of frequent antigen–antibody interaction.

Specific experimental work on the movement of eosinophils from the blood into tissue has been principally concerned with their entry into lymph nodes activated by antigen. It has been shown that eosinophils appear usually about the cortico-medullary junction of the node within a few hours of antigen being administered. In order to account for this accumulation, it is necessary to postulate some product of cellular activity associated with antigen–antibody union which presumably modifies local capillary endothelium so that circulating eosinophils are more or less specifically held and can pass by diapedesis into the tissues. The very rapid accumulation observed by Litt must, on current interpretations of the process of antibody production, result when antigen in the phagocytic reticulum cells makes contact with any reactive lymphocytes, well before there is proliferation and development of plasma cell clones.

The only positive suggestion as to the substance mediating the attraction of eosinophils to the site of an antigen–antibody reaction is derived from R. K. Archer's experiments in horses. Infiltration of skin areas with histamine caused a local accumulation of eosinophils within 30 hours, the intensity of the response being linearly related to the dosage of histamine used. Experiments with species other than the horse have not confirmed this, and general opinion is that some other product of antigen–immunocyte reaction must be involved.

Blood eosinophilia is characteristic of a variety of allergic states and of acute worm or protozoal infestations, all of which in human beings are associated with immediate skin reactivity of histamine type. There is also a fairly regular accumulation of eosinophils in the tissues containing metazoan parasites.

There is no accepted interpretation of eosinophil physiology but it must obviously be concerned in one way or another with the immune process. The fact that granules disintegrate when phagocytosis of antigen–antibody aggregates occurs with liberation of many enzymes suggests that the basic function of the eosinophil is as a mobile scavenger to assist in clearing up areas of subacute damage associated with immune responses. As in all such systems, one must postulate a system of controls and feedback to ensure an adequate production of eosinophils in the bone-marrow, to regulate the level in the circulation and to channel circulating cells to the tissues where they are required. There is no substantial information in regard to any aspect of these controls.

The striking association of eosinophilia with extensive metazoan infections may speculatively be ascribed (*a*) to the long persistence of antigenic material difficult for the body to dispose of, (*b*) a relatively high content of A-type antibody with its characteristic tendency to fix to tissues and react to circulating antigen with histamine liberation and (*c*) the common absence of G-type antibody with its blocking activity against sensitivity reactions.

## MAST CELLS AND BASOPHILS

It is rather depressing to find that the vast early literature on anaphylactic shock contained no reference to mast cells. Current opinion seems now to be firm that anaphylaxis is the symptom complex resulting from the discharge of the battery of pharmacologically active substances stored in the mast cell. These include heparin and histamine in all mammals, serotonin in mice and rats, and the slow-reacting substance of Kellaway and Trethewie in the guinea-pig.

Mast cells are common in all types of loose connective tissue and the standard experimental material is the omentum and peritoneal lining of the rat. They show only minimal evidence of mitosis as such and are usually assumed to arise by heteroplastic transformation of fibroblasts. In the course of a large number of histological examinations of mouse thymuses I found, not infrequently, massive areas of heteroplastic transformation of cortical thymocytes to mast cells. There is no indication as to the nature of the process causing the extraordinary transformation from fibroblast or thymocyte to mast cell nor is there any established function for the mast cell in the body's economy. It is the only known source of heparin in the body but there is no evidence that the anticoagulant action of heparin has any physiological function.

Since it is well accepted that histamine liberation is probably responsible for the first phase of inflammatory reactions, it will probably emerge that mast cells in some way facilitate local processes of defence and repair. Their capacity to discharge their granules and liberate histamine when stimulated by antigen–antibody reactions in various situations may be no more than a general response to a mildly damaging stimulus. There are a number of pathological conditions in which mast cells are conspicuous and some, such as urticaria pigmentosa, in which histamine liberation is responsible for symptoms, but none provides any significant clue as to physiological function. Anaphylaxis in its various forms seems always to be an experimental artefact—'a creation of the hypodermic needle'—or a rare accident in genetically abnormal individuals.

The only conclusion one can reach is that the mast cell and the basophil leucocyte are so widely present in vertebrates that they must have some function that is significant at the evolutionary level. Whatever that function may be it does not appear to have any aspects which can be meaningfully related to the general theme of adaptive immunity.

### COMPLEMENT

Since the classic studies of Bordet, Ehrlich and Wassermann, the concept of complement as an essential part of the mechanism of immunity has progressively been replaced by rather uncertain decision that the classic phenomenon of complement lysis of red cells is a laboratory artefact of no real significance for immunity.

The circulating blood plasma is an immensely complex mixture which, as R. G. Macfarlane has noted, achieves a miracle of 'peaceful co-existence with a vast surface area of the normal vascular endothelium'. When, however, trauma introduces the biological necessity of rapidly enforcing haemostasis in preparation for repair, a process involving at least twelve components leads to localized fibrin deposition; while at appropriate points platelet aggregation, kinin activation with modified endothelial stickiness and vascular contraction are co-ordinated with the fibrin formation to ensure haemostasis. When infectious processes with or without trauma are initiated, there are doubtless equally complex processes set in action and there can be little doubt that the variably large number of components, co-factors and inhibitors that have been described as being concerned with immune haemolysis by 'complement' represent a similar range of plasma factors, perhaps including some of those in the haemostatic group, whose co-ordinated function has as yet escaped recognition.

Research on complement seems to have failed to attract anyone with a capacity to look beyond finding a new 'component' but one can feel optimistic that an acceptable biological approach equivalent to R. G. Macfarlane's cascade view of the clotting mechanism will eventually emerge. As they stand, the $C^1$ factors $1p$, $1q$, $1r$, $1s$, $2$, $3a$, $3b$, $3c$, $3d$, $3e$, $3f$, $4$ and the co-factors $Ca^{++}$ and $Mg^{++}$ are biologically meaningless.

Certain strains of mice lack complement in the sense that, because of the absence of part of the $C^13$ complex, their fresh serum has no lytic activity. No one has demonstrated any inefficiency in meaningful immunological reactions in such strains.

Possibly all that can usefully be said is that when a complex of antigen and antibody is formed in the body it is liable to adsorb a variety of enzymically active components to produce an even more reactive aggregate. There are considerable differences in the ease with which antigen–antibody complexes adsorb complement in the diagnostic immunologist's conventional sense. A certain compact bulk of denatured globulin appears to be necessary. A single M-type antibody molecule attached to a red cell provides a nidus for the initiation of the sequence of enzymic actions that culminate in the production of a tiny hole in the surface membrane of the red cell through which haemoglobin can leak out. For lysis to be initiated with G (7S) antibody there must be sufficient antibody molecules for at least one pair to be attached adjacently to each other. In other systems involving the production of antigen–antibody aggregates there is equally an optimal size of the individual aggregates for complement fixation.

There is direct evidence that adsorption of complement components *in vivo* increases the activity of phagocytosis and the speed of killing of bacteria after ingestion. Under appropriate conditions, antigen–antibody–complement complexes can be shown to adsorb to certain red cells (the immune-adherence phenomenon). For the most part this is a laboratory artefact but it is closely related to the mechanism by which drug reactions producing haemolytic anaemia or purpura are mediated (see p. 284).

In the course of several types of autoimmune disease or other

6  BCI

types of immunopathology there is a sharp reduction in the measurable complement activity of the patient's serum. This has been observed particularly in systemic lupus erythematosus (SLE) and in kidney disease. Evidence that this results from fixation of complement in regions where auto-antibody has become attached to tissues can be obtained by the use of specific anti-human complement used in immunofluorescence tests. In practice, a serum against globulin $\beta 1$c which is a component of $C^1 3$ is used. With such methods, fixed complement can be detected in the kidneys in SLE nephritis and in the skin in SLE. A related finding is that synovial fluid from joints showing lesions of rheumatoid arthritis shows lower complement levels than normal synovial fluids. It is usual to ascribe a significant role to complement in the production of damage by antigen–antibody combination in both experimental and autoimmune situations. So little is known about the pharmacological activity of complement components that such statements are essentially meaningless.

Even if 'complement' appears to have little relevance as such to the interpretation of adaptive immunity, its importance remains unimpaired at the laboratory level as a reagent in complement-fixation tests for the measurement of antigen–antibody reaction.

# 8 The immunocyte and its response to specific stimulation

It is logical to adopt Dameshek's term 'immunocyte' for the cells directly concerned in immune responses but it must be recognized that a fully satisfactory definition of the term is not yet possible. It is a useful but perhaps oversimplified procedure to define an immunocyte as any cell that can be specifically stimulated by an antigenic determinant and that either itself produces antibody or is a potential ancestor or collateral of cells that can produce antibody. If a cell is to be specifically stimulated by an antigenic determinant it must carry a sensor, a receptor capable of recognizing and responding to the antigenic determinant in question. It would be unreasonable to look for such a specific receptor in any other terms than equivalence to the combining site of antibody globulins. This does not necessarily mean that the receptor is identical with A, G or M antibody as found in the circulating plasma, but since the light chain is common to all the immunoglobulins and appears to play an essential if minor part in the specificity of the combining site we should expect it to be present as part of any specific receptor. If only a limited portion of the Fd fragment of the heavy chain is needed to produce the combining site of a receptor there would be the further possibility that an immunocyte could conceivably be reactive to antigen without presenting evidence of the presence of any of the standard antigenic markers of the immunoglobulins.

At the operational level, a cell can be identified as an immunocyte (*a*) if it can be shown to be synthesizing antibody, (*b*) if it responds demonstrably and specifically to contact with antigen—and any effect due to adventitiously adsorbed antibody is appropriately excluded—and (*c*) if immunoglobulin can be shown to be present in the cell and, again, the possibility of its adsorption from the environment excluded. As in all immunological reasoning, only positive identification of an immunocyte is possible. As a rule, reactivity to one or, at most, a very small number of antigens is all that can be tested. Morphologically similar cells that do not react might, if they could be tested, react with one of the many thousand other possible antigenic determinants. Another characteristic difficulty arises from the fact that in a wide range of experimental manipulations a large population of cells, usually in the range $10^6$–$10^8$, is tested for immune activity of one sort or another. A positive finding means only that the population contains some of the immunocytes in question. It does not necessarily follow that because the population contains a large majority of one morphological type that particular cell is the immunologically effective one.

There are, however, methods by which the immunological reactivity of single cells can be recognized. Chapter 3 discussed the methods by which active production of antibody by isolated plasma cells can be demonstrated. Here we are concerned with the more interesting problem of detecting immunocytes at stages where no active production of antibody is taking place. In any form of clonal selection theory it must be postulated that as soon as a cell has been differentiated to an immunocyte it must have developed at least one of the immunoglobulin receptors that are within its genetic capacity to produce. There may well be a genetically predetermined predominance of certain immune patterns in the newly differentiated immunocytes but this has not been demonstrated as yet by any experimental technique. Where we are concerned with any standard foreign antigen we can expect only very small numbers of reactive cells in a population of millions. Highly sensitive methods are necessary for their detection. In practice the experimental approach is to look for cells reactive with a given antigen in animals which, as far as is known, have never been in contact with that antigen. An alternative is to devise methods of detecting very small amounts of immunoglobulin in lymphocytes without regard for its precise immunological reactivity. The presence of immunoglobulin, more correctly the presence of antigenic determinants specific for immunoglobulin, is prima facie evidence, but no more, that the cell has a specific immune function.

## THE PRESENCE OF IMMUNOCYTES IN UNIMMUNIZED ANIMALS

Reactions of immunological quality can be observed in unimmunized animals by three main types of technique:

(*a*) Graft-versus-host reactions; for example, that shown by chicken lymphocytes deposited on the chorioallantois of an unrelated strain.

(*b*) The Jerne–Nordin technique of detecting cells capable of producing haemolytic antibody against heterologous red cells by plating in agar containing the appropriate red cells. Discrete 'plaques' of haemolysis are produced around each antibody-producing cell.

(*c*) Stimulation of cells, typically small lymphocytes, to de-differentiate to blasts preparatory to proliferation, either by specific antigen or by antibody against immunoglobulins characteristic of the animal from which the cells are obtained.

The graft-versus-host reactions introduce many factors outside the scope of this book, and consideration of those that are relevant will be deferred until surveillance is discussed in chapter 13.

### *The significance of Jerne's antibody-plaque technique*

If a mouse is immunized with sheep red cells and four days later the spleen is removed, it can be shown that many cells in the spleen are actively producing haemolytic antibody. Jerne's technique is, in outline, to prepare a suspension of single cells from the spleen, mix them in warm liquid agar with an excess of washed red cells and pour the mixture on the surface of a plain agar plate. After 1 hour incubation the plate is 'developed' by exposure to guinea-pig complement for another 30 minutes. Small circular areas of haemolysis are each centred on a single cell often recognizable as a plasma cell. The reaction is specific for the type of red cell used for immunization.

Normal mice spleens contain small numbers of cells capable of producing such plaques. The number varies widely from one individual to another and the average value from strain to strain. It is very much lower in neonatally thymectomized animals. At least rabbit, sheep and pig cells have been used for test and appropriate experiments have shown that the number of normally reactive cells and the increase by antigenic stimulation are specific for the individual cell type. Suitable experiments have failed to show single cells haemolytic for two distinct red cell types.

The following interpretation follows that suggested by Jerne, but is equally compatible with the results from several other laboratories. Progenitor cells reach or arise in the spleen in very small numbers. These cells are probably subject to low-level nonspecific proliferation possibly, as suggested on p. 177, by a side-effect of the specific stimulation by antigen of other immunocytes. With the entry of the corresponding antigen perhaps after preliminary uptake and processing by splenic macrophages, a proliferation of plaque-forming cells occurs. Where the number of progenitor cells can be assayed by inoculation of normal isologous spleen cells into a heavily irradiated mouse, it can be shown that localized areas of 100–1,000 plaque-forming cells develop, each area from a single progenitor. If two different antigens are used the separate areas each contain only one type of antibody-producing cell. It appears still to be an open question whether the progenitors of these clones of proliferating antibody-producing cells are identical with the 'normal' plaque-forming cells or are non-antibody-producing but reactive to antigen. The fact that, in certain situations, lymphocytes can be stimulated by antigen to divide and to synthesize cytoplasmic protein but not immunoglobulin points to the second alternative. There is, however, nothing to negate the possibility that a plasma-blast proliferating slowly for nonspecific reasons may be *specifically* stimulated by antigen to accelerate its proliferation.

The antibody detected by Jerne's technique is wholly M in type and, as such, much more haemolytic than G-type antibody of the same specificity. As far as information is available it appears that all 'normal' antibodies are of M (19S) type. The Jerne technique has already been modified in various ways to detect other types of antibody. Anti-red cell Ig G antibody can be detected by treating with anti-mouse-globulin serum before the addition of complement. Other antigens can be used if, like many bacterial antigens, they can be firmly adsorbed to the red cell surface. In such experiments the red cells used in the test plates have been appropriately treated with the antigen.

In vitro *stimulation of lymphocytes to mitosis and blast formation*

As a by-product of attempts to produce rapid separation of red cells from leucocytes by agglutinating red cells in citrated or de-fibrinated human blood with kidney bean extract (phytohaemag-glutinin) it was found by Nowell that leucocyte cultures from cells isolated in this fashion showed enlargement and mitosis. Subsequently it became evident, to the surprise of most haematologists, that the cells involved were largely or wholly small lymphocytes. The use of phytohaemagglutinin for this purpose has greatly increased the ease with which blood cells can be karyotyped, but the immunological significance of the phenomenon is far from clear. There is adequate evidence that human small lymphocytes can be stimulated to undergo blast transformation to take up tritiated thymidine and to undergo mitosis *in vitro* by phyto-haemagglutinin and anti-human lymphocytic serum made in the rabbit; when two populations of lymphocytes from different individuals are mixed, a significant amount of blast transformation occurs, and when persons with a positive tuberculin test or with anti-streptolysin O in their serum are donors, their lymphocytes react respectively with tuberculin and streptolysin O *in vitro*. Claims that the reaction can be obtained with any antigen against which an individual has produced antibody, or that, on stimulation, specific antibody is produced have not been adequately confirmed.

It is probable that effective analysis of what is obviously a very important phenomenon will have to be carried out in experimental animals. So far it has proved much more difficult to obtain repro-ducible results in animals, and the present discussion is almost wholly based on work by Sell and Gell reported in 1965. In dis-cussing the nature of antibody, mention was made of the antigenic qualities of human immunoglobulins and their constituent chains. In experimental animals there are similar 'allotypic' differences between the immunoglobulins of different individuals. In rabbits, Ig G is made up of heavy chains which can carry any one of three antigenic determinants $A_1$, $A_2$, $A_3$, while the light chains also have three possible determinants $A_4$, $A_5$, $A_6$. Genetic studies indicate that qualities $A_1$, $A_2$, $A_3$ are co-dominant alleles of locus a, while $A_4$, $A_5$, $A_6$, are similar alleles at the locus b unlinked to a. In a heterozygous rabbit we may have the combination $A_1$, $A_3$, $A_4$,

A5, and in such an animal the Ig G will show the four types A1 A4, A1 A5, A3 A4, A3 A5. Everything suggests that when an anti-A4 serum is made by immunizing a rabbit with a genotype lacking A4, the antibody reacts only with the light chain of an immunoglobulin of the appropriate allotype.

TABLE 3. *Stimulation of lymphocytes by anti-allotype sera*

| | Lymphocytes from | | |
| --- | --- | --- | --- |
| Anti-allotype sera | A2, 3/5, 6 | A1, 1/5, 5 | A2, 3/4, 4 |
| Anti-A5 | +(x) | +(x) | — |
| Anti-A6 | +(y) | — | — |
| Anti-A5 + anti-A6 | +(x+y) | +(x) | — |

A schematic table based on Sell and Gell's results. *x* and *y* are the proportion of lymphocytes undergoing blast transformation.

When lymphocytes from a rabbit are exposed *in vitro* to an antiserum in a goat against rabbit γ-G globulin, the lymphocytes show blast transformation and thymidine uptake just as in the analogous human experiment. It is of much more interest that similar effects, but involving a smaller proportion of lymphocytes, can be produced by anti-allotypic sera produced in rabbits on lymphocytes from an animal of appropriate genotype. The effect is specific; an anti-A1 serum will only stimulate lymphocytes from rabbits with 1 in their genotype, and can be additive in heterozygotes (see table 3). Unlike the position with human cells, mixing of allotypically different lymphocyte populations does not induce mutual blast transformation. Further, none of the new protein synthesized during the blast transformation can be identified as immunoglobulin and, finally, lymphocytes from neonatal rabbits possessing only maternal allotypes in their immunoglobulins and making none of their own react with antisera corresponding to their genotype and the antigenic types of immunoglobulin that they will *subsequently* produce.

The implications of these facts are of first-rate importance for immunological theory and provide the first direct evidence for the presence of *genetically determined receptors of immunoglobulin character on lymphocytes* (including cells from very young animals) showing no other evidence of immunoglobulin production. They point, too, to the existence of phenotypic restriction allowing the production of only one type of light or heavy chain by any given cell in a heterozygous animal. The 'choice' as to which of two alternatives will be produced appears to be a random one. The next implication is that both heavy (Ig G type) and light chains are or may be present in the receptors. It does not, however, establish whether light and heavy chains are combined in the receptors in the same fashion as in immunoglobulins.

One might summarize by saying that experiments of this type confirm the lymphocyte as an immunocyte of which the immuno-globulin production is determined by genetic information, with phenotypic restriction to single chromosomes and that the first phase of specific stimulation is mitosis and proliferation, not immunoglobulin production. This is all precisely in line with clonal selection theory as formulated in 1958. There are, however, several areas still to be cleared up. It is not established that the antibody specificity of the receptor is present before the capacity to produce immunoglobulin develops, and there is no evidence for or against the contention that the combining site of antibody may have a genetic determination distinct from that of the antigenic determinants of light and heavy chains.

In the rabbit, mixtures of lymphocytes theoretically capable of reacting one with the other show no increase in blast transforma-tion and when lymphocytes from a rabbit hyperimmunized against a foreign allotype are exposed to the corresponding allotypic serum only a few cells show a response.

There is thus only a limited amount of direct experimental evidence to support the contention that cells can carry specific immune patterns without having had any contact with the antigens used to detect them. Nevertheless the evidence in its limited fields is unequivocal and there is no valid evidence that is equally un-equivocal in locating any source of immune pattern other than the genetic mechanism of the cell.

THE PHARMACOLOGY OF ANTIGENIC STIMULATION

*General principles*

Before considering the influence of antigen on cell receptors it is of interest to look at the modern approach to the nature of drug action. In general, there is recognition that a cellular receptor must be postulated for any response being investigated and that the action of drugs will depend on the effective concentration of drug in relation to receptor and the affinity as judged by $k_1/k_2$—respectively the association and dissociation constants for drug–receptor substance interaction. In terms of Paton's 'rate' theory, stimulant action is proportional to the rate of association between drug molecule and receptor; receptors become free for further stimulation by dissociation. If a drug is rapidly and almost irreversibly adsorbed, for example, in the action of nicotine on ganglion cells, there is a primary stimulation followed by a state experimentally indistinguishable from competitive inhibition.

Paton's point of view was used by A. Shulman to discuss the nature of pharmacological differences between drugs of the same general structure but differing in the nature of the substituents at appropriate positions in the molecule. The series he used were substituted glutarimides acting on the CNS (probably on cells of the ascending reticular formation) and giving a range of activity which, having regard both for drug structure and dosage, covered convulsant, analeptic, anti-convulsant, and hypnotic actions. Analysis of competitive actions between, for example, hypnotic and analeptic drugs indicated that common receptors were involved with results that could be quantitatively covered by appropriate elaboration of Paton's point of view. The dissociation constant $k_2$ is of primary importance in determining whether a drug functions as agonist, partial agonist or antagonist. On Shulman's view, a hypnotic is such by virtue of its power to block an endogenous stimulant.

This introduces a point of much importance both for pharmacology and immunology—the necessity of recognizing that any postulated cell receptor must in one way or another be associated with some normal cellular function. The clonal selection theory adopts the reasonable view that the immuno-receptors are in fact a specially evolved type of organelle integrated into the cell structure for the specific purpose of immune reaction. This does not, however, eliminate the possibility that such receptors may be stimulated in some nonspecific fashion or that there may be a 'final common path' for both immuno-receptor and some other type of receptor adapted, for example, to react to a thymic or other hormone.

Another feature which at present seems to be unique to the immunological field is the possibility of effectively replacing a genetically determined receptor with antibody present in the circulation. Such cytophilic antibody must always present a major difficulty in interpreting cellular reactions to antigen.

In interpreting immunological reactions, pharmacological principles are probably specially relevant in relation to the action of antigen on the genetically determined receptors of immunocytes. Basically similar considerations, however, will be equally relevant to the action of antigen on antibody more or less adventitiously attached to cells as exemplified, in all probability, by the liberation of histamine in passive anaphylactic experiments and from basophil leucocytes from allergic subjects following contact with the specific antigen. From this it is a simple further step to become involved in the cytotoxic action of antibody on cells carrying the corresponding antigen, and the reactivity and toxicity of antigen–antibody complexes.

*Application to immunocyte stimulation*

In interpreting the nature of cellular immuno-receptors, we must necessarily use soluble antibody as a conceptual model but be prepared to find the same sorts of relatively minor types of specificity difference that have been demonstrated between antibody reactions *in vitro* and the response to delayed hypersensitivity tests.

In regard to specific adsorptive capacity it is now well known that in an antiserum produced in response to immunization with an appropriate hapten–carrier combination there are antibody molecules for which, when tested with the hapten, association constants vary over a 100–1,000-fold range. In other words, for any given hapten there are many more or less similar configurations of combining sites with which reaction is physically detectable, but the chance of specific pharmacological response will presumably become increasingly rare as the association constant becomes smaller.

Other factors to be considered at the pharmacological level will be (*a*) the physical properties of the carrier of the antigenic determinant in so far as it influences the kinetics of association and dissociation and (*b*) the number and accessibility of the cell receptors. By hypothesis there are no immuno-receptors at all until the cell line has been differentiated in the thymus or elsewhere and it is reasonably certain that the number of receptors will be low until antigenic stimulation has occurred. (*c*) At the receptor–drug (antigen) reaction level we are only concerned with the initiation or blocking of stimulation. Once stimulation has been achieved, the result in terms of cell behaviour will depend on the physiological state of the cell, particularly as determined by the state of the internal micro-environment in the immediate neighbourhood of the cell.

At the present time it is not practicable to apply conventional pharmacological approaches to all the situations of interest. Attention may, however, be drawn to the following points elaborated either in the immediately following sections or in later chapters, all of which have pharmacological relevance.

(*a*) The influence of antigen dose (and age of the animal) in determining tolerance or positive immune response.

(*b*) The ability of specific antibody to prevent the initiation of a primary response to antigen in much smaller dose than is necessary to block a secondary response.

(*c*) The fact that in the presence of an adequate concentration of an immunosuppressive drug, a dose of antigen that would otherwise act as an active stimulus to antibody production can produce tolerance, presumably by destruction of the reactive immunocytes.

THE RESPONSE TO SPECIFIC ANTIGENIC STIMULATION

When we are concerned with the general problem of the responses of immunocytes to their specific antigen we have to work mainly on indirect evidence derived from the behaviour of large cell populations from immune animals. It is legitimate, however, to develop the implication of these experiments in the light of what has been drawn from the work we have just been considering on cells from unimmunized normal animals.

It is our contention that the result of contact of an immunocyte with the corresponding antigenic determinant will depend on a variety of factors involving the antigen, the physiological state of the cell and the internal environment in which contact takes place. The effect may be (*a*) mitosis and proliferation, (*b*) accelerated differentiation to the plasma cell series, (*c*) development to a 'memory' clone of lymphocytes, (*d*) surface damage with liberation of pharmacologically active cell products or (*e*) death with autolysis and/or pyknosis of nuclei.

As in so much of immunology, supporting evidence for each aspect of this statement is limited to a small number of well-worked out experimental situations plus fragmentary pieces of confirmatory evidence from a wide variety of sources.

*Mitosis and proliferation*
The primary postulate of any selective theory of immunity must be that, under some circumstances at least, contact of antigen with lymphoid cell of appropriate immune pattern must stimulate its selective proliferation. The evidence that this occurs *in vivo* will be considered later. Here we limit ourselves to *in vitro* studies, notably those of Dutton and his collaborators. Their work indicates that spleen cells from a rabbit immunized some months previously and 'primed' *in vitro* by a brief exposure to antigen show within 24 hours a rapid uptake of labelled thymidine. The technique is a versatile and convenient one and has allowed the recognition of some important additional points. The first is that stimulation of such cell suspensions is not necessarily specific. Phytohaemagglutinin, zymosan and staphylococcal extracts were all effective. If the cell suspension was irradiated, priming with antigen gave rise to only a minor uptake of label. Evidence of a specific effect could, however, be established by mixing such primed but inactive cells with normal unprimed cells themselves incapable of taking up more than trivial amounts of label. The mixture showed a greatly increased uptake. The irradiated cells had clearly been specifically stimulated to produce a transmissible agent, presum-

ably as a first stage to the initiation of DNA synthesis which, though inactive in the irradiated cell, was available to initiate the process in adjacent unstimulated cells. The simplest, though not of course the only interpretation, is that the primary effect is to liberate one or a number of pharmacologically active agents which provide the more immediate stimulus to mitosis. This would ascribe the nonspecific stimulation by phytohaemagglutinin, etc., to a similar type of activation and is in line with its known effects on human lymphocytes. This effect by which adjacent cells are stimulated as a result of specific antigen–cell interaction is likely to become important for understanding the interaction between thymus-dependent cells and immunocytes of other primary origin.

In several instances it has been shown that the initial protein synthetic activity induced by contact with antigen is not associated with increased immunoglobulin production but is mainly or wholly production of structural cellular protein. Most of the detailed work on blast transformation has made use of phytohaemagglutinin stimulation, and it remains to be proved whether this process has any close relationship to the proliferative activity induced in primed cells appropriately stimulated by antigen. As far as it is relevant it is clear that in the first stages by which a lymphocyte develops into a pyroninophil blast cell there is no more than trivial development of the endoplasmic reticulum which characterizes the plasmacytic series. If it is true that the progenitor immunocytes associated with delayed hypersensitivity should be differentiated quite sharply from those from which antibody-producing cells arise, some of these difficulties disappear. If the blast transformation involves only cells of the delayed hypersensitivity series no plasma cell formation would be expected.

## Accelerated differentiation in the plasma cell line

The data from experiments by Jerne's antibody-plaque technique and from Nossal's studies on the rat popliteal node seem to be best interpreted on the basis of slowly multiplying blast cells being stimulated to combined proliferation and progressive differentiation to mature plasma cells. There is still scope for eventual interpretation of the phenomenon of plasma-cell clone production in terms involving transfer of information from genetically determined cells to neutral 'nurse cells'. The pros and cons of this point of view are discussed in another section. From the present approach the alternatives are (*a*) the stimulation of a blast cell to give by proliferation a clone of plasma cells, or (*b*) the stimulation of an immunocyte to produce alone or, in addition to (*a*), transferable units of information (equivalent either to transforming principle (DNA) or to virus RNA) which can induce cells not in the same genetic line to give rise to a plasma cell clone.

*Development of a memory clone of lymphocytes*

The appearance of germinal centres is a characteristic sign of immunological activity, and there is evidence from immunofluorescence studies that germinal centres may be associated with antibody or immunoglobulin production at a low level. So far as one can deduce from indirect evidence, a germinal centre represents a centre of replication arising from a stimulated immunocyte. Germinal centres arise only in lymphoid tissue, and a large proportion of the newly formed cells pass to the blood and are widely distributed throughout the other lymphoid tissue of the body with the exception of the thymus. The simplest way to account for persisting immunological memory is to accept the production by each germinal centre, as it arises, of a large clone of lymphoid cells, a large proportion of which lodge as small or medium lymphocytes in various peripheral lymphoid tissues. In the absence of new entry of the corresponding antigen the survival of the clone (and of 'memory') will depend on (*a*) the numbers of lymphocytes produced during the immunizing episode and (*b*) the existence and extent of nonspecific stimuli to random immunocyte proliferation.

The relative absence of 'memory' in circumstances where antibody production is almost or wholly confined to Ig M has not been adequately explained, and interpretation has been complicated by the report of at least one example of a wholly Ig M secondary response. There is some basis for the suggestion that any immunocytes in which a significant proportion of receptors are blocked by union with antigenic determinant are incapable of giving rise to small lymphocyte descendants. They may become plasma cells, but these are end-cells and, once the stage of the mature plasma cell is reached, there is no evidence that this can by de-differentiation or otherwise ever become capable of mitosis and proliferation. In chapter 10 the effectiveness of specific antibody, particularly in the form of Ig G, in inhibiting further recruitment of Ig M antibody producers and perhaps accelerating the switch from Ig M to Ig G producers, is described. This may allow a proportion of the immunocytes to proliferate as memory cells which would otherwise have gone on to mature (end-cell) plasmacytes producing Ig M.

*Stimulation of lymphocytes to produce pharmacologically active products of cell damage*

Much of the recent discussion on cell damage associated with immune reaction has centred on the function of lysosomes. These are organelles visible in electron microscopic sections of a variety of cells. They are characterized primarily as containing acid phosphatase and a variety of other hydrolytic enzymes and seem to represent a heterogeneous group of structures. They are specially characteristic of polymorphonuclear leucocytes and macrophages but may also be found in lymphocytes and most other cell types. It has been suggested that they arise from pinocytosis vesicles and represent essentially organs for intracellular digestion.

A variety of cytotoxic influences can induce discharge of the lysosomes which may be either the immediate cause of cell death or one of the manifestations of the process. None of the defined components of lysosomes have specific pharmacological effects but it is probable that, associated with their disruption, proteolytic enzymes capable of producing peptide kinins, etc., will be activated.

Other aspects of cell damage possibly concerned in immune reactions are all somewhat uncertain. The part played by histamine is undoubted but general opinion is that the liberation of histamine is wholly a function of mast cells and basophils. The nature of the relationship, if any, of mast cells to immunocytes is problematical but there is no doubt that they can, in an immunized animal, be stimulated to liberate histamine and degranulate. The existence of the change from thymocytes to mast cells in some strains of mice may suggest that all immunocytes have some capacity to liberate histamine on stimulation but this has not been demonstrated.

Most of the information that bears directly upon the reaction of lymphocytes with antigens to which some of the cells in a popula-

176

tion are sensitized is clouded by the heterogeneity of all natural cell populations and the impossibility of obtaining more than a partial purification of any particular cell type such as small lymphocytes. Peritoneal exudates on which much of the work has been done are particularly difficult to interpret.

Further difficulties have arisen recently from the finding that nonspecific stimulation of lymphocytes by phytohaemagglutinin can result in a cytotoxic reaction against foreign cells, a reaction in which there is no overt evidence of antibody or immune patterns playing any part whatever.

Without being able to provide complete documentation for each component of the interpretation, the following account seems to be in accord with the experimental facts.

Any lymphocyte is a relatively vulnerable cell and when combination of antigen with antibody or immune receptor takes place on the cell surface, stimulation or damage results. In the presence of complement, destruction probably mainly by autolysis is common, provided the antigen–antibody reaction is a vigorous one. It is immaterial, in broad terms, whether the lymphocyte carries antigen on its surface or antibody (which may be either immune receptor or attached cytophilic antibody) or even whether a preformed antigen–antibody complex of the right physical character becomes attached to the cell surface. In every one of these situations there is evidence that damage may occur—often not readily demonstrable unless complement is present.

There are at least two types of experiment which indicate that antigenic stimulation of some of the cells in a lymphocytic population can produce similar results in normal cells. One is the uptake of tritiated thymidine; the other, adhesion of lymphocytes in culture when cells from a sensitized animal are exposed to the sensitizing antigen. In both cases it appears that specific stimulation by antigen produces, concomitantly with the effect being studied, liberation of active substances which confer presumably by simple pharmacological action the same effect on any normal lymphoid cells in the system. Probably closely related is the capacity of distinct populations of mouse lymphoid cells to react with mutual damage if they differ by major histocompatibility antigens even without any type of cross-immunization.

*Death of immunocytes following reaction with antigen*

It is obvious that in the body there is a high mortality of lymphocytes, particularly evident in the thymus but probably also occurring in all other regions. Death and autolysis is the normal fate of
probably 99 per cent of lymphocytes: antigen–antibody contact on
the surface will, under many circumstances, be the extra stimulus
which determines the actual time of cell death. Indirect evidence
suggests that once a clone of cells is developing to mature plasma
cells it has entered an irreversible path but, equally, there is no
evidence that an excess of antigen has any damaging effect on
plasmablasts or plasma cells actively synthesizing antibody. In the
discussion of tolerance in the next chapter it will be assumed that
tolerance represents either the elimination of all reactive immunocytes or their direction into the irreversible plasma cell development.

## GENERALIZATION OF THE RESULTS OF IMMUNOCYTE–
## ANTIGEN INTERACTION

At this stage it seems appropriate to try to provide a formulation
of the results of effective contact between immunocyte and the
corresponding antigenic determinant which can be used in the interpretation of such phenomena as delayed hypersensitivity and desensitization or antibody production, tolerance and auto-immune
disease. In all of these, antigen–cell contact is obviously a critical
step in the phenomena. In essence it is based on the central theme of
the clonal selection approach that the combining site of antibody
or cell receptor is a pattern produced by a random genetic process.

As in many other aspects of biology it is extremely difficult to
present the general characteristics of immune reactions in any sort
of a logical sequence. In this attempt to generalize the results of
immunocyte–antigen interaction it will be necessary to anticipate
some of the discussion in later chapters but it is felt that in its turn
the attitude schematized in figs 6 and 22 will be helpful in later
discussions.

*The importance of avidity*

If the pattern of the combining site as it emerges on differentiation
is wholly random, then, for any given antigenic determinant,

whether it is carried by a self-component or only on foreign macromolecules, there will be a range of newly differentiated immunocytes with avidity ranging from nil or negligible to a maximum determined by the still unknown potentialities of amino acid residue disposition in the combining site. One can be certain, however, that the distribution of immunocytes graded by avidity for a given antigen determinant will show a peak for those with minimal avidity and a progressive diminution of numbers as avidity increases (fig. 6). The highly avid combining site must always be very rare for the same reason that the chance of being dealt a complete suit in a hand of bridge is vanishingly small.

The process by which antibody against any antigenic determinant is developed will depend on the circumstances that allow cells with avidity in the higher ranges to be stimulated to multiply on contact with the antigenic determinant. Since the patterns by hypothesis are wholly random, capacity to combine with the antigenic determinant is the only quality which is relevant.

Our problem is to devise a means of predicting the result of combination between cell receptor and antigenic determinant under various defined conditions. From first principles, one could feel confident that both the concentration of antigenic determinant, and the avidity of interaction between combining site and antigenic determinant, would play a part. From our general knowledge of immune reactions we can also be certain that the response of an immunocyte is not simply proportional to the product of concentration and avidity. Both the way in which the antigenic determinant is presented to the cell and the physiological state of the immunocyte obviously play a part.

In an attempt to find a useful way to express the situation fig. 22 has been devised in which the ordinates and abscissae are the avidity of antigenic determinant–combining site union and the effective concentration of antigenic determinant respectively. Avidity between antigenic determinant and combining site has been extensively studied by Karush, Eisen and others using soluble haptens as antigenic determinant. It can be expressed as an adsorption coefficient, $K_a$, but as there is only limited quantitative experimental work it will often be necessary to use some qualitative index of higher or lower avidity. The term 'effective concentration of antigen' (or antigenic determinant) represents a complex function of concentration and time of exposure as qualified by the physical presentation of the antigen determinant to the cell. No single value can be attached to it in any physiological situation and information from *in vitro* studies can only be used cautiously and at the qualitative level in interpreting natural situations.

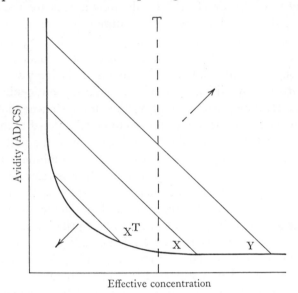

Fig. 22. To illustrate the concept that the result of specific contact of antigenic determinant and immunocyte may be either destructive or stimulating to specific activity, depending (*a*) on the affinity of receptor–A D union, (*b*) on the 'effective concentration' of the antigenic determinant and (*c*) on the physiological state of the immunocyte.

Destruction is more likely above and to the right of the diagonal line; proliferation and antibody production, below and to the left.

A wide variety of experimental approaches shows that small doses of antigen tend to promote proliferation and antibody production, while larger ones lead to tolerance and paralysis. On this basis, fig. 22 has been constructed.

It is obvious that if there is no antigenic determinant in the system or if the combining site has no avidity for it then there can be no response. There will therefore be a threshold for any response indicated by the rectangular hyperbola. On this can be drawn diagonal lines which can be taken as marking the zone on one side of which proliferative and antibody-producing responses to contact with antigen determinant will predominate while, on the other, destruction will be characteristic. The arrow indicates, upwards and to the right, increasing likelihood of destruction (producing tolerance and paralysis) or lesser degrees of damage associated with liberation of pharmacologically active agents (perhaps concerned with delayed hypersensitivity reactions). Downwards and to the left the arrow indicates that above the null thresholds the smallest concentration of antibody is likely to be the most effective in provoking antibody production and the production of memory cells.

## Changed cell reactivity

Irrespective of whether immunocytes undergo their primary differentiation in the thymus or elsewhere, there is general agreement that their first stage is one in which it can react with antigen but not by production of antibody. This has been called by various names: 'progenitor immunocyte', 'uncommitted cell', 'antigen reactive cell', or 'X cell'. For simplicity we shall use 'X cell', following Sercarz and Coons.

The effect of contact with antigen determinant is to convert an X cell into a Y cell (for which 'committed cell', 'plaque-forming cell' and 'memory cell' are approximately equivalent terms). Further antigenic stimulation will provoke Y cells to give rise to a clone of plasma cells (Z) or further generations of Y cells (memory cells).

There is evidence that X cells can be prevented from effective stimulation to Y cells either by the presence of preformed antibody of the same specificity or of an excess of the antigen concerned. Cells in the Y state are significantly less susceptible to either. In the diagram, therefore, line Y lies considerably to the upper right of line X.

It is an essential postulate of this approach that the general rules expressed by fig. 22 will hold under any circumstances where immunocytes and antigenic determinants react, with the necessary qualification 'other things being equal'. To a first simplifying assumption 'other things' will act by changing the position of the diagonal that lies at right angles to the double-headed arrow.

There are two sets of controlling factors which need special consideration, the local tissue environment and the effect of somatic mutation on an immunocyte line. The influence of drugs and hormones may also be important but is not immediately relevant.

In addition to factors which can modify the process that controls the response, there are implications of fig. 22 for the temporal course of antibody production and the relative avidity of antibody at different stages that are of great interest.

*The local environment*

The special features of the thymic environment have already been discussed. On general grounds, one would expect that somewhat similar situations can occur in germinal centres of lymph nodes and spleen, i.e. the standard position of the diagonal is displaced downward and to the left.

The other environment of special interest is the complex anatomical situation of the granuloma produced by antigen injected with Freund's complete adjuvant, and the draining lymph nodes. As will be discussed at a later stage, there are several pointers that favour the hypothesis that immunization with Freund's complete adjuvant results in a higher proportion of high-avidity immunocytes and antibody than is the case without adjuvant. This would mean that in the environment of the Freund granuloma and its draining lymph nodes the diagonal of our diagram is moved to the upper right. Despite what must be a relatively high antigen concentration, immunocytes of high avidity can proliferate to give memory cells and antibody-producing cells.

Any such statement has as a corollary that immunocytes of high avidity can arise and that they do not normally proliferate as a result of simple contact with antigen. On the random origin of pattern hypothesis we are using, in the uncommitted lymphocytes of the body there will be a range of immunocytes including some very rare ones with extreme avidity. One must assume that when a union of high avidity takes place in the germinal centre there is a strong probability that the cell will be destroyed and become visible as a necrotic pyknotic nucleus. We can reasonably assume that a small concentration of thymic hormone will be associated with primary nodules and germinal centres and that it will be absent in the Freund granuloma. Following this, our hypothesis becomes that in the complex antigen-containing environment of the Freund granuloma the most avid cells are allowed or even stimulated preferentially to multiply instead of being lethally influenced, as would be the case under more physiological conditions.

## Somatic mutation

Quite apart from the origin of immune pattern there are other possibilities of results arising from somatic mutation. If the local environmental changes or drugs can shift the diagonal line of our diagram, it is reasonably certain that there are also somatic mutations that could result in equivalent positive or negative cellular changes. If the mutation favours destruction of the cell concerned, nothing more will be heard of it but, as is the case in every evolutionary situation, if the mutation favours proliferation under the existent conditions, the mutant form will become evident. In the present circumstances only a mutation shifting the diagonal towards the upper right will have any significance. If, for example, such a mutation involved a cell capable of reacting with an autologous red cell antigen freely present in the thymic environment, movement of the line XT in the upper right direction would result in the escape from the thymus of *low avidity* cells capable of producing the equivalent incomplete antibody recognizable by a positive indirect Coombs test but not producing haemagglutination or haemolysis *in vitro* (see fig. 26).

It is immaterial how rare may be the concurrence of such a mutation with an immune pattern of the appropriate quality. Only one cell may be needed to initiate a pathogenic clone and there is evidence already that in a majority of cases of warm-type haemolytic anaemia the antibody that can be eluted from the patient's red cells is of monoclonal origin.

Combination of a similar mutation with immune pattern directed against a foreign antigenic determinant would rarely become evident but could be suspected whenever an individual being immunized gave an exceptionally active antibody response. There is a recent report which could be interpreted as a particularly striking example in which a burst of what appeared to be monoclonal antibody was seen in one of several rabbits immunized by a streptococcal antigen. No information as to the avidity of the antibody–antigen union in this case was provided.

Another possible field in which examples may be found is in the drug-induced purpuras and haemolytic anaemias in which binding of drug by antibody is demonstrable. Most of these diseases involve only a small proportion of those taking the drug and the likelihood is that only those persons who in one way or another can produce a highly avid type of combining site will be candidates for this type of drug reaction.

*The significance of avidity changes in the course of immunization*
Eisen has pointed out that, strictly speaking, 'affinity' is the correct term to use when one is discussing the intrinsic association constant for the reversible binding of simple antigenic determinants by antibodies. 'Avidity' is a looser term used by conventional immunologists to cover the overall binding activity of an antigen with perhaps many different types of determinant and a heterogeneous antibody population. It is convenient, however, when one must deal with a variety of situations to use 'avidity' at times with the strict meaning of affinity and at other times in the looser sense.

From the studies of dinitrophenyl (DNP) haptens and antibodies made in various laboratories and summarized in Eisen's 1964 Harvey Lectures, two points arise of special significance for our general approach. These are (*a*) that when a rabbit is injected with a single small dose of a DNP-immunogen in Freund's complete adjuvant, the avidity of the antibody present in the circulation rises about a hundred-fold in a six-week period and (*b*) that when 20–50 times that dose is given the rise in avidity does not occur.

This combined situation is of special interest in relation to the standard diagram of immunocyte response (fig. 22). With a very small circulating concentration of antigen, progenitor immunocytes of very high avidity for DNP-lysine will occasionally be stimulated by antigen to commitment and, at once, become more resistant against destruction by antigen contact unless a high concentration of antigen is encountered. If the concentration of antigen is very low the likelihood of effective stimulation of committed cells will be higher for cells with more avid immune receptors than for cells of low avidity. The figure shows that there will therefore be a build-up of high avidity antibody as long as the effective level of circulating antigen is maintained very low, as is the case where it is being slowly liberated from a Freund granuloma.

We have already (p. 121) adopted the point of view that as long as antibody is circulating the likelihood of a progenitor immunocyte being committed is very small. The subsequent history of a primed animal is therefore dependent almost wholly on the behaviour of committed cells and their descendants. New progenitor immunocytes that could react are prevented from doing so by the presence of antibody. As far as this particular reaction is concerned the thymus has no function. One might have predicted in fact the recent finding from Miller's laboratory that when a neonatally thymectomized mouse is supplied with adequate numbers of thoracic duct lymphocytes from normal syngeneic mice it produces antibody in normal fashion. Cells once committed are independent of any thymic control and do not require fresh recruitment of primary immunocytes.

### AVIDITY AND DELAYED HYPERSENSITIVITY

Much of the interest in the functional significance of avidity for immune reactions stems from the hypothesis of Karush and Eisen that delayed hypersensitivity represents the result of reaction between antigen and very low concentrations of highly avid circulating antibody. This was presented as an alternative to the dominant theory that delayed hypersensitivity and the closely related homograft immunity were cellular reactions. On the whole, the evidence points away from the importance of antibody, avid or otherwise, in delayed hypersensitivity but an extension of the concept of heterogeneity in avidity to the immunocytes could be highly relevant.

A quantitative study of desensitization of guinea-pigs using crystalline human serum albumin points strongly against the suggestion that delayed hypersensitivity is essentially a reaction between high affinity antibody and antigen. The central feature of the results is that when sensitized guinea-pigs are desensitized with human serum albumin intravenously they will remain reactive to large enough challenge doses even while significant amounts of antigen can be shown to be present in the circulating plasma.

All investigators are agreed that when the challenge dose of antigen is deposited intradermally the immunological reactant, be it antibody or cells, reaches the site from the circulation. The presence of small amounts of antigen in the circulating plasma would undoubtedly remove the postulated minute amounts of high affinity antibody. On the alternative theory which we favour, there are in the circulation lymphocytes derived from the lymph nodes draining the antigen deposit, which are immunocytes *vis à vis* human serum albumin. These cells will have variable avidity for antigen but, when effective contact is made, changes in the cell surface of a circulating cell will result in a very high probability that it is removed from the circulation, presumably in lymphoid tissue. This variable avidity or affinity may well involve both differences in the pattern of the combining site and of its relative accessibility to antigen in cells with the same combining site.

As Silverstein and Borek point out, their results point to a heterogeneity of affinity in the lymphocytes responsible for delayed hypersensitivity. If one accepts Spector and Willoughby's general interpretation of the tuberculin and similar reactions, movement through the local vascular endothelium is primarily nonspecific. When, however, specific immunocytes of high enough reactivity (avidity) enter the region where the antigen is present in relatively high concentration, they suffer a damaging reaction. This results in the release of a variety of pharmacologically active reagents. Among the effects of these on record are modification of the cell surface of other lymphocytes and monocytes leading to their diminished mobility and retention in the area of the reaction, and increased permeability of the local capillary endothelium allowing the entry of further specific immunocytes and other mobile blood cells.

On this view, the administration of a desensitizing dose of antigen intravenously will greatly reduce the proportion of circulating high avidity immunocytes and, to a progressively lesser extent, those of lower avidity for the antigen in question. A glance at fig. 20

will make it evident that in such partially depleted animals a higher local concentration of antigen will be necessary to produce a demonstrable damaging reaction.

## THE ORIGIN OF THE PLASMA CELL

Plasma cells are the major producers of antibody and the most striking histological feature of a secondary immune response is the rapid production of clones of plasma cells. The evidence is clear that plasma cells arise as clones derived from blast cells. They are characteristically found in close relation to small blood vessels and a number of histologists have claimed that they arise from perivascular cells. There is no doubt that when primed lymphoid cells from spleen or lymph node are placed in Millipore chambers in an isologous host, or injected into a heavily X-irradiated host, they respond to specific antigen by the production of numerous plasma cells. The precise origin of such cells is still unknown.

There is no shadow of doubt (*a*) that thoracic duct lymphocytes in which plasma cells and blasts are rare carry the information needed to allow antibody production, (*b*) that such cells can be converted into pyroninophil blasts which can undergo mitosis and (*c*) that most antibody is produced by cells of the plasmacyte series. There is no adequate evidence against the view that a small lymphocyte stimulated by antigen and in the right physiological environment will become a pyroninophilic blast from which proliferation and differentiation in the plasma cell series can develop. Equally it must be emphasized that there is no unequivocal positive evidence that this actually occurs.

A second possibility arises if there is any process by which genetic information capable of directing antibody pattern can be transferred to another somatic cell line genetically lacking that information. If antigen passes to a lymph node and there stimulates an appropriately patterned immunocyte to produce large numbers of information transfer units, we can conceive of these passing to any available blast cells and directing them to develop along the plasma cell line. There is no direct evidence in favour of this interpretation but a good deal of puzzling indirect evidence and nothing to veto it unequivocally.

187

At the present time it seems desirable to keep any ideas of inter-cell transfer of information in the background and to put the main emphasis on the experiments by Jerne's technique, which point rather directly to a direct proliferation to plasma cells of genetically determined lines. The process already mentioned on several occasions by which immunocyte–antigen reactions can modify adjacent cells may be relevant here. There are indications that a quiescent immunocyte of non-thymic origin needs to be activated in some way before it reacts with its corresponding antigenic determinant. The simplest way this could occur in a lymph node situation where a variety of antigens are likely to be available is for reaction by a thymus-dependent cell with *its* antigen to provide the nonspecific stimulus needed.

# 9 Antibody production

In this chapter we are concerned with the actual process of antibody production and only incidentally with the special character of antibody and the qualities that differentiate one antibody from another. It will be impossible to avoid some overlap with earlier and later chapters. What is being attempted is to bring together what has been discussed earlier in regard to the patterns of immune specificity and the cells concerned in immunity and to describe them in action. As everywhere in immunology it is impossible to provide a general interpretation that can be tested against each of the major areas available for study. Variability and heterogeneity are of the essence of the problem, there are 'good' antigens and 'poor' antigens in the sense of the amount required to provoke a measurable response and, for most antibodies to 'good' antigens, methods of assay are almost all still at the level of functional titration rather than truly quantitative estimation.

It follows that data obtained by the use of a good antigen like phage T2 or $\phi$X174 are very hard to compare with those obtained with a pure soluble antigen of relatively low antigenicity such as bovine serum albumin. The choice of what is to be taken as most significant in the observations must therefore be rather intuitive than logical—a choice in fact among possible formulations of those which require least distortion to cover phenomena distant from those on which they were primarily developed.

## ANTIBODY PRODUCTION FROM THE SELECTIVIST APPROACH

'The production of antibody' is a rather equivocal phrase particularly when one adopts a thoroughgoing selectivist approach. On this view, any immunocyte that can react with an antigen is a producer of antibody in the sense that the cell receptors have all the essential immune pattern of an antibody. Further, there is increasing evidence that all lymphocytes of the immunocyte series release

189

small amounts of immunoglobulins into their environment. The synthesis is of quite a different order of magnitude from the massive production of immunoglobulin by developing or mature plasma cells.

In general this chapter deals with the production of effective amounts of antibody by cells of the plasma cell series. This is a matter that can be looked at from several points of view, depending essentially on the type of experimental observation being considered.

1. For many types of experiment, what is convenient to follow is the amount or concentration of immunoglobulin or some physical fraction such as the conventional '$\gamma$-globulin' of simple electrophoresis. At this level we have two important questions to put: (*a*) What are the homeostatic mechanisms that keep the dynamic equilibrium concentration of $\gamma$-globulin in the blood approximately constant in health? (*b*) Is all $\gamma$-globulin antibody in the sense of carrying specific combining sites or are there wholly nonspecific globulins as well?

2. We have already described the physical differences among the three types of immunoglobulin having antibody function and indicated broadly the functional significance of the differences between A, G and M immunoglobulins in antibody function. There is a great deal to suggest that their specificity, their combining sites, are essentially identical or, more correctly, that they will show the same range of heterogeneity in each type. At this level we have three major questions: (*a*) What determines in a given immune response whether A, G or M is produced? (*b*) What changes in addition to M to G are possible within the same cell line? (*c*) Are there individual homeostatic mechanisms for each type?

3. It is axiomatic that, for the production of substantial amounts of antibody, an antigenic stimulus is necessary. This holds as much for a selection theory of antibody production as for an 'instructive' one. It is also evident that some antigens are much more effective stimulants to antibody production than others, and that the physical state of an antigen and its administration with or without adjuvant substance may also have great effect upon the antibody response. At this level we have the questions: (*a*) What is the process by which antigen initiates the sequence which leads to antibody forma-

tion? (*b*) What part do the cells which *par excellence* take up foreign material, macrophages and polymorphonuclear leucocytes play in the process? (*c*) What are the actual antigenic determinants concerned?

4. Finally, we have the striking difference between the primary and the secondary immune response, particularly when certain types of antigen are used. Since an active secondary response is the most characteristic form of antibody production and has been the most extensively studied, it will be convenient to begin the discussion of antibody production with this theme.

### THE SECONDARY-TYPE RESPONSE

If a rabbit or a rat has been immunized some months previously with an antigen, bovine serum albumin or *Salmonella* flagella for example, and now has a minimal or no detectable antibody titre, it will respond predictably and strongly to a second injection of the same antigen. If the injection is made locally, e.g. into a hind foot-pad, the regional lymph node will rapidly enlarge and, by three to four days, there will be a greatly increased proportion of plasma cells in the node. Antibody, almost wholly of Ig G type, will appear and show a rapidly rising titre in the blood.

A situation closely resembling this, though ostensibly a primary response, has already been described for the rapid increase in cells producing Ig M antibody in the spleen of a mouse immunized by sheep red cells. Analysis of the situation in the lymph node following a secondary stimulus leads to the same conclusion. The entry of antigen results in the rapid proliferation of cells to form clones, the members of which become progressively closer to the mature plasma cell type.

The plasma cell is unequivocally the major producer of antibody and its intracellular structure with complex endoplasmic reticulum is specialized for this function. The indications are that proliferation and antibody production take place concurrently until the descendant cells have reached the morphological character of mature plasma cells. Once that stage has been reached, proliferation ceases but antibody synthesis may continue for some time. The fate of the large numbers of plasma cells present at the height of a secondary immune response has never been quantitatively followed. There is a progressive diminution in numbers, presumably by death and autolysis, but individual plasma cells labelled as the product of a defined immune response may persist for as long as three to six months. Direct and circumstantial evidence both make it almost certain that the plasma cell is a differentiated end-cell incapable of giving rise to any type of descendant cell.

*The origin of plasmablasts*

What is still controversial is the nature and origin of the cells from which the plasmablast proliferation arises. There seem to be two formulations that cover a wide range of facts and represent two broad currents of opinion.

(*a*) In the previously immunized ('primed') animal, the antigen is rapidly taken up by a variety of phagocytic cells of which the most significant are the dendritic reticulum cells intimately associated with lymph follicles in lymph nodes, spleen and elsewhere. Here, a proportion of the antigen is converted to a more effective form of antigenic stimulant. A popular interpretation is that in the phagocytic cell partial proteolysis occurs and the antigenic determinant, with a few adjacent amino acids, becomes linked to a low-molecular-weight strand of RNA. This material is probably concentrated on the surface of the macrophage including its dendritic processes. A lymph follicle can be visualized as a moving three-dimensional crowd of lymphocytes from which large numbers are moving away mainly by efferent lymphatics to be replaced by approximately equal numbers arriving by the bloodstream and entering the follicle by passage through the thickened endothelium of postcapillary venules. With changing populations and as a result of their incessant active mixing movement, sooner or later lymphocytes capable of responding to a given antigen will make contact with that antigen presenting on the reticulum cell surface. Contact initiates proliferation and presumably influences both the mobility and reactivity of the cell. Either within the lymph follicle or in the course of its movement elsewhere the stimulated cell becomes a pyroninophil blast and the progenitor in the right environment of a clone of plasma cells.

This formulation can be summarized as the stimulation of a

small or medium pre-adapted lymphocyte, i.e. a committed immunocyte, by processed antigen to become a blast and initiate a plasma cell clone. Such a sequence has not been directly established but the importance of lymphocytes in allowing antibody production is undoubted.

(*b*) There is a need for a constant replacement of immunoglobulins in the circulation, irrespective of specific antigenic stimulation, and the second set of opinions would postulate that in some essentially random fashion some immunocytes are stimulated to take on the blast form, divide for one or two generations and produce small amounts of whatever antibody they are genetically capable of producing. The suggestion has already been made that when an immunocyte is specifically stimulated by the corresponding antigenic determinant the stimulus is mediated by the release of pharmacologically active material which could have a similar nonspecific stimulatory effect on adjacent lymphocytes. This would ensure that in any individual exposed to a natural environment there would be a random population of nonspecifically stimulated immunocytes, ripe as it were for specific stimulation. Just as in the spleen of the mice in Jerne's experiments, contact with antigen greatly accelerates the proliferation and differentiation of these cells. From this point of view it is immaterial whether the antigen is presented directly to the cells or after processing in macrophages. The most important evidence in favour of this approach is due to Nossal and Mäkelä, who found that when tritiated thymidine was given an hour before antigen most of the new plasma cells present four days later were labelled. This provides a prima facie case for the claim that the antigen was only effective in stimulating cells already in the stage of DNA synthesis for proliferation. The main weakness in this argument is that it does not wholly exclude early transfer of DNA or its molecular components from a disintegrating primarily labelled cell to others.

(*c*) A third possibility, that regularly or occasionally there may be a transfer of information from a cell that received it genetically to other cells capable of subserving a 'nurse' function, is better deferred.

### THE PRIMARY IMMUNE RESPONSES

There is a strong current tendency to minimize differences between primary and secondary responses and to look solely for quantitative reasons for the differences actually observed. On the simplest form of selection theory one might divide the type of primary response to be expected into three categories:

(*a*) in which, for some genetic reason, a relatively large range of combining sites can react with the antigenic determinant in question;

(*b*) in which previous experience with natural (bacterial or other) antigens has increased by a specific process the number of cells which can react with the chosen antigenic determinant;

(*c*) in which only rare configurations of the combining site will react significantly with the chosen antigenic determinant.

Only in the third circumstance is there likely to be a clear-cut differentiation between primary and secondary responses. In practice it will always be difficult to differentiate between (*a*) and (*b*) when we are looking for a reason why a given antigen provokes an active primary response only slightly less active than a secondary one. On the other hand there are instances where the primary response seems to have a different quality from the secondary. In my own early work I was greatly impressed by the contrast between a first intravenous injection of staphylococcal α-toxoid in a rabbit and the second, two to three weeks later. The first response is minimal and delayed, the second, after a lag of about 40 hours, rises exponentially for 3 or 4 days.

There are few examples in which the difference is so evident but there are many indications that cells are functionally changed after first contact with antigen. Strictly speaking, to obtain a true primary response, an antigen should be given in a single brief 'pulse' and in one way or another be immediately eliminated or removed. An intravenous injection of toxoid may come much closer to this ideal than, for example, any immunizing situation that involves a persisting deposit of antigen or in which the antigenic determinant is not susceptible to rapid breakdown by tissue enzymes. If, at a guess, to 'commit' an immunocyte requires rapid de-differentiation and perhaps two successive divisions within two or three days, then what is technically a secondary response could regularly be produced as a result of the primary infection or inoculation of antigen.

In line with our earlier discussion, the qualitative differences between primary and secondary response can be ascribed to an increase in the number and accessibility of immune receptors and a more 'robust' response to antigenic stimulation. Simple quantitative factors may well be even more important: there are simply more immunocytes available for stimulation by the antigenic determinants. This is probably the main or only reason for the

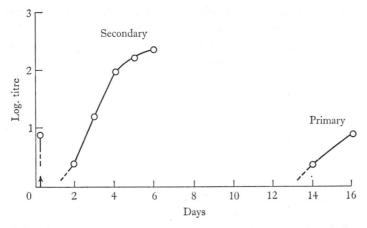

Fig. 23. Primary and secondary antitoxic response to α-staphylococcal toxoid. From F. M. Burnet (1941), *The Production of Antibodies*. Melbourne: Macmillan. Both injections given intravenously at day 0.

difference when the stimulus for the primary response is of such a character that significant amounts of antigen persist in the body for more than two or three days.

There is no reason to believe that the ways in which antigen can reach the receptor of the primarily reactive cell differ from those at work in the secondary response. The essential difference is that immunocytes of the right type are fewer in number and the requirements for stimulatory contact more strict. It is probable that the avidity of the cell receptor must be above a certain level for any stimulation to occur but also that any contact with more than minimal concentrations of antigen will have an inhibitory or destructive effect.

## Foetal and neonatal production

The former generalization, that no antibody could be produced in the embryo, is no longer valid. Recent work on sheep, pigs and cattle in which there is no transfer of immunoglobulins from maternal to foetal circulation has established this quite clearly. The period of gestation at which the embryo becomes competent to produce antibody varies very markedly from one antigen to another but with a phage antigen an antibody response was obtained in foetal sheep by injection at the earliest stage, 60 days, that was experimentally feasible.

In unimmunized foetal ungulates, very small but definite amounts of Ig M develop and foetal or neonatal antibody is also at first wholly Ig M. The rather patchy evidence is all consistent in indicating that the process of tooling up for antibody production goes on at least from the period when lymphoid cells appear in the thymus. Depending on the species, there is a transfer of maternal antibody to the offspring either *in utero* or after birth by way of colostrum and milk. On current information, the antibody that passes into the later human foetus is Ig G, while that concentrated in the ungulate colostrum is Ig A. This passively transferred antibody appears to act as a temporary shield against infection and under its cover the further development of the mechanism for active immunity proceeds.

There have been suggestions that nonspecific $\gamma$-globulin was produced during the foetal and neonatal period but there is no valid evidence that the situation is any different in principle from that obtaining in postnatal life.

## The change from Ig M to Ig G antibodies

Current interest in the primary response is concentrated largely on the characteristic switch seen, at least with predominantly protein antigens, from initial production of Ig M (19 S) antibody to Ig G (7 S). There is no evidence as to whether a similar switch from M to A may also occur.

In virtually every type of experimental immunization that has been reported the antibody first produced is of M type. This holds also for all 'natural' antibodies and for the antibody liberated by the splenic cells that produce Jerne plaques against sheep red cells. The characteristic methods of demonstrating the difference between M and G antibodies, susceptibility or resistance to the inactivating effect of 2-mercaptoethanol, are only applicable to antibodies which are conveniently demonstrable in tests made *in vitro*. This excludes Ig A antibodies and any Ig A or Ig M antibodies not demonstrated by the test being used.

There are some types of antibody which predominantly remain of M type; notably those produced in response to bacterial poly-saccharides. It is probably an important correlate that these are also antigens which do not give any evidence of immunological memory. An animal that has responded to pneumococcal poly-saccharide antigen and subsequently lost or almost lost its titre will respond to a second injection of the same type in what is apparently exactly the same fashion as a previously uninoculated animal of the same age. Recent work, however, indicates that this finding is, in part, a technical artefact. Following a first immuniza-tion, some Ig G antibody is produced appearing as usual a few days after the first appearance of Ig M antibody. Previous failure to recognize the Ig G production is due to the physical differences between the two types of antibody in relation to the distribution of somatic antigen on the bacterial surface. When compared on a molar basis, Ig M antibody may be up to a thousand times more effective than Ig G antibody in bactericidal action or as agglutinin. This depends apparently on the larger number (probably ten or twelve) of combining sites on the Ig M antibody compared with two on the Ig G molecule. On the bacterial surface, an Ig M anti-body is therefore able to establish two to four points of attachment to antigenic determinants and, following appropriate collisions, to a similar number on another bacterium. By contrast, Ig G anti-body can make only a single contact at each combining site, so providing a very much weaker bridge between two bacteria.

Following secondary immunization there is increased production of Ig G antibody, but the observed result is still dominated by Ig M. There is 'memory', even if it is not easily recognized, for Ig G production but the Ig M response retains the primary character.

The evidence suggests that the same cell line and probably the same individual cell can produce first M and later G antibody and this will be accepted as a provisional conclusion in our discussion. Direct evidence for this is limited to one set of experiments and many workers would still prefer to hold that Ig M and Ig G producers are more distinct than is accepted here.

The most important recent finding is that by administration of specific Ig G antibody to an animal producing M antibody there is induced a sharp change to G production and an almost complete inhibition of further M production. This phenomenon has some important implications. Since the effect is not given by any unrelated antibody the action must in some way involve the antigenic determinant. There is no doubt that most of the effect is a simple inhibition of the ability of antigen to act as a primary stimulus to what we have called 'progenitor immunocytes'. In addition there is presumptive evidence that the circulating Ig G antibody is able to switch some of the Ig M-antibody-producers to Ig G antibody production. This is doubted by some authors and it is not easy to provide a satisfactory explanation of the switch if it occurs. Whatever the details, there soon develops a situation in which the majority of active immunocytes are producing Ig G antibody and are committed as end-cells while an increasing concentration of G antibody will automatically prevent antigen from making initial contact with newly differentiated cells of appropriate pattern. These points follow fairly directly from the experimental data but to account for the existence of memory cells within the same framework requires some ingenuity.

'Memory cells' is a commonly used term for the cells in a previously immunized animal which, while not producing demonstrable antibody, are available for stimulation with the active antibody production characteristic of the secondary response. Always with the reservation that the actual situation is bound to be more complex than any hypothetical model, it is convenient to think of these as arising from uncommitted or committed immunocytes which, after stimulation by antigen, do not go on to form a clone of plasma cells. Either because they are cleared of antigen at an early stage or for other reasons, these cells in an appropriate internal environment can proliferate, perhaps in the form of a germinal centre, to give rise to large numbers of descendant lymphocytes with the ancestral immune pattern—reserving the possibility that a proportion may have undergone somatic mutation.

## THE POSSIBILITY OF TRANSFER OF IMMUNOLOGICAL
## INFORMATION BETWEEN SOMATIC CELLS

For many years there have been insistent suggestions that somatic cells might transfer immunological information from one to another. In bacteria and viruses there are several processes by which such transfer can occur, the first to be discovered and the one most potentially relevant to immunological discussion being type transformation in pneumococci. In this, segments of DNA pass to and can be re-incorporated into the genome of suitable recipient individuals. A second phenomenon, which has often been accepted as a possible prototype of information transfer between somatic cells, is the infectivity of nucleic acid from some RNA viruses. This has a legitimate claim for consideration since the phenomena are shown with many of the small RNA viruses which infect mammalian cells. The third approach is to accept the accumulating evidence that under some conditions hybridization and segregation can occur in somatic cells.

The reasons for postulating the existence of one or other of these processes are not very substantially based on experiment. The experimental observations that are reported have almost all been obtained in an attempt to establish a preconceived idea rather than to elucidate an unforeseen observation. Most have arisen in attempts to implant immune capacity on nonimmune cells by treating them with extracts of immune cells prepared in the way by which 'infectious RNA' was obtained from some of the smaller animal viruses.

Another reason for considering the possibility arose from efforts to explain within the limits of a rigid clonal selection theory the occurrence of a proportion of cells producing more than one type of antibody, the so-called 'double producers'. If a given clone of immunocytes can produce only one pattern of antibody or immunoglobulin, the existence of a small proportion of double producers in doubly immunized animals would call for some explanation in terms of transfer of immunological information between two independent lines of somatic cells. With the recognition that any acceptable present-day version of immunological theory is of necessity complex and flexible, the difficulties tend to disappear or at least become less urgent. One important possibility of understanding how nominally unrelated viruses used as antigens could on occasion react with a single type of combining region has already been raised. There are various other possibilities implicit in the interpretation of immune specificity developed in chapter 6.

In many ways it might be wise to postpone any consideration of transfer of immunological information until the experimental evidence for its existence becomes unequivocal and fully accepted. To discuss the three possibilities listed above can only be justified on the grounds of wide interest in the problem and the fact that there are recorded observations which appear to require explanation in terms of one or other process.

The most acceptable approach is probably to use the existence of somatic cell hybridization under certain conditions in tissue culture as a justification for assuming that immune pattern can be subject to recombination and segregation by fusion of somatic cells. The actual circumstances could only be guessed at and the overriding quality of phenotypic restriction makes any simplification highly artificial. We are forced to assume that of any alleles only one will be functional and if, for simplicity, we assume that only one gene in each cell is relevant to immune pattern, the cells taking part in recombination could be represented as *ao* and *bo* where *o* is a nonfunctional allele.

Suppose we have an antigen in the form of a single molecule containing two antigenic determinants *A* and *B*. This will stimulate cells of *ao* and *bo* clones respectively and in the appropriate lymphoid tissue we shall have proliferating cells of both clones in close proximity. If recombination and segregation can occur, these conditions could well be optimal. We assume segregation would be at random and would result in a ratio of 1 *aa* 1 *bb* 2 *ao* 2 *bo* 3 *ab* plus nonviable or inert *oo*. The phenotypic ratio at equilibrium would be 1 *a* 1 *b* and 1 *ab* but, as there will never be complete recombination there should always be something less than 33 per cent of *ab*. If, however, *ab* is more actively stimulated than *ao* or *aa* by *AB* antigen, which is not unreasonable, then a higher proportion of the double producers could be found.

There may be special limitations on the types of somatic cell that can undergo recombination and segregation and if the restriction of extensive double producers to antigenic determinants carried on the same molecule is valid, the restriction may well be to cells producing the same type of antibody G, M or A and perhaps to cells in equivalent physiological states. Since with a single molecule as antigen, antigenic determinants *A* and *B* must be handled in precisely the same fashion, these requirements would be fulfilled much more readily than when two unrelated antigens were inoculated simultaneously.

The two subcellular possibilities that have been canvassed can be expressed in more or less similarly simplified terms:

(*a*) Until much more is known about the genetic determination of the combining site, the form that any equivalent of pneumococcal transforming principle could take must remain completely unknown. Probably all that can be said is that unless the combining site or region is controlled by a single small genetic unit, a process of this sort is virtually inconceivable. The only qualification is that if there was a clear survival gain from the existence of cell-to-cell transfer of immune pattern, a process might conceivably have evolved for the immunocyte series alone. The lack of any biological analogy would then have less weight.

(*b*) If immune pattern could be transferred to another cell in the form of informational RNA this would imply that, like virus RNA, it fulfilled the dual function (i) of messenger-RNA in directing specific protein synthesis and (ii) of replicating to produce more RNA of the same pattern. It is also implicit that it should, in one form or another, provide information to enforce the development by the 'infected' cell and its descendants of the elaborate endoplasmic reticulum with its membranes and ribosomes characteristic of the mature plasma cell. By analogy with bacterial situations this would presumably demand a variety of information needed for the synthesis of necessary enzymes. It could, however, require no more than a signal to some 'operator' that could set the cell on the same road to differentiation as plasma cell, in a fashion analogous to what must occur in active antigenic stimulation.

None of the direct evidence for transfer of immunological information is both unequivocal (in excluding alternate explanations) and adequately confirmed. Most of what has been published as evidence of transfer by RNA is now regarded as representing the action of antigen modified in macrophages to units comprising antigenic determinant, peptide of perhaps minimal size, and low molecular weight RNA possibly of the type used for amino acid transport. An important recent experiment by Friedman has shown that a cold phenol extract of immunized mouse spleen has a minor effect in stimulating corresponding antibody production in normal mice but none in mice rendered specifically tolerant to the antigen.

It is probably best to maintain a sceptical approach to all suggestions of informational transfer between immunocytes and to seek other interpretations of awkward facts. The whole situation will probably remain obscure until a satisfactory means of growing pure clones of functional antibody-producing cells *in vitro* is developed and methods become available for full sequence-analysis of the immunoglobulins they produce.

THE ORIGIN AND HOMEOSTASIS OF IMMUNOGLOBULINS

There is no accepted doctrine in regard to the existence of immunoglobulins (as defined solely on physical and antigenic qualities) which lack combining sites. At the present time what evidence there is points very much against the possibility but it cannot be regarded as absolute. Until it was shown that at least three Ig G myeloma proteins were carriers of immune reactivity, it was possible to hold that these proteins lacked any combining site. It is still, of course, possible that this holds for some but not for others. The evidence to date, however, is that the light chains of myeloma proteins are structurally similar to each other and to the light chains of antibody. In addition, the finding that both light and heavy chains in mixed antibody populations can carry tyrosine molecules capable of specific reaction with hapten by the affinity-label method and that this corresponds to the variable portion of the light chain in Bence Jones proteins, gives solid reason for assuming that what is found in myeloma light chain can be transferred with only minimal reservations to the situation in antibody of normal physiological origin.

Our discussion will therefore adopt the viewpoint that all immunoglobulins conforming to the standard antibody structure are, in fact, antibodies in the sense that they carry two regions homologous with combining sites whether or not there is any antigenic determinant known or conceivable which would unite significantly with them. It seems, in fact, that if the structure of immunoglobulins that we have adopted is correct, 'nonspecific $\gamma$-globulin' could arise only as a result of gross error in the synthesis of the polypeptide chains. The failure of a given preparation of immunoglobulin to react with any available antigen is meaningless and a definite decision could only follow a demonstration (*a*) that the combining site had a specific genetic determination distinct from that of light and heavy chains and (*b*) that a physical difference could be shown between immunoglobulins which corresponded to one set having combining sites and the other lacking them.

On the interpretation we are adopting, non-antibody $\gamma$-globulin represents those immunoglobulin molecules which carry specific patterns nonreactive against the limited number of antigens that the experimenter has available for test. They can be regarded as a more or less random mixture of all the types of antibody a given individual can produce. The problem that arises is whether this random mixture results from a large number of small specific stimuli or is synthesized under the influence of some nonspecific process.

Our preference is for the second alternative although the evidence is probably inadequate to justify the choice. There is evidence, for instance, that when a specific response occurs there is a variable but often large increase in immunoglobulin which does not react with the antigen used. Freund's complete adjuvant given without additional antigen causes a rise in globulin not related to the antigens of the incorporated acid-fast bacteria. It has been claimed that 'immunization' of an animal heavily irradiated on the preceding day gives rise to an increase in $\gamma$-globulin but not of precipitable antibody. Some have also claimed that lymphocytes can be stimulated by phytohaemagglutinin to produce some additional $\gamma$-globulin.

A thoroughgoing genetic approach to antibody production would almost require that a proportion of the random patterns generated would not react with anything present in the body fluids and tissues nor with anything reasonably likely to be introduced into the tissues from outside. These, however, would be just the patterns required to help deal with any 'new' antigen which might appear and there is an easily recognizable biological requirement that a proportion of cells with such potentialities and a low concentration of the antibodies they produce should be retained in the body.

We are virtually forced, therefore, to assume that there is some way in which a nonspecific stimulus to proliferation can affect all clones of immunocyte without discrimination. There is a constant loss of lymphocytes across mucous membranes which is almost certainly nonspecific as well as the well-known losses in thymus and in lymphoid tissue which may have a specific element, and in one way or another these losses must be made good. One can hardly avoid postulating a series of homeostatic controls governing the total content of immunocytes in the body. On the positive side this would mean a continuing production of Ig M antibodies to maintain the natural antibodies of the blood. On the negative side the natural corticosteroids could well play the major role.

At several points the capacity of agents liberated by antigenic stimulation of a cell to activate adjacent cells has been described or suggested. If this activation is able to stimulate low-grade DNA synthesis and proliferation and to leave these cells temporarily more highly reactive to contact with their 'proper' antigenic stimulus, a number of experimental findings would be clarified, notably recent findings that an interaction between thymus-dependent and non-thymus-dependent immunocytes is necessary for the production by the latter of certain types of antibody.

It is an interesting point that, if the concentration of immuno-globulins acting on an appropriate sensor provided the necessary feedbacks to stimulate either lymphopoietic and antibody-producing activity or an increased corticosteroid output to diminish the immunocyte population, we should have the basis for a homeo-static control. There appears to be no evidence for or against such a concept which would provide a natural corollary to the specific feedback provided by Ig G antibody. Clearly there must be a complex process by which the A, G and M ratio is held constant or modified according to need, but no effective experimental approach has been described.

It is quite certain that mechanisms controlling the level of immunoglobulins in the body will be uncovered in the future. The existence of both congenital and acquired agammaglobulinaemia as pathological states in man points strongly to the presence of some unitary mechanism or substance necessary for the development of plasma cells and the production of immunoglobulins in more than minimal amounts. Elsewhere it has been suggested that this may be a hormone related to whatever tissue in the mammal corresponds to the bursa of Fabricius in birds. If such a hormone exists it presumably plays a major part in maintaining a standard level of immunoglobulins in the circulation.

### THE ROLE OF ANTIGEN

In any formulation of a selective theory of antibody production the function of antigen becomes wholly that of a specific stimulus which, acting through an appropriate receptor, initiates cellular response. This may be positive—for example, proliferation of the immunocyte to produce either plasma cells actively synthesizing antibody or memory cells—or it may yet be shown that direct conversion of lymphocyte to an antibody-synthesizing form can occur. Negative responses exemplified by destruction and perhaps non-destructive inhibition of immunocyte function are more relevant to the subject of tolerance (chapter 10), but must always be kept in mind.

### *The presentation of antigen as stimulus*

There are innumerable substances or organic structures which can serve as antigens and their effectiveness as stimulants to antibody production must depend on many factors. Bovine $\gamma$-globulin is a very 'poor' antigen in mice when it is completely soluble and undenatured and provokes unresponsiveness to the 'good' antigen that results when the bovine $\gamma$-globulin is partially aggregated. The physical condition of the antigen and the way in which it is presented are both obviously important. The intravenous route is standard when one wishes to induce tolerance. The production of a localized deposit of antigen by adding Freund's adjuvant to the inoculum can, in general, be relied upon to produce antibody in quantity.

Included in the 'presentation' of the antigen, and probably the main feature determining the outcome, are the relations of concentration and time of exposure to the antigen. Details are still to be worked out but the general picture that emerges, for example, from Mitchison's experiments, is that effective contact of an antigenic determinant with a cell receptor on a previously unstimulated cell will have a certain probability $(P)$ of initiating a positive response and the reciprocal $(1 - P)$ probability of initiating inhibition or destruction. In the fraction $P$ where positive action has been initiated, subsequent contact presumably with other receptors may have partial or complete inhibitory effects. In fig. 22 in the discussion of avidity in the previous chapter (p. 180) $P$ will increase at right angles to the diagonal in the lower-left direction and conversely the $(1 - P)$ probability of destruction in the opposite direction. The size of $P$ will almost certainly increase with the age of animal at least from mid-embryonic life till adolescence and with the relative maturity of the cell concerned. It will also differ greatly with the nature of the antigen concerned. A bacteriophage such as $\phi X 174$ seems to have a high value of $P$ in nearly all circumstances including embryonic life. Purified soluble plasma proteins on the other hand have a relatively low value of $P$ in mice and in young rabbits. The difference between the physical form of a complex virus and a soluble protein will, of course, be highly relevant.

In this discussion it is implicit that the characteristic negative response is functional and, probably, physical elimination of the cells concerned. These are therefore not subject to secondary effects. Positively stimulated cells are, however, activated to proliferation and differentiation. They remain susceptible to further stimulation by the corresponding antigenic determinant and the one thing we know about such cells is that with adequate exposure to high concentrations of antigen they can be functionally eliminated. It follows that in every situation where antigen is by one means or another maintained for a significant period of days or weeks in the body we have a highly complex dynamic situation in action. At all stages, a proportion of competent cells committed or uncommitted will be eliminated by reaction with antigen. Some uncommitted cells are stimulated to direct production of Ig M antibody, others in perhaps two or more sequential steps to pro-liferate either to antibody-producing plasma cells with eventual elimination of the line, or to 'memory cells' with continuing potentialities of reaction with the antigenic determinant.

If this analysis is correct we have an adequate theoretical basis to account for the dependence of the curve of antibody response on a host of factors: species, strain, sex and age of the animal immun-ized, physical and chemical characteristics of the antigen and the route concentration and time sequence in which it is presented. We may add to this two aspects discussed elsewhere: (*a*) genetic factors determining the readiness with which certain antigenic patterns can be produced by the individual and (*b*) the capacity of circulating antibody (7 S, G) to block the action of antigen as stimulus and to 'turn off' 19 S, M antibody production. Taken together, these factors provide an explanation of the need for quite extraordinary standardization of all experimental particulars if quantitatively uniform antibody responses are to be obtained.

When we look more closely at the function of antigen it soon becomes evident that the impact of antigen on the immunocyte is often, and as far as classic antibody production is concerned, possibly always, mediated by macrophages.

It has already been discussed in another context that macro-phages, especially the dendritic macrophages of lymph follicles, serve to prepare antigen to function as an effective stimulus to proliferation and antibody production. With the usual reservations about the possibility of alternative formulations, current opinion can probably be represented as follows. In order to be capable of stimulating antibody response the antigenic determinant with a pattern complementary to the receptor of the immunocyte must be in an appropriate physical form. The most effective conformation appears to be the antigenic determinant attached to part of a pep-tide chain, perhaps no more than one amino acid, to which in its turn a low-molecular-weight RNA, possibly s-RNA, is attached.

The major function of a dendritic macrophage is to maintain this complex in functional antigenic form on the surface of its projections for several days. The evidence suggests that less specialized macrophages quite rapidly destroy the antigenicity of any antigen which is basically protein in constitution.

At the evolutionary level the significant immunological needs are probably (*a*) the elimination of mutant somatic cells as they arise, (*b*) defence against micro-organismal invasion including both 'mopping up' of the first attack and prevention of re-infection thereafter and, possibly, (*c*) control of situations that arise in connection with pregnancy where two antigenically different organisms are in a close organic relationship. None of these is particularly close to the standard experimental situation in which a highly purified antigen is injected in relatively large amount by a single route. All results from such artificial situations must be carefully considered in the light of the realities that are relevant to the evolutionary process. So long as this is kept in mind the simplifications and abstractions of experimental work are fully justifiable.

*Hapten-carrier antigens*

The first type of study to be mentioned involves the use of simple haptens of known chemical structure such as dinitrophenyl (DNP) which are not themselves antigenic but become so when they are attached either to a suitable natural protein such as ovalbumin or to synthetic polypeptide. There are two important findings. If the hapten is attached to a synthetic polypeptide of L-GLU, L-ALA and L-TYR, the compound is antigenic and produces a corresponding antibody primarily directed against the hapten. If, however, the same hapten is attached to a synthetic polypeptide made up of D-amino acids it is not antigenic but when used as a test antigen it can be precipitated by antiserum made against the L-polypeptide complex. The second point is that when wholly synthetic antigens of this type are tested in guinea-pigs it is very rare to find all the test animals responding. If two sets of pure line guinea-pigs are used it is usual to find a certain pattern of response in one line and quite a different one in the other. Using non-responder guinea-pigs which produced no antibody to DNP-polylysine, Green and Benacerraf showed that complexing this to a carrier protein gave immunogenic material. Specific anti-DNP-producing cells could be demonstrated so that there was no genetic inability to produce the immune pattern.

Taken along with other evidence it seems that some preparatory process is needed to bring what is injected into a form that can directly stimulate antibody production. Following Benacerraf, the most likely interpretation is that the carrier peptide or protein must be broken down by intracellular enzymes in the macrophage that takes it up, so as to leave an antigenic determinant attached to an amino acid or small peptide to which low-molecular-weight RNA can be attached to give the 'directly antigenic' complex. The differences between individual guinea-pigs may well be due to differences in their capacity to break down the carrier, but genetically determined differences resulting in the presence or absence of appropriate patterns on the immunocytes may also be relevant in some instances.

## The effect of adjuvants

It has become almost a habit amongst immunologists to use Freund's complete adjuvant (F.C.A.), (a mixture of liquid paraffin, an emulsifying agent, 'Arlacel', and killed acid-fast bacilli) to enhance the action of any type of antigen used for experimental immunization. In general a longer lasting, higher titre, antibody response is given than when the antigen is injected alone. It is equally effective in increasing the ease with which delayed hypersensitivity is produced and its use is essential if 'autoimmunization' by homologous organ tissue is to be achieved. The action of F.C.A. is certainly complex. The water-in-oil emulsion induces a complex granuloma containing many cell types including macrophages, plasma cells and lymphocytes. Within this, the inoculum persists almost indefinitely, allowing a continuing release of antigen which, after the first day or two, continues at a low level for a very long period. There is experimental evidence that, at least as far as the course of antibody production is concerned, the effect of a single injection of antigen in F.C.A. can be reproduced by following up an initial injection of antigen alone with a prolonged series of small daily injections.

There are, however, qualitative as well as quantitative differences when F.C.A. is used and there are probably additional factors yet to be demonstrated experimentally. Raffel, for instance, has suggested that constituents of the tubercle bacillus may deflect the process of differentiation of the immuno-competent cell to allow the appearance of these qualities. This may be part of the truth but it does not fit easily into a selective approach.

The established qualitative differences after F.C.A. immunization which call for explanation are the emergence of immunocytes and antibody of higher avidity, and the appearance of unduly large amounts of 'nonspecific' immunoglobulin. Even in that statement there is a clear suggestion that there has been disorganization of normal homeostatic processes by this unnatural presentation of antigen. Under such circumstances it seems preferable to look for a change in the selective process rather than a meaningful deflection of differentiation.

The following suggestion may have experimentally testable consequences. The environment, either within the Freund granuloma or somewhere in the sequence of draining lymph nodes, is such that there is a movement of the normal balance between proliferative and destructive responses to immune stimulation toward the former. In terms of fig. 20 a proliferative response is obtainable when the product of avidity multiplied by effective concentration of antigen is higher than would be allowable in the absence of adjuvant. If, at the same time, antigen–immunocyte contact under F.C.A. conditions resulted in a much more effective or widespread nonspecific stimulus to proliferation of adjacent randomly distributed immunocytes, the observed conditions would be fulfilled.

Another point to be borne in mind is that long-continued release of very small amounts of antigen will induce a chronic proliferation of immunocytes under specific stimulation. Within these clones the possibility of (minor) somatic mutation, including movement to higher avidity, is always present. Again, the special local conditions might allow the proliferation of immunocytes with these secondarily avid receptors.

In one way or another, clones reactive with the deposited antigen may arise which would never become available under more physiological conditions. These may include clones with undue avidity for the antigen used and, when conditions are extreme, frankly pathogenic clones reacting with a variety of auto-antigens may be allowed to emerge.

## The behaviour of 'good' antigens

A comparatively recent development in academic immunology has been the study in relatively pure form of the antigens which for many years have been recognized by medical microbiologists as being particularly effective in producing high titre antibody. Typical examples of such 'good' antigens are bacterial flagella, influenza virus and certain bacteriophages. All these when used in the form of the natural organized material give a very rapid primary response, antibody being detectable in from one to three days. When, however, the same antigenic determinants are presented in molecular form in the two first examples there is a significantly longer period (7–10 days) before antibody is demonstrable. Similar experiments have not been recorded for bacteriophage. In both investigations the antibody produced by the molecular units was 7 S in character and, as far as currently incomplete data are available, only the molecular form, when given to neonates and continued at intervals, can provoke complete or almost complete tolerance.

In this response to a very good antigen, we may well be seeing the 'standard' form of the immune process developed to deal with pathogenic micro-organisms in the course of vertebrate evolution. The most direct interpretation may be as follows:

(*a*) Contact of the natural antigen with an uncommitted immunocyte, either directly or by way of a macrophage, will in a high proportion of instances result in stimulation of the cell to produce Ig M antibody, often of relatively low avidity.

(*b*) Contact of the monomer with immunocytes will commonly result in tolerance, especially when the animal is in the neonatal period.

(*c*) Taking up of the monomer by phagocytic reticulum in the lymphoid follicles provides a modified antigenic stimulus producing especially, but probably not exclusively, G-type antibody.

These results bring the findings into relatively close relationship to those observed with soluble antigens such as foreign $\gamma$-globulins. The behaviour of well centrifuged human $\gamma$-globulin in the mouse is discussed in chapter 10. It merely underlines the frequency with which contact of antigen with immunocyte must be lethal when large amounts of antigen are present and how much more effectively immunity is produced when the antigenic determinants are carried by organized or insoluble antigens. In this connection we can probably homologize denatured and partly aggregated foreign γ-globulin with the organized form of the microbial antigens.

# 10 Immunological unresponsiveness

Under the term 'immunological unresponsiveness' it is convenient to include all those phenomena in which, by one manipulation or another, an animal fails to produce the immune response regularly given by an equivalent untreated animal. It was first noted in regard to the failure of an animal to reject genetically distinct cells implanted neonatally, but much recent work has made use of simpler antigens. Unresponsiveness can be recognized by the use of any of the techniques used to demonstrate immune reactions and can be manifested as follows:

(a) Absent or diminished antibody production after a standard antigenic stimulus.

(b) Failure to reject a graft of allogeneic tissue.

(c) Failure to eliminate a viral infection, sometimes associated with abnormal absence of symptoms.

(d) Absence of the usual tissue reaction to challenge after a normally sensitizing injection.

Unresponsiveness or tolerance has now been recognized under a wide variety of conditions, the most important of which are:

(a) Physiological unresponsiveness resulting either:

    (i) from the presence of circulating antibody or

    (ii) from simple depletion of lymphocytes.

(b) Neonatal tolerance resulting:

    (i) from the implantation of living sources of antigen, allogeneic cells or certain viruses in embryonic or neonatal life or

    (ii) appropriate administration of non-living antigen in adequate amount at the neonatal period.

(c) Immune paralysis induced in postnatal life by doses of antigen, the amount needed varying widely with the type of antigen, the mode of administration and the species of animal used.

(d) The unresponsiveness of irradiated animals and of animals under the influence of immunosuppressive drugs.

Each of these will be discussed separately, but as a guiding line to the discussion it is important to remember that all immune processes are very complex and where there is an unexpected failure to produce antibody it is likely that more than one factor is at work. It should be remembered, too, that whenever purified or rare antigens are being used it is almost regular to find gross discrepancies in the antibody titre resulting from uniform immunization even in homozygous animals.

The approach to be adopted has necessarily been foreshadowed in earlier chapters and it may be convenient to summarize here the main points of the interpretation to be placed on the rather heterogeneous phenomena included within the general field of immunological unresponsiveness.

(*a*) Immunocytes under stimulation by antigen may respond positively or negatively, the decision being determined both by antigenic factors—concentration, sequence and mode of presentation—and by factors involving the immunocyte: maturity, local environment, presence of immunosuppressive drugs, recoverable damage from irradiation.

(*b*) Cells proliferating as plasma cell clones rapidly become irreversibly committed as end-cells and once their antibody-producing activity is ended they have no further existence.

(*c*) The presence of significant amounts of circulating antibody greatly diminishes or annuls any capacity of the corresponding antigen to act as a primary antigenic stimulus.

(*d*) Very small amounts of circulating antibody by opsonizing antigenic molecules or particles may greatly increase the primary response to a given antigen, particularly in young animals.

PHYSIOLOGICAL UNRESPONSIVENESS

*Inhibition by immunoglobulin G antibody*

The inhibitory effect of antibody, especially of 7 S Ig G type on primary-type stimulation, has been mentioned earlier, along with the probability that it may also have an effect in switching cells from Ig M to Ig G antibody production. One widely used experimental example of unresponsiveness is the repeated administration of sheep red cells to rats, starting immediately after birth. A detailed examination of this type of unresponsiveness by Rowley and Fitch in 1965 led them to conclude that the inhibitory effect of antibody and the end character of Ig G-producing plasma cells were the most important factors in the observed failure to produce more than small amounts of antibody.

This conclusion is based largely on the behaviour of plaque-forming cells in the spleen. These are cells capable of producing Ig M antibody against sheep red cells. If 10-week-old rats in three categories are tested (*a*) for serum antibody and (*b*) for the number of plaque-forming cells in the spleen, the results will be as shown in table 4.

TABLE 4. *Tolerance to sheep red cells in rats*

|  | Antibody | PFC per spleen |
| --- | --- | --- |
| Normal untreated | o | 40–100 |
| SRC twice weekly since birth | ± | 200–500 |
| 1 injection SRC at 7 weeks | + + | 100,000 ± |

SRC = Sheep red cells.     PFC = Plaque-forming cells.

The picture that emerges from such findings can be expressed as follows: There are probably several antigenic determinants in the sheep red cells but it is reasonable to assume that they behave in similar fashion and, for ease of discussion, refer to a single antigenic determinant, $S$. Immunocytes with the complementary $s$ pattern are probably present in the spleen from birth and are being steadily reinforced from the thymus. For the first two weeks they are highly susceptible to specific elimination by contact with $S$ but as the system matures a complex balanced condition develops. Antigen in various stages of processing by macrophages will be constantly present in the spleen and there will be a continuing entry of progenitor immunocytes. Some of these will be stimulated to produce Ig M antibody and become plaque-forming cells; more will be eliminated by antigenic stimulation at too high a level; others, for unknown reasons, will soon switch to Ig G production. Once a significant amount of Ig G antibody is present the number of plaque-forming cells falls rapidly and no fresh cells are recruited to this function.

Once an established production of Ig G antibody is under way the antibody will not interfere with its own production by cells

215

already in action. Depending on its concentration, however, it will shield a large proportion of any newly arising competent immunocytes from being committed by antigen to M → G antibody production. With cessation of antigen injections the content of both antigen and G antibody will steadily fall, a proportion of memory cells will persist and the situation will return to one equivalent to that which follows a primary immunizing injection. A new injection of antigen will give a secondary-type response. This is one form of partial tolerance, and similar factors will undoubtedly play a part in other manifestations of incomplete unresponsiveness.

*High-level and low-level unresponsiveness*

An example of different type may be taken from Mitchison's analysis of the response of mice to a wide range of doses of bovine serum albumin. He studied both the direct production of antibody and the subsequent response to a standard challenge by antigen (bovine serum albumin) in Freund's complete adjuvant. In this system there are two zones of partial tolerance or paralysis —a large dose (10–100 mg) given three times weekly gives a brief rise followed by a fall and partial paralysis. Very small doses give only a very slow, low-level response and leave the recipients strongly paralysed from an early stage. Interpretation of this will follow analogous lines. If we first consider progenitor immunocytes making initial contact with antigenic determinant, then at any level of antigen concentration there will be the two possibilities— of positive response on the one hand, committing the cell to proliferation and antibody production, and of negative response to damage and elimination, on the other.

The essential point to be remembered is that a cell once killed cannot thereafter be stimulated positively. On the other hand there is abundant evidence that even when an immunocyte has been set on the road to antibody production it can be overwhelmed by a large enough dose of antigen. The general impression from both the rat–sheep red cell and the mouse–bovine serum albumin systems is that, for progenitor immunocytes, primary contact with antigen present in more than minimal amount is always more prone to produce paralysis by elimination than immune commitment to antibody production or memory cell formation. There are undoubtedly circumstances, too, in which self-sterilizing antibody-producing end-cells are more readily produced than memory cells.

In an earlier section, the possible role of Mowbray's factor in facilitating specific elimination by contact with antigenic determinants in the thymus was mentioned. Mowbray's experiments have not been fully confirmed nor have they been extended by the author himself. It seems likely that there may be uncontrollable factors involved but it does appear that, under some conditions, injection of an $\alpha_2$-globulin fraction from bovine serum given before injection of antigen will prevent antibody production and leave at least a partial degree of unresponsiveness to the antigen used for some weeks afterwards. Search of various organ extracts for such an agent gave positive results only with the thymus.

It is biologically reasonable that there should be a hormonal factor which would raise the probability of destruction of immunocytes by specific antigenic contact and that this agent should be present in the thymus. Until the experimental results have been confirmed, however, no emphasis can be laid on this concept.

EMBRYONIC AND NEONATAL TOLERANCE

*Natural twin tolerance*

It is of the essence of the clonal selection approach that a Darwinian process of proliferation, mutation and selection is part of the natural history of the lymphoid cells of the body. To a large extent this concept is derived from a mass of experimental data that has been obtained by the use of antigens and of manipulations that are remote from any contingency that could be relevant to evolutionary development. The first development of ideas of self and not-self in an immunological connotation came, however, from a consideration of natural phenomena—the occurrence of twin chimeras in cattle and the persisting tolerated infection seen with lymphocytic choriomeningitis virus in an endemically infected population of mice.

When twin calf embryos arise each from a separately fertilized ovum, their blood groups will usually be genetically distinguishable. If they had developed separately calf *A*, when inoculated a few months after birth with blood from calf *B*, would produce agglutinins and haemolysins against *B* blood cells and any *B* cells

217

from the inoculum which might have persisted would be destroyed. In practice, however, the twin placentas *A* and *B* fuse at an early stage and there is a mixing of all the circulating cells and antigens of the two calves. The calves are born each with the same mixture of two genetically distinct types of red cell and, in general, the proportion of the two remains reasonably constant through life. Occasionally, progressive changes in proportion may occur but this is probably for other than immunological reasons.

The fact that the genetically 'wrong' cells persist without provoking immune response and, in fact, behave precisely as the animal's 'own' cells, provides the main evidence for the view—central to clonal selection theory—that normal unresponsiveness to the body's own potential antigens is not a directly genetic character but is generated by processes taking place during the embryonic phase and persisting at least till early adult life.

In all examples of persisting tolerance, whether to normal or chimeric antigens, it appears to be mandatory that the antigen should persist. Detailed study of the allotypes of human globulin and their antibodies has shown that when a mother is Gm($a^+$) and her child is Gm($a^-$), a considerable proportion of the children produce anti-Gm(a) antibody. This indicates that tolerance induced by the presence of the maternal Gm($a^+$) globulin in the early months of life does not persist long after the elimination of the maternal antigen. There is presumably a delicate balance of probability toward the end of the first year as to whether the remaining traces of maternal antigen will or will not find cells capable of being positively stimulated.

As a control to this result we may mention the findings in a large series of children who had had multiple transfusions necessarily resulting in the entry of $\gamma$-globulin of a variety of allotypes. A high proportion produced antibody, but never against an antigen present on their own $\gamma$-globulin. Tolerance to a persisting antigen obviously lasts indefinitely.

### Intrinsic immunological tolerance

Since Owen's demonstration of the double blood groups found in certain twin calves it has been implicit that an active process of similar quality must be responsible for the tolerance of the body for its normal components. I stated this more explicitly in 1954 by saying that 'recognition of self is something that needs to be learnt and is not an inherent genetic quality of the organism'.

It is only in 1968 that final unequivocal proof of that statement has been provided as a result of the work of Beatrice Mintz on the production of 'allophenic mice'. She has shown that when fertilized ova from mice are at the stage of 10 or 12 cells—early blastomere—they can be manipulated *in vitro* so that two such blastomeres can fuse and reorganize to form a single larger blastomere. This can then be transferred to the uterus of a suitably receptive foster-mother and with correct technique it will develop to a healthy mouse containing components of both the parental strains. This is just as practical with mice of different coat colour or of different major histocompatibility antigens as with syngeneic embryos.

The fact that an allophenic mouse which is a composite of two distinct H 2 types can be normally viable and fertile shows at once complete mutual tolerance. Whatever is incorporated into the structure of a new mammalian embryo and can persist, whether by hybridization or by this type of laboratory manipulation, will become completely tolerated by the organism as it develops. In the allophenic mice there is no suggestion of a choice between one or other histocompatibility antigen; both are demonstrable by appropriate experiment.

It has been rightly said that this is *intrinsic* immunological tolerance precisely equivalent to natural tolerance. It must replace the various demonstrations of *acquired* immunological tolerance as the experimental basis of the concept of the natural tolerance of self-components.

*Persisting tolerated infection*
The existence of persisting tolerated infection by lymphocytic choriomeningitis virus in mice has been extensively studied and the phenomena observed are of great interest. In all probability, serum hepatitis virus behaves very similarly in man and there is obvious relevance to the persistence of rubella virus from early foetal infection until after birth. In the context of theoretical immunology, however, these phenomena are not specially revealing and may be passed over with this brief mention.

## Homograft tolerance in mice

Since Medawar's team showed in 1953 that tolerance to homografts could be induced experimentally in mice, a very large amount of work has been published on this theme. The basic experiment is to allow a mouse of strain $A$ to accept a skin graft from strain $B$, tolerance being induced by implantation of $B$ cells of suitable type into newborn $A$-strain mice. In practice, for reasons to be mentioned later, the actual strains to be used for such an experiment must be properly chosen and an adequate dose of $B$ cells, preferably from bone-marrow or neonatal spleen rather than from adult spleen, given intravenously on the first day of life. When 'runt disease', that is, graft-versus-host reaction, is avoided by attention to these points, long-lasting tolerance with complete retention of the homograft is the rule. This tolerance is specific: an $A$ mouse tolerant to $B$ will reject skin from a third strain as readily as a normal mouse.

From experiments of this type, two main generalizations have emerged. The first is the relation between age and the dose of cells needed to produce tolerance. A very small dose given immediately after birth will sensitize or show no effect. An adequate dose is needed to give satisfactory tolerance, and a rapid increase in the number of cells needed takes place with each day of life. For mice over a week old it is usually impossible to induce tolerance by a single injection. A sufficient degree of unresponsiveness to allow retention of homografts can be induced in older mice but only by repeated intravenous injections of bone-marrow or spleen cells.

The second generalization is that persisting tolerance is always associated with colonization of lymphoid tissue by donor cells, i.e. chimerism. This does not necessarily imply that tolerance may not persist for a variable time after elimination of donor cells or that every chimera is fully tolerant to all tissues of the donor line. Split tolerance can be very readily shown by giving small numbers of spleen cells from male strain $B$ to neonatal females of strain $A$. Such mice become almost 100 per cent tolerant to grafts of male $A$ skin but a much smaller proportion will accept $B$ skin (either male or female) if, as will usually be the case, there is a major histo-compatibility difference between $A$ and $B$. Partial tolerance and split tolerance will presumably depend on the extent to which antigen is liberated both by the tolerigenic implant of spleen or other cells and by the test grafts, and on the still incompletely understood intensity of the destructive action of competent immunocytes against the graft. It is well known that a partially tolerant mouse may retain an already established homograft but reject a new transplant of skin of the same allogeneic character. The situation under such circumstances may be visualized as follows.

Relatively small numbers of *B* cells implanted in thymus and spleen, and the established *B* skin graft, are liberating small concentrations of the relevant histocompatibility antigens. Any newly differentiated *A* immunocytes which are highly avid, that is, closely complementary in pattern to the *B* histocompatibility antigens, will be destroyed on contact. Other immunocytes capable of less avid reaction will escape from the thymus and some will reach peripheral lymphoid tissues. At the time the new test homografts are made there will be, by hypothesis, no avid fully complementary immunocytes to any of the specific antigens of the graft. There are, however, some immunocytes capable of less avid reaction with the histocompatibility antigens. In the case of a major histocompatibility difference between graft and host, the proliferation of weakly reactive immunocytes will be adequate to bring enough of them into close enough contact with vulnerable graft cells to produce a mildly damaging effect. This may include the allogeneic inhibition effect described by Hellström and other workers at the Karolinska Institute. The specific tissues of an *established* graft of *B* skin are separated from the bloodstream by intact vascular endothelium and the graft is in a much less vulnerable position than one in the process of attachment.

Split tolerance in which an H2 difference leads to rejection, but a simple Y (male) antigen difference is tolerated, presumably depends either on the minor character of the antigenic difference in the antigen or on the small amount of it liberated from the tissue.

## Immunological capacity in the foetus

During the early years of work on perinatal tolerance it was accepted almost universally that the embryonic animal, mammal or bird was incapable of producing antibody or mounting any other type of immune response. It had, however, been known for many years that plasma cells were conspicuous in foetuses infected with congenital syphilis and it was shown experimentally in Australia as early as 1953 that a foetal sheep could reject a skin homograft in typical fashion. More recently, extensive work on foetal sheep at various stages of development has shown a curious progressive development of immune capacity. When a suitable bacteriophage is used as antigen, a foetal lamb will produce antibody from the 60-day stage onward. This is in fact the earliest stage at which it was found possible to give the immunizing injections *in utero*. On the other hand, diphtheria toxoid and *Salmonella typhi* vaccine are not immunogenic until at least 6 weeks after birth.

The lamb is born, proverbially well advanced in development compared with a mouse or rat, after a gestation period of 150 days. It is of interest, therefore, that there is a period before the 75th day of pregnancy when a homograft can be accepted and retained indefinitely. From 85 days onward, however, it is regularly rejected.

### Graft-versus-host reaction

The other main point of interest in neonatal tolerance is in regard to graft-versus-host reaction or its absence. In a large number of combinations, if strain $A$ is injected neonatally with spleen cells from strain $B$, the host animal will develop a characteristic runt disease with failure to gain weight, roughened fur and chronic diarrhoea. At an early stage there is enlargement of the spleen followed later by general atrophy of lymphoid tissues. The syndrome is ascribed to the proliferation of implanted $B$ immunocytes capable of reacting with histocompatibility antigens of the host strain $A$ and, in the process, damaging host lymphocytes and other cells. There are also combinations where adult spleen cells can be used to confer tolerance on neonatal hosts without risk of subsequent runting and many more combinations in which adult bone-marrow or embryonic spleen and liver cells can produce effective neonatal tolerance without subsequent harm.

The same contrasting behaviour with different combinations can be seen even more strikingly with mice or rats of different strains joined parabiotically. This is usually an unstable situation and the commonest result is for one partner to develop an immunological ascendancy over the other. If they are then separated the ascendant one will survive; the other will die of a graft-versus-host reaction, presumably of a highly complex pathogenesis. On occasion, however, both partners will survive and the combination remain viable indefinitely.

In these two related conditions, the tolerated chimerism without runt disease and the mutually tolerated parabiosis, one must postulate a double process by which both the donor immunocytes active against host antigens and host immunocytes active against donor antigens are eliminated. Such a process is of course not always successful. When the double elimination does take place the process is probably analogous to the unilateral tolerance obtainable in adult animals by the continuing intravenous infusion of antigen. Despite a number of theoretical formulations calling for the existence of 'tolerant cells' and one or two claims that they have been experimentally demonstrated, the established experimental results are compatible with the straightforward interpretation that an animal or a population of immunocytes is tolerant because it *lacks* immunocytes adequately reactive against the tolerated antigen.

## The role of the thymus

In chapter 3 the 'censorship' role of the thymus in preventing the development of immunocytes carrying patterns reactive against accessible self-antigens was postulated. If this is true, the thymus must also play a significant part in the development of neonatal experimental tolerance. The available evidence does in fact support such an approach. It is well known that when antigen is administered intravenously to adult animals or even to laboratory rodents more than a week or two old, very little antigen lodges in the thymus and practically none in the cortical zone. A number of authors have in fact spoken of a 'blood-thymus barrier'.

In neonatal rabbits, rats and mice it has now been shown that antigen labelled with radioisotope or fluorescent dye can pass freely into the substance of the thymic cortex. If, as we contend, to produce tolerance antigen must react perhaps more than once with *all* immunocytes in the body which bear an appropriate pattern, the chief advantages of using the neonate are, first, because the number of cells of any one pattern must be small, and second, because all immunocytes, particularly those newly differentiated in the thymus, are accessible to adequate concentrations.

Where the matter has been specifically studied, most animals which are fully tolerant to homografts have been found to be thymic chimeras. We should be on safe ground in postulating that tolerance to homograft needs for its establishment the development of persisting lymphoid cell chimerism involving the thymus. If this is the case, the effectiveness of the tolerance induced will presumably depend (*a*) on the relative proportion of donor cells present in the thymus and their continual reinforcement by stem cells of donor origin from the bone-marrow or elsewhere, and (*b*) on the concentration of the relevant antigens being liberated or otherwise made available within the thymus.

## UNRESPONSIVENESS ASSOCIATED WITH LYMPHOCYTE DELETION

### Drainage of the thoracic duct

One of the most unequivocal of immunological experiments is to deplete a rat of lymphocytes by chronic drainage of the thoracic duct. With this technique, Gowans not only established the mobility of the lymphocyte as between circulation and lymphoid tissue in the rat but showed that, with an adequate degree of depletion, power was lost to produce a primary antibody response to a normally effective dose of sheep red cell antigen or to reject a homograft from an animal differing only by minor histocompatibility antigens. These capacities could be restored by infusion of adequate numbers of lymphocytes from normal rats.

There were, however, considerable limitations to this approach. If a rat had been previously immunized with sheep red cells and was then heavily depleted of lymphocytes, it was still capable of giving a secondary-type antibody response to a new antigenic challenge. When a depleted rat was tested with a skin homograft from a rat with a major histocompatibility difference, the graft was rejected virtually as rapidly as from an undepleted rat of the same stock.

It is probable that at least part of the immunosuppressive effect of X-irradiation or corticosteroid administration is referable simply to the associated destruction of lymphocytes.

## Recapitulation of lymphocytic function

The lymphocyte is unique among body cells in its susceptibility to destruction by X-irradiation and cytotoxic drugs and both types of agent have been very extensively used in immunological investigation. Before summarizing the results of this work it is worth while restating briefly our earlier conclusions on the natural history of the lymphocyte in mammals. Under the morphological guise of the small lymphocyte there is a wide range of cells of varied potentialities, but all can be regarded as essentially mobile carriers of information. There are probably some 'lymphocytes' that have no concern with immunology, but from the immunological standpoint we can be reasonably certain that circulating lymphocytes include (*a*) stem cells of bone-marrow origin, (*b*) progenitor immunocytes as yet without experience of specific contact and (*c*) committed immunocytes with potentialities to take part in delayed hypersensitivity reactions or to produce antibody.

Only immunocytes in the small lymphocyte form are immediately relevant here. They are serving as carriers of information and each will be stimulated to function and sometimes to proliferation only when it meets the appropriate circumstance—usually antigen contact—calling for its specialized function. To make use of such a generalized mechanism of information transport there are three basic requirements. There must be a constant production of many cells of many types; the cells must be freely mobile, freely mixing and widely dispersed through the body and, finally, in order to allow a fully flexible response to transient needs and therefore a continuing rejuvenation of the population, there needs to be a high rate of destruction. This high turnover probably serves another important purpose. When there is need for proliferation of one functional type of cell there is a parallel requirement that the necessary building blocks for nucleic acid and protein synthesis should be accessible. As many have pointed out for other reasons, the lymphocyte (and the other leucocytes) can serve on autolysis as a very effective mobile store of nucleotide and amino acid.

*Lymphocyte deletion by irradiation and drugs*

The high vulnerability of the lymphocyte is essential to its proper functioning. At the physiological level it is vulnerable to an increase in concentration of adrenal corticosteroids and to contact with specific antigen. It is probably of the essence of the problem that, under different circumstances, hydrocortisone is essential for antibody production and specific antigenic contact provokes proliferation and antibody production. As well as by such physiological agents, lymphocyte levels can be lowered by a wide range of damaging stimuli—from acute infection to X-irradiation or the injection of toxic materials. It is very frequently an open question whether the action of the noxious agent is directly on the lymphocytes or is mediated by the liberation of corticosteroids.

When an animal is heavily but sublethally irradiated and given an immunizing injection approximately 24 hours later, there is often complete failure of antibody response. If very large doses of antigen are given, tolerance may be induced that is specific for the antigen. To obtain unequivocal results, the quantitative conditions seem to be critical. Usually there is a definite reduction in antibody response with a prolongation of the period during which Ig M antibody persists and sometimes loss of 'memory'. Irradiation two to three days *after* the immunizing injection will generally increase the antibody titre obtained and there is little or no effect on the secondary response in a primed animal. With some experimental models X-irradiation four or five weeks after the primary immunization will prevent the development of 'memory'.

In transplantation studies, the simplest means of producing tolerance in adult mice is to give mice of strain $A$ a lethal dose of irradiation and to save them by intravenous injection of bone-marrow cells from strain $B$. Such mice will accept skin homografts of $B$ and retain them indefinitely. In some instances it is reported that these chimeras will reject a syngeneic ($A$) graft but in other combinations the normal tolerance to the isologous skin graft is retained as well as the acquired tolerance.

What are essentially very similar results are obtained with immunosuppressive drugs such as 6-mercaptopurine, azothioprine ('Imuran'), and cyclophosphamide. It is easy enough to diminish antibody production, particularly if the drug is continued at the highest tolerated dose over the period of immunization, but persisting specific tolerance requires strict conditions depending on the nature of the antigen and the species of animal as well as the type and dosage of immunosuppressive drug. All cytotoxic drugs of this sort have a strong lympholytic effect, particularly evident in the thymus.

Another method of inducing tolerance (which has not been extensively studied) is to induce a temporary pyridoxine deficiency either by dietary restriction or by administration of deoxypyridoxine. Under these circumstances, administration of a dose of allogeneic spleen cells can induce specific tolerance which persists after the animal is returned to a complete diet.

### THE REQUIREMENTS FOR TISSUE AND ORGAN TRANSPLANTATION

From the human and medical point of view, an important current objective in immunology is to achieve effective and long-lasting transplantation of a healthy human organ, usually a kidney, to replace organs functionally destroyed by disease or injury. In the early 1950s surgical technique had advanced to the level where effective junction of arteries, veins and ureter could be done with confidence, blood dialysis by 'artificial kidney' was developing well and a number of kidney transplants were attempted. The results were as would be expected—some initial kidney function followed by immunological rejection. In 1954 an opportunity arose to transplant a kidney donated by an identical twin of the patient. Again, as might also have been expected, the operation was successful and the patient survived for several years.

The problem was clearly to find a means by which immune tolerance could be induced toward the foreign tissue or, alternatively, to find donors who, from the point of view of relevant histocompatibility antigens, were the equivalent of an identical twin of the proposed recipient. The current partial solution to the problem is the use of immunosuppressive drugs of which a derivative of 6-mercaptopurine, azothioprine (or 'Imuran') is the one that has so far proved the most effective.

Research on kidney transplantation has been intensive and widespread but few of the results have much bearing on the general problems of tolerance. Since the use of azothioprine with or without ancillary drugs including corticosteroids and actinomycin D, the proportion of patients surviving one year or more has been rising in a gratifying fashion. This is probably due to an increasing competence of surgical technique and biochemical control plus empirical experience in the handling of immunosuppressive drugs to deal with crises of rejection. There are still very few individuals bearing a transplanted kidney who have been able to maintain normal kidney function without the continuing use of immunosuppressives. Even where there has been a complete technical and immunological success with transplantation from an identical twin donor, the long-term results have not been happy. At least five recipients have died, after successful transplantation, from glomerulonephritis of much the same type as their original disease. Here the problem goes beyond the physiological and must be discussed with autoimmune disease.

*The action of 6-mercaptopurine*

The first immunosuppressive drug to attract the attention of experimenters was 6-mercaptopurine (6-MP) whose action was reported by Schwartz and Dameshek in 1959. More laboratory work has been reported with this drug than with any other and, as indicated above, its derivative azothioprine, commonly used in human transplantations, has a similar or identical activity but is less toxic in man.

In order to obtain suppression of antibody production and continuing tolerance to a given antigen by the use of 6-MP, several requirements must be fulfilled. In the first place the dose of the drug must be large and close to the toxic level for the species being used. The drug must be continued for some days after the injection of antigen and good results are not obtained unless large doses of antigen are given. The standard experimental animal has been the rabbit but mice can also be used; guinea-pigs do not develop tolerance. The tolerance, after such treatment, is specific and can be demonstrated either in relation to antibody production or homograft rejection. The secondary response in rabbits previously immunized can also be blocked by 6-MP with specific tolerance to a third immunization.

The findings are most readily brought together on the assumption that 6-MP, a mitotic poison, acts more or less selectively on immunocytes stimulated to proliferate by specific antigenic contact, destroying such cells and so eliminating or greatly reducing the size of the clones involved. As in all situations inducing toler-

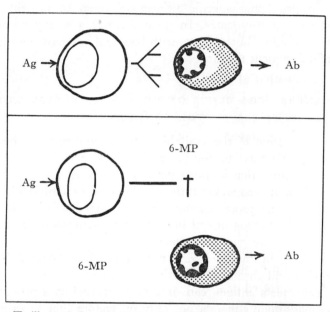

Fig. 24. To illustrate the antimitotic effect of 6-mercaptopurine on a stimulated immunocyte and its lack of action on mature plasma cells.

ance, it is necessary that *all* reactive immunocytes must be exposed simultaneously to antigen and mitotic poison. This is the explanation for the need of high concentrations of both antigen and drug persisting over some days.

It is reported that if 6-MP is given to rabbits for a week and 5 days later a dose of antigen (bovine γ-globulin), there is a strongly marked *increase* in the response compared to that in untreated rabbits. This is reasonably ascribed to secondary effects from the destruction of proliferating immunocytes reacting to 'normal' antigens at the time of treatment. This provides an increased

*Lebensraum* for stimulated immunocytes to proliferate and perhaps a higher local concentration of products of nucleic acid and protein breakdown which will facilitate synthetic activities in proliferating cells.

Of other drugs with immunosuppressive action, cyclophosphamide, a modified nitrogen mustard (and, again, a mitotic poison) is the most interesting. In guinea-pigs it is an effective agent in producing specific unresponsiveness to proteins and hapten-protein conjugates. In mice there is a strong immunosuppressive effect but at least, with bacterial antigens, no indication of subsequent tolerance.

GENERAL COMMENTS ON UNRESPONSIVENESS AND
TOLERANCE

A re-examination of the available experimental material bearing on tolerance seems to reinforce the basic postulate of clonal selection theory: that immune pattern is of genetic origin arising by a random process and that the development of natural tolerance or experimentally produced unresponsiveness is due to the complete or partial elimination of cell lines carrying patterns reactive with the antigenic determinants concerned.

Everything in immunological phenomena is soft-edged. There are no clear-cut absolutes—whether or not a given reaction occurs when cell meets antigen can only be expressed as a probability whose magnitude depends not only on factors that in principle might be expressed definitively, such as the structure of the antigenic determinant and its concentration in the environment of the cell concerned, but on many others that are virtually as complex at a cellular level as the genetic and ecological factors which determine the course of evolution at the macro-level.

In the most general terms, for effective inhibition of capacity to respond to a specific antigen, a situation must arise or be contrived by which sufficient antigen can make effective contact with *all* specifically reactive immunocytes which can have descendants. This means that all cells must either be set into irreversible antibody production as plasma cells or be inhibited or destroyed. Since there is good evidence that lymphocytes may remain inert but potentially capable of mitosis for long periods (several years in man) this may well require that antigen remains present for a long time in order to catch some of the cells in a vulnerable phase.

On the other hand, antibody production can theoretically result from contact of a single immunocyte with antigen. There are, in fact, several observations to suggest that populations of antibody molecules may be derived from a very small number of progenitor cells. In the following chapter the more definite evidence in regard to the origin of some pathological antibodies from a single clone will be discussed. The fact that a major population of antibody molecules may be produced by a single clone, and therefore in the last analysis from a single cell, does not of course mean that the appearance of an immunocyte with pattern $a$ and the existence of antigen $A$ in the body will necessarily result in the production of anti-$A$ antibody in demonstrable amount. A somewhat similar evolutionary situation arises when a mutant strain of a pathogenic micro-organism arises. It is not enough that the new strain should have qualities which would ensure the survival and wide dissemination of the strain if it ever became well established in a population of the susceptible species. It must also survive a thousand possibilities of extermination before it can succeed in that initial establishment of a viable 'beach-head' in the host species.

An indication of the importance of such factors in immunology can be drawn from the almost universal use of Freund's adjuvant to ensure a satisfactory antibody response. Whenever a group of rabbits or mice are injected with any 'rather poor' antigen, bovine serum albumin for instance, a considerable proportion of the animals will give feeble or negative responses. If an exactly equivalent group is given the same antigen with adjuvant of a number of types, a significantly larger proportion will give satisfactory antibody responses and the mean titre will be higher. If one can judge from the physical heterogeneity of all antibody populations that have been studied, in contrast to the homogeneity of almost all myeloma proteins from a single patient, antibody production usually occurs only when a fairly commonly occurring set of immune patterns can react to varying degrees with the antigenic determinant in question.

# 11 The integration and deployment of immune responses

It is an essential background to this book that experimental immunology, like any other special branch of biology, is justified (a) by providing material that can be integrated into the developing picture of the nature of living function in its evolutionary setting and (b) by providing information relevant to experienced human needs. The great majority of currently reported experiments concern highly artificial situations devised to supply valid information about individual facets of the natural phenomena. In many instances it would be more correct to say that the experiments are concerned with individual facets of laboratory phenomena encountered at one, two or more removes from any of the natural, that is, evolutionarily significant phenomena. It is therefore salutary to try to integrate the information obtained from the laboratories into an interpretation of the natural phenomena of defence or, as I would prefer, of the maintenance of bodily integrity.

Even for the purposes of such a discussion the situation must be simplified and three conditions only need be dealt with.

The first is a generalized infectious disease transmitted by a mosquito-borne virus—a contingency which must certainly have been relevant to the evolution of mammals and birds since their first appearance. Yellow fever can be taken as a prototype.

The second has been equally universal throughout the history of vertebrates—superficial, non-lethal trauma involving haemorrhage, exposure of normally sheltered tissues and infection by environmental micro-organisms. The third example, haemolytic disease of the newborn, represents an immunological accident whose more or less frequent occurrence is an inevitable consequence of the process of placental mammalian reproduction. It can be shown with reasonable certainty to have had an important influence on human population genetics.

The final example is an examination of the phenomena of delayed hypersensitivity from an unorthodox angle. The view adopted is that both delayed hypersensitivity reactions and the rejection of skin homografts are laboratory phenomena based on the natural processes of immunological surveillance by which any groups of cells with aberrant surface antigens can be removed from the body.

## HUMAN ARBOVIRUS INFECTION

The process of infection in a nonimmune individual is initiated by the injection of a small amount of virus in the mosquito's salivary fluid into the circulation. Virus particles are taken up by macrophages, particularly by the Kupffer's cells of the liver. The ability of the virus to multiply in these cells of its primary lodgment is probably a crucial step in the process. In mice it has been shown that inherited resistance of a mouse strain to one group of arboviruses could be correlated with failure of the virus to proliferate in tissue culture of macrophages of the resistant line in contrast to those from susceptible mouse strains. Once significant numbers of virus particles are being liberated from primarily infected cells, a rising viraemia will ensure infection of all susceptible cells in direct contact with the blood and, from these, secondary infection of other susceptible cells such as those of the liver in the case of yellow fever.

Large amounts of particulate and soluble antigens are present and, if the infection is to be overcome, any immunocytes capable of responding to the antigens will be stimulated to proliferate and produce antibody. This is initially of M type, but if animal experiments give an appropriate guide there is a rapid appearance of G- and probably A-type antibodies. Since it is easy to show that a mixture of virus with serum containing antibody has greatly reduced infectivity for the susceptible test system, whether this be mouse, chick embryo or tissue culture, it is easy to think of the process as a simple neutralization. Analysis of the situation, however, indicates that the effect of antibody must be a complex one interfering with one or more of the processes by which infection of the cell and viral proliferation occurs. Whatever the process, recovery from infection is associated with the appearance of 'neutralizing' antibody and substantial immunity.

The most characteristic feature of immunity to yellow fever in man is its extremely long duration, not only in the sense of immunity against symptomatic reinfection, but equally of very long (more than fifty years) persistence of measurable antibody in the blood despite residence in areas quite free of the virus. This has the implication that a large population of immunocytes developed during the infection including an exceptionally large number of memory cells. This in its turn implies that despite relatively large concentrations of viral antigen the concentration was not great enough (in recovering patients) to produce unresponsiveness. It is highly probable in fact that the concentration of antigen per kilogram was far smaller than the concentration of bland antigens needed to produce unresponsiveness in laboratory animals.

A second implication of interest from the theoretical angle is that in immunized individuals living in non-endemic areas there must be recurrent or continuous nonspecific stimulation of immunocytes to produce plasma cell derivatives which will maintain a constant or very slowly falling level of circulating G-type antibody. At one stage this was often ascribed to the indefinite persistence of antigen-producing virus in the body, a hypothesis which nowadays appears neither necessary nor credible.

There are still many points to be worked out in regard to how the presence of an immune response destroys the activity of virus and allows its elimination from the infected organism. Work that has been done with standard G-type antibody shows that union of antibody with virus receptors is reversible and is not in itself lethal to the virus. In view of the much firmer attachment of M-type antibody, each molecule of which has twelve potential combining sites, it is conceivable that M-type viral antibody may produce an essentially irreversible union which with subsequent action of complement might be effectively lethal for the virus. The 'neutralizing' activity of G antibody presumably depends on impeding attachment of the virion to any necessary cell receptors and distorting the initial intracellular changes. Simple thermal inactivation and the action of intracellular proteases and nucleases would then return the viral components to the metabolic pool. Damaged cells would be dealt with just as those damaged by trauma.

Workers with arboviruses have uncovered several immunological phenomena of interest concerned with matters other than the quelling of an established infection. From the practical point of view the most important is the long-lasting immunity that follows either a natural infection or appropriate vaccination, usually with an attenuated living virus. There is no reason to believe that the mechanism of this protective action is other than what is seen in the closing stages of primary infection, rendered much more effective by the fact that the whole process can be brought to bear on the infection at its earliest stage.

The main epidemiological feature of yellow fever in the days before its mode of transmission was known was its concentration of attack, very frequently lethal, on the newcomer while the indigene went unscathed. Every European army brought to the West Indies in the Napoleonic wars suffered disastrously. Men and women who had been born and raised in the endemic area were resistant, irrespective of their skin colour. For obvious reasons none of the regions studied for the distribution of yellow fever antibody when techniques became available between 1930 and 1940 were still in the hyperendemic state of West Indian cities in the eighteenth and nineteenth centuries. The results of these surveys, however, allow a reasonable interpretation of the former conditions. Infection-carrying mosquitoes were almost constantly present and children would be infected at an early age. In many there would be enough maternal antibody still in circulation to allow opsonization of the virus and a more effective immune response. Even in wholly unprotected children, however, a primary attack was milder than in an adult infected by the same strain; just as is known to be the case with polio-virus infection.

In yellow fever then—taken as a model of severe virus infection —the biological significance of the immune mechanism must be related not only to the capacity of a primary infection to be followed by resistance to the disease but equally to the fact that first infections in childhood are on the average much less likely to be lethal than those experienced for the first time in adult life.

It may be wise to stress that whether or not a primary infection with a given virus is lethal is much more related to metabolic and genetic factors than to the existence of the mechanism of adaptive immunity. The classically lethal virus disease is myxomatosis of rabbits, which in a virgin population will kill about 99·7 per cent of animals infected. Even with such a strain of virus, however, a proportion of rabbits will recover if they are maintained in a hot environment. Under conditions of the natural spread of myxomatosis in Australia during the 1950s there was a high survival premium on any development of genetically based resistance. Within ten years there was a highly significant increased proportion of survivors, given a standard inoculum of virus and a recrudescence of rabbit numbers in the field.

The evolutionary significance of adaptive immunity in this general field is presumably related to the possibility of transfer of maternal immunity and to a less extent the maintenance of post-infection immunity.

### TRAUMA AND LOCAL INFECTION

Twenty years ago there was much interest in the question as to whether hypersensitivity as expressed in the course of tuberculosis infection was relevant to the process by which infection was overcome. No decision on the point was ever reached, and current interest in delayed hypersensitivity is almost wholly at the experimental level, with the significance of the cytological changes in the site of the reaction the most frequent theme for study. Teleological or evolutionary considerations are out of fashion.

To introduce a discussion of the integration and effect of immunological reactions involving changes in the micro-circulation and movement of cells from the blood into tissues, one must have some natural phenomenon as defined above to maintain contact with biological realities. Even a very superficial consideration of significant evolutionary conditions makes one choice obvious— non-lethal trauma. In the wild, any serious trauma involving broken limb bones or severe haemorrhage is necessarily lethal through leaving the victim defenceless against predators. Minor trauma from accident or attack is frequent and the process of haemostasis, minimization of infection and repair is rapid and effective. The effective co-ordination of these three processes must have been a major consideration for survival and, if there is an evolutionary angle to immunological reactions involving blood vessels and cell migration, this is where it will be found.

236

It is a very interesting surgical principle that in the immediate neighbourhood of sites particularly prone to minor trauma such as the teeth, the anus and the skin generally, surgical procedures can be successfully carried through with a minimal degree of aseptic care. Any opening of the cranial cavity, peritoneum or major joint cavities—regions exposed only by lethal trauma in nature—must be made with far greater circumspection.

Many non-immunological factors are concerned in such local defence. It may be almost wholly a mobilization of nonspecific defence in the form of polymorphonuclears and monocytes assisted by responsive action of the micro-circulation in the area. Modern approaches to the nature of delayed hypersensitivity are more sophisticated than they were (see pp. 245–54) but as one who many years ago was intensely interested in staphylococcal infections, I still feel that one aspect of delayed hypersensitivity may be concerned with the facilitation of defence against traumatic entry of common saprophytes—semi-pathogens on the skin or in the body cavity. It is of the essence of my picture of the evolution of immunity that on top of a nonspecific capacity to resist invasion by pathogens which has been necessary since many-celled animals first arose, there developed a system of internal surveillance to recognize much finer differences between self and not-self. The two systems developed together to evolve into the complex and highly effective defence system of the mammal in which the mechanisms of adaptive immunity can be applied to increase the effectiveness of the older system.

There is a certain primitiveness about delayed hypersensitivity and it may have been in relation to surface infections, internal and external, that adaptive immunity was first linked to nonspecific defence processes. *Staphylococcus aureus* infection in man can be used as an example. The staphylococcus is a robust and ubiquitous potential pathogen. If one follows the course of a particular staphylococcal antibody ($\alpha$-antitoxin) at various ages, the curve falls to a minimum at 2–3 months and is rising at 1 year, average titres reaching a level of about 20 which is maintained for the rest of life. Infection clearly begins very early. It is also on record that young infants show no skin reaction to staphylococcal filtrates but increasing proportions react with age.

In many ways these staphylococcal findings belong to an earlier age of immunology but they suggest that low-grade exposure to bacterial products could gradually build up populations of immunocytes which could play a part in ensuring that entry of such bacteria through trauma is more rapidly and effectively contained. In part, this increased effectiveness may be associated with an accelerated accumulation of lymphocytes and monocytes, following the first impact of polymorphonuclear leucocytes, and mediated by what is essentially the mechanism of delayed hypersensitivity.

There is much to be said for Spector's point of view that whenever local irritation induces increased capillary or venular dilatation there is an escape of all types of cell into the tissues along with fluid from the plasma. In general, such cells are rapidly removed by lymph drainage unless there is a positive reason for keeping them *in situ*, such as:

(*a*) the presence of a chemotactic substance attracting leucocytes toward it;

(*b*) the existence in the tissues of an antigen which can stimulate the cell in question to become 'sticky';

(*c*) endotoxins or similar agents causing cell death.

The second of these is the one in which we are chiefly interested and it will be discussed in a later section where we are dealing directly with the significance of delayed hypersensitivity. Here we need only make the point that if the delayed hypersensitivity reaction has any direct teleological relation to infection it is to allow accelerated movement of functionally important cells from blood to tissues. If there is associated circulating antibody this also will pass more readily to the site of infection.

Before leaving the topic it is perhaps advisable to mention the very familiar phenomenon of accelerated and immune reactions to jennerian vaccination against smallpox. Here is a perfectly typical delayed hypersensitivity reaction preceding a mild or abortive infectious lesion which rapidly fades.

This approach to the nature of delayed hypersensitivity via traumatic infections is admittedly very incomplete. At most it is only of secondary importance and it may be that adaptive immunity has been of no real evolutionary significance at this level.

Where local trauma has been relevant to the course of

mammalian (or vertebrate) evolution has been in the field of haemostasis and repair. These fall outside the scope of this book, but it is relevant to discuss the nature of micro-circulatory adjustments both in relation to traumatic damage and to various aspects of hypersensitive reactions.

*Changes in the micro-circulation related to immune reactions*

An important aspect of many immune processes is seen in the part played by the vascular system in the phenomena of inflammation, acute and delayed hypersensitivity, passive cutaneous anaphylaxis, and the various local manifestations of allergic and autoimmune disease. In general the change from normal involves capillary or venular dilatation and increased permeability, surface changes in endothelial cells leading to stickiness for leucocytes, and migration from the vessel into the tissues by polymorphonuclears, monocytes and lymphocytes. A feature of special interest to the immunologist is that these reactions are to a considerable extent inhibited or reversed by appropriate concentrations of hydrocortisone or equivalent drugs.

There is no unanimity in regard to the detailed mechanism of these inflammatory and related responses but all are agreed about the importance of histamine in its initiation. One gathers, in fact, that in recent years there has been a considerable return to the classical views of Lewis on the importance of H-substance, histamine. In the body the main and perhaps only source of readily liberated histamine is the mast cells. The histamine is apparently held in some form of physical association with the heparin granules of the mast cells and is liberated by a variety of pharmacological agents and by appropriate antigen–antibody reactions involving the mast cell surface. No function of the mast cell other than the liberation of histamine, and in certain species other pharmacologically potent amines, is clearly established. One must therefore regard the widely distributed mast cells of the body as having primarily an emergency function to aid in the initiation of the local tissue conditions needed to deal with injury or infection.

In health the outstanding character of the micro-circulation from terminal arterioles to postcapillary venules is its adjustment of blood flow to local needs. This adjustment is necessarily a local one and is not seriously influenced by removal of nervous or systemic hormonal control. Schayer has postulated that histamine is also the basis of this local control largely on the basis of the distribution and changes in activity of the enzyme histidine decarboxylase.

Since the postulated action of intrinsically produced histamine is not readily inhibited by antihistamines the view is by no means universally accepted. It seems, however, that an intrinsic dilator with the general properties of histamine is needed to account for the natural processes and until a more satisfactory candidate appears it is reasonable to accept the histamine hypothesis.

On this view, vascular endothelium contains inducible histidine carboxylase capable of producing and building up concentrations of histamine as called for. This results in opening of precapillary sphincters and dilatation of postcapillary venules with partial opening of intercellular spaces between endothelial cells. Such action is antagonized by circulating catecholamines and glucocorticoids, both adrenal products, and when the local need is met these will lead to the gradual closing down of the local circulation to normal level. In traumatic and inflammatory situations the local situation is pushed beyond the 'normal' reversible state by the accumulation of active products of gross cell damage—bradykinin and other peptides, lysolecithin, and a variety of proteases and other enzymes. From the point of view of theoretical immunology these later changes are irrelevant.

DIFFICULTIES OF THE MATERNAL–FOETAL RELATIONSHIP

In seeking examples of immune phenomena which are natural in the sense of having possible significance in the evolution of the immune mechanism, the special problems of pregnancy come to mind. One of the major problems confronting the development of placental mammals must have been the necessity to render compatible the already established requirement that in nature there should always be histocompatibility differences between individuals of the same species including therefore, and perhaps especially, differences between mother and offspring, and the need for nutrition of the embryo from the mother's circulation. Clearly, in the placental mammals, foetus and the foetal components of the placenta represent a tolerated homograft in the uterine tissues of the mother.

There are interesting aspects of the function of the intermediate layer between foetal and maternal tissues and of the special immunological character of maternal tumours (chorioncarcinomata) derived from foetal cells, but the phenomenon of haemolytic disease of the newborn is a much better worked out example of immunological principles in action.

During pregnancy the plasma proteins of the mother, with some exceptions including Ig M, pass into the foetal circulation. Many of these proteins, including Ig G, will have antigenic differences from those of the foetus but for reasons discussed in chapter 10, the foetus makes no response against them. After birth the foreign proteins disappear gradually and tolerance persists for a variable period. Circulating maternal cells do not normally enter the foetal circulation, though there have been some rare cases of generalized disease in newborn infants which have been ascribed to a 'graft-versus-host' activity of maternal immunocytes which entered the foetal circulation *in utero*. There has even been a suggestion that the Burkitt lymphoma in African children may have such an origin.

### Haemolytic disease of the newborn

In what is popularly called 'Rh disease', the important happening is a leak of foetal blood into the maternal circulation. This is an accident that could conceivably occur at any stage of pregnancy, but detailed studies (based on the possibility of distinguishing red cells which contain foetal haemoglobin from those which contain haemoglobin of standard adult type) have shown that it is far commoner at the time of delivery. Significant amounts of foetal blood enter the maternal circulation in about 10 per cent of births and if conditions are appropriate the mother may produce antibody against any antigen in the foetal cells which she does not possess.

Among the very large numbers of genetically based differences in the antigenic qualities of red blood cells the only one of major practical importance is that between Rh + and Rh − or more precisely between the possession of antigen D as against d. When a woman genetically d/d is impregnated by a D/D man, her child will have both d and D antigens on its red cells. If the husband is D/d, half the children will be D/d, half d/d. The entry of D/d (or D/D) blood into a d/d individual is liable to immunize and result in the production of anti-D, initially Ig M and later Ig G. In addition, a mother so immunized will almost certainly develop clones of memory cells to be stimulated by any renewed contact with antigen D.

If she again becomes pregnant with a D/d foetus and her level of anti-D is high enough this antibody, with all other Ig G antibodies, will enter the foetal circulation. A dangerous haemolytic process develops which, in addition to causing severe anaemia, may have grossly damaging effects on the brain. The problems of this disease and its treatment by exchange transfusion immediately after or even before birth are now well known. Current interest in the possibility of the prevention of Rh disease, however, provides a story of much more interest for immunological theory.

### The prevention of Rh disease

It has been known for some time that only a small minority of pregnancies where the mother is d/d and the foetus D/d are followed by haemolytic disease. From what has been said it cannot happen with the first pregnancy unless the mother had inadvertently been transfused previously with the Rh + blood. In general, with each successive pregnancy the likelihood of disease increases, for obvious reasons, but there are still many who do not suffer.

Some years ago it was recognized that when the ABO blood group of the mother was incompatible with that of the foetus D/d cells failed to immunize. If we have an Rh − O mother and an Rh + A father and the d/D, AO blood of the foetus enters her circulation during delivery, the cells will be immediately opsonized by her anti-A, phagocytozed, and removed briskly from the circulation. No anti-D will be produced. If, however, the foetal cells were d/D,O they would circulate for a long time and find opportunity to stimulate immunocytes of appropriate pattern and more important to repeat the stimulation of any previously committed cells.

This phenomenon plus the progressive recognition of the power of Ig G antibody to block specifically the primary response to the corresponding antigen led Finn and others to suggest a means of eliminating Rh disease. If an adequate dose of anti-D can be given to a vulnerable mother who has just borne a D/d baby, the foetal red cells will be opsonized and removed from harm's way and no immunization should result. Anti-D can be obtained in the form of human Ig G prepared from a person immunized by pregnancy or transfusion and progress reports of its efficacy are highly favourable. For obvious reasons, strictly limited quantities of anti-D are available and its administration is only justified when proper blood tests of baby and mother show the infant's blood to be D/d and ABO compatible with the mother. The possibility of producing virtually unlimited amounts of anti-D by immunizing Rh − *men* with Rh + blood is already being explored.

*Selective effects of blood group incompatibility*
It is clear that these maternal–foetal differences should have some selective effect on the proportion of individuals with the various blood group antigens. There are very striking racial differences in regard to the proportion of Rh combinations, but it is not known to what extent these differences depend on the selective effect of haemolytic disease. Theoretically the effect of the disease should be to lead to the extinction of D or d, whichever was initially the rarer, apart from any reappearance by mutation. In this connection it is of interest that of the six genes (antigens) C D E c d e, D:d is the only pair which shows 100:0 ratio in any human groups. No individuals with d genes are recorded among South Chinese

TABLE 5. *Frequency of Rh genes in three races* (%)

|  | C | c | D | d | E | e |
|---|---|---|---|---|---|---|
| English | 43 | 57 | 59 | 41 | 16 | 84 |
| Australian Aboriginal | 71 | 29 | 87 | 13 | 22 | 78 |
| South Chinese | 76 | 24 | 100 | 0 | 20 | 80 |

Derived from G. A. Harrison, J. S. Weiner, J. M. Tanner and N. A. Barnicot (1964). *Human Biology*, p. 270. Oxford: Clarendon Press.

and New Caledonians, while in Australian Aborigines it is found only in the combination C d e. The absence of d in the South Chinese is at least suggestive of the action of what may be called 'immunogenetic selection'.

The only other area where immune factors may be important in selective survival of human genotypes concerns the ABO group. Haemolytic disease due to ABO differences is very rare but can occur, for example, when an O mother has an AO or BO infant. This is too rare to be significant in modifying blood-group frequencies. There is, however, statistical evidence that there may be a considerable prenatal loss associated with ABO incompatibility particularly when a group O mother has a group A child.

## DELAYED HYPERSENSITIVITY

### The tuberculin reaction

The phenomenon of delayed hypersensitivity was first induced in the course of Koch's experiments with tuberculin, and the Mantoux reaction in human beings infected with the tubercle bacillus is still the classical example of a delayed hypersensitivity reaction. Equivalent reactions are observed in experimentally infected guinea-pigs and other animals, so that there has been abundant opportunity to study the significance of the reaction. The essential features of a delayed hypersensitivity reaction in the skin of man or guinea-pig are its slow appearance and persistence beyond 24 hours. In man, a positive Mantoux reaction takes the form of a patch of erythema rather dull red in colour with moderate swelling and slight local tenderness. The reaction reaches a peak between 24 and 48 hours and fades slowly. The guinea-pig reaction is similar in all essentials.

The most generally accepted interpretation of what happens is probably that of Spector. At a rather superficial level the process can be described as follows. The deposition of the tuberculin antigen in the tissues produces minor endothelial damage allowing movement of polymorphonuclear cells, monocytes and lymphocytes from the capillaries into the tissue. When a sensitized cell, presumably a lymphocyte, reacts with the antigen a variety of pharmacologically active agents are produced. These have two important effects: fixation of the lymphocyte in the area in contrast to the speed with which other cells, notably the polymorpho-nuclears, are removed by the lymph flow, and fixation of other non-sensitized mononuclear cells in the region and variable further degrees of damage to the capillary endothelium. This secondary increase in permeability leads to the entry and fixation in the tissue of many more mononuclear cells, sensitized and non-sensitized, and the reaction builds up to its typical form.

It was obvious from the beginning that the reaction, whether invoked locally or generally, was harmful to the infected human patient or experimental guinea-pig. On the ground that any bodily reaction to a natural hazard is likely to be basically advantageous in dealing with the emergency, there has been a constant effort and debate to find some 'useful' function for the tuberculin reaction. The overall results have not been enlightening. Un-doubtedly, there is a significant degree of postinfective immunity in tuberculosis but there are many antigens concerned, with very little indication as to which are significant for immunity.

Neither is there any real indication that infections by acid-fast bacilli played any significant role in mammalian evolution or that the manifestations of delayed hypersensitivity in tuberculosis are of any relevance to the outcome of the infection. If neither tuber-culosis nor superficial trauma have been responsible for the evolution of delayed hypersensitivity we are almost forced to look at the possibility that non-infective processes must have been concerned.

*The range and character of delayed hypersensitivity reactions*
It seems worth exploring the hypothesis that the special character-istics of delayed hypersensitivity have arisen as part of a process which has been evolved to deal with body antigens modified either by environmental agents or by somatic mutation. In developing the hypothesis we shall necessarily equate homograft immunity with delayed hypersensitivity, an opinion which appears to be accept-able to most immunologists.

On this basis, delayed hypersensitivity-type reactions include, or are relevant to, the following phenomena:

(*a*) Sensitivity to tuberculin, coccidioidin and other reagents

245

associated with subacute infectious processes. There is direct evidence of delayed hypersensitivity in two virus infections, vaccinia and lymphogranuloma venereum, and in all probability other viral infections could be added.

(*b*) Skin hypersensitivity produced by reactive chemicals such as dinitrofluorobenzene and picryl, with poison ivy as a 'natural' agent of similar type.

(*c*) Experimental allergic encephalomyelitis may well be only the best-known example of a common process by which autoimmune damage to inaccessible organs may be produced. The lesion closely resembles a delayed hypersensitivity skin reaction in its cellular character.

(*d*) It is convenient to mention here three examples of immune processes related in one way or another to malignant change. These will be considered in chapter 13, but it is essential for the understanding of delayed hypersensitivity that they should be kept in mind:

(i) There are several instances in which virus-infected cells provoke a response of immunological character—either an inactivation of the modified cells or an inflammatory change based on the altered tissue—polyoma virus infection in young animals and Rous virus infection of neonatal rats may be cited.

(ii) Tumours produced in mice by methylcholanthrene and some other carcinogenic hydrocarbons are antigenic, and by suitable manipulations syngeneic mice can be immunized specifically against any such strain.

(iii) Evidence from human pathology suggests that many initiated tumours are destroyed, presumably by immunological means, before they become clinically evident.

(*e*) Experimental homograft immunity is mediated by delayed hypersensitivity-like activity.

There is a general consensus of opinion that delayed hypersensitivity reactions are mediated by cells, not by circulating antibody. The actual lesion seen in a test intradermal reaction is assumed to be the result of a damaging action of the antigen on sensitive lymphocytes with the release of material capable of affecting adjacent cells more or less harmfully. The sensitive cells may be specific immunocytes with appropriate receptors or, less probably, cells passively sensitized by attachment of cytophilic antibody.

The actual antigen and antigenic determinants are known in only a few of these conditions. It is possibly significant, however, that wherever they are known there is good reason for believing that the antigen is of host origin, modified either by the union of a hapten or as a result of somatic mutation or its equivalent.

Tuberculoprotein has a lower molecular weight than typical antigens while the encephalitogenic protein involved in experimental allergic encephalitis appears to be a polypeptide of mol. wt around 5,000. The effective antigen in both cases could well be a compound with a host component. Skin-reactive chemicals and certain drugs must combine with host protein to induce their immunological action. The intrusion of a virus genome into the genetic system of infected but resistant cells could be expected in one way or another to allow or enforce the synthesis of modified host proteins. The changes resulting from somatic mutation are probably of very similar character to the changes in histocompatibility antigens that become manifest when pure line strains of mice are developed and split into further sublines.

There are several reasons for considering the immunocytes concerned in these delayed hypersensitivity reactions as differing in some significant fashion from standard immunocytes. *In vitro* reaction of human lymphocytes with antigens to give blast transformation and mitosis is seen only in relation to delayed hypersensitivity. This holds also for the production by cell–antigen interaction of agents inhibiting macrophage migration *in vitro*, using guinea-pig material. Demonstration of specific antibody (or receptor) by immunofluorescence in small lymphocytes also has been successful only in relation to active delayed hypersensitivity. There is therefore considerable justification for exploring Raffel's suggestion that the antigenic stimulus is applied in some abnormal fashion and results in 'differentiation of a multipotential immunocompetent stem cell along a pathway which it ordinarily does not take in response to primary antigenic stimulation'. In line, however, with the point of view we have adopted, the phrase 'multipotential immunocompetent stem cell' would be replaced with 'progenitor immunocyte'.

## The special quality of delayed hypersensitivity immunocytes

There are two main suggestions as to the nature of the difference. The first is that the receptors concerned are of high avidity for the antigenic determinant and that delayed hypersensitivity develops when conditions favour the stimulation and proliferation of immunocytes carrying combining sites of high avidity.

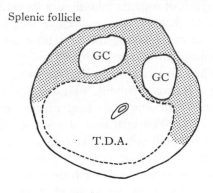

Splenic follicle

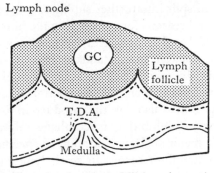

Lymph node

Fig. 25. Schematic diagrams of a splenic follicle and a portion of a lymph node to show the location of the thymus-dependent area (T.D.A.). GC = germinal centre.

The second (which may not exclude the first) is that the sensitizing antigenic determinants include an autologous component, and the stimulus to the immunocyte is of a different quality. Combining suggestions from a number of sources, we may consider the possibility that any antigen which is going to provoke delayed hypersensitivity must be held in some significant relationship to the lipoprotein histocompatibility antigens on the surface of some appropriate cell. *A priori* one would consider the dendritic phago-cytic cells of the lymph follicles to be the most likely, but recent experimental work on sensitization by simple chemicals indicates the primary involvement in lymph nodes of the 'paracortical' or 'thymus-dependent areas' where such cells are not normally seen. There is a real possibility that the effective antigen may reach the lymph nodes in the surface of mobile wandering cells, lympho-cytes or monocytes, from the area to which the chemical is applied.

Whatever the cells involved, the association of the antigenic determinant with the lipoprotein histocompatibility antigens would allow an entirely different type of presentation of antigenic deter-minant from that characteristic of ordinary antigens. The latter are, by hypothesis, held on the dendritic phagocytic cell surface in association perhaps with normal opsonin, perhaps with some form of RNA. The immunocytes capable of making effective stimulatory contact with the delayed hypersensitivity-type of antigen might well be characterized by some special quality, such as the need for high avidity toward the antigenic determinant and perhaps for a certain stable component within the different combining sites to facilitate union with antigenic determinants of this special type (see p. 140). If the surveillance function has the biological impor-tance ascribed to it in chapter 13 it is even possible (or probable) that a subpopulation of immunocytes—equivalent to the sub-populations A, G, M and D—may exist, specially adapted by evolution to mediate the surveillance function.

It would be quite in line with biological principles if such a sub-population had been differentiated evolutionarily to deal with contingencies arising through the production of abnormal body components. On this view the curious pathology of tuberculosis and leprosy may be very largely due to the fact that such infections were not significant factors in evolution and the response we see is, as it were, the best that could be done with available mechanisms. In 1950 I wrote about

a principle of very great importance in pathology—and there is an exact analogue of this principle in social history. It is that when a bodily mechanism has been evolved to favour survival it can be expected to function satisfactorily only in dealing with the standard common and significant type of situation which was (indirectly) responsible for its evolution. Once the mechanism has been established, however, it will be called into action by other more or less relevant situations to give responses which may be ineffective or positively harmful in dealing with the unusual situation.

## Delayed hypersensitivity in relation to infection

For many years there has been great interest in the relationship of hypersensitivity to immunity against tuberculosis and other bacterial or mycotic infections with more or less similar sensitizing proclivity. The experimental approach has been largely concerned with providing conditions which will allow effective immunization of animals against lethal or symptomatic infection and with the study of the capacity of macrophages to deal with ingested bacteria under various conditions.

The essential features stressed in Mackaness's recent review are that both protective immunity and the production of delayed hypersensitivity in mice against the 'facultative intracellular parasites' *Brucella abortus, Listeria monocytogenes* and *Salmonella typhimurium* can be produced only when there is infection by living bacteria. The delayed hypersensitivity is specific; resistance against death from infection is nonspecific, being associated with resistance against bacteria other than those used for immunization.

This has an interesting relevance to the hypothesis we are considering. It seems highly probable that non-lethal infection of cells would result in bacterial antigenic determinants accumulating to some extent in the lipoprotein surface of the macrophages involved. This would allow presentation of the bacterial immunogen to appropriate immunocytes in the fashion postulated as necessary for the induction of delayed hypersensitivity.

If this does represent the sensitizing situation it has obvious relevance to the nonspecificity of the resistance as tested by intraperitoneal challenge with heterologous bacteria. When the other strain of bacteria is inoculated there will be a movement into the peritoneal cavity of lymphocytes and monocytes from local sources and from the blood. These will include both specifically sensitive immunocytes and cells carrying surface bacterial antigen from past infection as well as other lymphocytes and monocytes with no specific qualities of either sort.

If, like most contemporary immunologists, we eliminate any active sensitization of monocytes, their increased capacity to deal with infection must result from activation by products of antigen–sensitive cell contact—a concept for which there is direct experimental evidence, and which is colloquially known in technical circles as the 'innocent bystander' effect. It follows, therefore, that when the heterologous bacterium is inoculated there is, by hypothesis, an almost equivalent opportunity for macrophage activation by interaction of the persisting homologous antigen and sensitive cells. The activated cells, perhaps by reason of increased lysosome production and activity, thus become available for ingestion and destruction of the organism against which the animal was *not* immunized.

Processes of this general character do not in any way interfere with the ability of the animal to produce circulating antibody of various types, including cytophilic antibody which has a special propensity for attachment to macrophages.

## The surveillance function

A discussion of the more specific aspects of this interpretation of delayed hypersensitivity in relation to surveillance and homograft immunity will be deferred until chapter 13. It is desirable, however, to elaborate one aspect because of its relevance to the understanding of autoimmune disease; namely, the nature of immune cytotoxic effects whether by antibody or by immunocytes.

If our general interpretation is correct the significance of the subpopulation of immunocytes concerned in delayed hypersensitivity reactions is to be sought in their surveillance function. They have been evolved to 'seek and destroy' aberrant cells carrying the antigenic determinant to which they are attuned.

At the experimental level there have been extensive studies of the damaging effect of specifically immune cells, obtained either from lymphoid tissue or in the form of a peritoneal exudate, against target cells. For reasons of technical convenience, malignant cells have usually been used as target. *In vitro* with a suitable pair of cell suspensions, the usual picture is for the immune cells to clump around the target cells which then show various degrees of vacuolation and disintegration. In some situations but not in others the immune cells are also damaged. The specific action of spleen cells or peritoneal exudate cells may also be demonstrated by *in vivo* experiments in which graded mixtures of immune cells and target cells are injected into an animal in which the neoplastic target cells can produce ascites or solid tumours.

The process by which the damage is produced does not seem to have been studied directly. It seems reasonable, however, to assume that the immune cell carries on its surface either fixed cytophilic antibody or its own characteristic receptor with the equivalent specificity. Union of this with the antigenic determinant on the other cell, and presumably a gradual increase in the number of such unions, will be expected to produce locally increased permeability with damage to the cell surface and lysosome breakdown. The possibility is very high that the proximity to a specific union between cells may result in a nonspecific increase in stickiness and reactivity of adjacent intrinsically non-sensitized immunocytes.

On the basis of this general approach the natural function of the sensitized immunocyte is to recognize cells which by mutation, damage or viral action have developed aberrant antigenic determinants as well as their normal surface antigens. When recognized they must be destroyed and eliminated. In the normal healthy animal the function would be recognizable only by the transient appearance of a few lymphocytes and a macrophage or two. It may be none the less important for having such insignificant manifestations.

## *Aggressiveness of immunocytes*
In various sections of this book I have discussed phenomena in immunologically modified animals in which passage of cells takes place from the blood to the tissues of a region where antigens or antigenic determinants were present. These include the standard ways by which delayed hypersensitivity reactions are demonstrated experimentally, positive reactions from simple application of sensitizing chemicals to the skin, the paralysing response in experimental allergic encephalomyelitis and the phenomena of autoimmune disease of specific organs. The mechanism of a straightforward test for delayed hypersensitivity has already been discussed in terms of Spector's interpretation. This, however, is quite inade-

quate to account for the phenomena of experimental allergic encephalomyelitis and it seems likely that in addition to the passive mechanism postulated by Spector for the entry of mononuclear cells from the blood there is another more active process. In some way it appears that small numbers of sensitized cells can be held in capillaries adjacent to a source of antigen and then pass into the tissue spaces. Once this has happened the pharmacological results of immunocyte-target cell or immunocyte–antigen contact will ensure the nonspecific entry of normal mononuclear cells, both monocytes and lymphocytes. The initial entry, however, appears to call for a special quality which can be called aggressiveness in the immunocytes concerned.

Aggressiveness is manifested against tissue antigens or their experimental analogues. It is specially characteristic when sensitization is by the use of antigen in Freund's complete adjuvant. It appears to be annulled with considerable regularity by circulating antibody, probably always Ig G, of the same specificity. There seems, therefore, to be another special quality of the immunocytes concerned with delayed hypersensitivity and related conditions which makes them specially prone to adhere to local endothelium carrying antigenic determinants, to pass into the tissues and perhaps undergo there lethal damage which allows changes leading to nonspecific entry of other lymphocytes and monocytes.

*A hypothesis of the cellular basis of delayed hypersensitivity*
None of the current hypotheses of delayed hypersensitivity is at the same time reasonably illuminating in regard to the experimental and clinical facts and consistent with the general clonal selection outlook. It is clear that any theoretical treatment must be wide enough to cover the other immunological phenomena which have been mentioned above as related to delayed hypersensitivity particularly contact sensitivity, homograft rejection and immunological surveillance. An attempt has therefore been made to summarize the situation in terms of a broad theoretical approach in the following terms.

The immunocytes concerned are derived from the thymus, and lodge primarily in the paracortical area of the lymph nodes. This is also the region where lymphocytes entering the node from the afferent lymphatics lodge. The delayed hypersensitivity lymphocytes represent a class of immunocyte analogous to but distinct from the classes producing G, M, A and D immunoglobulins. The uncommitted cell is reactive only to its corresponding antigenic determinant when this is incorporated in a special relation to the lipoprotein surface of another cell. On stimulation, change to the lymphoblast form occurs with subsequent active proliferation to committed lymphocytes. Secondary nonspecific proliferation of adjacent immunocytes is also probable, giving rise to a considerable population of recently 'born' lymphocytes which pass to the circulation. The committed delayed hypersensitivity immunocytes can react with the corresponding antigenic determinant when it is incorporated in a cell surface or free. In the first instance the result is to activate the metabolism of the immunocyte and produce some damage to the target cell. This will result in release of kinins, etc. and, in the case of capillary endothelium, opportunity to move into the tissues where the antigen is present.

As a secondary development of this point of view we need to look at the mechanism of immunological surveillance to be discussed in chapter 13. The primary assumption must be that histocompatibility antigens of tissue cells are labile and transferable to other cells, notably to any wandering lymphocytes in the area. Lymphocytes reaching the paracortical areas of the draining lymph node will therefore be carrying a sample of any aberrant surface antigens present in the tissues. Aggressive delayed hypersensitivity immunocytes of appropriate type will be able to proliferate and eventually deal with the focus of aberrant cells.

One of the outstanding conundrums of immunology is the nature of Lawrence's 'transfer factor' by which human leucocytes from a tuberculin-positive individual can confer a long-lasting reactivity on an initially non-reactive recipient. It would be inappropriate to elaborate this theme here beyond mentioning the possibility that the phenomenon may be based on the capacity of antigen incorporated in the cell surface to be passed on from one cell to another.

# 12    Autoimmune disease as a breakdown in immunological homeostasis

There are few people who at one time or another in their lives are not subject to some pathological manifestation of the immune mechanism. It is a highly complex mechanism of great significance for survival and in the course of evolution it has necessarily developed a series of controls, of homeostatic processes, to ensure that the potentially destructive reactions of the immunocytes are directed toward proper targets. The old analogy of the immune processes to national defence in fact takes on a modern flavour if we look on the whole function as a fail-safe system ringed around with controls to ensure that action against the 'enemy' does not damage the resources of the organism, whether that organism be a political one or a mammalian body.

The earlier approach to autoimmune disease was to look upon it simply as a failure of natural tolerance to develop. With the immense activity in the field of immunopathology during the last decade the approach at the theoretical level must be greatly broadened. Our approach now must be to consider each pathological condition or set of symptoms as a manifestation of some breakdown in one or more of the normal controls. Each disease and probably each individual example of autoimmune disease will have its own individual features. It is quite unjustifiable to seek a single formula to cover the nature of all autoimmune disease.

In immunopathology we are concerned, as in every other field of pathology, with environmental and genetic factors but here, more than in other fields, we must add a third qualitatively distinct set of factors—somatic mutation—which includes processes which might equally be called irreversible anomalies of differentiation. The working rule will be accepted that somatic mutation may take quite a similar form to mutation in germinal cells and that in any given immunocyte, somatic mutation will take place against a background of the genetic endowment of the cell. The influence of any mutational event occurring after the first division of the fertilized ovum will be expressed in relation to the other genetically controlled activities of the cell. The possibility that has arisen from results in tissue culture studies, that recombination and segregation may take place between somatic cells, has been referred to earlier (p. 200). There is as yet no evidence that this has any significance in relation to autoimmune disease but the possibility should be kept in mind.

At the environmental level we have first to recognize that it is a prime function of the immune mechanism to deal effectively with every type of 'natural' occurrence that introduces foreign material into the body. Immunologically speaking, local and general infections by micro-organisms are natural hazards providing stimuli to normal defence activities. The environmental agents relevant as such to immunopathology are of two groups:

(*a*) unnaturally reactive substances produced either by synthetic processes—dinitrofluorobenzene, picryl, etc., or evolved as specialized protective mechanisms by plant or animals, poison ivy being the classical example;

(*b*) potentially invasive micro-organisms which, apparently for wholly accidental reasons, carry antigenic determinants cross-reactive with potential antigenic determinants in certain organs of the susceptible individual. Streptococcal antigens in their relation to rheumatic fever and acute nephritis are the standard prototypes in this group.

There are of course antigens, pollens for instance, which are traditionally associated with allergic complaints but these represent an environmental impact common to everyone. Internal factors, genetic or somagenetic,* are responsible for the differences in response between allergic and normal subjects.

Autoimmune disease can be defined as a condition in which

---

* 'Somagenetic'—In the course of writing this book it became clear that on many occasions one would have to differentiate between cell characteristics genetically determined in the usual sense and others that had arisen subsequent to the initial development of the fertilized ovum by somatic mutation. The actual genotype and its phenotypic expression of any given cell will be influenced by both. In the interests of clarity the word 'genetic' and its derivatives will be used in its normal sense. For inheritable changes arising by somatic mutation, the word 'somagenetic' will be used with, in many places, the additional implication of recognizing that any change or character labelled 'somagenetic' is expressed on a background of the whole genotype of the cell.

structural or functional damage is produced by the reaction of immunocytes or antibodies with normal components of the body. This is wide enough to cover an extensive range of minor and major illnesses and disabilities. It will be best to sort out from these a small number of clinical patterns each of which seems to consist of a constellation of related conditions. It is of the essence of our approach to immunity that no two cases of autoimmune disease should be the same, but as long as this heterogeneity and the necessary existence of mixed and intermediate forms is borne in mind, classification into groups can be helpful.

The approach adopted is to consider first, three general auto-immune diseases: acquired (autoimmune) haemolytic anaemia (AHA), systemic lupus erythematosus (SLE) and rheumatoid arthritis (RA). Here we have three important sets of target antigens, respectively the red cell surface, nuclear and cytoplasmic constituents common to many cell types and, for RA, $\gamma$-globulin, all of which are readily accessible antigens in the sense of being available in adequate concentration in lymphoid tissues.

In a second group we have the organ-specific autoimmune diseases of which thyroid disease, as the most closely studied example, and myasthenia gravis, because of its relation to the thymus, are chosen for discussion. These are conventionally regarded as involving inaccessible antigens which are either absent from lymph and blood circulations or present in insignificant amounts. More information is required in regard to the amount of recognizable antigen liberated into lymph and blood from different tissues and at different stages of development. Current work suggests that the accessibility of such tissue antigens varies from one organ to another and according to the age of the animal. It is probable that the convenient division into accessible and inaccessible antigens is by no means sharp and may eventually be discarded.

Finally, it is necessary to consider the two groups of conditions mentioned earlier in which unusual immune responses to antigens from the environment inadvertently, as it were, give rise to damaging results. The examples to be taken are rheumatic fever and rheumatic carditis and the drug purpuras and haemolytic anaemias.

The application of the principles of clonal selection theory, particularly as they have been developed in relation to normal and experimentally produced immune tolerance, is best done in relation to the specific qualities of the separate autoimmune diseases. Throughout the central theme is the emergence of 'forbidden clones' of pathogenic immunocytes and the various ways by which these can arise and find ways of escaping the normal controls.

## AUTOIMMUNE HAEMOLYTIC ANAEMIA

Red blood cells have been favourite objects for immunological experimentation since the days of Ehrlich and Bordet, and as a basis for the discussion of autoimmune disease there is no better starting point than the disease in which a patient's own red cells are destroyed immunologically.

There are almost as many variants, minor and major, in the manifestations of autoimmune haemolytic anaemia (AHA) as there are patients with the disease. If we neglect minor points and concentrate on a 'standard' case of 'warm-antibody type' acquired haemolytic anaemia, there are several statements that can be made which bear directly on the general character of autoimmune disease.

(*a*) The destruction of the patient's red blood cells is not due to any abnormality in the red cells. If a healthy donor provides compatible cells for transfusion into a patient's veins, these are destroyed at the same abnormally rapid rate as the patient's own red cells.

(*b*) The destruction takes place for the most part in the spleen but is due primarily to the presence of antibody attached to the surface of circulating red cells.

(*c*) The antibody is of 7 S (Ig G) type and is characteristically incomplete, that is, it does not cause direct agglutination of normal red cells in saline nor haemolysis when complement is also present. It is detected by the use of antiglobulin sera in the direct Coombs test.

(*d*) In some 75 per cent of cases the antibody eluted from coated red cells can be shown to contain only one type of light chain antigen. This is the best evidence available that such antibody is of monoclonal origin.

(*e*) With only 10 per cent or less of exceptions the antibody is 'nonspecific', reacting with all types of human cells and not corresponding to any of the antigenic determinants defined by blood-group studies with natural or immune isoantibodies.

To these should be added two other statements in regard to other types of autoimmune haemolytic anaemia:

(*f*) In the cold-antibody type of human haemolytic anaemia the antibody is an Ig M globulin which does not attach to the target cells at 37 degrees, but at room temperature shows high titre agglutinating and often haemolytic activity.

(*g*) In the NZB mouse strain there is a model with virtually all the fundamental qualities of the warm-type human disease. Here the development of the disease between 3 and 9 months of age is a regular occurrence and appears to be a definite inheritable quality.

## *The origin of pathogenic clones*

If these qualities are accepted as typical of autoimmune haemolytic anaemia, any discussion of its aetiology in terms of the general clonal selection approach to immunity must start with point (*d*), the evidence that the antibody present in most patients is monoclonal in type. It is now generally accepted on precisely similar evidence that most cases of multiple myelomatosis are monoclonal and arise by the proliferation of a single pathogenic stem cell. Our contention, then, will be that AHA arises by the appearance through somatic mutation of a single immunocyte with the capacity to react by proliferation and antibody production when it makes effective contact with the nonspecific red cell antigen always present in the circulation, and to give rise to a continuing sequence of memory cells of the same quality. This formulation does not mean that the appearance of a progenitor cell of the right quality to produce AHA is a single unique episode. In both human patients and NZB mice it is probable that many such cells appear. The essential point is that of these it is only a rare individual cell that can escape all the controls and random obstacles and go on to establish a persisting clone with disease-producing capacity (fig. 26).

Acceptance of this hypothesis has the virtue that it immediately calls for discussion of many other questions. It has been emphasized, for instance, that clinical and immunological details vary widely from one case of the disease to another. A proportion, for instance, show a specific antibody most commonly directed against the Rh antigen e; some show active haemagglutination. This variability is a characteristic marker of somatic mutation. It is seen

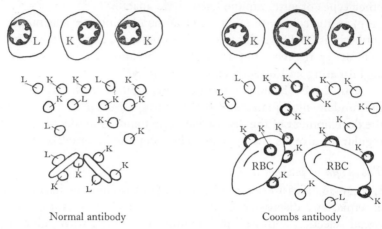

| Normal antibody | Coombs antibody |
|---|---|

Fig. 26. To illustrate the concept of autoimmune Coombs antibody as the product of a single clone in contrast to the heterogeneity of normal antibody.

even more clearly in multiple myelomatosis where virtually every case produces a single myeloma protein demonstrably different from that in any other patient. On our basic understanding of immune pattern there must regularly arise immunocytes carrying patterns that can react with one or other of the antigenic determinants of the individual's red cells. In the normal individual, none of these develops into an active antibody-producing clone. The primary aetiological question is to understand what allows a cell with such reactivity to do so in the individual developing autoimmune haemolytic anaemia.

The monoclonal character virtually forces us to look primarily for abnormality in the cell line. If the fault were in the internal environment, one would expect cells of varied character to escape control and produce antibody as varied in quality as that arising physiologically. The cell which is predestined to initiate the disease must differ genetically or somagenetically from its congeners. In the NZB mice with certainty, and by implication in human patients, there must be a genetic component. It is significant that in Dacie's series of cases of warm-type AHA, 9 out of 129 were associated with other autoimmune disease, six of them with SLE. It is at least evident that somatic mutation is more effective on some genetic backgrounds than others.

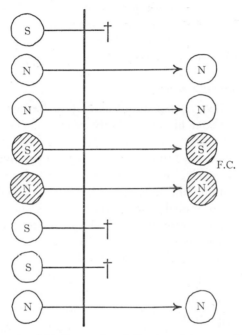

Fig. 27. The role of the thymus in relation to autoimmune disease.

Newly differentiated immunocytes capable of reacting with 'self' antigens (S) are normally destroyed on contact with the antigen in the thymus.

Non-reactive cells (N) escape as do S cells which have become resistant to the censorship function (shown by hatching) and can give rise to forbidden clones (F.C.).

Once the necessary somagenetic configuration has appeared it must, as it were, run the gauntlet of the controls provided by nature to prevent such cells from prospering. Red cells are ubiquitous and many of their antigens are shared with other cells. Whether the primary elimination of self-reactive immunocytes takes place in the thymus or elsewhere we must postulate that all the relevant antigenic determinants are present in the environment where differentiation to immunocyte takes place. Newly differentiated immunocytes reactive with antigenic determinants of the red cell surface would therefore normally be destroyed. The most probable somagenetic change responsible for the 'forbidden' quality is one which confers resistance or partial resistance to destruction when the newly differentiated immunocyte makes contact with its corresponding antigen in a thymic or equivalent environment. In previous discussions this has been taken as the basic abnormality leading to the development of a forbidden clone and the initiation of autoimmune disease (fig. 27).

## *Avidity and incompleteness*

There are, however, other requirements to be met, notably the incomplete character of the antibody produced. There are many differences between 'warm' and 'cold' types of haemolytic anaemia, but both have in common that the operative antibody is not overtly destructive *in vivo*. Since human beings can produce fully active immune antibody against foreign red cells, one must postulate some additional type of control which allows incomplete (that is, low-avidity antibody-producing) cells to flourish but not those capable of producing high-avidity antibody. There are two possibilities. The immunocyte with a highly avid receptor (and producing correspondingly avid antibody) may be overwhelmed by the impact of the abundant and readily accessible antigenic determinant. Except under quite exceptional circumstances, no somagenetic change is adequate to allow a fully avid immunocyte clone to escape destruction by antigen. The alternative is that immunocytes of high avidity that could escape the primary censorship and become committed cells could still fail to become an established clone if antigenic contact produced proliferation and antibody production but failed to allow any production of memory-cell descendants from such a cell. Unless memory cells *are* produced there is no possibility of antibody concentrations or immunocyte populations rising to a level which could be effectively pathogenic.

In line with the general Darwinian approach, one pictures within the population of immunocytes a vast repertoire of immune patterns arising within the labile segments of the somatic genome, but there is also scope for a lower level of somatic mutation in other regions of the genome. Following Burch's general approach it may well be that more than one independent mutation may be necessary before a cell has all the necessary qualities which will allow it to emerge as the progenitor of a pathogenic 'forbidden' clone. Such changes could be either genetic or somagenetic. For instance, when three independent mutations are needed to produce the necessary quality of resistance to homeostatic processes, if one or two are pre-existent as a result of genetic change the chance of the appropriate state being reached by somatic mutation will be increased correspondingly. In one way and another we must picture cells with an extreme variety of individual qualities, a proportion of which, depending largely on the genotype of the individual, will have a greater than normal chance of circumventing controls and giving rise to a pathogenic clone. As in equivalent evolutionary or epidemiological problems there are no absolutes, only varying probabilities, that a clone of cells, a new subspecies of mammal or a particular strain of micro-organism will develop to an effectively established population.

### SYSTEMIC LUPUS ERYTHEMATOSUS

SLE is the prototype of severe autoimmune disease involving a variety of tissues and associated with a wide range of abnormal antibodies in the circulation. It shows two theoretically important features lacking in AHA, a high concentration on the female and a characteristic age-specific incidence of the disease. Although the evidence is still incomplete and uncritical there is also a stronger indication of the part played by genetic factors than in AHA. There is further indication of the importance of the genotype in the existence of a mouse strain $F_1$ NZB × NZW which manifests with great regularity the essential features of SLE.

In the human disease there is a wide variation of symptoms and signs depending on the tissues predominantly attacked. The kidney is almost regularly involved and kidney failure is the usual cause of death. In the hybrid mice *all* deaths, except for accident, are due to kidney disease and life expectation is greatly reduced.

### The antibodies of systemic lupus erythematosus

In the features relevant to theoretical immunology the most important difference from AHA is probably the large number of different antibodies that can be found in a single patient's serum and their apparent concentration on antigenic determinants in DNA and other nuclear components. In looking over case histories of young women who develop SLE, there is an insistent suggestion of a preceding period of minor ill-health with a rather sudden onset of multiple symptoms that lead to the diagnosis. It is common for an emotional experience, an infection, administration of a drug or something else to be credited with triggering the onset.

Although there have been some very refined studies of the antibodies in a few individual sera, most serological work has been limited to following the LE cell test or the demonstration of antinuclear factor by immunofluorescence techniques. I am not aware of any careful study of sera taken before and during the development of the clinical condition. This would be of especial value in helping to clarify the pathogenesis of the disease. The simultaneous appearance (if it occurs) of a wide range of forbidden antibodies suggests the sudden collapse of a type of control not yet formulated.

In the full-blown case of SLE there is hypergammaglobulinaemia with auto-antibodies reacting with both nuclear and cytoplasmic cellular constituents. Kidney lesions are almost constant and of a type suggesting damage by soluble antigen–antibody complexes. Detailed study of sera with DNA and nucleotide combinations points to the existence of multiple antibodies. There is no substantial evidence that these antibodies are intrinsically pathogenic except in so far as soluble complexes with antigen in the circulating blood may be responsible for the kidney lesions. Current studies on the antibodies that can be released from SLE kidneys obtained at autopsy indicate that antibody specifically directed against components of the glomerular basement membrane is also present. As usual there are doubts as to how far such antibody is responsible for the actual lesion observed.

### The pathogenesis of the disease

In an attempt to picture the pathogenesis of SLE, one is compelled to assume an important genotypic abnormality, possibly, as Burch suggests, involving the X chromosome. It is reasonable that these genetic changes should make it easier for somagenetic change to increase resistance of a proportion of immunocytes to the destructive effect of contact with antigenic determinants of body constituents present in the thymus. SLE, however, differs sharply from any other autoimmune disease in the multiplicity of 'forbidden' immune patterns which emerge, and some explanation of this is called for. It seems as if there is at some point a massive breakdown in the control function, and an exceptional opportunity for immunocytes to be stimulated by products of nuclear autolysis.

A good case can be made for believing that this breakdown of homeostasis may be a result of the thymus becoming a target organ for the autoimmune process. At a certain stage of this attack the censorship function is lost or weakened so that there can be actual stimulation by antigens present in the thymus of any appropriate newly differentiated cells. All workers are agreed that the most striking metabolic activity of the thymus is the rapid proliferation of lymphocytes with a high nucleocytoplasmic ratio and their almost as rapid breakdown and presumed re-utilization as building blocks for the newly synthesizing cells. There must therefore be a high concentration of nuclear fragments and antigenic determinants of every sort. In the normal individual this merely ensures the elimination of all newly differentiated immunocytes capable of reacting with nuclear material. If, in the individual predisposed genetically to SLE, a relatively large proportion of immunocytes take on resistance of this type, then it will be particularly those cells reactive with nuclear determinants of various types which will emerge, since there is a higher concentration of these than of any other 'self-antigens' in the thymus. A process of this type would almost certainly give rise to germinal centres in the thymus in early SLE. It would also inevitably produce an intolerably unstable situation in the thymus, that could well lead to the disappearance of cortex and increase in epithelial cells that is in fact observed in the thymus even of untreated SLE.

Within this hypothetical frame there is scope for introducing other factors that might be concerned in determining when the disease is initiated. The balance between destruction and stimulation of cells carrying receptors of appropriate pattern within the thymus would be a very delicate one, and quite minor stress episodes with increased cellular destruction in the thymus could swing it toward the conditions allowing increased emergence of pathogenic clones. There is no suggestion that SLE is a monoclonal disease. Detailed serological study points to the involvement of many distinct clones of pathogenic immunocytes. No procedure exactly comparable to the elution of antibody from Coombs-positive cells in anti-nuclear factor is possible but it has been shown that ANF may be present in both G and M immunoglobulins and that each type contains light chains of both $\kappa$ and $\lambda$ specificity.

RHEUMATOID ARTHRITIS

Rheumatoid arthritis (RA) may well be the commonest of autoimmune diseases. Burch calculates that approximately 50 per cent of the populations in northern Europe are genetically susceptible and that, in women, most of that 50 per cent will show clinical signs if they live long enough.

The only fully established serological change associated with RA is the presence of rheumatoid factor, an M-type immunoglobulin strongly reactive with partially denatured immunoglobulins of human and, to a lesser extent, other mammalian origin. There is, as in every other well-studied autoimmune disease, a great heterogeneity of detail at both clinical and immunological levels, and an attempt to present a unitary picture of the pathogenesis is perhaps premature. Nevertheless there is nothing equivocal about the information that has been obtained serologically nor does there seem to be any reason to doubt that interaction between partially denatured immunoglobulin acting as antigen and corresponding antibodies or immunocytes plays a significant part in the pathogenesis of RA.

*Immunoglobulin as antigen*

The control of the antigenic potentialities of immunoglobulins is probably the most subtle of all the problems of immunological homeostasis in the body. The primary function of an antibody is to lay down on the surface of a foreign particle a coating of immunologically inert protein which is nevertheless effective in facilitating the phagocytosis of the foreign particle. When an antibody molecule is firmly attached to an antigen particle or surface it will undergo a variable degree of molecular distortion which may either result in the appearance of a new antigenic determinant or render an existing one more accessible. The standard approach to the antigenic differentiation of human Ig G has been to examine the capacity of human sera from suitable donors to agglutinate red cells coated with incomplete Rh antibody which acts as the antigen in the system. In this way a means has become available for categorizing the Ig G of the person who provided the Rh-antiserum. The tests can be elaborated in various ways by the use of cross-absorption and inhibition experiments to allow an easy classification of any human Ig G. The details have been of very great interest in regard to the genetic aspects of immunoglobulin production but are of no particular relevance to the pathogenesis of RA. The essential feature is that there are antigenic determinants by which individuals differ in the structure of either heavy chain (Gm factors) or light chain (Inv factors) of Ig G. Most rheumatoid factor antiglobulins are in antibodies not reacting with the patient's own immunoglobulin but some have also the quality of auto-antibodies.

Anti-$\gamma$-globulins are not unique to RA patients. If a Gm(a+) mother gives birth to a Gm(a−) child, her Ig G will be present in the child's circulation for some months after birth. A proportion of such children produce indefinitely antibody reactive with Gm(a+)-coated cells. In general, about 1 per cent of normal human sera can agglutinate one or other type of coated red cell. Such antisera always lack the factor they react with; in other words the antibodies are isoantibodies, not auto-antibodies.

Another finding which may be relevant to the pathogenesis of rheumatoid fever is the regular appearance of rheumatoid factor in cases of subacute bacterial endocarditis. Here there is a constant production and liberation into the bloodstream of bacteria from the damaged heart valves. Antibodies against the bacteria are also being produced and antibody-coated bacteria are therefore constantly being taken up by macrophages. The antiglobulin produced is an auto-antibody with most of the qualities of rheumatoid factor but, unlike the antiglobulin in an RA patient, it disappears when the infection is effectively removed by antibacterial therapy.

There is no evidence that rheumatoid factor as such has any pathogenic effect whatever and a good deal to suggest that it may have a protective effect against symptomatic expression of the genetic and immunological weakness.

## The pathogenesis of the disease

The pathological evidence that a part is played by immune processes related to the immunoglobulins has been obtained by the use of immunofluorescence studies of synovia from infected joints and draining lymph nodes. Many plasma cells in this situation are producing anti-human $\gamma$-globulin; there are phagocytic 'RA' cells which appear to contain complexed globulin, presumably antigen–antibody complexes, and there is an abnormally low amount of complement in the synovial fluid. Although persons can have a high titre of rheumatoid factor in their circulating blood without symptoms, it is interesting that several of the rare complications of RA in which cellular infiltrates are conspicuous are always accompanied by a high rheumatoid factor titre. These include subcutaneous nodules, Felty's syndrome of splenomegaly and generalized vascular disease.

In attempting an interpretation of RA in terms of the general approach we are using, we must consider first the almost constant presence within the body of partially denatured immunoglobulins. Harmless casual infection of the bloodstream is an everyday occurrence and more overt local or general infections will at frequent intervals be dealt with by processes in which fixation of antibody on micro-organisms or their products will result in its partial denaturation. In one way or another it is a necessity for survival that such material should not be antigenic. The actual way in which this form of natural tolerance develops is unknown. It may be that sufficient of the antigenic determinants are present on native circulating immunoglobulins to ensure that reactive immunocytes are destroyed soon after they differentiate. The anomaly in the potential rheumatoid individual would be, as in other autoimmune conditions, the emergence of mutants abnormally resistant to destruction by antigenic contact. All antiglobulins produced in man by any of the ways that have been mentioned are macroglobulins (Ig M) and as such are not significantly supported by the appearance of 'memory cells'. Normally, therefore, any reactive clone that emerged to produce rheumatoid factor would fade out fairly rapidly, just as happens when a patient with sub-acute bacterial endocarditis eliminates his infection. The anomaly in RA could well be a capacity for antibody-producing cells to go on proliferating, once specific stimulation has occurred for a longer period than normal. Such delayed maturation of committed immunocytes in which Ig M antibody fails to be replaced as normally by Ig G antibody is frequently seen experimentally when X-irradiation or immunosuppressive drugs are used. A pathological and persistent failure of maturation is the most direct interpretation of Waldenström's macroglobulinaemia, and in this connection it is of great interest that in some cases of macroglobulinaemia the abnormal immunoglobulin M has the reactivity of an antiglobulin.

So far we have been concerned only with the appearance of the abnormal antibody. Its presence implies proliferation of abnormal immunocytes and in all probability the symptoms and lesions of RA represent the cytopathic activities of immunocytes plus secondary changes. A certain 'aggressiveness' can be postulated as differentiating the antiglobulin immunocytes in the rheumatoid patient from those in the person with rheumatoid factor but no symptoms. The nature of this aggressiveness is unknown, but it can probably be equated with the similar increase in the cellular pathogenicity of immunocytes that is associated with the use of Freund's complete adjuvant. It is significant that a subacute arthritis can be produced in rats by injection of F.C.A. alone.

The abnormality in persons subject to RA is primarily genetic, allowing the appearance of somatic mutant cells which, in one way or another, have developed a capacity to enter tissues and react damagingly with any Ig G in some partially denatured form that may be present there. The joints are sites of election for serum sickness reactions and for the manifestation of postfebrile synovitis as seen in rubella and other infections. It is apparently a common site for the local fixation of antigen–antibody complexes. When this occurs the joint is in essentially the same position as any other local target organ. The specific antigen, in this case partially de-natured $\gamma$-globulin, is in the tissue and with the entry of an immunocyte with the needed aggressive quality, reaction with the antigen will allow further damage and greater opportunity for both immunocytes and antibody to reach the area. It is reasonable to consider Kunkel's suggestion that there may be a special type of pathogenic antibody, perhaps of abnormally high avidity, which is more likely to fix complement and mediate cell damage. Entry and accumulation of lymphoid cells (including plasma cells) is a feature, however, and it seems best to think of both immunocyte and its corresponding antibody as playing parts in the pathogenic process. The main result will be the inauguration of a vicious circle with fixation of globulin-antibody which, by local denaturation, becomes globulin-antigen.

If, as is the case in many analogous situations, contact of some types of antibody with cell-bound antigen is much less damaging than specific immunocyte contact, the presence of high levels of antibody may be beneficial to the patient by rendering antigen unavailable to react with immunocytes. In congenital agamma-globulinaemia, arthritis of rheumatoid type is common in the absence of rheumatoid factor or of more than minimal amounts of other immunoglobulins.

The variability in age and sex incidence, in intensity of the disease and in the occurrence of spontaneous, or drug-induced remissions of the disease, is in line with the factors concerned in the pathogenesis of RA. At the genetic level we may adopt Burch's formulation that susceptibles are homozygous at an autosomal locus and in addition must undergo one, two or more sequential mutations on the X chromosome, or the simpler interpretation of Maynard Smith that both the genetic and the somagenetic anomalies are on the X chromosome and that only one of each is needed. The primary provision of denatured immunoglobulin as a necessary stimulus must have a quantitative and probably a qualitative factor, and the old story of 'septic foci' may have a substratum of truth. Further development of the condition once potentially pathogenic cells have emerged will still be dependent on essentially random factors concerned with sequential mutation of varying degrees of effectiveness plus environmental factors, such as cold and minor trauma, rendering some joints more vulnerable than others. It is highly probable that genetic factors also play a part in determining the vulnerability of joints and the ease with which a vicious circle is initiated. As hinted earlier, the relative balance between titre of circulating antibody and numbers of active immunocytes may be important while, finally, the possibility must not be forgotten that a mutant clone of immunocytes may carry a new antigenic determinant that renders it subject to immune surveillance.

As in every other instance of autoimmune disease we have an immensely complex ecological and micro-evolutionary system at the cellular level.

### MYASTHENIA GRAVIS

From the angle of theoretical immunology, autoimmune disease is important only for the light it may throw on the controlling processes which allow the normal functioning of immunity. If the thymus is a key point in immunological homeostasis it is of obvious interest to pay special attention to its involvement in autoimmune disease. Reasons have been given for believing that the thymus might be of importance in the early stages of SLE but more attention has been paid to myasthenia gravis, the only generalized disease in which the thymus is manifestly and regularly involved.

Myasthenia gravis at the symptomatic level is a functional disease of the neuromuscular junction, shown by rapid onset of muscular weakness and fatigue, often limited to some muscle groups, and with drooping of the eyelids the commonest manifestation. The effect is temporarily alleviated by drugs of prostigmine type and at the pharmacological level represents ineffective transfer of acetylcholine across the neuromuscular junction. Biopsy of an affected muscle usually shows very little except for occasional small collections of lymphocytes. Very rarely the lymphocytic infiltration reaches the level of a chronic myositis. In nearly all cases the thymus shows striking changes, either moderate enlargement with conspicuous germinal centres in the medulla, or a low-grade non-malignant lympho-epithelial tumour which may be associated with germinal centres in the uninvolved portions of the thymus. The general consensus of clinical opinion is that thymectomy has a beneficial effect on the disease in the common group of young women with only germinal centre formation in the thymus, while patients with tumour usually show no benefit from removal of tumour and thymus.

No serious attempt to interpret the relationship of the thymic changes to the pathogenesis of myasthenia seems to have been made until we observed the consistent presence of germinal centres in the thymus of the autoimmune strain of mouse NZB. There had, however, already been a recognition that a substantial proportion of myasthenics had circulating antibody reactive with the A band of skeletal muscle and demonstrable by immunofluorescent technique. Later it was shown that, in about half the cases showing muscle reactivity, the same antibody reacted with thymic epithelial or myo-epithelial cells, but other sera reacted only with muscle.

There is no generally accepted interpretation of the pathogenesis of myasthenia gravis but it is not difficult to fit the observations into a pattern similar to that of other autoimmune diseases. The curve of age-incidence and the strong concentration of cases in adolescent and young adult females is in line with a number of other autoimmune conditions. Again, this suggests genetic and somagenetic factors with the initiating event being the emergence of a clone of immunocytes reactive against one or more of a range of antigenic determinants present within a variety of cells. None of these determinants has been chemically defined but they include structures in or functionally related to (*a*) the neuromuscular apparatus; (*b*) the A band of skeletal muscle; (*c*) cardiac muscle and (*d*) thymic epithelium or myoid cells. There is now a suggestion that the cells involved in the thymus are myoid cells with many of the characteristics of striated muscle cells. Their presence in the thymus remains unexplained. As in every other constellation of autoimmune disease we have again a Darwinian complexity of mutation and opportunity for cellular proliferation, survival and production of antibody. The part played in pathogenesis by antibody on the one hand and by direct cellular action on the other is obscure. In the absence of any record of cellular infiltration or other abnormality at the neuromuscular junction it is possible that the symptoms could be due to antibody becoming specifically attached to an antigenic determinant in neuromuscular junctions, but the possibility that the effect is due to a hormone-like agent rather than an antibody is not excluded.

From our point of view the thymic aspect is of much greater interest. Reverting first to the presence of areas of lymphoid cell proliferation in the thymic medulla, this is best seen in myasthenia gravis but it has also been reported in some of the relatively few biopsies that have been done or cases of other autoimmune diseases. In any extensive series of human thymic biopsies—for example, those obtainable in a variety of surgical procedures in the thorax—an occasional medullary germinal centre will be found. In animals showing autoimmune disease, only the NZB mice and their hybrids have been studied. All show variable but often extensive proliferative medullary lesions.

The least controversial interpretation is that normally the thymus is forbidden ground (because of some peculiarity of the internal environment) for the development of germinal centres or plasma cells and that when these appear they are the progeny of mutant cells resistant to the normal controls. If the antigenic determinant capable of specific stimulation of such cells is present in the thymus this would be an additional reason for the presence of the germinal centres.

## Lympho-epithelial tumours of the thymus

Of even more interest than the germinal centres are the lympho-epithelial tumours of the thymus. In general, these are not malignant. A proportion produce or are associated with no general symptoms and are recognized only because of pressure symptoms or from routine X-rays. There are, however, three important conditions associated to a significant degree with such tumours—myasthenia gravis, pancytopenia with aregenerative anaemia and acquired agammaglobulinaemia.

Myasthenia gravis is much the most common. Here several points deserve mention. The cases associated with thymic tumour are, in general, in older patients; are almost invariably associated with demonstrable antibody, and more often show frank myositis. Removal of the tumour has no influence on symptoms and does not change serological reactivity—in both these respects differing from simple myasthenia in young women without tumour. The tumours are always of mixed lymphoid-epithelial type; the two cell types varying in dominance from an almost wholly epithelial and spindled cell tumour to a histological appearance indistinguishable from Hodgkin's disease.

Various interpretations of the relationship of the tumour to the myasthenic syndrome have been suggested. The only one consistent with the approach I have adopted is that in these cases we are concerned with active and widely dispersed clones of reactive immunocytes with capacity to react at a low level with thymic epithelium. The result of that interaction is a chronic proliferative stimulus to the epithelial cells with, as a corollary, a more or less equivalent stimulus to lymphocytic proliferation. There is a real possibility that the type of stimulus is to some extent equivalent to the long-acting thyroid stimulator substance found in thyrotoxicosis.

If this point of view is accepted it is a logical extension to believe that in the target organs or substances previously mentioned are antigenic determinants common to the thymic epithelium. The thymoma would then be a variable concomitant of an autoimmune process, whose significant target was: muscle and neuromuscular junction—myasthenia gravis; haematopoietic stem cell—aplastic anaemia; 'gut-associated' hormone—acquired agammaglobulinaemia. It may be relevant that the branchial cleft epithelium from which the thymus is derived has wide potentialities and gives rise to thyroid, parathyroid and lung. The possibility that among the epithelial cells of the thymus there exists a variety of partially differentiated cells is suggested by the already mentioned myoid cells, and from the well-known fact that when small cysts appear in the mouse thymus the lining cells frequently show typical cilia with the structure of those in the respiratory tract. If this is the case it would have important implications for the censorship function of the thymus.

It would be in accord both with the facts and with this formulation that there should be occasional cases of mixed conditions or of the co-existence with thymoma of types of autoimmune disease outside this group.

## ORGAN-SPECIFIC AUTOIMMUNE DISEASES

It has been a curious development in the last decade to find an ever-increasing group of 'diseases of unknown aetiology' being brought into the autoimmune category. One can almost suggest that any disease not of clearly visible genetic, nutritional, infective or traumatic origin which involves a definable function or organ and comes on after a phase of normal health should be regarded as a candidate for interpretation as an autoimmune disease.

## *Thyroid disease*

One of the most interesting aspects of general medicine in the 1960s is the sudden swing of interest toward an autoimmune interpretation of thyrotoxicosis. Frank autoimmune disease of the thyroid, Hashimoto's disease, has been recognized for many years and need only be used here as the classical example of organ-specific autoimmune disease.

Of all the commonly looked-for auto-antibodies, those detected by thyroglobulin-coated tanned cells or complement fixation with thyroid extracts are the most frequently found. In many instances, thyroid auto-antibodies are found with no evidence of thyroid dysfunction but equally, examination of the thyroid *post mortem* in elderly females without history of thyroid disease will show many with significant minor infiltration by lymphoid cells in the thyroid. The serological tests used to measure anti-thyroid antibodies are numerous and several distinct antigenic determinants are probably concerned. The antigens used are thyroglobulin, a second colloid antigen, and a microsomal antigen extractable from thyroids from patients with thyrotoxicosis. Immunofluorescent studies using sections of thyroid tissue differentiate the location of these antigens. Complement fixation and tanned cell methods are also applicable with some, while the thyroid-stimulating antibody (LATS), mentioned below, is detected by *in vivo* tests.

There is quite extensive evidence to indicate a genetic component in thyroid disease. Close relatives of patients with Hashimoto's disease have a higher incidence of positive serological findings, particularly in women, than relatives of matched control individuals. Thyrotoxicosis and Hashimoto's disease are both much more frequent in women than in men but show quite different age incidences. Both patterns of age incidence have, however, analogies with standard autoimmune diseases: thyrotoxicosis with SLE and Hashimoto's disease with rheumatoid arthritis. It is reasonable, therefore, to explore the same general hypothesis of one or more somatic mutations on the background of a genetic deviation from normal. As before, the combination of genetic and somagenetic qualities is responsible for the resistance of the immunocytes involved to the normal homeostatic mechanisms.

What is less clear than in other instances is the distribution of the relevant antigens in the body and their role in controlling the numbers and type of reactive immunocytes. It is relatively easy to produce anti-thyroglobulin antibodies in experimental animals by injection of substantial amounts of thyroid extracts, even without adjuvant. There is therefore no significant tolerance to the specific antigenic determinants. In human adults there are minute amounts of thyroglobulin in the circulating blood and this amount is increased in association with surgical manipulations of the gland. Most individuals do not respond to these concentrations by demonstrable antibody production. It seems the most appropriate initial approach to the human situation to make the following assumptions: (*a*) there is insufficient circulating thyroid antigen of any type in the normal individual to produce tolerance or to induce antibody production, (*b*) small numbers of cells reactive with thyroid antigenic determinants are produced but the position is such that contact with antigen does no more than produce very small amounts of Ig M antibody and no memory cells and (*c*) in persons genetically predisposed, immunocytes of aberrant quality arise by somatic mutation which (i) escape contact with any organ-specific antigen in primary sites of differentiation, (ii) are avid enough to react with thyroglobulin, producing Ig M, Ig G, and memory cells and (iii) have 'invasive' or similar qualities allowing entry into thyroid tissue with production of diffuse lymphocytic infiltration and germinal centres. It is highly probable that such populations of pathogenic immunocytes are heterogeneous and with different functional potentialities.

In earlier discussion the suggestion was raised that an automatic protection of inaccessible antigens from involvement in auto-immune processes could be provided if movement of lymphocytes through the tissues was confined to committed cells, that is, those derived from cells which had already made effective contact with an antigen. Virgin cells would be limited in their movement to entering and leaving peripheral lymphoid tissues. It is difficult to obtain evidence for or against this hypothesis, and it should probably be kept in reserve until a technical approach to testing it is available.

There are considerable numbers of cases on record in which the sequence of thyrotoxicosis, Hashimoto's disease and myxoedema can be recognized. All are strikingly commoner in females. The sequence, however, is by no means invariable and *a priori* it seems unlikely that all cases of thyrotoxicosis have an autoimmune origin. Nevertheless, from our present approach, what is required is simply an interpretation of the part which can be played by auto-immune processes in the sequence.

Interest in thyrotoxicosis as an autoimmune condition arose mainly because of the recognition that LATS present in the blood of some patients with thyrotoxicosis was not, as originally thought, a pituitary hormone of abnormal character but an Ig G. In addition it had long been known that, next to Hashimoto's disease, the highest proportion of sera with antithyroid antibodies came from thyrotoxicoses and, as already mentioned, the age and sex incidence is that to be expected of an autoimmune disease.

Unequivocal demonstration of LATS is possible only in a pro-portion of cases, but is characteristically present in severe cases with ophthalmic symptoms and pre-tibial oedema. As yet it is not known whether in such cases there is lymphocytic infiltrate in unusual amount in thyroid substance, or whether germinal centres in the thymus are regularly present. Both could be expected and actual demonstration that this was the case would increase our confidence in ascribing thyrotoxicosis to an autoimmune process. An Ig G antibody might well be the primary agent in stimulating excessive activity of the thyroid. The actual antigen or antigenic determinant involved is not yet known and, by implication, it is different in specificity from both thyroglobulin and cytoplasmic antibodies. As in other autoimmune situations the initiation of the process could, by increasing the amount and accessibility of the antigen, generate a vicious circle and also increase the likelihood that other types of immunocyte reactive with thyroid antigenic determinants would be brought into active proliferation.

In all forms of autoimmune disease we find a basically similar situation—the appearance of a wide range of immunocytes reactive against the antigenic determinants involved and a complex and variable ecological situation dependent on the local availability of the antigens concerned and any mutant characters influencing the immunocytes' response. Admittedly this is a very flexible approach which can rather readily be adapted to meet any anomalous circumstance that may arise, but the actual facts of autoimmune disease *are* so individual and only broadly reproducible that any approach must of necessity be a flexible one.

Some special features of the thyroid autoimmune complex call for brief comment. The thymus is regularly enlarged in thyrotoxicosis and when its upper isthmus area is biopsied during surgery of the thymus, about one-third of the specimens show germinal centres. It is unfortunate that there is no safe and convenient way of obtaining biopsy material from the thymus. The sequence of thymic changes associated with a severe case of thyrotoxicosis could be very enlightening and it could well emerge that some of the antigenic determinants were common to thyroid and thymic epithelia.

*Other conditions involving specific organs*

Another organ of special interest is the adrenal. Non-tuberculous Addison's disease is usually autoimmune in character and organ-specific antibodies are commonly demonstrable. At least half such cases also show thyroid antibodies and there is one reference to the occurrence of germinal centres in the thymus. On the other hand, the great majority of sera positive for thyroid antibodies are negative when tested with adrenal antigens. Of all the organs of the body the thyroid is the most prone to autoimmune disease. In the absence of any reason to the contrary, we can assume that if some human individuals can produce a particular organ-specific antibody others in principle can also give rise to cells with this capacity and, further, that there is no *a priori* reason for believing that any one type of autoimmune pattern will arise with greater frequency than another. It is axiomatic in our approach that any of the mutant forms of resistance to immunological controls arise wholly independently of the immune pattern(s) carried by the cell line. If this is true, then the frequency of different local autoimmune disease types in women must be related to antigen factors, presumably concentration and accessibility. The pre-eminence of thyroid disease would presumably be related to the marked changes in physiological activity associated with emotion, pregnancy, etc., and perhaps an associated leak of antigens into the circulation on a relatively larger scale than other organs.

The next most commonly found antibody is directed toward an antigen present in the gastric parietal cells. This antibody is found in nearly 100 per cent of patients with pernicious anaemia and there is a strong positive correlation with the presence of thyroid antibodies. Using the same type of argument, the accessibility of gastric antigen presumably depends on local trauma and infection, particularly in association with low acid secretion.

The antibodies concerned are each specific so that the positive correlation of antibodies directed against thyroid, gastric cells and adrenal must be more deeply based and at the theoretical level represents the main problem of localized autoimmune disease. Expanding slightly the above interpretation, we assume that in certain individuals, notably females, there is a predisposition for mutant immunocytes to appear in which there is a heightened capacity to induce local autoimmune disease. The exact character of the *cellular* expression of the genotype is unknown, but one or more of the following differences from the normal might be involved: (*a*) resistance to destruction by antigen in the phase of first differentiation, (*b*) enhanced capacity to swing from Ig M production to Ig G and (*c*) capacity to move through tissues in the uncommitted state. Such mutant character will be randomly distributed amongst immunocytes and will be expected to have survival advantage only when the associated immune pattern is reactive with an organ-specific antigenic determinant and when it can meet an adequate concentration of the determinant. With a constant very small influx of mutants of the necessary type, many subjects may escape disease indefinitely unless some initiating trauma or infection allows an abnormally high concentration of accessible antigen. In general the proportion of antibodies for each organ will depend on the various factors relevant to this accessibility of antigen. In fact, the various figures in the literature are not greatly discrepant from what would be expected, on the very simple assumptions that 10 per cent of females in a standard population will produce significant numbers of mutants and that the chance of these initiating antibody production is 1:3 for both thyroid and gastric parietal cell and 1:200 for adrenal. Then the proportion showing antibody in four different populations would be as in table 6.

TABLE 6. *Occurrence of tissue-specific antibodies in human populations*

| Population | Thyroid | Gastric parietal cell | Adrenal |
|---|---|---|---|
| Standard | 3·3 (4) | 3·3 (7) | 0·01 (—) |
| Hashimoto's disease | 100 (97) | 33 (27) | 0·1 (—) |
| Pernicious anaemia | 33 (33) | 100 (83) | 0·1 (—) |
| Adrenal atrophy | 33 (52) | 33 (—) | 100 (60) |

Percentages to be expected from the simplified hypothesis given in the text with, in parentheses, approximate percentages observed in series recorded by Mackay and Burnet or Doniach and Roitt.

These fit the published data reasonably well, except for idiopathic adrenal disease, where 60 per cent gave adrenal antibody but 52 per cent gave thyroid antibody. This merely accentuates the point being made by suggesting that a rather higher grade of anomaly is needed to give adrenal damage.

IMMUNOPATHOLOGICAL CONDITIONS ASSOCIATED WITH THE ENTRY OF ENVIRONMENTAL FACTORS

## Rheumatic fever

Rheumatic fever has many of the features of an autoimmune disease but it is quite clearly related to streptococcal infection. The connection is currently ascribed to the accidental resemblance or identity of a streptococcal determinant and an inaccessible tissue determinant in cardiac muscle and possibly in synovial tissue. This has been taken by a number of authors as a prototype for all or most autoimmune diseases, so allowing the postulate of an external cause (usually an unidentified virus) instead of the less easily comprehended concept of somatic genetic origin.

The pathogenesis of rheumatic fever is not yet fully understood

although Kaplan's demonstration of common antigens in streptococci and cardiac muscle makes it highly probable that the carditis
is largely a result of damage by the antibodies and immunocytes
that have emerged as a result of the streptococcal infection. The
origin of the rheumatic nodule (including the Aschoff body) and
of the polyarthritis is controversial, but it seems likely that they
represent sites of lodgment of dead or lethally affected streptococci
and associated antibodies. The situation is clearly very complex and
has a distinct resemblance in some aspects to delayed hypersensitivity. The whole picture suggests that acute or subacute
infection of the tonsils provides a situation where antigen, toxic
substances and immunocytes are intimately related in a fashion
reminiscent of the Freund adjuvant granuloma.

By the time the infection has been present for from 10 to 20 days,
a large population of immunocytes, including plasma cells and
'aggressive' lymphocytes, has accumulated and passed to other
regions of the body. In one of two possible ways the synovial cells
are vulnerable to immunocyte and/or antibody attack. They may
have taken up streptococcal antigens circulating in the early stages
of the infection or the vicinity of the synovial tissues may be a
preferred site for lodgment of antigen–antibody complexes. There
is no evidence that an antigenic determinant cross-reacting with a
streptococcal product is present in synovia but it cannot be ruled
out. Possibly in its favour is the migratory character of rheumatic
polyarthritis, indicating that there is some resistance to starting
the inflammatory process but that, once started, it involves all the
susceptible cells in the joint concerned. If a blood-borne antigen
from the streptococcal nidus in the tonsils was concerned it should
be readily available to antibody also arriving by the blood. If,
however, the antigen is inaccessible until the cell containing it is
damaged, the independent build-up of a destructive process in
each joint cavity is to be expected.

The process can be pictured as starting with the deposition of
antigen–antibody complex in the perivascular region of the joint.
This will result in some local damage with leak of antigen from
synovial cells and increased local permeability with entry of cells
and antibody from the blood. Some of the entering immunocytes
will be reactive and the destructive process will continue.

## The general problem

This type of action can be considered in a more general form. Inaccessible antigenic determinants as discussed in relation to myasthenia gravis and autoimmune thyroid disease do not provoke tolerance or antibody production in the normal individual. Small numbers of pre-adapted cells are available as for any other random antigen, and when infection by a micro-organism carrying a major antigenic determinant equivalent to one of the inaccessible body antigens occurs, the normal response takes place giving antibody and an increased population of immunocytes. This does not, in itself, give rise to attack on the cells carrying the equivalent body antigen any more than simple immunization of a rabbit with a thyroid extract provokes damage to its thyroid. In both cases some additional 'aggressiveness' of the immunocytes is needed either genetic in origin or invoked by the use of Freund's adjuvant in the experimental situation. It may also be highly relevant how the immune response is distributed among immunoglobulins A, G and M and the various functional forms of immunocyte including those associated with delayed hypersensitivity. A major function of Ig G antibody is to moderate or annul responses in other immunological categories.

It is clear that if a commonly present extrinsic antigen has an accidental cross-over with an inaccessible body antigenic determinant, any tendency of the individual to autoimmune disease will be biased toward the organ in question. Rheumatic fever and glomerulonephritis may well be determined in part by the fact that the region of proliferation of immunologically competent cells is probably at the site of infection where the infected tonsil has many of the qualities of a Freund granuloma. Any anomalous mutants of high survival value and potential pathogenicity will have an abnormally favourable situation to proliferate, a fact that may be responsible for the acute character of rheumatic fever.

In rheumatoid arthritis we have the interesting probability that low-grade streptococcal infection in older people may provide an ever-present opportunity to initiate an immunological process which develops as an autocatalytic one by the appearance of denatured $\gamma$-globulin as the bearer of target antigen determinants. This, however, can only happen when the individual has (*a*) the necessary genetic constitution and (*b*) can allow the appearance by somatic mutation of the abnormal clones which can react pathogenically with denatured γ-globulin.

*Non-bacterial antigens, including drugs*
The next possibility to be considered is based on the incorporation of a foreign antigen in the cell by a low-grade viral infection. It is essentially immaterial whether the new antigen is a viral protein, or a host protein that has been rendered foreign by the incorporation of a viral episome in the cell genome or some equivalent process. Three experimental examples may be mentioned:

(*a*) It is generally accepted that polyoma virus fails to produce tumours in animals injected after the neonatal period because antibody or a cellular immune response against a polyoma-induced antigen eliminates neoplastic cells by a process analogous to the homograft reaction (see chapter 13). Here there is nothing equivalent to autoimmune disease since the only cells damaged are neoplastic ones.

(*b*) Lymphocytic choriomeningitis infection is harmless if there is no immune response normally demonstrable as perivascular cuffing with lymphocytes in the CNS. This can be prevented either by embryonic induction of tolerance by irradiation or by the use of cytotoxic drugs. The actual antigenic determinants concerned are unknown and the effect is rather remote from autoimmune disease, but it is an example of damage by an immune response.

(*c*) There is available a strain of Rous sarcoma virus which, when injected into neonatal rats, multiplies and gives rise to a curious multicystic disease of the lymphatic system. With complete neonatal thymectomy immediately before inoculation with the virus the rats remain healthy. The lesions affect tissues in which the virus can be shown to multiply and presumably represent a cellular immune response equivalent to delayed hypersensitivity. Here the resemblance to an autoimmune condition is more striking.

There are no established examples of human conditions fitting into any equivalent categories, but there are perhaps hints that hepatitis virus infections may eventually have to be considered as a possible source of something equivalent to group (*c*). It is known that hepatitis virus infections in man may be long lasting and the range of cells that can suffer non-damaging infection may be more widespread than the liver. Suggestions have been made that lupoid hepatitis and the New Guinean disease kuru (a cerebellar degeneration) may be autoimmune reactions against tissues carrying a new or newly accessible antigen for which hepatitis virus was in some way responsible. Once satisfactory ways of culturing the hepatitis viruses are available, these questions should be readily resolved.

Quite apart from micro-organisms, non-living, potentially noxious material can enter the body and there are two practically important types of immunologically based disease which can result—hay fever and similar allergies and a variety of drug reactions. Like autoimmune disease, these are the reactions of a minority of people though sometimes a large minority. The syndromes cannot be regarded as physiologically normal ones and some intellectual framework, flexible as always, must be found to fit them.

In many ways the best studied and most interesting are the drug reactions which produce haemolytic disease or thrombocytopenic purpura. Here the damage seems to result from the interaction of antigen (hapten) with antibody, producing a complex which attaches to red cells or platelets and with complement leads to their functional destruction. The process can be clearly demonstrated *in vitro*. In a typical case, for example, of stibophen haemolytic reaction, haemolysis of the patient's or anyone else's red cells can be produced in a mixture containing appropriate amounts of drug, patient's antibody, red cells and complement. In addition to the red cells all three components must be present and analysis of the reaction shows that the first effect is the firm union of drug to antibody. This union changes the quality of the immunoglobulin and perhaps after an intermediate stage of aggregation it is 'nonspecifically' attached to red cells if these are present. With complement, haemolysis follows.

In general, if the antibody is principally Ig M, the complexes attach to red cells; if Ig G, to platelets, with production of thrombocytopenic purpura. N. R. Shulman and others have suggested that what are normally regarded as autoimmune haemolytic anaemias or thrombocytopenic purpuras are in fact the response to an undetected virus or other antigen.

Perhaps a more illuminating way to approach the phenomena is to regard them as being based on the same foundation as auto-immune disease in general, the appearance of unusual types of immunocyte. Only a small proportion of persons given the drugs in question develop haemolytic disease or purpura; in some way they differ from the normal. What is probably the simplest answer is that in the susceptible individual there are a few immunocytes with immune receptors of unusual pattern avid enough to bind the hapten directly. Such cells would also produce small amounts of similarly avid antibody. It is possible that the hapten is itself immunogenic in the sense of being able to stimulate such unusual immunocytes directly, but other explanations are possible. In any case, ingestion of the drug in the susceptible individual results in immunological stimulation and significant amounts of the same abnormally reactive antibody are produced and liberated into the bloodstream. The more important factor is the abnormal immunocyte and antibody, not the nature of the drug.

Hay fever and other allergies can be looked at similarly. The main abnormality seems to be in the production of an abnormally high proportion of antibody in the Ig A (Ig E) category. The overwhelming importance of genetic (including somatic genetic) factors is shown by the occurrence of highly allergic subjects who produce almost all their diphtheria antitoxin in the form of Ig A. As yet we seem to have no suggestion as to the nature of any normal homeostatic mechanisms which maintain the ratio of Ig A to Ig G in the plasma, and until this is remedied there is nothing to be said at the theoretical level about the mechanism by which a genetic predisposition to allergic disease is expressed.

# 13 Immunological serveillance and the evolution of adaptive immunity

In the last decade there has come into being, without either flourish of trumpets or serious controversy, a general current of belief in what I have come to call 'immunological surveillance'. Somatic mutation is constantly occurring in the mammalian body, and on current genetic theory all mutations in structural genes will result, when the gene is activated, in the production of some protein in which the amino acid sequence differs in some respect from the form genetically characteristic of the individual. Theoretically, at least, immune response against the new protein is therefore a possibility. If a sufficient amount of the new antigen is liberated there will appear a clone of immunocytes which could be expected to mount an attack equivalent to a homograft reaction against the mutant cells and eliminate them from the body. One can therefore picture a form of surveillance by which the body is being continually patrolled, as it were, for the appearance of aberrant protein (or perhaps also polysaccharide) patterns.

The idea has arisen from consideration of a variety of immune phenomena associated with clinical or experimental malignant disease. It is in the field of experimental carcinogenesis that we find experimentally verifiable extensions of the idea.

IMMUNOLOGICAL ASPECTS OF NEOPLASTIC DISEASE

*Experimental carcinogenesis*
When a tumour is provoked by a virus or chemical carcinogen it is reasonable to assume that a process operationally equivalent to a sequence of somatic mutations has occurred, giving rise to at least one protein antigenically distinguishable from any in normal cells. With growth of the tumour, enough of such protein should be produced to allow a demonstrable immune response. This has now been fully established. Tumours produced by injection of polyoma and other 'tumour viruses' in newborn mice or hamsters

contain a new antigen specific for the virus used and currently interpreted as representing the work of a viral episome incorporated into the host cell genome. The immune response to this antigen confers resistance against polyoma tumour cells which are presumably eliminated by the standard method of homograft immunity. Sarcomas produced in mice or rats by intramuscular injection of methylcholanthrene are also antigenic in a similar sense, but here each tumour seems to have a different new antigenic pattern which presumably arises by some form of somatic mutation. It is possible to amputate the leg bearing the primary sarcoma and establish the tumour cells in tissue culture or by passage through syngeneic animals. They can then be used to show that the mouse in which the tumour arose has an active capacity to reject the tumour cells when they are injected, although control mice of the same strain develop tumours.

The most striking characteristic of the tumour viruses is that, except under such special conditions as neonatal thymectomy, tumours are produced only if the virus is administered to newborn mice or hamsters. Injection of virus into older hosts produces antibody to the virus and no tumours. From virus-induced tumours of both polyoma and SV 40 viruses, strains of tumour cells have been obtained that are free of virus and can be transplanted like any other tumour in syngeneic animals. There are two ways by which animals can be rendered resistant to such a tumour. The potential host can be inoculated when more than a week or two old with the cancer virus $X$ or it can be inoculated with a transplant of a tumour induced by virus $X$ in an allogeneic strain of host. In the second case the tumour graft is rejected as a homograft but a few weeks after the immunizing manipulations such mice show a significant resistance to syngeneic tumour which is specific. Tumours produced by polyoma virus have a common antigen which is not demonstrable in the virus but is generally thought to arise by the activity of a viral episome in the malignant cell genome. Similarly, tumours arising after SV 40 virus infection have a common antigen unrelated to that of the polyoma tumours.

The necessity for neonatal infection if tumours are to be produced is currently interpreted as a manifestation of the relative immunological incompetence of very young mammals. Whatever

287

the process of malignant transformation, it takes place rapidly and is associated with a definable antigenic change or addition in the transformed cell and with active proliferative growth. In the neonatal animal the proliferative process is well established before there can be sufficient stimulation by the new transplantation or histocompatibility antigen to produce the immunocytes that could inhibit the growth of the tumours. When virus is given later there is, by hypothesis, transformation and initiation of neoplastic growth but the response is more rapid and enough specifically active immunocytes are mobilized to nip the process in the bud.

The outcome, tumour or no tumour, is dependent on a balance of factors which can be tipped by minor circumstances. An occasional tumour appears when virus is injected into older animals and many more if such animals have been neonatally thymectomized. If, after neonatal injection of virus, a series of further injections is given, either of virus or of irradiated tumour cells, the incidence of tumours can be sharply reduced. Both findings point indubitably to the importance of immunological factors.

In some ways these virus-induced tumours provide an ideal experimental model of the process of immunological surveillance. New tumours with a definable new antigen begin to proliferate. If the immune response is effective they may fail completely to give any symptomatic or pathological manifestation. This is precisely what is assumed to occur in the process of immune surveillance.

*Clinical phenomena in man*
Malignant disease is proverbially progressive and lethal unless the whole mass of neoplastic cells can be removed by surgery or its equivalent. In fact, however, there are well-known phenomena which point to a significant impact of immunological processes. There are rare but well-documented instances where an established malignancy proven by biopsy and histological examination has retrogressed spontaneously. Most clinicians would agree that unless there was a significant resistance, presumably immunological, against the proliferation of malignant cells and their spread through the body, the results of surgery, radiation and drug therapy of cancer would be much worse than they are.

There are several other aspects of neoplastic disease in man

288

which are or may be relevant. Pathologists concerned with cancer biopsies are aware that there are considerable differences in the extent to which there is a round-celled reaction at the edge of the tumour mass. In some, plasma cells are conspicuous among lymphocytes and histiocytes. Two studies to correlate the intensity of lymphocytic and plasma cellular response with survival showed an inverse correlation. The more extensive the infiltrate the longer the patient was likely to survive. It is a sound rule that where lymphocytes and plasma cells congregate, an immune process of some sort is going on. The correlation with prognosis suggests that the process here is a homograft response against cells which have been recognized as antigenically distinct from normal cells.

It has been of obvious interest to pathologists to seek for early stages in the development of neoplasia by examining tissues that are prone to malignant change but which have been taken from autopsies on persons dying from unrelated conditions. In several such investigations, systematic histological study has shown that areas of apparently malignant change are much more frequent than would be expected from the known age incidence of clinically diagnosable cancer of the organ concerned. There are, of course, uncertainties about the histological diagnosis of cancer, and it may not be wholly legitimate to assume that a nodule with the histo-logical character of a carcinoma would inevitably become clinical cancer if it were not actively inhibited by an immune process. As far as they go, however, results for prostate, thyroid and adrenal point in this direction. Of rather special interest is the quite differ-ent shape of the curves of age-specific incidence of histologically diagnosed carcinoma of the prostate and of death from prostatic cancer. The latter rises very sharply with age and only becomes significant 15–20 years after 'malignant' areas begin to appear in the autopsy sections.

Analogous findings are on record in regard to neuroblastomas of the adrenal, which are forty to fifty times more frequent in autopsies on newborn children than would be expected from the later clinical incidence of the tumour. Thyroid nodules classed as malignant on histological criteria are also vastly more common than overt thyroid carcinoma.

One of the recurrent themes of cancer research has been the

search for antigens in human tumours which were specific either for one class of tumours or for malignant cells as against normal cells. In the light of modern work on histocompatibility differences, most such studies were based on naïve ideas and no clear positive conclusions emerged. There are, however, a number of clinical phenomena which, though rare, are directly associated with malignant disease and seem undoubtedly to be mediated by immunological processes. If this is so it is worth while to look at the ways by which abnormal antigens could be produced by malignant cells.

A tumour cell, like any other somatic cell, could in principle produce any antigen to be found in any tissue of the organism in which it developed. It is also in principle susceptible to mutation or its equivalent by which aberrant proteins with new antigenic determinants may be produced. As a tumour develops, 'progression' to greater malignancy with increasing aneuploidy must mean possibilities of gross disturbance of function in the genome. Mutation will be common and, in addition, if de-repression of 'wrong' loci should occur, this could result in the synthesis of proteins normally produced only by specialized cells in particular organs. If a tumour cell developed a metabolic anomaly of this type and concomitantly proliferated freely so as to produce a substantial mass of tumour, abnormal proteins and their characteristic antigenic determinants could be released in significant amounts.

This point of view can be used in an attempt to interpret some of the rarer concomitants of cancer in man.

(*a*) The abnormal protein may be a potent hormone producing its characteristic effect. The best known example is the occurrence of hypercalcaemia in cancer patients, associated with the presence of parathyroid hormone in the tumour. In a number of cases, removal of the tumour corrected the hypercalcaemia. Other hormonal effects have also been reported. According to Lebovitz these include (in addition to parathyroid hormone) ACTH, antidiuretic hormone and thyroid-stimulating hormone. Tumours of the liver have also been recorded from which excessive symptom-producing amounts of porphyrins were liberated.

(*b*) The protein may have the antigenic quality of a substance normally present as an 'inaccessible' tissue-specific antigen in some distant tissue. According to the general rule discussed in chapter 12, this antigen can provoke the proliferation of immunocytes and antibody. Under appropriate genetic and perhaps other types of individual circumstances, this could provoke localized autoimmune disease. The standard example which may call for such an interpretation in the carcinomatous neuromyopathy which in one form or another is fairly often associated with oat-celled (anaplastic) bronchial carcinoma. It may be suggested that the symptoms and lesions in the CNS represent an autoimmune attack by immunocytes (and perhaps antibody) directed against antigenic determinants specific for CNS cells and 'inaccessible' in the sense of not being capable of producing normal tolerance. Different antigenic determinants are probably responsible for the anatomical distribution of the lesions. The effect of this fortuitous synthesis and liberation of normally inaccessible antigens will only be visible (i) in individuals capable of giving rise to pathogenic clones of immunocytes, that is, genetically susceptible to autoimmune disease (one component of this susceptibility is female sex), (ii) when the target organ is such that immunological damage produces significant symptoms in a debilitated patient or (iii) when, for one reason or another, antibody is produced which can be detected by available techniques. The hypothesis accounts satisfactorily for (i) the relationship to a specific histological type of tumour, (ii) the limitation to a small proportion of those with the tumour, (iii) a higher incidence in females than males, (iv) a wide variety of localization within the CNS and (v) the presence of tissue-specific antibody in a proportion of cases. Another important condition usually associated with malignant disease, and with all the qualities of an autoimmune process, is dermatomyositis. The same general hypothesis could be applicable here.

(*c*) The protein is a normal body component and has no detectable effect.

(*d*) The protein is of aberrant structure and therefore treated as a foreign antigen. In most instances there would be no indication of the existence of either the new antigen or the antibody. It is possible that one particular kind of aberrant protein is produced by a considerable range of malignant tumours. If so, it might be found that a particular tumour extract might react in immunologically demonstrable fashion with a proportion of sera from cancer patients but not with normal sera.

Part at least of the characteristic age incidence of cancer probably depends on the progressive weakening of the surveillance function with age. It is well known that elderly patients with cancer will often fail to reject an inoculum of a standard tissue culture line of human carcinoma cells.

There is thus substantial evidence for the contention, first, that new antigens arise in association with the changes which characterize malignant cells and second, that immune reaction to the new antigens occurs under suitable conditions. These are important points when it comes to deciding how the immune mechanism developed in the course of evolution.

## MUTATIONAL ORIGIN OF HISTOCOMPATIBILITY ANTIGENS

In this book I have paid much less attention to homograft immunity and organ transplantation than the current volume of work and surgical interest in the field would seem to warrant. This is primarily because its interest for the central theme of the biological significance of immunity is only peripheral. There is no counterpart in nature of blood transfusion or tissue transplantation but it is the object of this chapter to show that two basic mammalian plenomena, placentation of the fertilized ovum and malignant disease, are better understood if the immunological aspect is included in their consideration. In the first instance it is necessary to consider the significance of what might superficially seem to be a surprising finding, namely, that in every type of bird or mammal that has been examined, any two individuals of the same species will reject each other's skin grafts. There are, of course, exceptions, but it must be firmly emphasized that pure line strains of mice or rats are biological monstrosities that could never survive in nature. Nevertheless it has only been by the use of pure line animals that most of our recent understanding of immunity has come, and for the discussion of immunological surveillance and its evolutionary significance the nature of histocompatibility differences is vital.

There are now some hundreds of registered pure line strains of mice, many of which had common ancestors. It is regarded as a reasonable rule that if from a pure homozygous line one develops two parallel lines which are thenceforward kept quite distinct, the two lines will probably begin to show lack of reciprocal acceptance of skin grafts by the time they are separated for ten generations. These histocompatibility changes are genetically based and presumably arise by mutation. Several complex loci are involved and in mice the H2 locus is one responsible for major differences correlated with parallel differences in the agglutinogens of the red blood cells. A number of other loci may also be involved, and there is a particularly interesting set of phenomena based on a histocompatibility gene carried by the small male Y chromosome. In many homozygous strains of mice, male skin grafted to a female will be rejected, since the Y-based antigen is foreign to the XX host. Female skin is, however, accepted by a male host since it contains no antigen foreign to an animal with both X and Y chromosomes.

It seems therefore that the H2 and other histocompatibility loci must be relatively highly mutable and that there is either a premium for survival on diversity or there is no selective advantage of any particular pattern. In discussing the genetic situation one must be careful not to forget that artificial brother–sister mating is specifically designed to avoid conferring any advantage on heterozygosity, so that if a line with histocompatibility genes *AA* undergoes a mutation by which one allele of *A* becomes *B*, then even if *AB* is more vigorous than *AA* or *BB* there will be around a 12 per cent chance that the *AA* line will, in a few further inbred generations, become *BB*. However, it is known that wild or pen-bred stocks are highly heterozygous in histocompatibility characteristics and there is at least prima facie evidence of some advantage for survival when the members of a species are heterogeneous in histocompatibility characteristics. The first reason for this that comes to mind is that the absence of such differences would allow the transmission from old to young of malignant disease. If all human beings were antigenically identical, malignant cells, for example, from carcinoma of the skin in an elderly individual, could implant in any abrasion on the skin of an infant. A little thought will show, in fact, that once any sort of malignant tumour developed it would spread probably with steadily increasing invasiveness throughout the community.

To prevent such a calamity we believe that nature, to speak teleologically, invented the mechanism of vertebrate immunity. In some way the body even of the infant must recognize and reject any cell that is not its own. For this there must be differences to recognize and a means of recognizing such differences. It is the central contention of clonal selection theory that the mechanisms of signal and sensor, stimulus and receptor, arose in characteristically biological fashion by making use for a specialized purpose of the already established phenomena of germinal and somatic mutation.

In developing this approach we assume that the regions of the genome coding for histocompatibility antigens are for some reason either more highly mutable than other regions or, perhaps more likely, a much smaller proportion of mutations result in lethal or nonfunctional results. It is logical to assume that this holds for these loci as much in somatic cells as in germinal cells. Since there is evidence that most or all somatic cells produce histocompatibility antigens it follows that the corresponding structural genes must be active in transcription as well as replication with perhaps greater liability to mutation. We can also assume that if historically a certain strain of mouse has given rise to mutant forms carrying histocompatibility antigens *A*, *B*, *C*, *D*, then it is reasonable that in an *A* mouse, *somatic* mutation is likely to produce similar results so that occasional cells have antigens *B*, *C* or *D* as well as or in place of *A*. We should expect, therefore, to find in any homozygous mouse a few cells of alien character, bearing foreign antigens which would in general correspond to the sort that over the history of the species had arisen by germinal cell mutations. In addition, the existence of the foreign antigen would sooner or later allow the proliferation of complementary clones of immunocytes with or without some production of antibody. In this way we should have a highly efficient mechanism to ensure complete protection against the potential calamity of free transmission of cancer cells amongst the population.

There is a good deal of evidence to suggest that this formulation is in fact very near the truth. Simonsen used a fairly elaborate graft-versus-host assay to test the power of strain $A$ to provoke splenic enlargement in $F_1$ mice in which one parent was $A$ and the other differed in minor or major histocompatibility characters. Where there was only a minor difference it required large numbers of spleen cells from $A$ to produce the standard degree of splenic enlargement in $(A \times A^1)F_1$s, but if $A$ mice were hyperimmunized with $A^1$ cells and then used as donors, very much smaller numbers were required. When $(A \times B)F_1$s were used in which $B$ differed strongly from $A$, very few cells from either normal or immunized donors were necessary to produce the standard effect. It was as if in $A$ there were already plenty of immunocytes active against $B$ antigen, while any $A^1$ antigen which had appeared normally in $A$ was apparently not sufficiently distinctive or in too small amount to provoke a reaction.

Even more impressive results are obtained using the CAM reaction when adult fowl leucocytes are deposited on the chorioallantoic membrane. This is also a graft-versus-host reaction mediated by differences at the B locus. It can be shown that the number of foci produced on the chorioallantois is a measure of the number of cells in the inoculum capable of reacting with the major antigens *not* found in the donor but present in the embryo. Immunization of the donor with antigens of the recipient strain has no significant effect. The number of reactive cells is sometimes quite high (as much as 1 per cent of the large lymphocytes in the inoculum). The most likely interpretation is that this is the result of stimulation by an antigen not present in the genotype of the donor and therefore presumably or necessarily arising by somatic mutation. This, however, has not been properly established and it would be unwise to exclude the alternative possibility that in the region of the genome coding for the combining site there are short stable sequences which ensure that the appearance of a related group of specificities follows change in adjacent labile nucleotides.

A possibility of special interest is that abnormal lymphoid cells (immunocytes) may differ antigenically from normal and therefore be in themselves subject to the surveillance function. This has been suggested on mathematical grounds by Burch and at the clinical level provides a reason, otherwise not easy to find, why autoimmune diseases do not always persist indefinitely. I know of no direct experimental evidence to substantiate the suggestion, but it has been well established that thymic lymphocytes differ anti-genically from the majority of circulating lymphocytes.

The picture that emerges then is that in two complementary areas the mammalian organism has made a virtue of mutation as such to evolve, on the one hand, antigenic heterogeneity within the species and, on the other, to create the diversity of immune pattern in the fashions discussed in chapter 6. The existence of antigenic heterogeneity and of a cellular mechanism to recognize and react with it are both necessary to ensure that foreign cells can be recognized and destroyed. This is an adequate answer to the problem of protection against cancer contagion but it has other implications which may be more important.

The whole process of evolution depends on the liability to error of any mechanism for the replication of information and of the indefinite perpetuation of such errors unless they can be corrected. The evolution of genetic systems has given rise to an indescribably elaborate process of chromosomal control to ensure that the organ-ism that arises from the fertilized ovum shall be a fully functional viable being. The processes by which differentiation of, and sub-sequent maintenance of, bodily function and structure take place make use of the same cellular mechanisms that are concerned with the transmission of inheritable qualities in the germ cells and, in so far as they are exercised in replication and transcription, must be similarly subject to error.

The likelihood of error is usually calculated in terms of the number of replications or generations and the figure of $10^{-8}$ per nucleotide has been suggested. In the human body there are of the order of $10^{15}$ cells of which at least $10^{13}$ are proliferating cells in blood and lymph systems and in expendable epithelial surfaces. Over the whole period of growth and maintenance an average generation time for any cell line of about one day is probably of the right order of magnitude. It must follow, in any large long-lived animal in which maintenance is dependent on a large turnover of cells, that somatic mutation is a highly significant source of potential disaster. For reasons discussed earlier in chapter 6, somatic muta-tion is only significant when it can be expanded by preferential proliferation of descendant cells and, from the point of view of danger to the individual, the most significant abnormalities are those which initiate or predispose to malignant disease. There are, however, marginal conditions sometimes of great clinical importance which cannot be called malignant but probably depend equally on somatic mutation. From our point of view the most important is autoimmune disease, but there are also Paget's disease of bone, polycythaemia vera, Hodgkin's disease and the various paraproteinaemias.

### THE ALLOGENEIC INHIBITION PHENOMENON

It has been recently suggested from the Karolinska Institute that although surveillance of the tissues for 'nonconformist' cells is a reality, the mechanism is not an immunological one.

The primary observation is that a cytotoxic effect on mouse tumour cells can be produced by lymphocytes from a mouse of a different strain which has been immunized either against the tumour or against normal cells of the strain in which the tumour arose. Normal allogeneic lymphocytes have no such effect but if, by the addition of phytohaemagglutinin, aggregation results, cytotoxic action basically similar to that mediated by immune cells is produced. Syngeneic lymphoid cells have no action on the tumour cells under the same conditions. The suggestion is clear that it is the close apposition of two cells differing in histocompatibility antigens which is responsible for the damage. The function of the immune state of the sensitized cell on this view is primarily to ensure intimate contact between the two cells.

The nature of the reactivity of the wandering cells of invertebrates toward foreign structures will be discussed in a later section (p. 302). The existence of such reactivity in the absence of any specific antibody or specifically reactive cells makes it easy to accept Hellström's point of view, but it by no means follows that the immunological component is unnecessary for surveillance. It is a perfectly arguable case that where there is a major histocompatibility difference between a cell in a tissue and all the adjacent cells, allogeneic inhibition could result in the death and removal of the alien cell. With minor antigenic differences such as are present in methylcholanthrene-induced murine sarcomas, however, all the indications are that for these to be differentiated from normal cells an immunological process is essential. Even the immune response may be initially small and require reinforcement to be effective. In general, the mouse or rat in which a tumour arises and is removed is not significantly resistant to the tumour line until it has been immunized with doses of irradiated cells.

Any necessity for serious consideration of the allogeneic inhibition phenomenon seems to have been removed by Mintz's recent description of composite 'allophenic' mice produced by embryological manipulation and fusion of early embryos of two distinct histocompatibility types. These show complete tolerance to and between both types of cell.

## SOMATIC CELL HETEROKARYONS AND THEIR SIGNIFICANCE

In connection with this question of the mutual incompatibility of vertebrate cells it is of very great interest that by suitable artifices this incompatibility can be overcome and hybrid cells or, more correctly, heterokaryons produced in which nuclei of quite different origin function normally in a composite cytoplasm. The possibility arose from studies of Japanese workers who found that when Sendai virus, a distant relative of influenza virus, was grown in tissue culture the cells were not obviously damaged but fused into syncytial masses without cell wall demarcation. Subsequently it was found that by inducing infection or even adding heat-killed Sendai virus to a mixed tissue culture of two types of malignant cells, one from a mouse tumour and one of human origin, mixed heterokaryons containing one or more of the two distinctive types of nuclei could be produced. This has been extended in many directions and a number of instances have now been reported where when simultaneous mitosis of two different types of nuclei has occurred the chromosomes have mingled and the two descendant nuclei contained a full complement of chromosomes of each of the two species. Sometimes a viable clone of cells of this double type has developed.

Even allowing for the fact that virus is present in the system this set of phenomena has implications of great importance for many aspects of biology. In a recent review, H. Harris points out that from the immunological angle the important implication is that *there are no intracellular mechanisms for the recognition of incompatibility*. A foreign nucleus appears to be quite at home in the mixed cytoplasm and synthesizes DNA and RNA in normal fashion.

A further observation of great immunological interest is made when human small lymphocytes are added to a culture of HeLa cells. In the absence of PHA or some equivalent agent, small lymphocytes show no DNA synthesis or mitosis in tissue culture. They can, however, produce heterokaryons with HeLa cells and in these the lymphocytic nuclei enlarge, become less heavily stained and actively synthesize DNA. The same enlargement of nucleus and appearance of DNA synthesis is seen even with nucleated erythrocytes from fowl or frog.

Harris, perhaps moving faster than is fully justified, finds that these results suggest that the orthodox view of differentiation as involving a very elaborate process of repression and de-repression in the nucleus is mistaken. In his view the nucleus is subject only to a 'coarse' control of its general synthetic ability, the fine adjustments of protein synthesis and assembly being controlled in the cytoplasm. Should this heterodox approach find experimental support it is going to be of obvious importance to immunological theory. For the present it is the only possible choice to adopt the ruling opinion that vertebrate cells function according to the rules derived from the study of bacteria. This will always be subject to the qualification that conditions are much more complex than in bacteria and that the model we apply may be basically correct yet from its nature divert our interest from other important aspects of the control processes in vertebrate cells.

In the present context of the interaction of body cells with mutant or allogeneic types, it is clear that the process is at a surface level and must involve the generation of signals by the impact of external pattern on surface pattern acting as receptor. This would, of course, be compatible with either of the two alternatives of allogeneic inhibition or immune surveillance.

## THE EVOLUTION OF THE IMMUNE PROCESS

### Defence processes in invertebrates

Vertebrates evolved from invertebrates, which from the time of the first multicellular organisms must have developed adequate mechanisms to protect them from micro-organismal infection. There are in mammals many signs of defence mechanisms unrelated to recognition of foreignness or antibody production, which represent direct developments from the primitive invertebrate system. It is our contention, however, that as vertebrates evolved, grew larger and more active, the need to counter the internal dangers of somatic mutation provided the frame within which the immune system we know in mammals had to evolve.

There is still much to be done in comparative immunology but it seems possible to make a series of general statements based largely on the writings of Good and his colleagues. The characteristic features of the mammalian immune system are: (*a*) capacity to reject homografts and accept autografts, (*b*) *specific* immune responses either by cellular activity or antibody production, (*c*) the thymus as the initial organ of differentiation, (*d*) lymphocytes and plasma cells as effector cells and (*e*) the immunoglobulins as the vehicle for antibody.

None of these five characteristics is found unequivocally in any invertebrate, though it would be hard to rule out the possibility that a vehicle for antibody, if it existed, could well have very little in common with the mammalian immunoglobulins or that a wandering mononuclear cell might not be equivalent to a lymphocyte. Fairly extensive studies have been made on insects as the most easily handled of the invertebrates. No antibody production has ever been detected and experimental zoologists have no difficulty in transplanting hormone-producing tissues from one individual to another. A striking instance of the absence of immune reaction in invertebrates can be seen in the way certain nudibranch molluscs when feeding on sea anemone tentacles transfer the nematocysts in functional form to their own tissues.

There is of course a fairly elaborate protective response against the entry of foreign organisms or material into invertebrates. After all, it was the classic observation of Metchnikoff on the phagocytic cells of the freshwater crustacean *Daphnia* that initiated the cellular approach in immunology. The evidence appears to be adequate that nothing recognizable as antibody could be detected after the injection of material that would be highly antigenic in mammals into representative insects and marine worms. Nor is there any evidence of homograft rejection although, as would be expected, if one moves too far away there is rejection. A *Cecropia* pupa will accept a tissue from a variety of related species but not from other orders of insects.

There can, however, be no doubt that there is some recognition of foreignness in invertebrates even if it is of a much cruder character than occurs in warm-blooded vertebrates.

In the first place, several authors have found that the body fluids of invertebrates may contain relatively high-titre agglutinins against a range of mammalian red cells and, just as would be the case with 'normal antibodies' of vertebrate origin, appropriate absorption experiments showed that in the coelomic fluid of a Californian lobster a relatively large number of different molecular forms of protein must be present in the fluid. There is no evidence that these 'normal antibodies' are increased by immunization and their existence may mean no more than that any soluble globular protein is likely to carry sequences of amino acids, perhaps in duplex or multiplex form, which have a physical adsorptive action of similar quality to that of an antibody combining site. It is axiomatic that in an organism which has reached a dynamically stable situation, cells must tolerate cells adjacent to them and that mobile cells and proteins must have no harmful effect on cells that are accessible to them. These imperatives would eliminate organisms in which mutation led to a pattern which had a specifically harmful effect but would have no bearing on simple polymorphisms with no significant impact on survival. It means nothing to a lobster whether or not its body fluid will agglutinate rabbit red cells or whether several different (but all harmless) patterns may be present amongst such proteins. But there is material here that could be worked up by mutation and selection till a vertebrate immune system evolved.

At the cellular level the arthropods and most other invertebrates have haemocytes in their body fluids which accumulate around foreign spicules or parasites. At most there is no more than a certain nonspecific increase in the rapidity of such mobilization after 'immunization'. The nature of 'recognition as foreign' still presents a problem. A partial answer may be to adopt the relatively naïve but still legitimate hypothesis that the cell surface has been evolved to be dynamically stable in relation to any other surface or soluble component that it encounters normally. In contact with 'anything else' the consequences have no evolutionary significance. The presence and extent of functional damage to a complex self-renewing surface will depend on straightforward physico-chemical factors impossible to detail but which one can expect to be specially active when the foreign material is remote from normal and especially if it has enzymatically active groupings. Partial denaturation of the surface of contact could well have as a concomitant adhesion to the foreign surface and the development of the typical foreign-body reaction. In the short-term the essential result will be to coat the foreign material, a parasite for example, with a continuous layer of haemocytes which eventually insulates it from providing any further damaging contacts.

The wandering cells of invertebrate body fluids are highly phagocytic and capable of breaking down organic material so that the combination of recognition of grossly foreign character plus phagocytosis and intracellular digestion should be sufficient to provide the basis of an effective protective system against bacterial infection.

### Analogies with the invertebrate system in vertebrates

Many of the characteristics of the generalized invertebrate system can be recognized in the mammal and other vertebrates. One of the important *non sequiturs* in much discussion of natural antibodies is to assume that any effect produced by serum in a situation analogous to that used for titrating standard antibody is due to antibody. To be meaningful, that statement requires that the protein responsible should be known to have the standard structure of immunoglobulin M, G or A and that attachment to the substance it reacts with must be by the standard combining site. Many natural antibodies may depend on regular but biologically accidental configurations on a plasma protein that are unrelated to the combining site on immunoglobulins. In reading an excellent recent review on natural antibodies by Boyden, one's chief impression is of the heterogeneity and unrelatedness of the reported facts and the inability to obtain consistent experimental support of any proposed hypothesis relating natural antibodies in a specific fashion to classical immunoglobulin antibodies.

Instead of seeking any unitary explanation of the phenomena I should be content to leave the situation in the plasma of a young normal animal as a basically unanalysable situation. In addition to a vast variety of Ig M molecules with a wide range of specific pattern mostly present in low concentration, Ig G and Ig A antibodies mostly from specifically stimulated clones—there are all the other globulins, haptoglobins, etc., each with their own physiological function. Any of these may show individual reactivity with some arbitrarily chosen reagent, red cell or bacterium, for example, for reasons which will probably never be expressible in more than the general statement that the whole evolutionary function of proteins depends on the capacity of one or other sequence of amino acid residues to combine selectively with almost any conceivable surface configuration. There must be a vast number of biologically meaningless reactions to be detected by any industrious apprentice to immunology which conceivably are related to *any* protein in serum or to other aspects of immunoglobulin than the combining site.

The behaviour of macrophages in mammals has also many resemblances to that of invertebrate cells. Although the results of experiments on the standard source of macrophages, peritoneal exudates in rodents, are often hard to interpret from the presence of large numbers of immunocytes (lymphocytes) as well as of phagocytic cells, the evidence is clear that basically the activity of macrophages is not an immunologically specific one. Macrophages from immune animals are no better at breaking down the substance of the bacteria against which the immunity is directed than those from normal animals. When the macrophages of a mouse develop power to destroy *Listeria* they concomitantly become able to destroy other unrelated organisms such as *Brucella*.

As has been discussed at various places in this book, macrophages of one sort or another are intimately concerned with immune processes in the strict sense, but they are ancillary rather than essential. Opsonization, which may mean the specific or non-specific coating of a particle with partly denatured immunoglobulin often with attached complement components, is a potent and often indispensable adjunct to phagocytosis. Mouse macrophages from peritoneal cavity readily adsorb cytophilic antibody and are so enabled to phagocytose antigenic particles such as red cells. Finally, there is the capacity which may be present in all macrophages but is particularly evident in the dendritic phagocytic cells of the lymph follicles to prepare and retain antigenic determinants in a particularly immunogenic form.

However, there is still evidence that substances such as *damaged* red cells and foreign particles like carbon and starch can be phagocytosed without first being coated with serum protein. Even a pneumococcus can be phagocytosed without opsonization if the macrophage is supported by tissue or some mechanical substitute. There is evidently still present some of the crude ability to recognize foreign material that is essential in the invertebrate.

## Immune responses in primitive vertebrates

The most primitive existing vertebrates are the hagfishes and lampreys which appear to be specialized and in part degenerate descendants from the stock that gave rise to the earliest marine vertebrates, the ostracoderms, which appeared in the Upper Silurian. The hagfishes (*Myxinidae*) are parasitic on fish and anatomically appear to be more primitive than the lampreys. Extensive studies made by Good's group in Minneapolis have failed to show any evidence of antibody production or any of the other 'markers' of immune capacity. Lampreys are more advanced anatomically and in their biphasic life history. They have undoubted immune capacity although they produced antibody against only one of the five antigens tested. There are groups of 5–20 cells in the epithelium of the peripharyngeal gutter which could represent a thymus and there was a definite rejection of homografts although autografts were well retained. All the higher fishes show immune responses although elasmobranchs in general give poorer responses than teleosts. Immunoglobulins can be recognized and there are highly specific homograft reactions demonstrable in scale grafting experiments. In amphibia and reptiles the most interesting feature is the great increase in speed and effectiveness of the immune responses when the temperature is raised.

From the evolutionary point of view then we can probably decide that the evolution of the immune system, essentially in the form known in the higher mammals, took place in the Silurian, perhaps more or less coincidentally with the first development of relatively large ostracoderms or their unarmoured ancestral forms. The application of modern techniques to the structure of the immunoglobulins in the lower vertebrates is just beginning and should eventually provide information of much importance to the understanding of the genetic origins of immune pattern. As yet the most interesting finding is that the immunoglobulin of an elasmobranch, although present in both 7 S and 17 S form, is antigenically uniform and corresponds almost certainly to Ig M in mammals. In an amphibian, both Ig G and Ig M can be recognized and there is the usual Ig M, Ig G sequence in the standard immune response to an antigen.

Light and heavy chains can be identified in elasmobranchs and all higher vertebrates, so that the need is urgent to look at the fine structure of immunoglobulin in the available cyclostomes if its early evolutionary history is to be unravelled.

## A GENERALIZED SKETCH OF THE EVOLUTION OF ADAPTIVE IMMUNITY

In a contribution at present in the press I have tried to draw a self-consistent picture of the evolution of adaptive immune mechanisms, starting with the premise that the process was initially concerned not with defence against infection but with the maintenance of the cellular integrity of the body.

Such an attempt is necessarily more speculative than the generalized approach to the experimental facts with which this book is concerned. I have felt, however, that to present a summary of this picture of how adaptive immunity evolved is the most logical way to round off an exposition of theoretical immunology which claims to have been based throughout on evolutionary principles.

When animals became large enough and lived long enough for somagenetic changes in their cells to become a significant factor in survival, the immune mechanism was initiated as a two-sided exploitation of genetic lability. Histocompatibility antigens became highly modifiable by genetic processes; cell globulins with potential adsorptive power for a variety of organic configurations became increasingly subject to accelerated somatic mutation or some equivalent somagenetic process.

At all stages the immune process is concerned only with the cell surface. As I have already indicated, when cell surface characteristics become inoperative, as when mouse and human cells are grown in mixed tissue culture with Sendai virus, heterokaryons form without any evidence of disability to either nucleus. Histocompatibility differences and the possibility of recognition of foreignness are strictly functions of the cell surface. One can epitomize the immunological situation as a reaction by which one cell, the immunocyte, carries a modified globulin which can 'recognize' an unfamiliar pattern carried on the surface of another cell. That unfamiliar pattern can be a new histocompatibility antigen or a wholly foreign molecule or configuration. Recognition is by reversible union based on steric complementarity of the two patterns.

As has been discussed at some length in chapter 11, delayed hypersensitivity is a relatively primitive reaction in which the antigenic determinant can function as an immunogen only when it is incorporated into the lipoprotein of the surface of one of the mobile cells, lymphocyte or monocyte. Contact of immunocyte and antigen-carrying cell has a stimulatory effect producing as standard response blast transformation of the immunocyte and limited proliferation to a small clone of lymphocytes. Subsequent contact of such sensitized cells by antigen in solution results in more drastic stimulation and the liberation of pharmacologically active substances as described earlier.

Antibody production is a further development on the same essential theme with the casual mobile carrier of antigen replaced by a specialized fixed cell, the dendritic phagocytic cell of the lymph follicles. Recent work suggests also that the differentiation of the stem cells to immunocytes that are potential active producers of antibody is not a function of the thymus but of bursa or bursa-equivalent tissues. The immunocyte, too, has evolved for a more specialized group of functions. There are cell types restricted to the production of one only of the immunoglobulins M, G or A and, for each of these, the corresponding mature plasma cell develops an elaborate lamellate expansion of the endoplasmic reticulum. The standard induction of the immunocyte to antibody production is by contact of its immune receptor with antigenic determinant on the surface of a dendritic phagocytic cell. There are many complexities, diversions and anomalies such as those discussed throughout the body of this book in relation to antibody function and the behaviour of antibody-producing cells. The same central theme of cell surface to cell surface can, however, be discerned almost as clearly as in relation to delayed hypersensitivity.

As we see the picture in mammals, including ourselves, we can summarize immune function against such an evolutionary background as follows.

The primary immune function of surveillance to destroy mutant cells within the body is directly demonstrated in the laboratory models of homograft immunity and graft-versus-host reactions. It provides, too, the master-key to the understanding of delayed hypersensitivity and related phenomena.

Immunoglobulin is synonymous with antibody; this can now be taken as a dogma of the modern approach.

Ig M is the earliest type of antibody produced to any antigenic stimulus and may be the only form which results from nonspecific stimulation of immunocytes. Its main function is as an opsonin to allow phagocytic removal of micro-organisms and to facilitate further antibody production against their antigenic components.

Ig G may have as its main function to damp down the antibody response to a widely present foreign antigen. In mammals with haemochorial placentae it is the main vehicle by which maternal immunity is passively conferred on the newborn.

Ig A is specialized for secretion in glandular products. In ungulates it is concentrated in the colostrum and is the sole vehicle of passive maternal immunity. It appears also to have special qualities in protecting mucous surfaces from infection or in becoming the vehicle of allergic sensitization.

The evolution of adaptive immunity conforms to the classic pattern. We see a simple theme seized on by the 'master constructors' mutation and selection and progressively moulded and elaborated. From primitive beginnings we reach in the higher mammals a finely tuned process exquisitely adapted to function against what is foreign or abnormal, but equally ringed round with devices to ensure that it will not offend by attacking the normal.

The immune system is a creation of evolution. It has arisen by the exploitation of error in nucleotide replication and it is no more infallible than any other aspect of biological function. It is sound biology to remember that autoimmune disease and myelomatosis are as much part of the universe of immunology as recovery from yellow fever or pneumonia.

# Epilogue

A book, even a scientific text, is always liable to develop in a fashion that diverges subtly from what was originally conceived. This one ends without providing that 'definitive' form of clonal selection theory which it was its primary objective to present. Doubts and soft edges continue to abound. I grow more and more doubtful whether we are capable ever of understanding mechanisms that have taken a billion years of evolution to design and prove. Our theories remain crude and naïve.

Any summary can represent only a tentative halt along the road from which we can look forward hopefully or despondently, according to temperament, at the mounting complexities in front of us. Since the days of Ehrlich's side-chain theory there has been a progressive increase in sophistication of the requirements of immunological theory. With each major new development in biology, existent theory necessarily becomes inadequate and with the current applications of molecular biology to the immunoglobulins, the whole theoretical basis of immunology awaits recasting.

The process will go on indefinitely but at every step, as now, it must always be compatible with current understanding of biological processes generally. Now and perhaps permanently there are five levels or categories that have relevance to immunology.

1. The stochastic processes of error in replication which provide the raw material, through mutation and somatic mutation, for all biological change. This applies both to progress or degeneration at the species level and to a significant proportion of physiological and pathological changes in the individual.

2. Stochastic processes by which, from among the many available patterns of biological information, selection is made for some only to reach expression in organism or cell. These include sexual recombination at the level of the organism and phenotypic restriction at the cellular level.

3. The impact of environmental factors in allowing selective

309

survival and proliferation of certain forms at both the individual and the cellular level.

4. The molecular basis of the storage, replication and expression of biological information as it has been developed from viral and bacterial models, plus what indirect evidence can be gained of its applicability to vertebrate cellular activities.

5. The determinate aspects of biochemical processes as governed by stereochemical patterns in protein structure.

What I have written about the cellular and molecular basis of immune phenomena is a deliberate attempt to keep within those categories to the limit of my knowledge and understanding. Immunology from its very nature, its unique blend of stochastic and determinate processes, is probably better fitted to exemplify the realities of biology than any other major field in vertebrate physiology and biochemistry. Perhaps it is not too immodest to hope, even, that this book, by providing such an example of principles in action, may help toward a broader understanding of biology amongst a group of scientists now over-inclined to limit their interest wholly to the molecular approach.

# Bibliography

As a guide to the literature in English bearing on the more academic aspects of immunology, the following notes may be helpful:

*Primary publication* of new research may be in any one of a hundred journals but a considerable proportion will be found in:

*Immunology, Clinical and Experimental Immunology* (U.K.), *Journal of Immunology, Transplantation, Immunochemistry, Journal of Experimental Medicine* (U.S.A.), *International Archives of Allergy* (Swiss), *Australian Journal of Experimental Biology and Medical Science.*

*Review articles* appear in:

*Progress in Allergy, Advances in Immunology, Annual Review of Microbiology, Bacteriological Review, Annals of the New York Academy of Sciences.*

Recent *textbooks* include:

Humphrey, J. H. and White, R. G. *Immunology for Students of Medicine*, 2nd ed. Oxford: Blackwell Scientific Publications (1964).

Kabat, E. A. *Experimental Immunochemistry*, 2nd ed. Springfield: Thomas (1961).

Raffel, S. *Immunity*, 2nd ed. New York: Appleton-Crofts (1961).

The following are probably the key references for the development of immunological theory:

Metchnikoff, E. *L'Immunité dans les maladies infectieuses.* Paris: Masson (1901).

Ehrlich, P. On immunity with special reference to cell life. *Proc. R. Soc.* 66, 424 (1900).

Landsteiner, K. *The Specificity of Serological Reactions*, 1st ed. Cambridge, Mass.: Harvard University Press (1936, rev. ed. 1946).

Breinl, F. and Haurowitz, F. Chemische Untersuchung des Präzipitates aus Hämoglobin und Anti-Hämoglobin-Serum und Bemerkungen über die Natur der Antikörper. *Hoppe-Seyler's Z. physiol. Chem.* 192, 45 (1930).

Pauling, L. Theory of the structure and process of the formation of antibodies. *J. Am. chem. Soc.* 62, 2643 (1940).

Burnet, F. M. and Fenner, F. *The Production of Antibodies*, 2nd ed. Melbourne: Macmillan (1949).

Jerne, N. K. The natural selection theory of antibody formation. *Proc. natn. Acad. Sci. U.S.A.* 41, 849 (1955).

Burnet, F. M. *The Clonal Selection Theory of Acquired Immunity.* Cambridge University Press (1959).

# BOOK TWO

# Introduction

In this volume I am concerned to provide as clearly as possible the detailed justification for those statements in the first volume which are not self-evidently true to anyone with a reasonable knowledge of biology and some special interest in immunology. The only reasonable excuse for writing a book on immunological theory is the hope that it can present a broad and flexible concept that has significance for the whole range of phenomena conventionally classed as immune. It is convenient to refer to the approach I am using as 'clonal selection theory' simply because I cannot recognize any point at which the interpretation of past and current knowledge required a sharp break in the way my theoretical ideas have been developing since the first full statement of clonal selection theory was made. Its current presentation differs a great deal from that of 1957–8, and the need is now to demonstrate that it has evolved in such a fashion that it is still applicable to the whole field of immunology. If it were physically possible either for writer or reader, every technically competent paper in the literature could be discussed for its relevance pro or con. I think I can claim that in fact every paper in immunology that I have read or listened to since 1957 has been looked at in this fashion and an abstract made of each one that either contributed a significant point in favour of the theory or appeared difficult to accommodate within the current confines. This meant that a vast number of good papers which appeared to me to have no bearing one way or another on theory or merely underlined a point already established were not recorded. The bibliography is therefore unbalanced and will often cite for a given point only the papers which initially attracted my attention to the topic. It has also been necessary to rely rather largely on secondary sources for references before 1957, but having regard to the well-defined objective of the book, this appears to be quite legitimate. As the project has developed, I have become impressed with some aspects of experimental immunology familiar to most experienced workers but which for obvious reasons are rarely commented on in print. The first concerns the way in which the relevance of certain areas of investigation to the general picture varies with the development of the science. I am impressed, for instance, with the immense importance of the immunology of pneumococcal infections in the 1930s and the virtual absence of any scholarly interest in the field in recent years. There are also fields which grew up as an essential part of the study of the immune process and are still being actively exploited, yet have only the most marginal relevance to the present approach. Two noteworthy examples are the components and significance of complement and experimental anaphylaxis. Neither is given more than cursory treatment.

A second observation in much the same context is that some observations may be accurately made and reported and on the surface be incompatible with a vastly larger internally consistent body of established data and logic. This is the proverbial situation of a beautiful theory destroyed by one little fact. In relation to theories of immunity the significant instance is Fishman's work on the transfer of immunological information by RNA. Perhaps it is advisable to discuss this general point in the light of current ideas because it seems intensely relevant to the problem of justifying acceptance of any generalization covering more than a very restricted biological field.

Briefly the thesis is that because under special laboratory conditions some reproducible phenomenon can be demonstrated, it does not necessarily follow that the phenomenon has any biological significance in the sense either that it has any relevance to the evolutionary development and survival of the organism that shows it, or that it occurs at all under natural conditions or, finally, that it has any but the most remote bearing on the field of biology to which it appears to be relevant. Virtually all the examples one has in mind of such real but misleading phenomena are brought to light by making use in one way or another of the uniquely biological quality of replication of specific pattern. Organic evolution is only possible (*a*) if all biological replication methods are highly consistent in producing exact replicates and (*b*) if the replication mechanism has a low but definite liability to error and a considerable capacity to tolerate and perpetuate that error. This flexibility

316

and tolerance is such that one can expect wholly unphysiological manipulations occasionally to interfere with the genetic mechanism producing an effect which does not preclude replication of the unit concerned and therefore can allow an experimentally recognizable change.

We can accept the current dogma that m-RNA moves from the nucleus to find an appropriate group of ribosomes to allow synthesis of the translated message in a polypeptide chain. It would follow that exposure of cells in an appropriate physical state to large concentrations of many different types of intact and damaged RNA might, on occasion, result in the synthesis by a cell or a number of cells of small amounts of a protein in which the experimenter was interested. This would not, in the least, mean that the normal way in which this protein was produced was by the transfer of m-RNA from another cell.

Similarly, an oncologist interested in cancer viruses finds a tumour of the usual spontaneous origin which has allowed the growth of a casual low-grade mouse virus. During passage of the tumour the only virus variants which will survive in the laboratory are those with a nicely balanced capacity to live in tumour cells without, on the average, either causing necrosis or being eliminated from the system. When now a filtrate is passed to a neonatal animal the temperate virus will have many opportunities to interfere with somatic cell genomes. Any particular interference that sets malignant proliferation of a cell under way will automatically provide another nidus for virus persistence. Under persistent and intense laboratory selection pressure to produce an oncogenic virus, a helper virus or an autoimmune virus, the starting material, however innocent, will sooner or later produce what the experimenter desires. From the point of view of defining the natural history of cancer the work on rodent and chicken oncogenic viruses is quite valueless, whatever interesting light it may throw on the possibilities of modifying the genome of somatic cells.

Medicine is, however, not concerned only with normal biological processes of evolutionary significance. This becomes only too evident when one looks over the genetic catastrophes that fill the idiot asylums and the prisons for the criminally insane. The trisomic mongols and the homicidal criminals with XYY chromosomes are the results of accidents that have a curious resemblance to the efforts of the virologist to insert genetic material from one genome into another. Somewhere there is on record an individual whose blood group mosaicism was such that two maternal gametes and one male gamete must have contributed to the zygote. None of these anomalies has more than the most marginal bearing on the understanding of normal genetic processes.

There are excellent arguments for regarding any type of reproducible biological phenomenon as a perfectly valid object for detailed scholarly investigation. It may well be said that without years of study of evolutionarily meaningless or positively harmful laboratory mutations in *Drosophila*, it would have been impossible to carry out the work by Dobzhansky and others on the population genetics of wild-living species and variants of the genus. Similar things could be said about much of bacterial genetics. Molecular biology has essentially been based on the study of variants of a few laboratory cultures of *Escherichia coli* and of a chosen group of bacterial viruses. Probably none of the bacterial variants and very few of the phages used could ever survive in nature and in that sense the work has probably contributed nothing to medicine, agriculture or industry. Yet there is no one who would not admit that, as a feat of scholarship, nothing calls for more admiration than the achievements of the molecular biological approach to bacterial genetics in the last decade.

The legitimate grounds for quarrel with artificial biological experimentation only arise when results are unintelligently or disingenuously transferred to other fields. The understanding of cancer has not been helped by the oncovirologists, and I expect that in twenty years' time three major areas of experimental immunology will have had to be written off as blind alleys— anaphylaxis, the use of complex adjuvants, and the experiments on intercellular transfer of immunological information. They are blind alleys in the sense that findings within the corresponding field of experimental study cannot be transferred without the greatest circumspection into a general context.

The general rule is, I believe, completely straightforward. Biological process is wholly the creation of organic evolution and biological generalization is only significant in its evolutionary context. It is this rule that I have tried to follow in writing this book. There are occasions when it must be broken. Scholarly work in blind alleys can still be of first-rate quality and bring prestige to the investigator. Modern medicine feels as much responsibility for the individual crippled by a biologically meaningless genetic anomaly as for a victim of trauma or infection. Industrialized animal husbandry may demand animals or birds which in nature would be nonviable monstrosities.

Any piece of investigational work done with good technique, good planning and complete integrity is worth doing but if its results add significantly to the evolutionary understanding of the group of organisms involved, it has a much greater reason to be remembered.

# 1-2 The history of selective theories of immunity

## THE DEVELOPMENT OF SELECTION-TYPE THEORIES (1955–1959)

The first formulation of a selective theory of antibody formation is unquestionably due to Jerne (1955). As he pointed out, up to that time there had been two views with significant support: (*a*) the 'antigen-template' theory, now usually referred to as 'instructive' and associated with the names of Breinl and Haurowitz (1930), Mudd (1932), Alexander (1932) and Pauling (1940) and (*b*) the attempts to use analogies with adaptive enzyme production for which Burnet and Fenner's monograph (1949) was the chief support.

The essence of Jerne's theory was:

> The antigen is solely a selective carrier of spontaneously circulating antibody to a system of cells which can reproduce this antibody. Globulin molecules are continuously being synthesized in an enormous variety of different configurations. Among the population of circulating globulin molecules there will, spontaneously, be fractions possessing affinity toward any antigen to which the animal can respond. These are the so-called 'natural' antibodies. The introduction of an antigen into the blood or into the lymph leads to the selective attachment to the antigen surface of those globulin molecules which happen to have a complementary configuration. The antigen carrying these molecules may then be engulfed by a phagocytic cell. When the globulin molecules thus brought into a cell have been dissociated from the surface of the antigen, the antigen has accomplished its role and can be eliminated.

The stimulus to Jerne's theory was his study of normal antibody capable of inactivation or modification of bacteriophage in unimmunized animals. In 1955 the pattern of protein synthesis had not been elucidated and Jerne's postulated mechanism was that 'spontaneously produced' globulin carried by antigen into a phagocytic cell would enforce cellular RNA to produce more globulin of pattern identical with that brought in. Even in 1955 this was very hard to accept and it was, in fact, the only point that prevented my adopting the theory almost immediately.

In 1957 Talmage and Burnet independently asked why Jerne's idea of selection should not be applied in more direct fashion to the antibody-producing cells.

Talmage (1959*a*), in discussing the mechanism of the antibody response, strongly supported Jerne's natural selection ideas against instructive and modified enzyme hypotheses. He, incidentally, suggested that Ehrlich (1900) had produced the first selection theory of antibody production. This is literally true, but Ehrlich's point of view at that time had no possibility of experimental verification or disproof and was in fact automatically discarded by all immunologists, with the realization, largely due to Landsteiner, of the immense variety of antibody patterns that can be produced.

Talmage's (1959*a*) views were still strongly influenced by Jerne's interest in circulating normal antibody. Talmage's modification was to regard the 'normal' situation as resulting from the production of antibody by antibody-synthesizing units (ASU), the activity of which was controlled by a negative feedback based on antibody concentration. Each ASU is presumed to produce one antibody and to go on doing so spontaneously until a certain concentration is reached. With injection of antigen there is a sharp fall in antibody concentration in plasma with release of the ASU to more activity. This is clearly inadequate to account for production of antibody beyond the 'normal' level. Talmage does not use the concept of cell receptor responsive to antigen but implies, like Jerne, the existence of one responsive, either positively or negatively, to *antibody*.

My own modification of Jerne's theory (Burnet, 1957*a*) can be visualized from the following quotations:

> It is believed that the advantages of Jerne's theory can be retained and its difficulties overcome if the recognition of foreign pattern is ascribed to clones of lymphocytic cells and not to circulating natural antibody.
>
> Each type of [antibody] pattern is a specific product of a clone of mesenchymal cells and it is the essence of the hypothesis that each cell automatically has available on its surface, representative reactive sites equivalent to those of the globulin they produce.
>
> [Antigen] will attach to the surface of any lymphocyte carrying reactive sites which correspond to one of its antigenic determinants... [then] the cell is activated to settle in an appropriate tissue... and there undergo proliferation to produce a variety of descendants.

321

> We have to picture a 'randomization' of the coding responsible for part of the specification of γ-globulin molecules...any clones of cells which carry reactive sites corresponding to body determinants will be eliminated...in the late embryonic period with the concomitant development of immune tolerance.

In the following year I used the general topic of clonal selection for the 1958 Abraham Flexner lectures at Vanderbilt University Medical School. In the published expansion of these lectures (Burnet, 1959*b*) the 1957 sketch was filled out in detail, the most important additions being the adumbration of the forbidden clone approach to autoimmune disease and the claim that multiple myelomatosis was a legitimate model of (pathological) clone production by cells closely related to antibody-producing plasma cells.

Lederberg spent two months in my laboratory in late 1957 and was keenly interested in the new approach. It was his suggestion that an experimental attack, using *Salmonella* antigens, would be a practical one, and this initiated the productive studies of Nossal and collaborators (see Nossal and Lederberg, 1958). As a geneticist he was also deeply concerned with the genetic implications of the theory and in an article in *Science* he (1959) clearly contrasted what he called 'instructive' and 'elective' theories of antibody formation. He also raised the possibility that each potential immunocyte might carry a very large armoury of genetically determined patterns and that, in some way, contact with antigen *A* would ensure that only the corresponding antibody and not anti-*B*, -*C*, -*D*, etc., would be produced by the cell. From the point of view of one who was at that time firmly committed to the clonal selection view with one cell, one antibody (or having regard to diploid cells, two at most) this could be regarded as a return to a pseudo-instructive viewpoint.

In the same number of *Science* Talmage (1959*b*) favoured the general selective approach but was mostly concerned with finding a way to reduce the number of genetic patterns that must be postulated. He endeavoured to replace 'the classical concept of unique globulin molecules for each possible antigen...[by] immunological specificity based on a unique combination of natural globulins'. This, however, seems to have been in part based on a confusion of the two concepts, antigen and antigenic determinant (AD).

322

The immediate result of this round of exposition of selection theories was to stimulate experiment. The first point to be tested was whether a single immunocyte produced more than one type of antibody in an animal simultaneously immunized with two unrelated antigens. Nossal and Lederberg (1958) made the first report, using cells isolated in microdroplets and inhibition of motility in Salmonellas as test for antibody production. Of 456 lymph node cells from a doubly immunized rabbit, 33 produced antibody to one antigen, 29 to the other, while no cells produced both. Using a fluorescence technique with two different stains, Coons (1958) had found that in doubly immunized animals, plasma cells stained as if they were producing either one antibody or the other, not both.

This clear-cut evidence did at least indicate that, in general, a given cell produced only one type of antibody; it did not prove that there were no exceptions to the rule. Real or apparent exceptions were in fact reported quite soon and have been important in bringing a more sophisticated elaboration of the naïve 'one cell, one antibody' approach. These will be discussed at length in later chapters and need only be mentioned briefly here. As Nossal and his colleagues continued their work, occasional 'double producers' were encountered. Some of these, in which only traces of a second antibody were detected, could be discounted for technical reasons but amongst their doubly immunized rats there was a very small proportion (2 out of 2,600, or 0·08 per cent) of indubitable double producers (Nossal and Mäkelä, 1962b).

Attardi et al. (1959), using rabbits doubly or triply immunized with serologically distinct phages over a long period, obtained a much larger proportion, approaching 20 per cent, of double producers amongst lymph node or spleen cells. The interpretation of these results is still controversial. The phages used were serologically distinct in the conventional sense but it is a pity that experiments were not devised to see whether the antibody from a double producer cell was in fact made up of two populations. This could in principle be tested by absorbing the antibody with phage A and seeing whether, concomitantly with the disappearance of anti-A, anti-B also disappeared. The possibility of double immunization producing antibody of a different quality from that obtained by either antigen singly is well known to virologists. The significance of these apparently discordant findings is discussed elsewhere in this book.

The overall result was rather to minimize the importance of theorizing. Most workers got on with experiments and made an occasional perfunctory remark in their discussion as to whether or not their results favoured instructive or selective theories. One might almost summarize the period since 1958 as a time when all the set experiments designed to disprove clonal selection seemed superficially to do so, while virtually every new advance arising wholly from growing edge experimentation with a minimum of theoretical preconceptions fitted far more easily into the clonal selection pattern than any other.

It was perhaps to be expected that the instructive theories should continue to be supported by workers with predominantly chemical interests at a time when biologically minded ones were moving toward sympathy with selection ideas. Pauling (1940) expressed his classical modification of the Haurowitz–Mudd–Alexander view in the following words:

> antibodies differ from normal serum globulin only in the way in which the two end parts of the globulin polypeptide chain are coiled, these parts...having accessible a very great many configurations with nearly the same stability; under the influence of an antigen molecule, they assume configurations complementary to surface regions of the antigen, thus forming two active ends.

In Pauling's paper no mention was made of the process by which this intramolecular three-dimensional folding could be stabilized. The most important development in recent years in this aspect of protein structure has been the recognition of the importance of the formation of disulphide links in stabilizing protein structure. Karush (1957b, 1962) noted that in the Porter fragments I and II which carry the antibody specificity there were the equivalent of sixteen half-cystine residues in I and thirteen in II. If these residues were free to pair in all possible fashions there would be many millions of different configurations possible. Karush's suggestion was that as the nascent polypeptide chain oriented itself against the antigen template, the opportunity arose for adjacent cysteine residues to form S—S bonds which would then stabilize the configuration of the globulin in the antibody form. This suggestion has gained few supporters. Most protein chemists, largely as a result of Anfinsen's work, now favour the view that the tertiary structure of a protein is implicit in its amino acid sequence and that part of the function of the sequence is to determine when the disulphide bonds will form.

Haurowitz (1965) still supports a modified form of instructive theory but accepts the probable existence of amino acid differences which are significantly related to immune pattern. He would attribute these differences to

> a disturbance of the coding mechanism by serologically determinant fragments of the antigen molecule resulting in a change in the cellular phenotype but not the genotype...cells would have the ability of forming antibody only in the presence of antigen fragments but would not be able to transmit the ability to its daughter cells in the absence of the antigens or its fragments.

Haurowitz, however, concedes that selection on a subcellular (intracellular) basis is still a possibility and cannot be excluded.

Most immunologists probably regard the instructive theories as highly improbable and would look on Haurowitz's support as a rather hopeless rearguard action. For the majority it accords both with theory and with the experimental data that the immune pattern should be determined by a genetic mechanism. It is equally evident, however, that this genetic mechanism differs in important respects from that concerned with the production of a standard protein by a secreting cell.

Present-day discussion of immunological theory seems to have reached the stage of accepting the essential features of a modernized clonal selection theory which can be expressed as follows:

(*a*) Antibody-producing cells are members of clones of immunocytes; members of each clone produce only one type of immunoglobulin and antibody.

(*b*) Immunocytes can be defined as cells which carry receptors of specificity equivalent to the antibody pattern characteristic of their clone; specific cellular reactions of immunological character are initiated by a drug-type contact of antigenic determinant (AD) with such receptors.

(*c*) The diversity of immune pattern in immunoglobulins

depends on genetic processes taking place in the individual subsequent to the formation of the zygote, that is, in somatic cells. 'Somatic mutation' may be used in a broad sense to cover such processes.

(*d*) The only function of antigen is to react with patterns of appropriate steric configuration which have been produced by these 'random' processes of diversification.

The residual problems at the fundamental level that require solution within the general framework are:

(*a*) The nature of the genetic processes responsible for diversification of immune pattern and for the phenotypic restriction to a single pattern.

(*b*) The part played by the accumulation over the course of vertebrate evolution of duplications of a primitive gene, each subsequently susceptible to independent mutation. This is specially relevant to the structure of the different immunoglobulin types.

(*c*) The relationship of immune pattern to the chemical structure of the immunoglobulins.

(*d*) The possibilities of transfer of immunological information to a cell of a different clone either by somatic recombination or by some subcellular process.

If this formulation of the current theoretical problem is a valid one, the main experimental approach will have to be a genetic one combined with a detailed specification of the amino acid residues making up the combining site (CS). This is not yet practicable but should become so within a few years.

The rest of this chapter is primarily concerned with the objections that have been raised at various stages against the concept of clonal selection as it was presented in 1959. In any discussion of this sort there will always be the difficulty that the background of relevant knowledge has been progressively changing over the whole period of controversy. As is often the way in science, what seemed at first to be a clear incompatibility between fact and theory has often taken on a different aspect with the passage of time. It will probably be noticed that there is an old-fashioned air about many of the points made pro and con in this discussion of the objections to clonal selection theory.

PROBLEMS OF CLONAL SELECTION THEORY

*The production of antibody against artificial haptens*

It is certainly true to say that at no time in the course of evolution did any animal encounter a combination of dinitrophenyl with a synthetic polypeptide. Yet laboratory animals can readily produce antibody against such substances. To many immunologists, including Haurowitz (1965), the principal objection against selective theories is based on this ability of animals to make antibodies against quite unnatural antigens.

This objection is based on a complete misinterpretation of the concept of randomization. To simplify the situation, let us take a 20-letter alphabet (A–T) to represent the 20 amino acid residues and assume that the CS is of the size suggested by Kabat and consists of two identical sequences of 10 amino acid residues. Even if, as seems likely, there are some invariant requirements for the CS such as a tyrosine residue at a certain region, there will be an extremely large number of different 10-letter words ($20^{10} = 10^{13}$ if there were no restrictions) that could be produced as a result of point mutation within the controlling cistron. By analogy with what is known about species differences in such proteins as haemoglobin or cytochrome C, some single-residue changes will have virtually no detectable effect, while others will produce major functional differences. A selection theory would picture the production of perhaps 100,000 such patterns distributed amongst the cells of an individual man or an individual rabbit but in each cell only one pattern can be functional. Having regard, however, to the diploid character of somatic cells, one must also assume that a nonfunctional allele potentially responsible for a second pattern would be present in each immunocyte.

In the body any such cell, immediately it has differentiated, will have to run the gauntlet of meeting perhaps 10,000 patterns (potential AD's) carried by body components with which it might be capable of reacting 'by chance' with an adequately firm linkage. If so, it will by hypothesis be eliminated. This could happen to a substantial proportion, perhaps 90 per cent of the random patterns carried on immune receptors. There will remain 10 per cent of the original cells carrying, say, 80,000 antibody patterns which are

327

random, apart from the fact that none will react strongly with any AD in the body producing them. Those 80,000 patterns are just as likely to react with a synthetic macromolecule as with a biological one, only provided that the synthetic substance has, like most organic and many inorganic materials, some capacity to be adsorbed on proteins generally and has the sort of physical qualities needed if it is to be antigenic in the strict sense.

One must not forget that *all* ADs are just as 'unknown' to the animal reacting immunologically as the most unbiological synthetic hapten. It is the great virtue of a selective theory that it allows this freedom of approach. An antigen, synthetic or natural, does not instruct the animal to produce a custom-made antibody or receptor. A response occurs only if the potential antigen can find immunocytes carrying patterns with which it can react. This is a rule as easily applicable to synthetic as to biological materials.*

## The immunochemical approach

An important group of immunochemists whose spokesmen are Eisen and Karush have resisted the general trend to concentration on cellular activity by seeking explanations for delayed hypersensitivity (DH) and tolerance strictly at the antibody level. They are concerned with the point which has now been fully established that there are very great differences in the avidity of union with the AD amongst the antibody molecules of an antiserum or of a semi-purified globulin preparation. Sometimes the association constant $K_A$ may be very high and their hypothesis of the nature of delayed hypersensitivity (Karush and Eisen, 1962) is

> that the affinity of an antibody for its homologous antigen can be sufficiently great for the antibody to form a stable union with the antigen at concentrations of the uncombined antibody which are too low to be detected by current methods... [and allow] the accumulation of a sufficient quantity of antigen–antibody complex to yield observable tissue damage.

To account for tolerance, Eisen and Karush (1964) suggest that in relation to 'susceptible cells', which could be either uncommitted immunocytes already restricted in their reactivity or, on an

---

* The significance of the report by Eisen *et al.* (1968) of a myeloma protein with the specific quality of an anti-dinitrophenyl antibody is discussed on p. 440.

328

instructive theory, 'virgin' cells capable of being instructed to produce the appropriate antibody, the effect of antigen will depend on its relationship to existent antibody. Antigen alone is regarded as being incapable of action on the susceptible cell, the bi-molecular complex antigen-antibody is the *only* stimulus to antibody production while the triple combination antigen–antibody–antigen is ineffective (table 7). Tolerance is therefore established whenever the concentration and avidity relationships ensure that any available antibody is wholly in the form of inert antigen–antibody–antigen complexes. A major difficulty recognized by the authors is that there is no way by which antigen can initiate antibody production *ab initio*, so that one must fall back on 'natural antibody' of the right type.

TABLE 7. *Eisen and Karush's (1964) approach to immunogenicity and tolerance*

| Complex | Biological activity |
|---------|---------------------|
| Ab alone | Inert |
| Ab–Ag | Immunogenic |
| Ag–Ab–Ag | Tolerance |
| Ag alone | Inert |

The attitude of a biologically minded immunologist can be summed up in the opinion that these views are founded on an illegitimate abstraction and simplification of the system in which the reactions occur. There can be no doubt that the variable avidity of antibody—and of the receptors on the immunocytes, a point not considered by Karush and Eisen—is of great theoretical importance. Equally, it is well established that antigen lightly opsonized is more immunogenic than if no antibody is present. Every point made by the authors is a worth-while one but it is impossible to picture immune reactions as taking place in a uniform fluid medium in which uniform and highly abstracted cells are suspended. Even when one is concerned only with immunological phenomena, the organization of the internal environment of the living vertebrate is intensely complex and, for the present, interpretable only in concepts almost all of which are at a less abstract level than those of chemical dynamics. Differentiation, homeostasis, somatic mutation, phenotypic restriction, selective survival and morphogenetic control—none of these is expressible wholly in terms of chemical dynamics. All seem to be required to present a provisional picture of immune phenomena in terms appropriate to the state of biological science in the 1960s.

*The production of multiple antibodies by a single cell*

For reasons of technical difficulty, no more than two different sorts of antibody have been simultaneously sought in single cells from doubly immunized animals. A great deal of work can be summed up by saying that in the majority of cases only one antibody is produced but, on occasion, a second. It is well known that when an animal is immunized with any conventional antigen, the antiserum obtained shows a wide heterogeneity of the antibody molecules which can demonstrably react with the antigen. In the first place there is a wide range of avidity distributed, according to Karush (1956) as a Gaussian error-function. If the antigen is a bacterium or other convenient particle, absorption of the antiserum with more or less closely related antigens will allow the partial separation of many populations of antibody according to their range of cross-reaction. This heterogeneity can be regarded as resulting (*a*) from the wide range of CS patterns which have some reactivity with the ADs concerned, (*b*) the intensity and quality of the stimulus and response associated particularly with the avidity of union between receptor and AD and (*c*) the concentration of antigen and the initial distribution of the types of cell subject to proliferative stimulation.

As Fazekas and Webster (1964) point out (in somewhat different words), the cells stimulated by antigenic determinant *A* will include a proportion that can be stimulated also by a related determinant *B*. Where the difference between *A* and *B* is considerable, there may be only one clone, present initially in very small numbers, which is strongly avid for both *A* and *B*. It will be hardly represented in antiserum *A* or antiserum *B* but it may dominate an antiserum derived by immunizing with *A* and later with *B*. The immune reactions of viruses provide a number of such examples. Theiler and Casals (1958), for instance, have shown

330

that as far as neutralization tests are concerned, the three arbovirus B species, yellow fever, Ilheus and dengue I, are distinct. Serum from a man infected by one of these will neutralize the homologous virus but not the other two. If, however, a man who has become immune to yellow fever is infected with Ilheus, his serum now neutralizes all three. Generalizing from this and other similar studies, it is in fact probable that most or all of the Group B arboviruses will be neutralized by such a serum.

If these enhanced cross-reactivities depend on the progressive selection of initially rare patterns for cell proliferation and antibody production, cells carrying such patterns could readily be taken as double producers in standard single-cell studies. As far as I am aware, no one has devised a practical means of attacking this possibility at the single-cell level.

As is discussed later, the other possibility of obtaining double producers while still preserving the essential features of phenotypic restriction to a single emergent pattern is to postulate a special type of somatic recombination and segregation for which the analogies at present are scanty, or to look for some means by which transfer of genetic information at a subcellular level may take place.

There is current discussion which suggests that where two distinguishable ADs are carried on the same molecule, double producers are much more common. This is intrinsically reasonable since it is the only situation in which completely similar situations must be met by each determinant. The most recent report on this topic by Benacerraf's group (Green *et al.* 1967*b*), gives, however, no support for this thesis and virtually removes any necessity for going beyond the standard clonal approach.

*The presence of 'too many' cells of a given specificity in unimmunized animals*

The main reason why at one period I found it necessary to swing from the straight clonal selection approach to an acceptance of multiple potentialities within each immunocyte came from the study of the Simonsen–CAM phenomenon. If blood leucocytes from adult fowls are placed on the chorioallantoic membrane of commercially obtained chick embryos, a number of focal lesions develop. By the use of inbred strains of fowl it can be shown that the appearance of the focus depends on an antigenic histocompatibility difference between the adult donor and the recipient embryo. The focus is in fact a graft-versus-host reaction. This has been clearly established by Cock and Simonsen (1958), Burnet and Burnet (1961) and shown to be related to the B blood group series by Schierman and Nordskog (1963). A detailed study of the numbers of foci produced in relation to lymphocytes in the inoculum by Szenberg *et al.* (1962) gave four findings which at the time were regarded as incompatible with the clonal selection theory. They were, first, that in all adequate experiments there were some embryos showing large numbers of foci corresponding to one focus per 40–100 large lymphocytes instead of the usual ratio of 1 to 2,000+. This suggestion that 1 per cent of cells had one special specificity without any immunization procedure seemed quite incompatible with the numbers of specific clones needed to cover all antigens. The second difficulty was that when random blood cells were plated on pure line *AA* membranes, the mean numbers were the same as on random embryos, instead of showing the relation one would expect to 1:2. The third was the failure to increase the number of foci on *AA* eggs by immunizing random fowls with *AA* cells. The fourth objection was the finding that when *AA* cells were passed through a non-inbred chick embryo, the cells harvested from the spleen of the recipient produced foci on *AA* embryos.

It was therefore concluded that our original explanation of the Simonsen–CAM reaction was untenable although the more limited studies on pure line material, both by Burnet and Burnet (1961) and by Schierman and Nordskog (1963), showed no significant discrepancies. In a review paper (Burnet, 1961), therefore, I withdrew from my previous position and at a symposium at Houston (Burnet, 1963*a*), I presented a very qualified account of clonal selection and accepted the claim by Trentin and Fahlberg (1963) that clonal selection in its original form was 'out'.

Subsequent work by Warner (1964) and a re-examination of the position of graft-versus-host reactions in the light of Simonsen's (1960) results in mice allowed a return to the original point of view that there is nothing in the experimental results incompatible with a clonal selection interpretation of the Simonsen–CAM results. The key finding was that our source of 'random' fowls was in fact largely inbred and closely related to our *AA*. They could be regarded as *AX* where *X* could be one of several alleles. Further, Warner was able to show that the apparent change of specificity of active cells on passage was due to the fact that if *AA* cells and embryonic *BC* spleen cells (both incapable, when alone, of producing foci on *AA* membranes) are mixed, the *BC* antigens can stimulate *AA* cells to proliferation and focus production on *AA* membranes. The significance of the high counts on a small proportion of random membranes is not yet clear, but the failure of immunization to increase the count is paralleled by Simonsen's (1960) results in mice plus Warner's (1964) finding that, as in mice, if one deals with a *minor* histocompatibility factor the number of foci *can* be increased by immunization.

By 1964, therefore, it had become unnecessary to make any *ad hoc* change to the clonal selection theory to account for the phenomena of focal cellular growth by adult immunocytes on the chorioallantoic membrane.

*The use of the Till–McCulloch technique to dispute the validity of clonal selection theories*

For the present it is accepted that pure clones of lymphoid cells in tissue culture have never been established as a continuing line, so that a direct test of the validity of the clonal selection approach has not been possible. The obvious method of disproving the clonal selection theory is to start with a single lymphoid cell and place it in a situation where it can develop a small clone. The clone is then divided into three and these groups of cells are each exposed to a different antigen. If all three populations produce the antibody corresponding to the antigen used as a stimulus and if the number of cell generations between the first duplication of the progenitor cell and the initiation of antibody production is small enough to ensure that even accelerated somatic mutation could not produce a sufficient number of patterns, one would have to accept the ability of an AD in one way or another to instruct the immunocyte what antibody to produce.

In the absence of an *in vitro* approach, several groups have

333

considered the possibility of an analogous cloning experiment done *in vivo*. The technique is derived from experiments by Till and McCulloch (1961), in which they observed that when a lethally irradiated mouse was injected intravenously with bone-marrow cells from a normal mouse, and the spleen was examined 7–10 days later, this showed white nodules a few millimetres in diameter whose number was linearly proportional to the number of normal cells inoculated. These nodules are never lymphoid in character but according to Lewis and Trobaugh (1964) they may be erythropoietic (42), granulocytic (21), megalokaryocytes (21), or mixed (16), the percentage of each type being shown in parentheses. If a nodule of erythropoietic character is tested in the same fashion for its content of stem cells (defined as cells initiating splenic foci) none is apparent until the fourth day after inoculation, after which they double about once each 24 hours. The second-generation colonies show the same distribution of cellular types as are seen with the first generation from bone-marrow cell inoculation.

There is good reason to believe that, as in the not greatly dissimilar case of the Simonsen foci on the chorioallantoic membrane, each nodule is initiated by a single donor cell lodging in an immunologically incompetent host. The application of this phenomenon to immunological investigation can only be indirect, as the proliferating cells are primarily of non-antibody-producing types and there must always be the possibility of movement of cells produced elsewhere into the nodule.

The earlier experiments of Trentin and Fahlberg (1963) as well as those of Feldman and Mekori (1966a, 1966b), gave apparently positive results in the sense that passage of 'cloned' material eventually allowed antibody production by the final irradiated recipient. All such experiments suffer from two almost insuperable difficulties. The first is to exclude the possibility that actively proliferating donor cells may reactivate immunocytes of the host otherwise paralysed by the irradiation (cf. Taliaferro and Jaroslow, 1960); the second is the likelihood of movement from one site of clonal proliferation to another, so that by the time the nodule is taken for test it may contain representatives of any other nodule and of any host cells which may have been stimulated to multiply. The most effective way to overcome the first of these difficulties is to use as donor a strain of mouse which has a karyotype clearly differentiable from that of the recipient. This is the approach used by Trentin *et al.* (1967) in an elaborate experimental study.

In brief, they used bone-marrow cells from $F_1$ CBA-T6 to initiate splenic nodules in lethally irradiated CBA recipients. These first-passage mice showed the presence of 4–13 spleen nodules of Till–McCulloch type, composed of haematopoietic cells. The spleen nodules were harvested at 10–12 days and cells from them given to a second-passage series of lethally irradiated CBA mice. A small proportion of these survived, and 100–120 days later spleen and bone-marrow were transferred to a third series of irradiated mice. In both second and third passage, three standard antigens were given to those mice surviving 30 days. The essential result was that in both second and third passages at least two of the three antibodies were produced, despite the fact that virtually all the mitoses observed were of the CBA-T6 type and, by implication, all of the proliferating and antibody-producing cells were of donor type.

At first sight this is a formal disproof of clonal selection theory, but there are two important qualifications. The first is that even if there are only 4–13 colonies in the primary spleens, these would in all probability also receive a few cells from the 25–100 other clonal proliferations in the mouse.

The second qualification is more important; it concerns the number of cell generations between the cloned cell and the initiation of antibody production. For the time being we can look at the experiment from the angle of interpreting it in terms of the hypothesis that diversity of pattern arises by processes analogous to somatic mutation starting from the two allelic patterns in the fertilized ovum. Suppose we take the spleen nodules at their face value as pure clones averaging 10 per spleen and assume that the 10 progenitor cells exhibited the sorts of differences equivalent to a week's proliferation and mutation. When harvested then, the cells had proliferated very actively for 10 days and were at the equivalent of 17 days from the beginning of the process responsible for diversity of pattern.

If we take a single initiating cell, the resulting colony (clone) will have proliferated very actively for 10 days. It is then given a further opportunity to proliferate freely in bone-marrow, spleen, etc., of the second-passage mouse and by 30 days the clone is 40 days old, during which time it has proliferated at least as actively as it would in normal development. In discussing the results, Trentin *et al.* claim that they eliminate the possibility of clonal selection unless a remarkable degree of somatic mutation is postulated.

Perhaps one should point out that the whole population of cells in a man or mouse is in a perfectly legitimate sense a pure clone derived from the fertilized ovum. The gestation period of a mouse is 18–20 days and the Trentin experiment does no more than underline that for a normal 10- or 20-day mouse to produce antibody, an intensive process of somatic mutation is required on the original clonal selection theory.

The essence of all the currently acceptable alternatives is the emergence in one way or another of phenotypically restricted cells potentially subject to stimulation by antigen. The diversity of their immune patterns may be produced in ways only remotely similar to point mutation in somatic cells. What the Trentin experiment underlines is the need for an active and random diversification process to ensure that a thousand or more types of immunocyte are available in a mouse forty days after its conception.

## THE CONCORDANCE OF NEW DEVELOPMENTS
### WITH THEORY

In 1956 I predicted that 'immunology seems to be ready for a new phase of activity that may be even more fruitful than its first flowering in the hands of Ehrlich and Bordet half a century ago'. In the intervening decade, this hope has been abundantly fulfilled. There has been an almost steady stream of new advances, many of them of completely unexpected character. What has convinced me more than anything else of the validity of the clonal selection theory has been the consistency with which new experimental discoveries have found a natural place in the theoretical framework.

The significance of the new developments is discussed elsewhere but a few comments may be made here. If we take as a starting point the two important symposia held in Prague and de Royaumont, France, during 1959, the important new developments have been:

(*a*) The production of specific tolerance in adult rabbits by combined treatment with antigen and 6-mercaptopurine (Schwartz and Dameshek, 1959).

(*b*) The recognition that small lymphocytes can be transformed to blast cells, both *in vivo* (Gowans, 1962) and by phytohaemagglutinin (PHA) *in vitro*, which was observed by Nowell (1960) but probably clearly established as a response of the small lymphocyte first by Marshall and Roberts (1963).

(*c*) The demonstration of the effect of neonatal thymectomy in the mouse (Miller, 1962).

(*d*) The origin from bone-marrow of the stem cells which become thymocytes (Micklem *et al.* 1966).

(*e*) Jerne and Nordin's (1963) technique of antibody plaques with its demonstration of the independence of the cells producing antibodies against different erythrocyte species.

(*f*) The demonstration that when synthetic antigens were used, clear-cut differences in immune response could be shown between pure line strains of guinea-pigs and mice (Kantor *et al.* 1963; McDevitt and Sela, 1965).

(*g*) The analysis of the classic example of immune paralysis with pneumococcal SSS in mice by Brooke and Karnovsky (1961) and Brooke (1966).

(*h*) The progressive recognition of the *monoclonal* character of the myeloma proteins and their use for the elucidation of the chemistry of the immunoglobulins.

(*i*) Koshland and Englberger's (1963) demonstration of amino acid differences between antibodies of different specificity in the same rabbit.

(*j*) The progressive recognition of phenotypic restriction in immunocytes in relation to type of immunoglobulin, allotypes and specific reactivity.

(*k*) The discovery by Hilschman and Craig (1965) and Titani *et al.* (1965) of the variable amino acid sequence in the N terminal half of the light chains from K-type human myeloma proteins.

(*l*) The finding by Eisen *et al.* (1968) of a myeloma protein reacting as an antibody with DNP-hapten.

This is admittedly a selected list of topics and there are probably a substantial number of other discoveries which have an equal right to be included. Even if they were included, the claim can be made that all the outstanding developments of the decade have fitted into the picture of clonal selection theory, and allowed its progressive development in detail and in depth.

# 3    The thymus in relation to the origin and differentiation of lymphoid cells

In discussing the basis for present opinions on the function of the lymphocyte and particularly its identification as the primary form of the immunocyte, it will be convenient to take as a background the presentation in Yoffey and Courtice's second edition (1956) of *Lymphatics, Lymph and Lymphoid Tissue.* In general, only work later than this will be cited and, as in all sections of the book, the main objective will be to provide the pros and cons for any new or controversial interpretations included in Book I.

## THE ULTRASTRUCTURE OF THE IMMUNOCYTES

In the interpretation adopted, the key cells are the small lymphocyte and the mature plasma cell. Both are derived from large pyroninophil cells which we have called 'blasts'. There is no conclusive evidence whether or not a single blast has the potential to initiate either a lymphocytic or a plasmacytic clone. It is in fact extremely difficult to conceive of an experimental approach which could give a direct answer to the question and for the present the indirect evidence will be accepted that lymphocytes, plasma cells, blasts and intermediate forms may all be members of a single clone.

The main features of the structure seen in electron micrographs are as follows:

*Small lymphocyte.* The cytoplasm contains only a small number of mitochondria and an occasional vesicle. There is no endoplasmic reticulum or Golgi body and only isolated ribosomes (Bernhard and Granboulan, 1960). The nucleus often shows a narrow invagination and there is usually a small nucleolus.

*Mature plasma cell.* The striking feature of the cytoplasm is the profuse development in lamellar form of rough endoplasmic reticulum and a marked development of the Golgi apparatus. This is responsible for the typical clear area of the cytoplasm near the nucleus that is seen in sections stained with Unna-Pappenheim.

339

Moderate numbers of mitochondria are present. The nucleus shows a well-marked nucleolus.

*Blasts and intermediate types.* Electron micrographs of sections of the cells in thoracic duct lymph show, among the larger cells, some containing Golgi apparatus with mitochondria and numerous vesicles but no endoplasmic reticulum, while others show quite definite development of endoplasmic reticulum both in the rat (Marchesi and Gowans, 1964) and in human thoracic duct lymph (Zucker-Franklin, 1963). The only difference between lymphoblasts and plasmablasts is the significant amount of endoplasmic reticulum in the latter, and there are good reasons for doubting whether this is any more than a minor functional difference (Harris, in discussion of Bernhard and Granboulan). Amano and Maruyama (1963), however, maintain that plasma cells in lymph nodes are derived from vascular adventitia.

In many recent studies of the active cells produced in the early stages of the immune response (Hall *et al.* 1967), or producing antibody at the centre of Jerne plaques (Hannoun and Bussard, 1966; Harris *et al.* 1966) only minimal development of endoplasmic reticulum has been noted in most cells although a range of intermediates toward typical plasma cells can usually be found as well.

### THE BEHAVIOUR OF THE SMALL LYMPHOCYTE

From the point of view of the immunological functions ascribed to the lymphocyte the important aspects of behaviour are its motility, its capacity for conversion to a pyroninophilic stem cell and its susceptibility to damage.

### Motility

Lymphocytes are found in all tissues of the body, being specially numerous in the intestinal mucosa where they are very frequently seen intracellularly in the intestinal epithelium. Following Marchesi and Gowans (1964) it appears that a basic feature of the small lymphocyte (in which it differs from other types of leucocyte) is the ability to pass freely from the blood *through* endothelial cells of the postcapillary venules in lymphoid tissue. It is now generally accepted that a large proportion of small lymphocytes recirculate, leaving the blood in this fashion and rejoining the circulation via efferent lymphatics and the major lymphatic ducts. Hall and Morris (1965), studying lymph nodes in sheep with cannulation of both afferent and efferent lymphatics, showed a surprisingly active turnover of lymphocytes and calculate that about 10 per cent of the lymphocytes reaching the lymph node in the blood pass into the tissue of the node across the endothelium of the capillaries.

There is much to suggest that the special morphology of the endothelial cells of the postcapillary venules is concerned with this behaviour of the small lymphocytes (Gowans and Knight, 1964). Such venules are characteristic of lymph nodes but are also conspicuous in Peyer's patches. They are not seen in thymic cortex but I have observed essentially similar vessels in the neighbourhood of proliferative lesions in the thymic medulla of NZB mice and their hybrids. Most adventitious accumulations of lymphocytes show very little evidence of blood vessels, but thick-walled vessels of postcapillary type are sometimes seen.

The movement of the lymphocyte is associated with a single cytoplasmic protrusion—the 'hand mirror' appearance of Yoffey *et al.* (1958)—the direction of movement being away from the handle. In speeded-up cinematographic studies of tissue cultures containing both lymphoid and epithelial cells, e.g. from human thyroid, the lymphocytes move actively and randomly except for a striking tendency to move toward any mitosing cell when telophase begins (Humble *et al.* 1956). No other type of tropism has been established (Harris, 1954). In a number of *in vitro* situations, lymphocytes are observed to congregate around a macrophage, sometimes making contact with the single foot process, 'uropod', of the lymphocyte (McFarland and Heilman, 1965). Such appearances have been regarded as representing the morphological equivalent of a transfer of immunological information either from lymphocyte to macrophage or from macrophage to lymphocyte (Fishman *et al.* 1963).

## Vulnerability of the lymphocyte

*In vivo*, the small lymphocyte is the most vulnerable of all cells to X-irradiation and to a variety of cytoxic drugs. Trowell (1958) summarizes his observations by saying that lymphocytes are

341

unduly sensitive to ionizing radiation, cortisone, nitrogen mustard, colchicine and barbiturates. They are relatively resistant to changes in pH and $CO_2$ tension, SH-poisons, urethane and alcohol. There is some basis for considering that most of these destructive effects are mediated *in vivo* by cortisol which, in relatively large dose, is highly lympholytic. It is of great interest in this connection that very small concentrations of cortisol are necessary if *in vitro* production of antibody from primed cells is to be obtained (Ambrose, 1964).

## THE CONVERSION OF SMALL LYMPHOCYTES TO PYRONINOPHIL CELLS WITH PROLIFERATIVE CAPACITY

The experiments responsible for the general recognition that small lymphocytes are not necessarily end-cells are undoubtedly those of J. L. Gowans. Before these experiments, large numbers of workers felt confident that small lymphocytes could give rise to descendant cells (see Yoffey and Courtice, 1956). This was the whole basis of the clonal selection theory in its first form (Burnet, 1959). Gowans (1962) used the purest available source of lymphocytes, thoracic duct cells from the rat, labelled them isotopically and used them to provoke a graft-versus-host reaction. He made use of two inbred strains of rats and their $F_1$ hybrids and injected parental strain lymphocytes intravenously into $F_1$ recipients. With an adequate dose, $2 \times 10^8$ cells, a fatal graft-versus-host reaction occurred. Examination by autoradiographic methods of various tissues showed at 24 hours considerable numbers of heavily labelled pyroninophil cells with large active nucleoli in the regions which 4 hours after the injection had contained large numbers of labelled small lymphocytes. It was in these same areas of white pulp in spleen, and in the primary follicles of lymph nodes, that the damaging effect of the graft-versus-host reaction was most evident.

Nowell in 1960 showed that extract of kidney bean (PHA) capable of agglutinating human red cells of group O could initiate mitosis in cultures of normal human leucocytes. This was confirmed by Cooper *et al.* (1961) and Carstairs (1961) and the latter suggested that the dividing cells must be derived from small lymphocytes. Marshall and Roberts (1963) showed conclusively that this was in fact the case. There is now a general consensus of opinion that the enlarged and dividing cells seen in lymphocyte cultures treated with PHA are equivalent to those obtained *in vivo* in Gowan's experiments.

The next step was due to Pearmain *et al.* (1963) who showed that tuberculin could induce similar changes in lymphocytes from persons with a positive tuberculin reaction but not from non-reactive individuals. It is still very difficult to gather from the literature how far such antigenic stimulation can be generalized. There appears to be unequivocal evidence that normal human lymphocytes can be stimulated to enlargement and mitosis by:

(*a*) anti-leucocytic sera (Gräsbeck *et al.* 1963);

(*b*) mixing with lymphocytes from an unrelated individual (Bach and Hirschhorn, 1964);

(*c*) by streptolysin S (Hirschhorn *et al.* 1964);

and that lymphocytes from a tuberculin-sensitive individual can be stimulated by tuberculin. It is reasonable to extend this to other sensitizing agents, and there are many isolated reports that persons immunized by various antigens give lymphocytes which react specifically. There are so many technical hazards and so many negative results, however, that it is wisest to adopt a sceptical attitude.

A similarly cautious attitude must be maintained in regard to reports that such stimulated cells produce immunoglobulins or specific antibody. Johnson and Roberts (1964) made electron micrographs of stimulated cells and found only traces of endoplasmic reticulum, although mitochondria were increased and nucleoli large and active. There is plenty of evidence that leucocytes can take up antibody from serum, e.g. Ridges and Augustin (1964), and with sensitive tests it will always be very difficult to distinguish between such cytophilic antibody, particularly Ig A, and antibody being actually synthesized by the cell concerned.

This topic will be further elaborated in chapter 9.

OTHER POSSIBLE LYMPHOCYTE TRANSFORMATIONS

There is a general reluctance at the present time to go further than Gowans in regard to possible transformations of lymphocytes. It has often been suggested that macrophages and fibroblasts can be derived from lymphocytes. There seems to be no doubt that when thoracic duct cells are cultured, either *in vitro* or *in vivo* in Millipore chambers, a proportion of macrophages appear (Shelton and Rice, 1959; Holub and Říha, 1960). Howard *et al.* (1965) produced graft-versus-host reactions by injecting parental $C_{57}BL$ mouse cells into $F_1$ hybrids with $T6$ and showed that the active Kupffer's cells present in the liver were of donor origin, i.e. they had been derived from some cell present in the thoracic duct lymph. Volkman and Gowans (1965 *a*, *b*) showed conclusively that the phagocytic mononuclear cells which migrate into an area of inflammation in the rat are derived from cells recently produced in the bone-marrow and not from circulating lymphocytes in the usual sense. The possibility is not excluded that the cells as they left the blood in these experiments had the form of small lymphocytes but it is equally probable that they would have been identified as circulating monocytes.

Recent claims to establish the transformation of lymphocytes into fibroblasts have been made by Shelton and Rice (1959) and Petrakis *et al.* (1961), but general opinion seems to favour a local origin of the fibroblasts seen in areas of inflammation. Nettesheim and Makinodan (1965), for example, found no development of histiocytes or fibroblasts in diffusion chambers containing thoracic duct cells from the rat.

The suggestion that lymphocytes in general could function as bone-marrow stem cells from which granulocytic and erythrocytic cells could be derived has been fairly conclusively disproved by showing that even very large numbers of lymphocytes from thoracic duct lymph will not protect lethally irradiated mice which could be saved by very much smaller numbers of bone-marrow cells. Another point of considerable significance, due to Whang *et al.* (1963), is that in cases of myelogenous leukaemia associated with the Philadelphia chromosome, mitoses of lymphocytes in circulating blood, taken during a remission of the disease, showed no evidence of the abnormal chromosome although it could be seen in almost all mitoses in bone-marrow material taken simultaneously.

All the findings are compatible with the hypothesis that most of the circulating lymphocytes are immunocytes which are incapable of differentiating into forms outside the lymphocyte–plasma cell series, but that in both thoracic duct lymph and blood there is a proportion of small round cells which are not differentiated immunocytes and retain a wider capacity for differentiation.

## THE EMBRYOLOGY OF THE THYMUS

There appears to be general agreement that lymphocytes in the mammalian body are first seen in the thymus. This lends special interest to the embryology of the thymus, particularly as it concerns the origin of the lymphoid cells. There are striking differences in the anatomy and embryological development of the thymus in different mammals. In reptiles (Fraser and Hill, 1915), the thymic primordia arise from gill pouches I, II, III and in some, IV. In most placental mammals the thymus comes from III. In marsupials there are two and sometimes three sets of thymuses. A superficial cervical thymus is present in the kangaroo, Australian possum and wombat. This is derived from the cervical sinus I and II. In addition there is one or, in the possum (*Trichosurus*), two thoracic paired thymuses. All have similar histology. One Eutherian mammal, the mole (*Talpa*), has a cervical thymus and in embryonic life a thoracic one from pouch III which atrophies before birth. There is as yet no evidence of any functional differences between thymuses arising from different gill pouches.

The first lymphocytes in the thymus have been variously ascribed to conversion of epithelial cells or entry of mesenchymal cells which in the thymus differentiate to lymphocytes. Auerbach (1960) has applied an experimental approach to the problem and concludes, on the basis of experiments in which thymic epithelial rudiments from mouse embryos of 12 days were separated by a Millipore membrane from various types of mesenchymal tissue (Ball and Auerbach, 1960), that lymphocytes arise by transformation of epithelial cells induced by the proximity of mesenchymal cells which also serve to provide the stromal elements of the gland.

345

Most of the earlier workers on the subject, e.g. Grégoire (1935, 1958), have preferred the view that mesenchymal cells from elsewhere in the body infiltrate the primordial epithelial mass, proliferate as lymphocytes (thymocytes) and transform the epithelium into the loose sponge of reticulo-epithelial cells characteristic of the developed thymus.

It is extremely difficult to be sure that in Auerbach's experiments no potential mesenchymal ancestors of lymphocytes were present in the 12-day anlagen, and in view of the evidence that bone-marrow cells colonize the adult thymus one hesitates to accept Auerbach's interpretation as adequate. A modification of Grégoire's approach that appears to be compatible with both the experimental facts and the evolution of the thymus is to accept a complex interaction of the epithelial primordium with primitive mesenchymal cells. By their mutual inductive effect, a milieu develops in which undifferentiated mesenchymal cells are differentiated to immunocytes with the morphology of the lymphocytic series. The thymus is thus regarded as necessary for the differentiation of lymphocytes even if they are not linear descendants of thymic epithelium. In view of the immunological function of the bursa of Fabricius in chickens, the possibility that a gut-associated source of primary lymphopoiesis may be found in mammals seems to be real. Good and collaborators (Sutherland *et al.* 1964; Cooper *et al.* 1967) have supported the view that Peyer's patches and the more concentrated lymphoid tissue of similar character in the appendix have this function. The human tonsil, they believe, may also be part of the same system.

### THE STRUCTURE OF THE THYMUS AND CELLULAR CHANGES

The essential features of the histological structure of the mouse thymus have been extensively discussed by Metcalf (1964, 1966) while the electron microscopic appearances have been reported by Clark (1963).

Recent observations relevant to the general theme of this book are:

(*a*) The concept of cortical packets of multiplying thymocytes enclosed by a tenuous 'reticulo-epithelial' sheath (Clark). This is modified by Metcalf (1964) to admit the presence of 'PAS cells', phagocytic cells staining positively by the periodic-acid-Schiff method, in the perivascular wall.

(b) Metcalf regards the reticulo-epithelial cells of both cortex and medulla as representing a single tissue, the distribution of proliferating cortical lymphocytes being determined by hormonal and vascular mechanisms, which in the normal young organ ensure a relatively constant relationship of medulla and cortex. Thymic grafts do not develop unless a portion of medulla is included (Metcalf, 1963).

(c) In mammalian thymuses including those in man, the progressive decrease in relative and absolute size with ageing involves cortex rather than medulla. The classical evidence on this point is due to Hammar (1936), while Metcalf (1966) has provided quantitative data for the mouse strains AKR and C3H.

(d) In my own laboratory (Burnet and Holmes, 1964; Holmes and Burnet, 1964) the thymic appearances of the NZB autoimmune strain of mice and its hybrids have been extensively studied.

Changes in the thymus of the NZB mouse and its hybrids recorded by Burnet and Holmes (1964) were primarily concerned with the possible relationship of the germinal centres to the autoimmune process. Incidentally, however, a number of observations were made which seem to throw some light on the general nature of cellular changes in the organ. A characteristic feature is the appearance in the thymus of medullary centres of lymphoid cell proliferation, usually from 5 or 6 months of age onward. These serve as local centres of stress influencing the adjacent cortex. Wherever a germinal centre develops it soon shows a roughly circular area of 'medulla' surrounding it instead of the irregularly angular shapes of normal medullary areas. It is obvious that the visible circle results from local depletion of the cortical areas on which it impinges.

Observations on the nature of this local depletion are necessarily limited to strains which develop medullary lesions spontaneously, such as NZB. We regard one completely regular finding as of special significance, viz. that mast cells which are relatively frequent in the substance of NZB thymuses occur almost wholly in two situations: (*a*) in the cortex adjacent to the cortico-medullary margin and (*b*) in areas of medulla which have clearly encroached recently on cortex. Close examination will also show that at the margin where medulla for any reason is expanding at the expense of cortex, both mast cells and a variety of other cells are responsible for the rather mottled appearance of the cortical margin. Irregularly associated with a majority of normal thymocytes are pyroninophil cells with an increased proportion of cytoplasm, foamy phagocytic cells with small nuclei and a slightly pyroninophil cytoplasm, epithelial cells, and a relatively large number of nondescript mononuclear cells with less chromatic nuclei than lymphocytes. In fully developed normal medulla we have, apart from cells immediately associated with blood vessels, a majority of the epithelial and nondescript mononuclear cells, occasional pyroninophils and a moderate number of diffusely distributed small lymphocytes but no mast cells.

An interpretation of these findings on the basis of the Clark packet is necessarily somewhat speculative. For the present we can consider only the nature of depletion processes, leaving any question of the entry of cells from extra-thymic sources till later. The lymphocytes (thymocytes) in the cortical environment are obviously highly vulnerable; the immediate stimulating or destructive agents being presumably diffusible pharmacologically active agents released locally in the course of cell damage or, in the case of hydrocortisone liberated by the adrenals, in response to general stress.

When the immediate stimulus is sufficiently intense, the thymocytes are necrosed and the picture seen histologically within 24 hours of the action is of large numbers of pyknotic nuclei, fragmented nuclei and active phagocytosis. There is often hardly an intact lymphocytic nucleus to be seen in the cortex. The process of autolysis and removal is rapid and, as the cortex shrinks, the relatively unaffected epithelial and macrophage cells become much more conspicuous. If a cytotoxic agent like cyclophosphamide is administered at intervals, one can in a week or two find an atrophic thymus which on section seems to be almost wholly composed of epithelial cells. This picture of acute depletion is well known and the interpretation offered is probably generally acceptable.

348

Our interpretation of the changes on slow expansion of the medulla is based wholly on the NZB group of mouse strains. The impact of the hypothetical diffusible agent on thymocytes within the packet is assumed in the first instance to be similar to the effect *in vitro* of PHA or human small lymphocytes. Some of them enlarge to become pyroninophil cells, generally resembling immature plasma cells but without any significant development of antibody-secreting endoplasmic reticulum. In sections of thymuses in which the medulla appears to be enlarging rapidly, mast cells are often closely associated with pyroninophil cells and in some expanding medullas the distribution of pyroninophil cells corresponds to the periphery of the area, i.e. to the area which was recently part of the cortex. With something a good deal short of certainty we believe it likely that the change to pyroninophils is a movement to a labile form which under appropriate stimulation can move to mast cells or plasma cells as well as to lymphocytes. We believe, too, that the evidence from mast cell changes in the NZB thymus also provides important additional evidence that *all* thymocytes in the mouse are potentially capable of conversion to mast cells.

## SEROLOGICAL DISTINCTION BETWEEN THYMOCYTES AND CIRCULATING LYMPHOCYTES

Two groups of workers using isoimmune mouse sera by cytotoxic methods have shown that thymic cells may carry an antigen not regularly present in lymphocytes from other sources but rather commonly found in mouse leukaemia cells. There are some inconsistencies between the two sets of results. Old *et al.* (1963, 1964) describe an antigen, TL, present in many leukaemic cell lines and in the thymus of A and C58 strains. Thymic cells of C3H, C57BL and AKR were not affected by their test immune sera (TL−) but several leukaemias from TL− strains were sensitive to the cytotoxic effect.

Reif and Allen (1964) used cytotoxic sera produced by immunization of C3H mice with AKR thymus cells and vice versa. They produced in this way cytotoxic sera which differentiated between cells according to whether or not they contained antigens which they designated $\theta$-AKR and $\theta$-C3HeB/Fe respectively. The first was present in AKR and RF thymic cells and in several (four out of six) AKR leukaemias. The second was found in the thymuses of all other mouse strains tested.

If, for the present, we take both sets of results at face value and assume that comparison of material between the two groups will remove inconsistencies, there are several points of great interest. It appears that there are a small number (perhaps two groups, each with minor variants) of antigens characteristic of thymic as against other lymphocytes. The thymic antigen is also found on a proportion of leukaemic cells, the rules for distribution of the different antigens amongst various leukaemias not yet being fully established.

Absorption studies by Reif and Allen showed that brain (AKR) tissue absorbed as actively as thymus and that peripheral lymphoid tissues had a low but definite absorptive power for antibody, about 1/20 that of thymus. There is therefore a likelihood that a proportion of the circulating lymphocytes have thymic antigenic quality. In view of the probability that most peripheral lymphocytes in young animals are only a few somatic generations from thymocytes, this suggests that the antigenic quality is lost in a generation or two outside the thymus. It is of interest in this connection that Boyse *et al.* (1963) found that C57BL mice immunized with a TL+ leukaemia of foreign origin were still fully susceptible to a TL+ leukaemia strain of C57BL origin, but the cells developing in the immune host had lost the TL antigen. A single passage in a nonimmune host restored *in vitro* susceptibility to the cells. It is clear that the antigen is associated with some labile cell constituent whose presence or absence has very little biological significance.

Other evidence of the existence of a specific thymic antigen has been obtained by immunizing rabbits with rat or mouse thymus homogenates. Nagaya and Sieker (1965) compared the activity of such an anti-thymic serum with similar serum prepared against lymph node cells. The anti-thymic serum (AT) differed from the anti-lymph (AL) in being much more toxic, but when the dose was reduced to a nontoxic level there was complete acceptance of allogeneic skin grafts. The AL serum had a minor effect in prolonging graft survival. Russe and Crowle (1965) produced a similar

AT rabbit serum using washed cells without adjuvant and found lymphopenia and a marked decrease in the number of germinal centres in the spleen. They found no change in the histology of the thymus. In such mice there was a diminution in the development of hypersensitivity to protein antigens but only after a prolonged course of antiserum administration.

These findings suggest that some interesting results could be derived from an elaboration of this approach, making an attempt to assess the cells or antigens in the thymus responsible for stimulating the effective antibody. Until it is clear that epithelial cells or products are not involved, the results may be irrelevant in the present context.

Although not strictly relevant to the present topic it is of interest that claims have been made that thymocytes differ from lymphocytes elsewhere in the body in that they stain for alkaline phosphatase by Gomoris's method (Beauvieux, 1963) and that, following antigenic stimulation in rabbits, significant numbers of such cells appear in the blood. Metcalf *et al.* (1962) have noted that a small proportion of thymocytes in mice show this reaction but when lymphoma develops in the thymus in AKR mice the lymphoma cells are uniformly positive for alkaline phosphatase.

### THE IMMUNOLOGICAL ACTIVITY OF THYMOCYTES

The thymus is not by any means a simple accumulation of cortical thymocytes, and in assessing the results of tests of immunological function we must keep this in mind, particularly in mice whose thymus has been in any way damaged by previous experimental manipulations. In most older thymuses, human or murine, there are lymphoid cells in perivascular regions running into the medulla which often include a proportion of plasma cells. In a thymus damaged incidentally to experimental immunization, Landy *et al.* (1965) and G. Möller (1965) both find that at the fourth or fifth day there is a transitory high count of plaque-forming cells in a proportion of thymuses tested. Some show none and in samples taken on the seventh or eighth day, when spleen showed a high count, none was present in the thymus.

Direct extracts of thymus from an immunized animal later

351

than the fifth day contain no more than traces of antibody or of antibody-producing cells and one can be confident that the great mass of cortical cells play no direct part in antibody production.

All tests of the immunological capacity of thymic cells must therefore be done by transfer of cell suspensions to other animals. The method by which thymic cell suspensions are prepared experimentally makes it probable that well over 90 per cent are cortical thymocytes plus small lymphocytes from medulla and perivascular lesions. With due caution, therefore, any major immunological effects of such cells can be ascribed to cells which are probably from the cortex of the thymus. Minimal effects may well be ascribed to the small proportion of 'normal' lymphocytes, plasma cells and macrophages present in relation to the vessels of the medulla. Thymus cells from an immunized rabbit were found by Dixon *et al.* (1957) to have no capacity to transfer antibody-producing capacity to an irradiated recipient. In mouse experiments, Stoner and Bond (1963) found that transfer of thymus cells from immunized mice to irradiated recipients gave some secondary-type production of antitoxin, but only 2–10 per cent of that obtained from equivalent numbers of splenic cells.

Holub *et al.* (1965) studied the response of lymphoid cells from normal young adult rabbits when they were implanted with antigen (BSA) in Millipore-type chambers in the peritoneal cavity of rabbits less than a week old. The response of thymic cells was negligible but if small lymphocytes were separated from the whole population and tested they gave a significant response in the general range of about 1/8 the titre of small lymphocytes from lymph nodes.

If the normal structure of the thymus is disrupted by the direct inoculation of antigen into its substance, germinal centres and plasma cells appear with antibody production (Marshall and White, 1961). The importance of normal structure is also exemplified in Stutman and Zingale's (1964) experiments on the effect of early autografting of rat thymus to a subcutaneous site. When they were immunized with diphtheria toxoid 60 days later, an equivalent response was obtained in both intact and autografted animals— approximately four times that in neonatally thymectomized animals. The autograft in the immunized animals showed a very high content (24 per cent) of plasma cells including specific anti-toxin producers in contrast to thymuses in their normal anatomical relationships where about 1 per cent of plasma cells was found.

Studies for the presence of immunoglobulins in the thymus have given interesting results. In human foetal thymus, no immuno-globulins are present at a time when the spleen shows immuno-globulins G and M (van Furth, 1964). In the same author's study of thymus from surgical material he found that G and A globulins were produced in thymus cell cultures but no M. He drew atten-tion to the fact that the results with thymus were quite similar to those with lung and thyroid and unlike those of spleen and lymph node, where G-, A- and M-producing plasma cells were always present and a proportion of small lymphocytes produced Ig M. The competent lymphoid cells in the thymus and thyroid are simply interpreted as cells picked up from the bloodstream and not produced *in situ*.

Thorbecke and Cohen (1964) make the brief but very interesting comment that in the immature guinea-pig thymus the only immunoglobulin component detectable is the light chain—the element common to all the immunoglobulin types.

The combined findings suggest that most of the observed immune competence of the thymus is due to such more or less adventitious cells but that the most differentiated of the cortical cells, the small lymphocytes, can, when transferred to a suitable peripheral environment, develop demonstrable activity.

As far as the immunochemical findings go they are consistent with the view that only the earliest stages of the process leading to the establishment of receptors take place in the thymus. Stimula-tion by antigen or by some nonspecific peripheral stimulus is necessary for evidence of immunoglobulin to be observed.

In adequate amount, thymic cell suspensions can reconstitute immune capacity in syngeneic neonatally thymectomized mice (Hilgard *et al.* 1964; Yunis *et al.* 1965), the competent cells in these experiments being shown to be of *donor* origin and associated with tolerance to donor skin grafts. Congdon and Duba (1961) showed that isologous mouse thymus cells prevented the normal capacity of heterologous (rat) bone-marrow to protect lethally irradiated mice. Presumably amongst the thymic cells there were some able

353

to settle in spleen and respond immunologically to the heterologous antigen.

The other indirect approach to the immune capacity of thymus cells is to test their ability to evoke graft-versus-host reactions in appropriate systems. The results vary with species and strain. Billingham and Silvers (1962) found that rat thymus was almost completely inert. Thorbecke and Cohen (1964) on the other hand observed some effect with newborn thymus cells from rabbits. Using C 57 BL donors and newborn A recipients, Cohen *et al.* (1963) found that both neonatal and adult thymus gave a significant graft-versus-host reaction as judged by Simonsen's spleen-index method.

## INDIRECT EVIDENCE OF THE IMMUNE FUNCTION OF THYMOCYTES

At this point it is worth enumerating briefly the less direct evidence for the view that cells differentiated in the thymus subsequently play a part in immunological function. Some of the points have been or will be referred to in other contexts.

(*a*) Dispersed thymus cells in large enough number will reconstitute the immune competence of a neonatally thymectomized mouse. The competent cells are of donor origin and the animal is tolerant of donor skin homografts (Yunis *et al.* 1965).

(*b*) Syngeneic radiation chimeras saved by spleen cells from neonatally thymectomized mice accept homografts but if the spleen cells are from normal syngeneic mice the homografts are rejected (Miller *et al.* 1964*a*).

(*c*) When mice are thymectomized in adult life, lethally irradiated and injected with syngeneic marrow, lymphoid tissues are not recolonized and the animal is immunologically deficient (Miller *et al.* 1963*a*). In their opinion: 'There is in the marrow inoculum a population of cells which has immunological potential but can only express this if the thymus is present.'

(*d*) In rats (Sainte-Marie and Leblond, 1964) and guinea-pigs (Ernström, 1965) venous blood from the thymus contains more lymphocytes than blood taken at the same time from the carotid artery.

354

(*e*) When a mouse is treated with tritiated thymidine long enough to allow extensive labelling of small lymphocytes, subsequent examinations show a rapid disappearance of all labelled small lymphocytes from the thymus (Bryant and Kelly, 1958; Mims, 1962; Craddock *et al.* 1964).

(*f*) Thymectomy prolongs the period of tolerance to antigens such as BSA administered neonatally or the duration of immune-paralysis after injection with well-centrifuged bovine $\gamma$-globulin in mice (Claman and Talmage, 1963; Taylor, 1964).

### THE EFFECTS OF NEONATAL THYMECTOMY

Neonatal thymectomy in the mouse was introduced as an immunological technique by Miller in 1960–1 and most of the essential changes that can be observed as a result have been recorded by him. Miller has published several reviews (Miller, 1963, 1964; Miller *et al.* 1962) that are easily available, and discussion here will be limited almost entirely to consideration of the recent work on the relative importance of deficiency in cellular and humoral factors in producing the characteristic immune crippling of the neonatally thymectomized animal.

The features in which a neonatally thymectomized mouse of a suitable strain differs from a sham operated control are:

(*a*) High mortality in first 4 months of life.

(*b*) Prolonged acceptance of allogeneic skin grafts.

(*c*) Inability to give a normal immune response to standard antigens.

(*d*) Increased susceptibility to a variety of disease conditions, tumour induction, fungal infection and endotoxin intoxication.

### THE BEHAVIOUR OF GERM-FREE ANIMALS

It is now established that complete neonatal thymectomy in germ-free animals is not followed by lethal runt disease (McIntire *et al.* 1964; Wilson *et al.* 1964). When asepsis broke down in some of the latter's experiments, 'runt disease' appeared in four to eight weeks. It is evident, therefore, that some infectious factor, probably a virus or toxic bacterial products, is responsible for the character-

355

istic disease and it is a natural deduction that this is a secondary result of the inadequacy of immune response to the pathogenic agent. It should be added that under germ-free conditions (*a*) thymectomized mice showed retention of homografts and (*b*) inoculation of lymphoid cells from a strain capable of giving graft-versus-host disease in conventional mice was equally effective in germ-free recipients. In the absence of knowledge of the pathogenic agents it is impossible to analyse the differences in the mortality experience seen with various strains of mice, though in general one can assume that the more rapidly the thymectomized animals die the greater the degree of immunological crippling. Dukor *et al.* (1966) found that in outbred Swiss mice neonatal thymectomy is followed by minimal secondary disease and the animals rapidly recover immunological competence, including ability to show a normal plaque-forming cell response in the spleen.

### DIMINISHED IMMUNOLOGICAL RESPONSIVENESS IN NEONATALLY THYMECTOMIZED ANIMALS

Apart from the delayed mortality already referred to, Miller's (1962) early findings in neonatally thymectomized mice were (*a*) severe depletion of the lymphocyte population; (*b*) impairment of H-agglutinin production to immunization with *Salmonella typhi* vaccine; and (*c*) prolonged survival of allogeneic skin grafts (fig. 28).

The immune response as judged by immunoglobulin and antibody production is diminished in neonatally thymectomized animals, but the changes are irregular and may vary considerably from one mouse to another (Humphrey *et al.* 1964) (table 8). Normal immunoglobulin levels are common but are reached more slowly than usual. Several observers (Parrot and East, 1962; Azar *et al.* 1963) have recorded abundant plasma cells in spleen and lymph nodes. Immunoglobulins A, G and M are all present. Wide variation of antibody titres within any group of neonatally thymectomized mice may be the most important theoretical finding from these experiments. Humphrey *et al.* (1964) discuss various possibilities to account for the apparently random depletion of antibody-producing capacities. By far the simplest interpretation

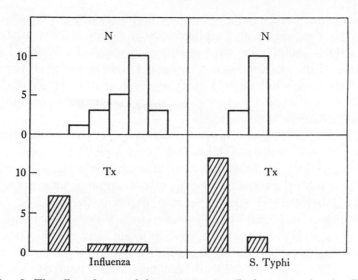

Fig. 28. The effect of neonatal thymectomy on antibody response in mice. Data from Miller and Dukor (1964). Antigens were killed vaccines of influenza virus and *Salmonella typhi*. Titres are shown at log. 2 intervals and the blocks indicate the numbers of mice at each level of antibody. N = normal.

TABLE 8. *The variable response of neonatally thymectomized mice to different antigens (Humphrey* et al. *1964)*

|  | Antigens | | |
|---|---|---|---|
| Mouse | Sheep red cells | Pneumo III SSS | Haemocyanin |
| 1 | 15 | 120 | 10 |
| 2 | 0 | 20 | 0 |
| 3 | 40 | 40 | 240 |
| 4 | 160 | 15 | 240 |
| 5 | 40 | 160 | 200 |
| Controls | 120–2,560 | 0–80 | 80–1,280 |

Titres from five completely thymectomized mice are compared with the range of titres in control mice.

357

is to assume that the range of immune patterns in the clones seeded through the peripheral lymphoid system before thymectomy is incomplete compared to the average complement of a young adult mouse. If the development of any given clone is essentially by a random process influenced in part by the genetic character of the individual, this patchy quality of immune incapacity in this field is just what would be expected.

Any alternative source of randomness, e.g. in varying concentration of thymic hormones, must find it very difficult to explain the individuality of specific depletions and the fact that in some thymectomized mice the response to one antigen, a pneumococcal poly-saccharide (SSS III), was higher than in any of the controls.

Ability to retain first-set allogeneic grafts also varies somewhat from one individual mouse of a group to another, and in relation to the 'strength' of the antigenic difference between graft and host. Thus Miller (1962) found with neonatally thymectomized (AK × T6) $F_1$ mice the following percentages of (a) prolonged acceptance and (b) early rejection with the grafts named:

|  | (a) | (b) |
| --- | --- | --- |
| C3H | 70 | 9 |
| C57BL | 54 | 23 |
| BALB/c | 58 | 25 |
| DBA/2 | 57 | 14 |
| Rat | 15 | 62 |

### RECONSTITUTION OF IMMUNE CAPACITY IN NEONATALLY THYMECTOMIZED MICE

The most physiological means of reconstituting capacity is to graft a neonatal thymus, either syngeneic or allogeneic (Miller, 1962). When an allogeneic thymus is used it is capable of establishing tolerance if grafted early enough. Such mice will accept skin grafts from their own strain and from the strain supplying the graft but reject any 'third party' graft.

Injection of syngeneic spleen or lymph node cells in doses of $10^7$ cells will reconstitute the immune reactivity (Miller, 1962) and large numbers of thymus cells have also been found effective (Hilgard *et al.* 1964). Bone-marrow cells, however, have failed even

358

in very large doses (Cross *et al.* 1964). This is an important finding in showing the inability of the bone-marrow 'small lymphocytes' to replace thymus.

Mitchell and Miller (1968) have recently reported a complex series of experiments to show that in CBA mice reconstitution of capacity to produce plaque-forming cells against sheep red cells in thymectomized and irradiated mice requires both thymic cells specifically stimulated by antigen and cells of probable bone-marrow origin. The nature of the interaction between the two sets of cells is still problematical and it remains to be seen how widely applicable these findings are to other strains and species.

The effect of humoral factors in allowing reconstitution is discussed under a separate heading.

### HUMORAL FUNCTIONS OF THE THYMUS

In 1956 Metcalf provided evidence that extracts of mouse and human thymus injected into baby mice produced a temporary lymphocytosis in the circulating blood. With this technique he was able to show that similar activity was present in the plasma of pre-leukaemic mice and of human patients with chronic lymphatic leukaemia (Metcalf, 1956*a*, *b*).

Little recent work with extracts has been reported but Osoba and Miller's (1964) experiments with Millipore chambers impermeable to cell passage are assumed to be equivalent, in that any cellular activity in the chambers can only affect the host through the mediation of some soluble agent.

The facts are well established that thymus tissue from foetal or neonatal donors placed in Millipore chambers in the peritoneal cavity of mice recently subjected to neonatal thymectomy will allow normal weight gain, absence of wasting disease and ability to produce antibodies and reject homografts at approximately the normal level of activity (Osoba and Miller, 1964; Levey *et al.* 1963). If such chambers are examined histologically after some weeks, all lymphocytes have disappeared and the contents of the chamber are essentially epithelial cells.

It has also been shown that the additional disability that adult thymectomy confers on lethally irradiated mice can be removed by using chambers containing thymus in addition to bone-marrow transfusion (Miller *et al.* 1963*a*).

In addition to the experimental evidence, Metcalf (1964) has summarized indirect evidence which leads him to believe that medullary epithelium and 'PAS-cells' of both cortex and medulla produce the agent responsible for the high mitotic rate of cortical lymphocytes in the thymus. The identity of Metcalf's (1956*a*) 'lymphocytosis-stimulating factor' (LSF) with the 'competence-inducing factor' (CIF) demonstrated by Osoba and Miller, has not been shown but there seems no reason why they should not be identical. The statement by Yunis *et al.* (1964), that a well-established graft of rat thymus showing healthy epithelial growth failed to restore immune competence to thymectomized mice, suggests that a careful study of the species specificity of thymic humoral effects might be of great interest.

Finally, brief mention should be made of Mowbray's (1963) report that the thymus (of the dog) was the only organ in which he could recognize the factor inhibitory to antibody production that he had previously isolated from bovine serum. I suggested (Burnet, 1963*b*) the possibility that this could be responsible for the 'censorship' function of the thymus in eliminating cells capable of reacting with accessible body antigens.

It is clear that although the existence of diffusible substances produced by some of the non-lymphocytic cells of the thymus has been adequately established, we are a long way from knowing the number of such agents and their chemical nature. From the immunological point of view the CIF of Miller is the most important and in the present state of knowledge it will be best to refer to this as a single agent, the thymic hormone.

### THE EVIDENCE FOR THE EXISTENCE OF BONE-MARROW STEM CELLS WHICH CAN POPULATE THE THYMUS

It has been recognized for many years that effective 'rescue' of a lethally irradiated mouse is more readily effected by injection of syngeneic bone-marrow cells than by any other type of cell. This has led to the definition of stem cells as the cells which, in adequate number, would allow colonization of the haematopoietic and lymphoid tissues of the irradiated animal and allow its survival. To say that the bone-marrow or the circulating leucocytes contain adequate numbers of stem cells does not, of course, identify the morphological type or types actually responsible. The nearest approach to this is the finding by Cudkowicz *et al.* (1964*b*) that bone-marrow cells passed through glass wool which removed virtually all but cells with the morphology of small lymphocytes can still protect against radiation injury. Cells of similar appearance from spleen and lymph node do not.

The bone-marrow contains large numbers of cells with the morphology of small lymphocytes. Osmond and Everett (1964) deduce from autoradiographic studies that these are being constantly produced and rapidly exported elsewhere. A single injection of tritiated thymidine will show 40 per cent of bone-marrow lymphocytes labelled 72 hours later and there is good evidence that all but a small proportion of these are derived from local precursors in the bone-marrow. Cronkite (1964), in a recent review, speculated that the stem cell of the erythropoietic series had the morphology of a small lymphocyte, that in the bone-marrow it was imprinted to differentiate along the erythropoietic line and after taking up erythropoietin, perhaps in the kidney, returned to the bone-marrow to settle down as an erythroblast. If a karyotypically distinguishable graft of neonatal spleen or thymus is placed in a host that will accept it, within three weeks all the cells showing mitosis in either type of graft are of host origin (Metcalf and Wakonig-Vaartaja, 1964) (fig. 29). Similarly in a lethally irradiated animal rescued with allogeneic bone-marrow, most animals become chimeras in which lymphoid cells are all of donor origin. In such animals Goodman (1964) found that all peritoneal fluid cells (monocytes, macrophages) were of donor type as also were the mast cells. Finally, Ford *et al.* (1959) observed in a mouse lethally irradiated and saved by rat bone-marrow, that eventually the rat cells were eliminated. Examination of bone-marrow, thymus, spleen and lymph node showed that, in all, 95 per cent of the karyotypes showed a similar chromosomal anomaly. This could only indicate that at least erythroid, lymphoid and granulocytic cells were derived from an initial clone from a host stem cell modified at the time of the irradiation.

This, however, appears to be a rather exceptional situation. It is more usual to find a rather sharp discontinuity between erythroid, granulocytic and megakaryocytic cells on the one hand and lymphoid cells on the other. According to Lewis and Trobaugh (1964), when either splenic or bone-marrow cell suspensions

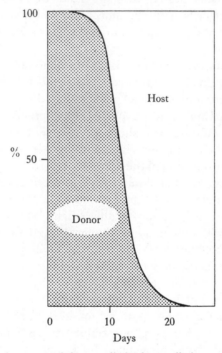

Fig. 29. The replacement of donor cells by host cells in a compatible thymus graft in mice. The percentage of donor and host karyotypes of dividing nuclei is shown. Drawn from data of Metcalf and Wakonig-Vaartaja (1964).

or circulating leucocytes are injected intravenously into lethally irradiated mice by Till and McCulloch's (1961) technique, a mixture of clonal foci appear: roughly two of erythroid cells to one of granulocytes, one of megakaryocytes and one mixed. If erythroid foci are examined at intervals by the same technique, no stem cells are present before the fourth day; thereafter there is an approximate twofold increase per day of cells which can induce splenic foci. From a well-developed erythroid focus the derivative foci show the same distribution of types as the original material. If we can exclude entry of cells from elsewhere in the body this would indicate that a proportion of cells differentiated to produce erythroblasts can be switched presumably after de-differentiation to produce granulocytes or megakaryocytes. In none of these experiments are lymphoid clones observed. To produce them it is necessary to use lymphocytes which have been treated with PHA (Mekori *et al.* 1965).

Another piece of evidence in the same direction comes from the finding that in certain cases of myeloid leukaemia nearly all mitoses in the bone-marrow, including erythroblastic and granuloblastic cells and probably also megakaryocytes, may show the Ph chromosome but it is almost or wholly absent in the blood cells, small lymphocytes, stimulated to mitoses by PHA (Whang *et al.* 1963) in a period when the leukaemia is in regression.

Barnes *et al.* (1964) find that foetal or neonatal liver, which is the major hacmatopoictic organ in mice at that period, also has a high content of stem cells as defined above. In the neonate the blood is exceptionally rich in such cells, apparently moving from liver to bone-marrow.

This represents only a small sample of the papers which point in the same direction. Very few immunologists would quarrel with the statement that among the 'lymphocytes' of the circulating blood and in the bone-marrow are a proportion of cells which in an appropriate internal environment can differentiate to give clones of any of the several types we have mentioned.

The best evidence that such cells do actually enter the thymus is derived from Harris *et al.* (1964) who worked with parabiotic mice, one of which had the T chromosome marker. In a pair of animals which had been united for five weeks, the thymuses contained 12 and 13 per cent of mitoses due to the partner's cells, while in the lymph node the percentages were 35 and 36. Here the conditions were very close to the physiological and we can be certain that the partner cells were derived from the blood. It may turn out to be relatively unimportant where such circulating stem cells come from, but such evidence as there is points to the bone-marrow. When chromosomally labelled bone-marrow cells are given intravenously to mice (Loutit, 1963) they are first recognized in the thymus and only later in spleen and lymph nodes. Lymphoid cells on the other hand are first found in spleen and lymph nodes.

Foetal liver is an actively haemopoietic tissue which can be used as an equivalent to bone-marrow. Using a rather indirect method by which foetal liver of strain A was injected into sublethally irradiated $F_1$ (A × B) mice, this primary host being either intact or thymectomized, Tyan *et al.* (1966) tested spleen and lymph node cells from primary hosts 60 days later for their capacity to produce lethal graft-versus-host disease in a second series of sublethally X-irradiated $F_1$'s (A × C). No manifest disease was seen in the primary hosts but in the intact irradiated group the development of A cells capable of reacting against C antigen was shown by the death of 43 of 48 secondary hosts. If the primary host had been thymectomized this development failed to occur, there being only one death in the secondary $F_1$ (A × C) hosts. The experiment is shown schematically in table 9.

TABLE 9. *Tabulation of Tyan* et al.*'s* (*1966*) *experiment*

| A cells (foetal liver) into | |
| --- | --- |
| $F_1$ A × B normal irradiated | $F_1$ A × B thymectomized irradiated |
| ⌈New A cells in thymus give⌉<br>⌊ rise to anti-C, anti-D etc. ⌋ | ⌈Anti-B only proliferate⌉ |
| Injected into irradiated $F_1$ A × C | Injected into irradiated $F_1$ A × C |
| ⌈ Anti-C proliferate giving ⌉<br>⌊ + + graft-versus-host ⌋ | ⌈ No anti-C cells ⌉<br>⌊No graft-versus-host reaction⌋ |

The importance of the thymus was underlined by showing that when the source of A cells was a late foetal thymus graft, it made no difference whether the primary host was thymectomized or not. Spleen cells from both were equally effective in killing about 80 per cent of the recipients.

There is very little evidence that lymphoid cells from spleen lymph nodes or thoracic duct ever enter the thymus (Gowans and Knight, 1964) and none that they can establish themselves in the cortex. Most of the findings of definite but minimal immune competence of thymic cells may well be related to the small perivascular accumulations of lymphocytes and plasma cells that are quite commonly seen in the medulla and along the entering arterioles.

THE SIGNIFICANCE OF THYMIC PALLOR

One of the most striking features of the thymus in any normal young mammal is its pallor (Metcalf, 1966) and this pallor may serve as a useful central theme around which to present our current theoretical picture of thymic function.

The macroscopic appearance indicates that there is relatively little blood passing through the thymus and, histologically, blood vessels are by no means noticeable, particularly in the cortex. When tritiated thymidine is administered intravenously, a high proportion of thymocytes are labelled, but at a threefold lower grain count than the cells labelled elsewhere in the body. On the other hand, the mitotic activity of the thymic cortex is only exceeded by the epithelium of the small intestine, which has a very abundant blood supply. For some years it has been difficult to account for the enormous numbers of new thymocytes being produced by the assumption that all cells not needed for growth are liberated into the circulation and settle in other lymphoid tissues. There is, in fact, little evidence that more than a very small proportion of the cells produced there actually leave the thymus. This difficulty is accentuated by Metcalf's (1966) recent work on multiple thymus grafts. He finds that a mouse may carry twelve times the normal mass of thymic tissue, each graft showing normal structure and high mitotic activity, without more than a small increment in the number of circulating lymphocytes or in the mass of other lymphoid tissue. It seems inescapable that the majority of the lymphocytes arising by mitosis in the thymus, or in thymus grafts, either do not leave the thymus or, if they do, are destroyed or shed from the body so rapidly that they are not detectable in the tissues. What evidence is available suggests that both processes occur. In very young animals there is a large movement of cells from the thymus but rapid destruction of most of the emigrants (Mims, 1962). In older animals, such as mice six to twelve weeks of age, the observations fit best with the hypothesis of destruction of the great majority *in situ* with re-utilization of their constituents within the thymus.

Some of the best evidence for active emigration comes from Mims's (1962) findings in young mice given tritiated thymidine. In one of his experiments a seven-day mouse was heavily treated over nine days and then examined by autoradiography ten days after the last injection. Throughout the lymph nodes, 70 per cent of cells, mostly small lymphocytes, were labelled but the thymocytes were now completely unlabelled. Mims agrees, too, with Bryant and Kelly (1958) that a single pulse of thymidine labels a large proportion of large and medium lymphocytes in the thymus but that labelled cells are gone in 2–3 days. The suggestion, from both sets of experiments, is that there is in fact a high level of movement of lymphocytes from the thymus but that a large number of them are destroyed, almost at once, in extrathymic areas. Mims's experiments are done with immature animals and one can be certain that a high export of cells does occur at this age (Weissman, 1967). Even in young mice, however, there is evidence of breakdown of cells *in situ* and there is much to suggest that as the animal ages there is a progressive increase in the proportion of cells destroyed in the thymus itself.

In a thymus which has reached an approximately stable size, the production of new cells by mitosis must be balanced by the sum of those lymphocytes leaving the thymus plus the fraction which dies and autolyses *in situ*. Autolysis would mean that cellular components are broken down to metabolites small enough to be utilized for the synthesis of nucleic acids and proteins in new cells. If this were the case, one might expect in the thymus (and in the bursa of Fabricius in birds) a very active mitotic rate with an abnormally low demand for metabolites from the circulating blood. The picture then emerges of a physiological situation in which many new cells are being produced but a high proportion, perhaps 95 per cent or more, are being immediately scrapped and 'cannibalized' in the interests of continued production. In addition, Mims's findings agree with all others in showing that the disappearance of label from the thymus is not balanced by the recognition of an equivalent number of labelled cells in the blood or tissues. Inside or outside the thymus, a very large proportion of newly produced cells fail to reach a functional state. The situation appears to be unique in the mammalian body and there is a clear challenge to find what determines the eventual emergence, as functioning lymphocytes, of a small, perhaps a very small, proportion only of the cells which arise by mitosis.

On the basis of the chemical and genetic considerations discussed in chapter 6, a stem cell entering an environment where it will be differentiated to an immunocyte carries a complex range of genetic information from which, on differentiation, the cell will adopt one particular form of immune pattern. This has been generated, in part over the period of vertebrate evolution, in part by somatic mutation in the sequences coding for the combining sites. The 'choice' of which cistrons will contribute to the construction of the chains involved in the combining site is of the same random quality as the Lyon phenomenon—which somatic mutations have occurred in the chosen sequences must also be a wholly random matter.

In the course of differentiation in the thymus these randomly generated immune patterns will emerge. It is simplest but not necessarily correct to assume that from the moment of differentiation the phenotypic restriction to one type of immunoglobulin and one combining region configuration holds. Having regard to the immense variety of potential antigens present in body fluids and in cell components it can be assumed with confidence that if immune patterns are random a very large proportion of those that emerge will be capable of reacting with one or another of the accessible body components. By hypothesis, cells carrying such immune patterns will be destroyed either in the thymus or soon after leaving it. Although other factors may also be involved there seems no reason why the very high mortality ($\pm 99$ per cent) envisaged by Metcalf should not depend on this process of immune censorship.

Only cells certified, as it were, not to react with body components are released to become ancestral to part of the body's working complement of immunocytes. The rest are broken down and re-utilized for the growth of the new clones that are constantly being developed.

Such a point of view is consistent with all that has been discussed of the traffic of cells into and from the thymus and provides a satisfactory interpretation of the humoral aspects. If the active agent or agents are present in relatively high concentration within the Clark packets this should provide the necessary highly specialized environment for:

(a) stimulation of a high level of lymphocytic mitosis;

(b) progressive differentiation to immunocytes;

(c) destruction of differentiated or partially differentiated immunocytes by contact with the corresponding AD.

In this way the massive turnover of cells in the thymus is achieved with only minor contribution of metabolites from the circulating blood, the source of new building-blocks being 'cannibalization' of cells which fail to pass the test for survival.

## THE EVIDENCE FOR THE IMPORTANCE OF CELLS SEEDED FROM THE THYMUS FOR THE DEVELOPMENT OF IMMUNOLOGICAL CAPACITY

The first interpretation of the effects of neonatal thymectomy was that the thymus was associated with the provision of cells differentiated to immune function (Miller, 1962; Auerbach, 1960) and that on leaving the thymus these cells were responsible for 'colonizing' the various peripheral lymphoid organs. I found this approach extremely attractive in supporting a clonal selection theory and gave a fairly detailed description of its implications (Burnet, 1962). Brought slightly up to date, the interpretation offered was that the normal thymus is constantly being re-populated with stem cells, probably with the morphology of small lymphocytes from the bone-marrow. Each such cell undergoes a series of mitoses producing small lymphocytes and, in some period of the order of 20 days, exhausts its reproductive capacity in the thymus. In the course of multiplication in the thymic environment, cells of the developing clone undergo differentiation to immunocytes with minimal production of immunoglobulin receptors. Contact of cells with the corresponding antigen while they are within the thymus will result in destruction of the cell. As a result of the random emergence of immune patterns and the very large number of potential ADs in the thymic environment, a large proportion of the lymphocytes are destroyed either within or soon after leaving the thymus. Those cells passing this censorship move to the circulation and lodge in lymphoid tissue until stimulated to proliferate by antigen.

Since then, the field opened up by Metcalf (1956b) of the humoral functions of the thymus has been expanded by Osoba and Miller (1964), Levey *et al.* (1963) and others and there has been an increasing tendency (see Metcalf, 1966) to see the whole function of the thymus in relation to lymphocytes and immune function as hormonal.

Direct evidence that thymic cells colonize the peripheral lymphoid organs under physiological conditions is technically very difficult to obtain. The requirement is to label thymic cells with tritiated thymidine without labelling the other regions of the body and without disturbing the delicately organized processes of cell proliferation and differentiation going on in the thymus. Nossal (1964) used the superficial thymus of the guinea-pig for such experiments and obtained heavily labelled small lymphocytes in spleen and lymph nodes 3-4 days later. This approach has been elaborated by Linna and Stillström (1966) who compared the distribution of labelled lymphocytes in guinea-pigs after injection of tritiated thymidine (*a*) into the thymus and (*b*) intraperitoneally. There was a highly significant movement of cells labelled in the thymus to the spleen. Even more interesting are the experiments described by Weissman (1967). He inoculated tritiated thymidine through microneedles into the thymus in mice of various ages and at the same time maintained an infusion of unlabelled thymidine intravenously. This ensured that no significant labelling occurred outside the thymus. Autoradiographs showed considerable colonization of thymic cells in the spleen, the highest proportion found being in very young animals.

Less direct evidence may be sought in many directions but, in all, the experimental situation differs grossly from the natural. It is easy to label most of the lymphocytes in the thymus by a short series of injections of tritiated thymidine and to inject such cells intravenously in syngeneic recipients. Most of the cells that can subsequently be recognized lodge where lymphocytes from thoracic duct, lymph or lymph nodes also lodge, in lymph nodes in or at the periphery of primary lymph follicles or in the spleen. Extremely few are found in the thymus (Harris and Ford, 1964; Fichtelius, 1953; Mims, 1962).

A somewhat different approach is due to Parrott *et al.* (1966). They injected syngeneic thymus or spleen cells, labelled *in vitro* with tritiated adenosine, intravenously into thymectomized and

369

intact mice. They were specially interested in what they called 'thymus-dependent areas' in spleen (surrounding the central arterioles) and lymph nodes (in mid- and deep cortical areas). These were markedly depleted in mice 6–7 weeks after neonatal thymectomy but shortly afterwards tended to fill up with pyroninophil cells.

The important finding was that, at 24 hours, labelled thymus cells were five times as frequent in these dependent areas as in other regions of the spleen, while labelled spleen cells showed the reverse with only one labelled cell in the dependent area for four in other areas. There was no significant difference in the distribution in intact or thymectomized hosts.

Another approach is to follow the disappearance of labelled cells from the thymus after a course of injections of tritiated thymidine sufficient to label the great majority of thymic cells. Bryant and Kelly (1958) found that after a single pulse all labelled thymic cells had gone from the thymus in 3 days. Mims (1962) described one very young mouse heavily labelled during the 6th to the 17th days of life. When it was killed 10 days later the lymph nodes still showed 70 per cent of labelled cells but there was none in the thymic cortex. This does not tell more than that all cells newly produced in the thymus have left it (or been destroyed *in situ* without significant local re-utilization of DNA) and that circulating small lymphocytes do not lodge in the thymus.

Metcalf (1966) uses a different approach by enormously increasing the bulk of thymus in a mouse by a large number of grafts of syngeneic neonatal thymus. He finds that the grafts show the same histological structure and mitotic activity as normal thymus but in labelling experiments he can find no evidence of any more seeding of lymphocytes to peripheral organs than from mice with only the normal thymus. This is perhaps the strongest evidence against any large-scale distribution of thymic cells in adult mice.

There is adequate confirmed evidence that neonatal thymic grafts implanted 7–10 days after neonatal thymectomy liberate cells which colonize spleen and lymph nodes, and can be recognized there by karyotypic individuality. The first such experiments by Miller (1963) made use of $F_1$ hybrids with T 6 as the thymectomized host and thymic grafts from the parental strain with the normal karyotype. Thirty out of thirty-one of these animals showed the presence of donor mitoses in the spleen. A more technically satisfactory experiment due to Harris and Ford (1964) made use of CBA and CBA-T6T6, which are fully histocompatible. In thymectomized and grafted animals tested between 3 and 5 weeks after grafting, donor cells were regularly present in lymph nodes, the largest proportion being 65 per cent of donor cells in a mouse killed when 27 days old. As has been the universal experience, the thymus graft is eventually re-populated completely by thymocytes of host karyotype. Harris and Ford then showed that when such a thymus graft was transferred to a neonatally thymectomized secondary host—the $F_1$ (CBA × CBA-T6T6)—cells from the primary host via the thymic graft were numerous in lymph nodes and present in smaller numbers in the spleen. In none of the two sets of experiments were donor cells found in bone-marrow.

It must be admitted that anatomical conditions in the graft differ greatly from those in the normal thymus. Dukor *et al.* (1965) divide the changes seen into four phases. In the first there is general necrosis of all but a thin rim of thymocytes and reticulo-epithelial cells at the periphery of the cortex. During the third to the eighth day, phases II and III, there is intense donor-cell mitotic activity leading to the local reconstitution of the cortex and finally (phase IV) to the re-establishment of normal cortico-medullary relations. In the period from the eighth to the fourteenth day the graft appears to be anatomically and physiologically normal and *all mitoses are of donor type*. It is presumably at this period that maximal seeding to lymph nodes takes place. From now on, however, host cells begin to populate the graft and by 21 days 100 per cent of karyotypes are of host type.

Evidence of functional activity of cells derived from a donor thymus in spleen or lymph nodes has, however, been provided by Leuchars *et al.* (1964) using cells identified by karyotyping. In their fuller report, Davies *et al.* (1966) describe experiments in which they used syngeneic CBA radiation chimeras which are immunologically incompetent but can be reconstituted by grafting a neonatal CBA-T6T6 thymus. They showed that donor cells in the spleen were stimulated to proliferate when the mice were given sheep red cells intravenously and that a sharper increase (to nearly

50 per cent) in donor karyotypes appeared when a secondary antigenic stimulus was given 44 days after transplantation. Equivalent findings in regional lymph nodes but not in the spleen were obtained when the antigenic stimulus was a homograft of skin.

This indicates that in addition to passing from thymic graft to spleen as had been shown earlier by Miller (1962) such donor cells in the spleen had shown the characteristic reaction of immunocytes in proliferating in response to antigenic stimulation.

Taylor (1963) implanted neonatal thymic grafts from T 6 mice into histocompatible 7-week-old CBA mice which had been thymectomized at 2–3 days of age. Three weeks later lymph node cells from these mice were used to provoke graft-versus-host reactions in lethally irradiated $F_1$ CBA × C 57 BL mice. At 2–3 days after injection the spleens were karyotyped, essentially to determine the origin of cells proliferating under the stimulus of the foreign C 57 BL antigens. In these mice 23 per cent of the mitoses were T 6 in type, indicating that considerable numbers of donor cells had established themselves in the primary recipients from the neonatal thymus graft.

All these experiments, as the authors point out, are made on animals which have been subjected to grossly unnatural manipulations, and the results may be only indirectly relevant to physiological happenings. However, taken with the other evidence in regard to thymic function, everything on record is consistent with the interpretation of the thymus as the primary centre for the differentiation of stem cells to immunocytes and for 'censorship' of the differentiating cells.

None of the apparently contradictory indications of the relative importance of cellular and humoral factors in the immunological function is incompatible with our general approach. If we accept as the crux of the matter the special significance of the perinatal period, the best interpretation of the humoral function of the thymus would probably be as follows.

At the immediately prenatal period there is active dissemination of newly formed immunocytes which lodge largely in areas which will subsequently have the character of lymphoid tissue. We assume they are small lymphocytes and that they have a fairly long lifetime if not subject to accident. For a given lymphocyte to proliferate it requires (*a*) presence of thymic hormone (CIF) and (*b*) antigenic stimulation or a nonspecific equivalent. In the absence of either, these cells will remain inert and eventually be lost by one form of accident or another. Once a clone is initiated by the co-existence of (*a*) and (*b*) its subsequent survival is unrelated to the necessity for thymic hormone. If the thymus is removed at three weeks of age or later there is no evidence of any immunological inadequacy for the rest of the animal's life, provided no major stress such as lethal irradiation and recovery by bone-marrow therapy is encountered.

The test animal thymectomized at birth and given the active diffusion chamber or the empty control can be thought of as having a large number of very small clones widely distributed and a steady entrance of small amounts of many antigens from the environment. In the controls many clones will lose all their representatives within a few weeks. Only those cells which have been stimulated while there was still circulating thymic hormone will be able to go on to produce the rather large clones of plasma cells character-istically found in these mice. If the circulating hormone is renewed the previously unstimulated clones are in a position to be effectively stimulated and take on (as clones) potential immortality.

Before concluding this section it needs to be emphasized that work on the immunological function of the thymus in mice has made use predominantly of only two tests of immune capacity— the ability to reject allografts of skin and the production of haemo-lysin against sheep red cells especially as estimated by counts of plaque-forming cells in the spleen. It is now widely held that there is an alternative organ to the thymus by which immunocytes can be differentiated and, presumably, censored. In the fowl the bursa of Fabricius obviously has such a function and, according to Cooper *et al.* (1967), Peyer's patches and other gut-associated lymphoid tissue may have the same role in mammals. If this is the case we shall have to be prepared to find that the thymus can be wholly responsible for immunocytes with DH and homograft function, require the co-operation of another system in establishing antibody analogous to red cell haemolysins and be quite unrelated to the production of immunocytes ancestral to cells producing antibody against bacterial polysaccharides.

### EVIDENCE FOR THE DIFFERENTIATION OF STEM CELLS
### TO IMMUNOCYTES IN THE THYMUS

In a lethally irradiated animal all the evidence suggests that thymocytes are completely destroyed, but a large proportion of the
reticulo-epithelial cells and macrophages are still viable in the
thymus. Such an animal can be restored to full physiological and
immunological capacity by the injection of an adequate dose of
syngeneic bone-marrow cells.

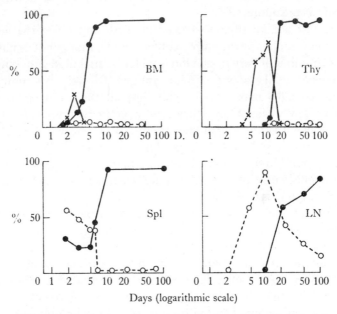

Fig. 30. To show distribution of host and donor karyotypes after administration
of cells from bone-marrow or lymphoid tissue. (Redrawn and simplified from
Micklem *et al.* 1966.) ●——— = cells from donor bone-marrow; O - - - - =
cells from donor lymphoid tissue; ×——— = host cells of 'viable' karyotype not
seen in spleen or lymph node.

In an animal which has been completely thymectomized a few
days before lethal irradiation, a similar dose of bone-marrow cells
will save its life but will not reconstitute its immunological capacity
(Miller *et al.* 1963*a*).

The presence of a thymus is necessary therefore for bone-
marrow cells to reconstitute immune capacity in an irradiated

mouse. After intravenous injection into irradiated mice, thymus cells settle preferentially in spleen and lymph nodes but not in the thymus. Bone-marrow cells settle readily in the thymus (Ford and Micklem, 1963).

Later experiments of the same type by Micklem *et al.* (1966) show that after bone-marrow transfusion there is no significant donor mitosis in lymph nodes until 21 days, i.e. at a time when the thymus has almost completed recovery. On the other hand, lymphoid cells from the donor can give rise to proliferation in the spleen and lymph nodes forthwith.

If we neglect, for the time being, conditions in the perinatal thymus the evidence is complete (*a*) that cells of bone-marrow origin readily establish themselves and multiply in the thymus; (*b*) that neither thymocytes, thoracic duct lymphocytes (Gowans and Knight, 1964) nor lymph node cells can lodge and multiply in the thymus; (*c*) that thymocytes do not recirculate in the rat in the way that thoracic duct lymphocytes do; (*d*) that thymocytes injected in adequate dose can (i) reconstitute immune capacity in a thymectomized mouse, (ii) give rise to a graft-versus-host reaction and (iii) give an immunological response against injected heterologous cells and (*e*) that none of these faculties is shown by normal syngeneic bone-marrow.

Cells derived from the circulation and almost certainly from stem cells in the bone-marrow are constantly entering the thymus and replacing clones of cells which have presumably exhausted their capacity for continued multiplication (Harris and Ford, 1964). Thymocyte suspensions have the capacity to induce a variety of immunological responses at a relatively low grade of efficiency.

Unless some new technique is developed it seems unlikely that any more direct evidence than this will become available. As in nearly every phase of immunology we are working with such a complexity of conditions that the *experimentum crucis* is very rarely forthcoming. No one can ever prove that the picture of differentiation and censorship in the thymus is true, but if such an hypothesis cannot be disproved by the wide range of experimental work done on the thymus in the last five years, it is at least a useful hypothesis.

IMPLICATIONS OF CLINICAL THYMUS DEFICIENCY
IN INFANTS

There are now considerable numbers of reports on severe immunological deficiencies in infants which appear to be related to failure
or abnormality of development of the thymus (reviewed by Peterson *et al.* 1965; Good *et al.* 1967). Included amongst a fairly
heterogeneous group are several cases on record which have
common features of significance for the interpretation of thymic
function (Tobler and Cottier, 1958; Hitzig *et al.* 1958; Harboe *et
al.* 1966; Gitlin and Craig, 1963; Schaller *et al.* 1966). This has
been called 'Swiss type' thymic deficiency.

Broadly speaking, the infants were well for their first 3 months,
then developed a variety of infections and died, usually around
6 or 7 months, although they sometimes survived up to 2 years.
The pathological and haematological findings vary but are approximately as follows:

Lymphopenia but not usually of extreme degree, lymph nodes
small without lymph follicles or germinal centres but plasma cells are
usually seen. The level of circulating immunoglobulins varies and
both agammaglobulinaemia (Gitlin and Craig, 1963; Hitzig *et al.*
1958) and hypergammaglobulinaemia (Schaller *et al.* 1966) have
been reported. Where the evidence has been sought, Ig M is very
low or absent in all. The one case with hypergammaglobulinaemia
had autoimmune haemolytic anaemia and glomerulonephritis with
a sharp spike of Ig G in the electrophoresis pattern. There was a
moderate increase in both Ig G and Ig A concentration but very
little Ig M, and none on immunoelectrophoresis.

At autopsy the thymus is very small or absent and free of
lymphocytes. Since in all cases there had been prolonged and
serious illness, this does not necessarily represent the condition
before the onset of symptoms. The absence of Hassal's corpuscles
and other features have led most authors to conclude that there
had been a primary histological defect in the thymus.

In all cases there has been complete or almost complete failure
to produce antibody after standard antigenic stimulation or to
become sensitive to skin-sensitizing chemicals.

On the interpretation of thymic function that we have adopted, the primary defect is a failure of colonization of thymic cortex with stem cells from bone-marrow, so eliminating normal differentiation to immunocytes and censorship. If this is the case, the circulating lymphocytes in these infants are not immunocytes but stem cells, a variable proportion of which are subject to abnormal differentiation in regions other than the thymus. It is possible to provide a reasonable interpretation of the variation in immunoglobulins on the assumption that only stem cells which had undergone appropriate somatic mutation could be differentiated and gain opportunity to proliferate as plasma cells. On the other hand, in children with these conditions, there may well be associated primary or secondary malfunction of the system, analogous to the avian bursa, which is responsible for differentiation of stem cells to plasma cell precursors. Understanding will be incomplete until the part played by gut-associated primary lymphoid tissue is clear. There is certainly inefficiency of both systems and it could be expected that aberrant cells capable of autoimmune proliferation would flourish in such circumstances if the appropriate mutant arose.

### THE FUNCTION OF THYMUS AND BURSA IN BIRDS

In this book, attention is concentrated almost wholly on mammalian immunology, but the function of the bursa of Fabricius in chickens and presumably all other birds is so important that it can hardly be omitted. There is, too, the real possibility that in mammals there may be an equivalent function perhaps more diffusely related to the lymphoid tissues along the gastro-intestinal tract.

As with the thymus in the mouse, removal of the bursa, or of both bursa and thymus, more than 15 days after hatching has no easily demonstrable effect. Many years ago Woodward (quoted by Glick, 1964) showed that combined bursectomy and thymectomy at ages 10 to 60 days had no detectable effect on development or subsequent laying capacity of chickens. The discovery by Glick *et al.* (1956) was due to the purely fortuitous use of healthy-looking birds previously bursectomized at an early age for a class demonstration in antibody production.

Elaboration of this finding showed that an influence on antibody production was evident after bursectomy up to 12 days of age but that after this it became negligible. The effect of sexual maturation in bringing about atrophy of the bursa was well known and Meyer *et al.* (1959) followed a variety of endocrinological experiments by showing that treatment of young embryos with a single dose of 9-nortestosterone could completely inhibit the development of the bursa. Warner and Burnet (1961) found similar activity of testosterone. Workers in both groups showed that Glick's findings in surgically bursectomized chickens held also for those hatching from treated eggs.

It is now established that hormonal bursectomy (with minimal damage to the thymus) completely inhibits antibody production (Mueller *et al.* 1960; Szenberg and Warner, 1962) but leaves virtually uninfluenced homograft rejection (Mueller *et al.* 1962; Warner *et al.* 1962; Aspinall and Meyer, 1964) and the capacity of blood leucocytes to produce focal lesions on the chorioallantoic membrane (Warner *et al.* 1962). Warner *et al.* (1962) also found that sensitization to tuberculin was lost in such birds. In chickens surgically bursectomized 2 days after hatching, Janković and Išvaneski (1963) found the same results in regard to antibody production and homograft rejection. In their experiments, however, bursectomy had no effect on sensitization to tuberculin or brain lipids and allowed normal production of experimental allergic encephalomyelitis.

Cooper *et al.* (1965) considered that more uniform results were obtained if surgical thymectomy or bursectomy on neonatal chickens was combined with whole-body irradiation (600 r). With such material they found that thymectomy was associated with failure of antibody production in about 40 per cent of chickens, and bursectomy with 100 per cent failure. Splenic lymphoid nodules were present in thymectomized birds but, in flat contrast to Warner *et al.*'s results with unirradiated chickens, none was present in those bursectomized.

Experiments on procedures to reconstitute the immune reactivity of bursectomized chickens have been on a relatively small scale. There is one report that a subcutaneous bursal graft allowed retention of antibody production against a red cell antigen (Isaković and Janković, 1964) and two claims that bursal cells in diffusion chambers will partially return antibody-producing capacity (St Pierre and Ackerman, 1965; Janković and Leskowitz, 1965). In neither set of experiments was anything like full return of immune capacity found.

The process by which the bursa influences the immune response has hardly been investigated. Dent and Good (1965) found no evidence of antibody production by the bursa in chickens immunized with red cells and no cells capable of producing antibody plaques were present in either bursa or thymus. In very limited experiments, Lind and Burnet (1962, unpublished) found significant low production of antibody to BSA in chickens after instillation of antigen in soluble form into the lumen of the bursa. Alum-precipitated antigen was inert. The spleen is undoubtedly the major antibody-producing organ in birds and there are significant differences in the histology of the spleen after hormonal bursectomy (Warner *et al.* 1962) or after bursectomy plus heavy irradiation (Cooper *et al.* 1965). There are differences between the published findings but both groups agree (*a*) that when both thymus and bursa are absent or nonfunctional, no lymphoid nodules appear in the spleen and (*b*) when bursa is absent, plasma cells are absent or extremely rare.

There are still serious gaps in our knowledge of cellular dynamics in the bursa. In the young bird, very active mitosis is taking place far beyond what is needed for growth of the organ. Cooper *et al.* (1967) found that the ultrastructure of germinal centre cells in the chicken resembled that of bursal cells, but it is no more than a reasonable hypothesis that the bursal cells pass to the spleen and on appropriate antigenic stimulation initiate germinal centres and give rise to plasma cells. There is definite evidence (Moore and Owen, 1966) that circulating cells (? stem cells from bone-marrow) are constantly entering the bursa, and the conditions are likely to resemble those of the thymus in most respects. It is noteworthy, however, that the avian thymus is an active lymphocyte-producing organ when the bursa is still wholly epithelial.

The possible importance of 'gut-associated lymphoid tissue' in subserving functions in the mammal equivalent to those of the bursa cannot yet be based on direct experimental work. Indirect approaches are referred to in chapters 4 and 11.

# 4   The immunocyte: its definition, recognition and distribution

Any discussion of cellular immunology must of necessity be centred on the lymphocyte. If one adopts any type whatever of selection theory the statement still holds good that the lymphocyte 'is the obvious and in fact the only possible candidate for the role of the responsive type in the clonal selection theory' (Burnet, 1959). It is equally evident, however, that lymphocytes morphologically similar may be functionally different. According to Gowans (1962), rat thoracic duct cell preparations can be freed of large lymphocytes by a few hours' incubation at 37°, giving a cell suspension in which > 98 per cent of the cells have the morphology of small lymphocytes. Such suspensions can be used:

(a) to reconstitute the capacity of a syngeneic animal to produce antibody to a primary stimulus when that capacity has been lost either by prolonged depletion of lymphocytes or by irradiation (McGregor and Gowans, 1963);

(b) to break down a state of immune tolerance (Gowans et al. 1963);

(c) to provoke a lethal graft-versus-host reaction when an adequate dose of cells from a parental strain is injected into young $F_1$ recipients (Gowans, 1962).

The same cells, however, have no power to prevent secondary disease in lethally irradiated recipients (Gesner and Gowans, 1962). Such protection is easily accomplished by bone-marrow cells. Conversely, studies of the morphological cell types in bone-marrow that could protect against lethal secondary disease in mice have led Cudkowicz et al. (1964a) to identify marrow lymphocytes as responsible. This is a conclusion also reached earlier by Yoffey (Yoffey and Courtice, 1956) from a variety of less direct indications. In the population of 'small lymphocytes' in bone-marrow are cells which are presumably pluripotent stem cells capable of giving rise to several functional types and so protecting against secondary disease. Till and McCulloch's (1961) procedure of producing cellular nodules in the depleted spleen of lethally irradiated mice by intravenous injection, in the first instance of bone-marrow cells, has been extensively used to analyse the situation. These nodules are almost always composed of granulocytic or erythropoietic cells and there is good evidence (Becker *et al.* 1963) that the nodules are essentially clonal in character. Many of the nodules are mixed and may contain also megakaryocytes. Sub-inoculation into further animals of the cells from a nodule will give a similar range of granulocytic and erythropoietic nodules but no nodules containing lymphoid cells.

Feldman (1963) reported some success with thymic cell suspensions but has recently obtained much more regular results by treating lymph node cells with PHA before injection into the depleted recipient (Mekori *et al.* 1965; Feldman and Mekori, 1966*a*). Such results, however, in no way detract from the conclusion that small lymphocytes in a lymph node suspension lack many of the potentialities of cells of similar appearance from the bone-marrow.

In chapter 3 the evidence was summarized that the 'small lymphocytes' of the bone-marrow were responsible for the steady entry of new stem cells into the thymic cortex.

## STUDIES ON SINGLE CELLS

The heterogeneity of all cell populations that can be shown to be immunologically active makes it of special interest to devise ways by which single cells can be identified (*a*) as carrying some immunologically significant property and (*b*) as of some specific cellular type in terms of classical histological cytology or of some electron microscopic or histochemical criterion.

There are now rather large numbers of techniques which can provide information about the status of individual cells. These methods can be enumerated as follows:

1. Methods requiring physical isolation of individual cells for functional study, e.g. Nossal and Mäkelä's (1962*b*) method of detecting flagellar antibody production in microdroplets.

1 *a*. Methods by which isolation of individual cells is achieved by dilution followed by fixation of cells at definite points in a suitable gel, e.g. Jerne and Nordin's (1963) method of antibody plaques.

2. Methods in which cells in sections, smears or monolayers are treated so as to differentiate a proportion as functionally significant. Methods include:

2 *a*. Tests for cytoplasmic immunoglobulins using appropriately prepared fluorescent antisera against globulins or globulin fractions of the species being studied;

2 *b*. Tests for the presence of specific antibody by the 'sandwich' technique of Coons, by treatment with fluorescent antigen or by using electron microscopic methods with a recognizable micro-particulate antigen such as ferritin (de Petris *et al.* 1963).

A great deal of the experimentation in this field has been directed toward the problem of the specificity of the antibody response. How many types of antibody can a given immunocyte produce? Can more than one form of immunoglobulin come from the same cell? Discussion of these features will be left to chapters 5 and 7. Here we are concerned with recent approaches to the definition of the immunocyte.

## THE IDENTIFICATION OF ANTIBODY PRODUCED BY SINGLE CELLS IN MICRODROPLETS OR UNDER GEL FIXATION

By 1955 it had become universally accepted that the plasma cell was the major, perhaps the only, producer of antibody. Fagraeus's (1948) studies on different fractions of splenic tissue from immunized rabbits pointed very strongly in this direction and the immunofluorescent staining experiments of Coons *et al.* (1955) had virtually clinched the matter. An actual demonstration that a single plasma cell could liberate sufficient antibody to be demonstrable in the fluid of a microdroplet was, however, not made until 1958. In that year Nossal (1958) published studies primarily concerned with whether plasma cells from doubly immunized rats produced one or two antibodies. Almost incidentally this established that all the cells identified as antibody producers had the morphological appearance of cells in the plasmacytic series from blasts to mature plasma cells.

In basically similar experiments, Attardi *et al.* (1959) found that some of the antibody-producing cells had to be classed as small lymphocytes; the majority were of the plasmacyte series.

The development of the 'antibody plaque' method by Jerne and Nordin (1963) has brought a more elegant and quantitative method of studying the numbers and type of antibody-producing cells in a given cell population. It has already been widely applied and will need to be referred to in a variety of contexts. The method is applicable to any type of antibody which, with the addition of complement, can induce haemolysis of red cells. In its classical form, lymphoid tissue from an animal immunized with sheep red cells is dispersed to give a single cell suspension which is washed and, after suitable dilution, mixed with a relatively dense suspension of the appropriate red cells in melted agar at 40°. This is immediately poured to give a thin layer opaque with red cells above a base of clear agar gel previously prepared. After 30 minutes at 37°, guinea-pig serum diluted to a standard complement activity is poured over the surface for another 30 minutes. The plaques take the form of small circular clearings, 0·2–1 mm in diameter, each centred on an antibody-producing cell. The method is not ideal for identifying the histological character of the central cells but they are regularly larger than small lymphocytes and most are obviously plasma cells.

In Bussard and Hannoun's (1965) experiments, using a different supporting gel, a proportion of the antibody-producing central cells deviated considerably from typical plasma cells but all those studied showed a well-defined endoplasmic reticulum. Probably the most critical set of experiments are those of Harris *et al.* (1966), who prepared electron microscopic sections of central cells, the whole plaque being removed with the agar and dehydrated. They concentrated on well-defined plaques with good clearing of red cells and with an unequivocal central cell. These cells were highly pleomorphic and included what would have to be termed small lymphocytes, large lymphocytes and plasma cells. The lymphocytes differed from classical small lymphocytes in showing a definite Golgi body and some endoplasmic reticulum. In the

383

smaller lymphocytes this took the form of sparse small vesicles rimmed with ribosomes; in the medium lymphocytes there were fairly numerous tubules cut in various fashions but no lamellae. The presence of an organized endoplasmic reticulum in lamellar form was taken as diagnostic of plasma cells. On these criteria, about 50 per cent would be classed as lymphocytes, 50 per cent as plasma cells. The authors regard this as strong evidence that plasma cells develop from antibody-producing lymphocytes and that the main difference may be in the greater storage capacity of plasma cells for antibody rather than in their speed of synthesis.

It should be added that in these experiments those concerned with antibody against flagellar or bacteriophage antigens were dealing essentially with 7S,G antibody. The antibody plaque technique is concerned only with 19S,M antibody. Neither in these, nor in other investigations has any morphological difference between the plasma cells producing the two types in normal animals been detected.

## THE VISUAL IDENTIFICATION OF IMMUNOGLOBULIN OR ANTIBODY IN CELLS

A method which is in a sense intermediate between the use of single isolated cells and the various techniques using immunofluorescence depends on the adhesion of antigenic particles to immunocytes which can be seen in wet preparations by phase-contrast microscopy. Bacteria have been used as particulate antigen by Mäkelä and Nossal (1961) and Nossal *et al.* (1964*a*), as well as by a number of earlier workers (Reiss *et al.* 1950; Cooper and Pillow, 1959). Other workers have used red cells both for immunization and as test object. Much use has been made of this technique by workers interested in the behaviour of peritoneal macrophages, e.g. Boyden (1964) and Perkins and Leonard (1963), but most of the reactions which take the form of a rosette of red cells around a single macrophage are apparently due to the take-up of cytophilic antibody by the cells. Biozzi *et al.* (1966), using sheep and pigeon red cells, were able to show a striking contrast between lymph node cells, which did not become positive, and 'peritoneal macrophages', which showed a high proportion of reactors following intravenous immunization with maximal development of antibody-producing cells in the spleen. No statement of the morphological character of the positive cells in these experiments has been published but all past experience indicates that they are lymphoid cells predominantly of the plasmacyte series but including a proportion of small lymphocytes. Analysis of the number of double reactors indicates that lymphocytes and plasma cells are nearly all single reactors so that a passive effect of cytophilic antibody can only be effective in a small proportion. The proportion is much higher amongst peritoneal cells but there is still a majority of single reactors, suggesting that a relatively large number of 'macrophages' are in fact modified immunocytes derived from lymphoid cells.

With modern methods of immunofluorescence it is possible to determine in fresh human material whether a given cell is producing A, G or M immunoglobulins and whether the corresponding light chain is of K or L antigenic character. Appropriate antisera are prepared usually by immunization of rabbits with purified monoclonal immunoglobulin from patients with multiple myelomatosis of A, G or M type. Monovalent antisera against each light-chain type can similarly be produced against fractions from appropriate myeloma proteins.

The findings have been consistent that a given plasma cell is capable of producing one type of immunoglobulin only (Mellors and Korngold, 1963; van Furth, 1964) except for an occasional plasma cell which appears to contain both G and M. This is in accord with the finding by Nossal *et al.* (1964*c*) that at a certain stage after immunization some plasma cells liberate both M- and G-type antibodies.

Working with an anti-A serum, Carbonara *et al.* (1963) found that plasma cells producing Ig A could be found in varying stages of maturity in the spleen but were far less numerous than G-producing plasma cells. They are, however, the predominant form among the numerous plasma cells in the intestinal wall (Crabbé, 1967).

The only other situation where immunoglobulins are regularly observed by immunofluorescence is in germinal centres, where it is characteristically located intercellularly.

385

Studies using allotypic antisera against $\kappa$ and $\lambda$ antigens of human light chains by Pernis and Chiappino (1964) showed that in any human spleen there are K plasma cells and L plasma cells but none showing mixtures. In germinal centres, however, there is clear evidence of both types. Using corresponding reagents suitable for work with rabbit allotypes, Pernis *et al.* (1965) obtained the same type of result. The significance of these findings with germinal centre cells is discussed on p. 475.

The use of fluorescent label on antigen to detect cells carrying the corresponding antibody has been much more limited. Mellors *et al.* (1959) used labelled $\gamma$-globulin partly denatured to detect auto-antibodies in rheumatoid arthritis patients. This is such a special situation that it is better left for treatment in relation to autoimmune disease (chapter 12).

Various modifications of Coons's sandwich technique (Coons *et al.* 1955) have been used in which cells are first treated with antigen, then washed and stained with fluorescent antibody against the antigen. The most interesting recent finding is from Raffel's laboratory (Martins *et al.* 1964; Rausch and Raffel, 1964), where it was shown that 6–20 per cent of lymphocytes from the regional lymph nodes of guinea-pigs sensitized with BCG show a rim of bright fluorescence after such treatment using unheated tuberculoprotein as antigen. Very similar results were obtained with lymphocytes from guinea-pigs sensitized with spinal cord extract and Freund's complete adjuvant, the antigen used to coat the cells being a purified 'encephalitogenic protein' of mol. wt 16,000. Most of the animals developed encephalomyelitis but a proportion of positive cells was also present in sensitized animals showing no symptoms. Cells showing specific staining in both series of experiments were nearly all small lymphocytes.

These findings are of high significance for the interpretation of DH (see p. 579) but in the present context they can be taken as establishing the existence in animals showing DH of a subpopulation of lymphocytes carrying specific immune receptors, in all probability of intrinsic character, although the possibility of their representing adsorbed cytophilic antibody is not wholly excluded.

THE PRESENCE OF IMMUNOCYTES IN CONVENTIONAL
SOURCES OF CELLS
## Thoracic duct lymph
This represents the purest source of lymphocytes, and since the contents of the thoracic duct discharge directly into the bloodstream all the cell types found in the lymph must also be present at some time in the blood.

Approximately 95 per cent of the cells in thoracic duct lymph from the normal rat are small lymphocytes, but with continued depletion the proportion of medium and large lymphocytes increases (Gowans, 1959). A proportion of thoracic duct cells show endoplasmic reticulum in electron microscopic sections (Zucker-Franklin, 1963) and would therefore be classed as immature plasma cells. Using immunofluorescence with appropriately specific antisera, van Furth (1964) found that cells reacting with sera against Ig G, Ig A or Ig M could be recognized. Most of these were medium-sized cells. Small amounts of M are seen on some small lymphocytes. Small lymphocytes with pyroninophil cytoplasm have been noted in rabbits (Hulliger and Sorkin, 1965). This presence of antibody-producing cells can be confirmed directly (Wesslén, 1952; Hallander and Danielsson, 1962) by culture of thoracic duct lymph *in vitro*.

Thoracic duct cells, and specifically small lymphocytes, can produce severe graft-versus-host reactions in suitable parent-into-$F_1$ combinations both in the rat (Gowans *et al.* 1961) and in the mouse (Hildemann, 1964). Similarly they can provoke a homograft reaction against heterologous tissue as shown by Gesner and Gowans (1962), who found that while lethally irradiated mice could be saved with rat bone-marrow the addition of thoracic duct cells from normal mice completely prevented the protective action.

Finally, thoracic duct cells cannot protect a lethally irradiated syngeneic animal (Gesner and Gowans, 1962) or produce nodules in the depleted spleen.

## Circulating leucocytes
Most of the small lymphocytes of the blood are involved in a recirculation (Gowans, 1959) of which the most important phases are apparently escape to lymphoid tissues through postcapillary

venules and return via the lymphatic trunks. These results are due primarily to Gowans and collaborators working with the rat and cannulating the thoracic duct. Evidence for similar active recirculation has been obtained by cannulation of afferent and efferent lymphatics of the popliteal lymph node in the sheep (Hall and Morris, 1964).

Since the only possible channel connecting bone-marrow, thymus and peripheral lymphoid tissues is the general blood circulation, it follows that all cell types which by direct demonstration or by logical deduction pass from one of these situations to another must spend part of their time as circulating leucocytes.

Experimental studies have in fact been able to establish the presence of immunocytes with these various functions in leucocytic or lymphocytic preparations from circulating blood. Many positive results attest to the presence of cells capable of inducing graft-versus-host reactions in suitable recipients. These include the production by chicken lymphocytes of splenomegaly (Simonsen, 1957) or focal lesions on the chorioallantoic membrane (Boyer, 1960; Szenberg and Warner, 1961); and analogous results in mice (Cole and Garver, 1961; Hildemann *et al.* 1962); and guinea-pigs (Brent and Medawar, 1963).

Antibody-producing cells are regularly present in the blood during the height of an immune response and they may be seen as typical members of the plasmacyte series—5 to 10 per cent of the leucocytes in Hulliger and Sorkin's (1965) experiments with rabbits. Landy *et al.* (1964) used Jerne's plaque method and showed that the maximum count of plaque-forming cells was on the fourth day after immunization. Indirect evidence to the same effect can be seen by the sudden appearance of plaque-forming cells in the thymus on the fourth day after immunization (Landy *et al.* 1965; G. Möller, 1965). It is well known that human blood at the height of a measles attack contains blasts and plasma cells.

In mice, cells capable of protecting lethally irradiated mice are present in the blood, particularly in neonatal animals when there may well be an extensive movement of cells from the liver, the main haematopoietic organ in foetal life, to the bone-marrow (Barnes *et al.* 1964).

## Immunocytes in the peritoneal cavity

The question of the immunological competence of peritoneal macrophages or, more precisely, of the population of mobile cells washed from the peritoneal cavity under standard conditions, has been extensively discussed. Undoubtedly, a large part of the phagocytic activity, or capacity to hold red cells, of such cells from immunized animals is due to cytophilic antibody or opsonization by associated antibody. Such phagocytic cells conform to all definitions of macrophages and are dealt with in chapter 7. The possibility that they may include a proportion of antibody-producing cells has been raised but it could only be substantiated by a demonstration that they are actively synthesizing antibody, and this has not so far been obtained. In addition to frank macrophages, peritoneal exudates contain a proportion of cells which do not adhere to glass and resemble lymphocytes morphologically. It seems probable that most of these are immunocytes.

Weiler and Weiler (1965) used a rather complex technique to give what is perhaps the clearest picture of the situation. Mice were heavily irradiated (800 r) and injected intravenously with spleen cells from syngeneic mice primed with a phage antigen. They were then injected intraperitoneally with the same phage in Freund's complete adjuvant. A heavy cellular exudate was obtained which was separated by adsorption to glass and simple aggregation into populations (L) predominantly of small lymphocytes and (MN) predominantly of macrophages and polymorphonuclears. Washed L suspensions injected into irradiated mice treated with bone-marrow gave rise to high antiphage titres of the appropriate type; MN suspensions of macrophages and polymorphonuclears, similarly used, were inert.

Related studies by Weiler (1965) showed that these lymphocytic cells from the peritoneal cavity, washed and placed in Millipore chambers, gave rise to antibody and large numbers of plasma cells within the chambers.

The technique in these experiments involves very extensive departure from physiological conditions and perhaps no more should be said than that, under these conditions, large numbers of immunocytes enter the peritoneal cavity in lymphocyte form.

## THE EVIDENCE THAT DIFFERENTIATED CELLS INCLUDING IMMUNOCYTES CAN MAINTAIN THEIR DIFFERENTIATED CHARACTER ON PROLIFERATION

An essential feature of clonal selection theory is that, once a clone has been established, descendant cells of the clone retain the capacity to continue to produce the same immune pattern, subject only to the possibility of somatic mutation.

From general impressions of the behaviour of cells within the body this seems a perfectly reasonable assumption, but there have been suggestions that the long-term persistence of immunological memory is due simply to the physical survival of individual cells over many years, and not to any transfer of information to successive generations of descendant cells.

Until recently it appeared that the use of tissue culture as an approach to testing the persistence of differentiated qualities in somatic vertebrate cells was wholly unsatisfactory owing to the necessary absence of the interaction of the organism with the single cells multiplying in pure culture. Recent work, however, suggests that it may be possible to abstract from the organism as a whole simple factors which may allow a differentiated somatic cell line to maintain its quality. A striking example is the work by Hauschka and Konigsberg (1966) on cloning of chicken myoblasts. This had previously (Konigsberg, 1963) been achieved only by growth on media 'conditioned' by fibroblastic cell growth. It was found, however, that the conditioning effect was simply due to the laying down of a layer of collagen on the agar surface. This could be imitated by pre-treatment with purified collagen and on such a medium active clonal growth of differentiated muscle cells occurs.

In somewhat similar fashion, pigmented retinal cells and cartilage cells from chick embryos can be cloned at least for several generations (Cahn and Cahn, 1966; Coon, 1966) provided due care of the conditions of culture is maintained. Deviations are liable to result in the de-differentiation which has previously been regarded as inevitable. The important conclusion from this work is that 'cells may stably inherit the ability to express specific differentiation for many generations, often without cell contacts and interaction'.

No similar work has yet been reported at the immunological level but there seems no reason why refined experimentation should not allow the cloning of lymphocytic memory cells with subsequent test of the immunological competence of descendants by transfer to a syngeneic irradiated host. In the absence of any such direct evidence we are compelled to concentrate on more peripheral approaches.

TABLE 10. *Selective proliferation of anti-A and anti-B clones of immunocytes initially present in parental population P (based on Elkins, 1966)*

| | | | | | | | |
|---|---|---|---|---|---|---|---|
| | | | P cells | | | | |
| | F₁ P × A | | | F₁ P × B | | | |
| | + + + G-v-H | | | + + + G-v-H | | | |
| P | P × A | P × B | | P | P × A | P × B | |
| − | + + + | − | | − | − | + + + | |

The production of graft-versus-host reactions in different strains by cells of the primary graft-versus-host reactions is shown.

A recent study of very great interest is by Elkins (1966), who was dealing with the infiltrating kidney lesion induced by injecting allogeneic lymphocytes under the kidney capsule. In essence, the experiment was carried out by injecting lymphocytes of rat strain P into hosts which were $F_1$ P × A or $F_1$ P × B. The massive graft-versus-host reaction in P × A reached its peak in 6 or 7 days and cells from the reacting tissue taken at that time were tested for their capacity to produce graft-versus-host reactions in the three rat types under study. The results were quite definite; these cells (a mixture of proliferated donor cells and many host lymphocytes) produced graft-versus-host reactions in $F_1$ (P × A) but no reaction in either P or $F_1$ (P × B). Here we have evidence that in the original P inoculum there were cells capable of initiating a graft-versus-host attack on antigen $A$. Equally, there were cells similarly directed toward $B$. When proliferation of donor cells took place in (P × A) these cells retained their specificity of attack on (P × A) and retained their innocuousness for P, but they had lost capacity to act on (P × B). The only conclusion is that in the graft-versus-host reaction, immunocytes reactive with *A* multiplied actively with retention of their immune character. Any cells reactive with *B* failed to proliferate and were lost by dilution. The donor cells remained antigenically P but only that subpopulation pre-adapted to proliferate on contact with antigen *A* increased in number.

It is now well accepted that once a clone of plasma cells is initiated, the descendant cells produce antibody of a uniform specificity, although it still seems possible that a cell initially producing Ig M antibody may give rise to descendants which finally produce only Ig G. Elkins's finding is of special interest because there is general agreement that graft-versus-host reactions are mediated by lymphocytes and not by plasma cells.

COMMITMENT OF IMMUNOCYTES

Immunology refuses to be cast into a consecutively logical story. One cannot discuss first the behaviour of cells and then pass on to discuss wholly *ab initio* the nature of antibody specificity. Antigenic stimulation is an essential feature of the traffic in immunocytes.

In earlier pages I have given reasons for using the hypothesis that, in addition to a large proportion of newly differentiated immunocytes that are destroyed in the thymus, others leave the thymus, probably by the veins, and lodge in the developing or developed peripheral lymphoid tissues of the body. A proportion of these initiate continuing clones of immunocytes, and pressing the hypothesis to its logical conclusion *all* clones of immunocytes can be assumed to derive from cells which either before or after birth were differentiated to immunocytes in the thymus. At the present time the possibility cannot be excluded that there are other 'lympho-epithelial' organs where differentiation to immunocytes can take place, but there can be no doubt that in standard laboratory mammals the thymus is much the most important site.

Evidence has also been given which suggests that immunocytes which have recently left the thymus require the continuing presence of a humoral agent produced by the epithelial cells of the thymus if they are to develop normal immunological effectiveness. The next logical step in the discussion of cellular aspects of immunity is to sort out the evidence on the various steps or 'compartments' that a cell line must pass through before it gives rise to a proliferating plasma cell clone producing Ig G antibody, and to latent memory cells which can ensure a rapid response to any subsequent entry of the appropriate antigen.

The primary difficulty in such an attempt is the need to differentiate either experimentally or in discussion between cells all of which have the morphology of a small lymphocyte.

On this hypothesis we shall expect to find in an adult mouse (or any other mammal) a number of functionally different types of cell with the morphology of the small or medium lymphocyte. In order to clarify discussion it will be desirable to define and name the groups that are postulated.

*Stem cell.* The rapidly proliferating small lymphocytes of bone-marrow with potentiality to give rise to immunocytes (Harris and Ford, 1964), monocytes and macrophages (Volkman and Gowans, 1965 a, b; Goodman, 1964) and mast cells (Goodman, 1964). Such stem cells (SC) are not immunocytes and in the absence of a thymus or equivalent organ cannot confer immune capacity on an animal.

*Progenitor immunocyte.* This is the newly differentiated cell liberated from the thymus and initially requiring the presence of a small concentration of thymic hormone. One must probably postulate that progenitor immunocyte (PI) cells can be stimulated to proliferate by nonspecific stimuli without changing their character except possibly removing the necessity for thymic hormone. These are equivalent to Coons's X cells (Coons, 1965) and to the PC 1 cells of Albright and Makinodan (1965). From Šterzl's (1966) studies of young animals using Jerne's antibody plaque technique it seems probable that contact with specific antigen at high concentration (or possibly as a result of repeated stimulation) will induce antibody synthesis in a PI cell without proliferation and with eventual death.

*Committed immunocyte.* When a PI makes effective contact with the corresponding antigen (which may require special conditions, for example, antigen associated with a phagocytic cell), blast cell formation and proliferation to a small clone of committed immunocytes (CI) occur. These will usually be small lymphocytes morphologically, and correspond functionally to Coons's Y cells.

393

*Plasmacyte-producing cell and memory cell.* Contact of specific antigen with CI again is assumed to induce blast formation and, depending on the local circumstances, development of an antibody-producing clone of plasma cells or further lymphocytes which have probably characteristics equivalent to CI (Coons's Z cells).

This is a speculative classification and is quite obviously inadequate to cover all the relevant phenomena such as the special quality of the response to Freund's complete adjuvant, the association of a given antibody with A, G or M immunoglobulins and the apparent association of memory-cell formation with concomitant appearance of cells producing G-type antibodies.

This method of summarizing the facts (see fig. 31) owes much to the work of two groups of workers, those associated with Makinodan at Oak Ridge and Coons's group at Harvard.

The former group have developed their method of *in vivo* culture in irradiated mice to examine the cell dynamics of antibody production: Albright and Makinodan (1965) speak of progenitor cells, PC 1, as pluripotential and possibly having capacity for giving descendant erythropoietic cells as well as lymphocytes. They would therefore be equivalent in some ways to our PI and, in others, to SC or some even less differentiated type. As long, however, as the discussion is wholly at the immunological level, PC 1 is equivalent to PI. The main point of contention is their claim that such cells are pluripotential. For this, two types of experimental evidence are provided. Perkins and Makinodan (1964) showed that when a population of normal spleen cells was injected into irradiated mice and antigen (sheep red cells) given at variable intervals thereafter, the antibody production was highest if antigen followed immediately. When the delay was anything from 2 to 11 days the amount was around one-fifth of that obtained when the antigen was given without delay.

The second approach was to show that in normal mice pre-injection of a large dose of sheep red cells produced a competitive reduction in antibody production to a smaller inoculum of rat red cells. The maximal effect was seen with a delay of 10 hours between the two injections and it was almost gone by 5 or 6 days. In both types of experiment the evidence is capable of many other interpretations, and in view of the very clear evidence from Jerne and Nordin (1963) that the populations reacting with sheep or rabbit red cells in both normal and immunized mouse spleen are quite distinct, we prefer to hold to the view that PI already carry a specific immune pattern.

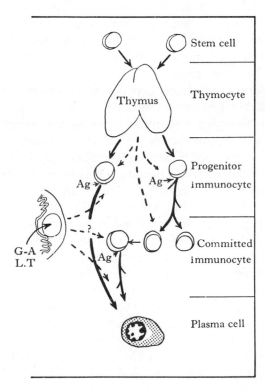

Fig. 31. To summarize cellular and hormonal processes in the differentiation and commitment of immunocytes. G-AL.T. = gut-associated lymphoid tissue; solid lines = cellular development; broken lines = hormonal influence.

In their analysis of commitment they find that there is a linear $\log_2$ relationship between activity, as judged by antibody production, and numbers of nucleated spleen cells. The lower end of such a curve can be used to give an 'all or none' result capable of defining an active unit in terms of cells. In normal mouse spleen 1 cell per $10^6$ was found to be capable of responding to sheep red cells;

for primed animals the spleen contains approximately 100 times as many responsive cells.

A finding of special interest is the progressive increase in capacity to produce antibody using the *in vivo* culture method with age (up to 40 weeks) of the normal mice from which spleen cells were obtained (Makinodan and Peterson, 1964). The rise is steepest in the period 0–12 weeks and thereafter changes are relatively minor. This can be most readily interpreted by the increasing number of clones accumulating during the earlier part of life and held in peripheral lymphoid tissues.

Coons's (1965) approach depends considerably on the work with his collaborator (Sercarz and Coons, 1960), which showed that immediately following a secondary response spleen cells could not convey capacity to give a secondary response to a suitable recipient. This is better discussed in relation to antibody production (p. 544).

Studies on the process of priming, i.e. the change from PI to CI in our terminology, and from X to Y in Coons's, are more relevant. With Butler and Cruchaud, Coons found that a number of cytotoxic drugs given in near maximal doses not only prevented a primary antibody response but also prevented priming. Since the primary effect of such drugs is antimitotic, this is indirect evidence that a necessary part of the change is blast formation and proliferation (Butler and Coons, 1964; Cruchaud and Coons, 1964).

Following Gowans *et al.* (1961) we can presume that under most circumstances the progeny of a blast produced by primary antigenic stimulation take the form of lymphocytes, but they are now committed lymphocytes, with the same specificity but with an increased and qualitatively changed reactivity for the appropriate antigen.

An important clue as to the nature of the change from uncommitted to committed lymphocyte may perhaps be drawn from Levey and Medawar's (1966) work on anti-lymphocytic serum (ALS). In one experiment, mice of strain CBA were grafted with skin of strain A. As soon as first-set rejection was complete they were given a 12-day course of anti-lymphocytic serum from a rabbit immunized with CBA mouse thymus cells. At intervals of 7, 10 and 14 weeks, groups were tested for capacity to reject a second graft. In controls without serum treatment, there was normal accelerated second-set rejection. The experimental mice tested at 7 weeks still showed the effect of ALS with a prolonged retention of the graft. At 10 and 14 weeks, however, the grafts were treated as first-set grafts would be on a completely 'virgin' animal of the same age.

The effect of ALS will need to be discussed again in relation to tolerance. Levey and Medawar have made a number of suggestions as to how the antibody affects lymphocytes, of which the most interesting is that they are 'blindfolded'. The coating of antibody is not lethal to the cell but prevents access of antigen to receptors. One of the most interesting features of the experimental situation is that the mice do not develop antibody against rabbit serum antigens that are present in the ALS. Slightly modifying the authors' conclusions we may suggest that, in the uncommitted cells carrying receptors reactive with *A* antigens, the masking of receptors is complete in the sense that no stimulatory contact with antigen *A* can be made. When the rabbit antibody is eventually metabolized, a proportion of the uncommitted cells will have survived. With committed cells, however, perhaps simply because larger numbers of surface receptors are present, the masking is inadequate. It is easy to believe that an immunocyte already hampered by a partial coat of ALS antibody will find reaction with the corresponding antigen to be lethal. If significant amounts of antigen persist long enough, all committed cells will be eliminated although uncommitted ones of the same specificity may survive. This seems almost the only way to account for the quite unusual type of semi-tolerance that is described.

THE LYMPHOCYTE IN RELATION TO THE SPLEEN AND
LYMPH NODES

One of the most striking changes that the immunological approach to lymphocyte function has made is in the interpretation of the structure and function of lymph nodes and spleen.

In 1956 it was generally believed that the lymph nodes were self-subsistent entities in which lymphocytes, plasma cells and macrophages were derived from primitive reticular cells. There was, and to some extent still is, confusion about the significance of germinal centres and, arising largely from the work of Drinker and his school, most interest was taken in the function of the nodes as filters for bacteria and other foreign material.

To interpret the lymph node purely from an immunological point of view will undoubtedly leave much more to be said. The present attempt is based almost solely on the experimental and immunological literature, and no attempt has been made to survey the recent anatomical literature on the subject.

There is little controversy in regard to the actual structure of lymphoid tissues nor is there now any doubt about the major importance of spleen and lymph nodes in antibody production. Even ten years ago it was recognized that the appearance of germinal centres and of increased numbers of plasma cells were both signs of antigenic stimulation. Perhaps the greatest change in approach can be exemplified by quoting from Loutit (1962). After a mouse receives 1000 r of whole-body irradiation, 'its capacity to recognize and reject foreign cells is permanently lost—when "rescued" by another animal's bone-marrow its immunological potentiality is that of the donor'. In other words, the whole immune capacity of a lymphoid organ can—and in the course of ontogeny presumably does—reach it by way of mobile cells in the bloodstream.

Pointing in the same direction is the mobility within the body of immunological capacity. It requires only a small region of the body containing lymphoid tissue to be shielded against X-irradiation to protect the animal against death or against loss of immune capacity. In an early investigation, Wissler *et al.* (1953) showed that while adequate whole-body irradiation would prevent antibody production in the rat, lead shielding of the spleen would allow maintenance of immune capacity even if the spleen were removed shortly after irradiation and before antigen was given. Similarly whole-body irradiation of the rabbit can be rendered ineffective if the appendix is shielded (Sussdorf and Draper, 1956; Sussdorf, 1960). The most direct evidence comes from Hall and Morris's (1965) study of cannulated popliteal lymph nodes in sheep. This allowed administration of tritiated thymidine by the afferent route and analysis of the degree of labelling in the cells leaving by the efferent trunk. From the results they concluded that less than 4 per cent of the cells leaving the lymph node are actually produced in it.

Calculations suggest that on the average the whole complement of lymphocytes in a quiescent node is replaced every 70 hours. Gowans's studies with labelled thoracic duct cells in rats point equally to a massive circulation of lymphocytes (see Gowans and McGregor, 1965). Another striking experiment by Hall and Morris (1964), still using the cannulated sheep lymph node, is to show that heavy local irradiation, presumably capable of destroying all lymphocytes actually present in the node, has virtually no demonstrable effect on the function of the node 48 hours later.

Basing the interpretation largely on Gowans and McGregor's review we can offer the following picture of the circulation of immunocytes. In the blood there are, in addition to small lymphocytes of immune character, morphologically similar cells with qualities equivalent to the small lymphocytes of bone-marrow which can be regarded as stem cells and, by hypothesis, a few newly differentiated immunocytes from the thymus. In speaking of lymphocytes it will be convenient to regard them all as immunocytes, always with the qualification that the other functional types of lymphocyte may behave differently.

Lymphocytes in the blood enter the lymph nodes by passage through the postcapillary venules of Marchesi and Gowans (1964). These have cuboidal endothelium which appears to have a positive attraction for lymphocytes. These are found in the primary follicles and other cortical areas of fairly densely packed lymphocytes. There is a suggestion from Parrott *et al.* (1966) that the perifollicular areas of the cortex have a special relation to the thymus since a very much larger proportion of labelled thymic cells lodge here than is the case with cells of splenic origin (see fig. 25). It is possible that some cells from lymph nodes enter the blood directly but there is a consensus that the great majority leave by the efferent vessels and enter the blood by the major lymphatic trunks. Rats can be very extensively depleted of lymphocytes in all parts of the body by chronic drainage from the thoracic duct (McGregor and Gowans, 1963) and the same holds for calves in which circulating lymphocytes in the blood are destroyed by irradiation of the blood in an extracorporeal circuit (Cronkite *et al.* 1962).

An important aspect of lymphocytic circulation which emerges from the work by Hall *et al.* (1967) concerns the movement of

399

lymphocytes from the tissues to the primary draining lymph node and thence to other lymph nodes that intervene before the main lymph trunks are reached. There is something to suggest that the antigens concerned with the induction of DH may reach lymph nodes carried on the surface of mobile cells. From an actively stimulated node there is a very active flow of large basophilic cells, and if these are diverted by draining the efferent lymph flow, antibody production fails to occur. The most likely interpretation is that favoured by Hall *et al.*, that the basophilic cells settle to some extent in intermediate lymph nodes and via the blood in spleen and other lymphoid tissue where they initiate plasma-cell production. There is still much interesting detail to be learnt as to where cells entering by the afferent lymph vessels lodge in the node. The paracortical areas would seem to be the most likely.

It is well known that large numbers of small lymphocytes and plasma cells are present in the mucous membrane of the bowel. Many lymphocytes appear to be actually within epithelial cells and many must be shed into the gut lumen. In dogs with a Thierry intestinal loop given labelled thoracic duct lymphocytes this has been shown directly by Ambrus and Ambrus (1959). It is equally evident to any histologist that, particularly under conditions of stress, large numbers of lymphocytes are being destroyed within the tissues. Some source of new production is therefore required, the obvious one being the germinal centres of lymph nodes, spleen, Peyer's patches and probably other lymphoid areas such as the tonsillar and pharyngeal lymphoid tissue in man. The older interpretation that the rim of small lymphocytes around a germinal centre is the mature product of proliferation in the germinal centre has been proved incorrect by autoradiographic experiments, but this by no means excludes the germinal centre as the major source of new lymphocytes. In addition there is a growing opinion that the initial stage of the production of a plasma cell clone takes place in germinal centres (Thorbecke *et al.* 1962; Fernando and Movat, 1963).

There is no reason to believe that germinal centres differ seriously in structure or function from one organ to another, and what holds for lymph node probably holds also for Peyer's patch and spleen. In this chapter we are concerned only with the lymphocyte, and the special characteristics of the different lymphoid organs probably depend on their relation to the corresponding drainage field and the type of antigenic material characteristically derived from it.

## A SUMMARY OF LYMPHOCYTIC MOVEMENT IN THE BODY

The lymphocyte population of the body, and more particularly the large proportion with immune function, is highly mobile. We can accept implicitly the circulation defined by Gowans by which a cell can pass via the thoracic duct to the blood and thence into the lymph follicle of a lymph node through the cuboidal epithelium of the capillary venules. In due course it may then move via efferent lymph through more central lymph nodes to the main lymph trunks to the circulating blood again. There are many alternative pathways and a constant loss in one way and another of cells from the circulating pool. Lymphocytes are being constantly destroyed or lost from mucous surfaces. Their numbers are maintained by proliferation within lymphoid tissue of various types predominantly within germinal centres. Newly formed lymphocytes are prone to pass rapidly into the circulation and few of them stay in the vicinity of the centre producing them. In a quiescent peripheral lymph node there are no germinal centres and in all probability germinal centres are always the result of antigenic stimulation. Leaving detailed discussion of this to chapter 9, we can merely suggest here that each germinal centre represents the proliferation of a clone derived from an antigen-stimulated lymphocyte which found all the circumstances necessary to allow implantation and proliferation. In a healthy animal not subject to experimentation most of the germinal centres are related to the bowel—Peyer's patches and mesenteric lymph nodes—and presumably represent stimulation by bacterial antigens from the gut.

## THE FUNCTION OF GERMINAL CENTRES

Shortly after the first draft of this chapter had been completed, abstracts of the papers presented at the conference, 'Germinal Centers in Immune Responses' (Bern, June 1966) were received.

These have now been published *in extenso* (Cottier *et al.* 1967 *a*). These indicated quite strikingly that there is now virtual unanimity in the broad interpretation of the germinal centre. Since this interpretation is that which has been presented it will be summarized in this section in very brief dogmatic form but with references only to this recent work. Earlier references are included in the preceding pages.

Germinal centres are induced by antigenic stimulation (Keuning and Bos, 1967) and arise *de novo* by the localization of 'germinoblasts', i.e. stimulated immunocytes derived from other regions of the body in the substance of primary follicles (Cooper *et al.* 1967; Hinrichsen, 1967; Cottier *et al.* 1967*b*). The cells of the germinal centre are actively proliferating as judged by their uptake of tritiated thymidine but do not give rise to the accumulations of small lymphocytes which surround them (Everett and Caffrey, 1967; Hinrichsen, 1967; Parrott, 1967). In the rabbit tonsil there is evidence that a cap of small lymphocytes is derived from the proliferating germinal centre (Koburg, 1967). The tingible bodies so characteristic of germinal centres are the result of phagocytosis of cells which have died in mitosis or immediately thereafter (Odartchenko *et al.* 1967; Fliedner, 1967). There is indirect evidence that germinal centre cells pass freely into the blood and may localize in the spleen in the form of small lymphocytes (Wakefield *et al.* 1967). There is a close association of germinal-centre production with the appearance of plasma cells (Cooper *et al.* 1967; Simar *et al.* 1967; Cottier *et al.* 1967*b*; Good *et al.* 1967) and some evidence that germinal centre cells are actually transformed via pyroninophil blast cells to typical plasma cells (Cottier *et al.* 1967*b*; Micklem and Brown, 1967). The long-held possibility that plasma cells might be derived from endothelial or reticular cells can be eliminated by studies of the regeneration of lymphoid tissue in X-irradiated animals (Everett and Caffrey, 1967; Keuning and Bos, 1967).

There is some conflict in regard to the relationship of Ig M- and Ig G-antibody production to germinal-centre formation but it appears that Ig M can be produced in the absence of germinal centres (Fitch *et al.* 1967; Good *et al.* 1967) though some antigens, such as pneumococcus III, which produce relatively large amounts of Ig M antibody are active stimulants to germinal-centre production (Turk and Oort, 1967), and the early production of anti-red cell haemolytic antibody (Ig M type) in mouse spleen is in the region of lymph follicles and germinal centres (Young and Friedman, 1967). There appears to be unanimity that the common association is germinal-centre production, appearance of plasma cells and of Ig G antibody (Cooper *et al.* 1967; Balfour *et al.* 1967; Fitch *et al.* 1967; Good *et al.* 1967). It is highly probable that memory cell production, i.e. the development of immunocytes which can later respond specifically to contact with AD, is a function of germinal centres (Congdon and Hanna, 1967; Thorbecke *et al.* 1967). Turk and Oort (1967) found that DH might become established without the appearance of germinal centres, and Micklem and Brown (1967) found similarly that allografts could be rejected in a first-set reaction before their appearance.

Germinal centres do not develop in neonatally thymectomized mice (Grundmann and Hobik, 1967; Rogister, 1967) nor in the spleens of irradiated and bursectomized chickens (Cooper *et al.*). In human pathology there is a similar absence in sex-linked congenital agammaglobulinaemia and in some dysgammaglobulinaemias (Good *et al.* 1967) in which Ig M is produced without Ig G or Ig A.

THE ORIGIN OF PLASMA CELLS

Indirect evidence suggests very strongly (most would say conclusively) that small lymphocytes can be converted into pyroninophil cells capable of proliferation either to plasma cells or to small lymphocytes. There is, however, no reasonably direct evidence of any transformation of small lymphocytes to plasma cells.

The position of the plasma cell as the principal producer of antibody is well established, but recent work on the nature of antibody-producing cells at the centre of 'antibody plaques' has destroyed any contention that it is the only producer of antibody. Harris *et al.* (1966) examined such cells by electron microscopy and found three types. Two were medium lymphocytes, A having sparse endoplasmic reticulum in the form of small vacuoles rimmed with ribosomes and B with more abundant endoplasmic reticulum in the form of tubules. The others were definable as plasma cells from the presence of a lamellar endoplasmic reticulum but the cells were very pleomorphic and showed widely differing degrees of development of the endoplasmic reticulum.

Bussard and Hannoun (1965) found a similar pleomorphism of these central cells. The results obtained by both groups would fit best with the view that antibody-producing cells may be at various stages of development, in a sequence from the primary pyroninophil blast with no definite endoplasmic reticulum to the fully developed lamellar plasma cell.

Probably the first substantial claim for direct conversion was that of Roberts and Dixon (1956) who injected lymph node cells from immunized donors into X-irradiated recipients and excised the area 3–6 days later. They found aggregates of plasma cells at the periphery of the injection site. No such appearance was seen when immunologically inert or inactivated cells were injected, and there was good correlation with antibody response. Sainte-Marie and Coons (1964) analysed this situation in more detail. They observed active proliferation of plasmablasts and antibody production but calculated that with the rate of multiplication observed (a generation time of 6–7 hours) the plasma cells must have been derived from less than 1 per cent of the cells injected. They could have arisen just as well from one of the minority forms in the inoculum as from the predominant small lymphocytes.

In papers by Thorbecke and collaborators (Thorbecke *et al.* 1962; Cohen *et al.* 1966) evidence is collected in favour of the view that the plasma cells which form such a conspicuous feature of the medullary cords of an active lymph node arise from lymphoblasts produced in a germinal centre. These then migrate to the medullary cords, dividing and maturing as they go. A comparative study of the number of plaque-forming cells in separated cultures of red and white pulp, from spleens of rabbits at various stages of immunization with sheep red cells, indicated strongly that antibody-producing cells were multiplying in germinal centres (Cohen *et al.* 1966).

Older histologists considered that plasma cells in the medullary cords and elsewhere arose *in situ* by development from perivascular adventitial cells. This view has in recent years been strongly supported by Amano (Amano and Marumana, 1963). He believes that cells of the plasma cell series can be recognized by their association with adjacent collagen fibres, a mark of their origin from adventitial cells.

From autoradiographic studies of immunocytes in the popliteal node of the rat, Nossal and Mäkelä (1962a) concluded that the antibody-producing plasma cells in a secondary response were descendants of cells which had been actively synthesizing DNA *before* the antigen was administered. There are a number of possibilities of misinterpretation in such experiments, but the results with their hint of a transfer of specific information to a 'nurse cell' line should be kept in mind. They may become important if and when the possibility of transfer of genetic information between somatic mammalian cells is established.

On the basis of current information there appears to be no valid evidence against the view that plasma cells arise as one of the modes of secondary differentiation within an immunocyte clone. A committed lymphocyte, after making effective contact with antigen, lodges in a primary lymphoid follicle of a lymph node and initiates a germinal centre, the first stage being necessarily its conversion to a blast cell. As the germinal centre enlarges, descendant cells move away to the circulation via the efferent lymph trunks, or to the medullary cords. The morphological and functional form taken by these newly produced cells will probably depend on the nature of their immediate tissue environment.

### THE REQUIREMENTS FOR PLASMA CELL DIFFERENTIATION

There is little direct evidence as to the requirements which allow a stimulated immunocyte to proliferate to form a clone of plasma cells. In following recent work, particularly in immunopathology, I have been struck with a number of observations which suggest two possible factors of importance in the tissue environment. The first is a relatively high oxygen tension; the second the presence of a humoral factor derived from 'gut-associated lymphoid tissue'. The existence of such a factor has already been hinted at in relation to the function of the bursa in birds, and it will be convenient to preface a discussion of the situation in mammals with an account of recent work and speculation about bursal function.

*Function of oxygen tension*

Dealing first with the question of oxygen tension, we will take as a working hypothesis that where there is adequate oxygen tension a proliferating immunocyte will develop as a plasma cell clone while, if it lodges in packed lymphoid tissue with an associated low oxygen tension proliferation, if it occurs, will be along the lymphocytic series and may take the form of production of a germinal centre.

In any standard lymph node reacting to an antigenic stimulus, germinal centres develop in the primary follicles and contain only rare plasma cells. These become more frequent just beyond the periphery of the germinal centre where cells seem to be passing down into the medullary cords, and it is in the cords themselves with their free blood supply that the plasma cells predominate. Many appear to be applied directly to the venules. This is perhaps the reason which has led Amano and others to state that they are derivatives of perivascular endothelium. One frequently finds enlarged lymph nodes in which there is a highly cellular mass of plasma cells unrelated to the medullary cords. Characteristically, however, these are quiescent cells without mitoses and often include hypermature forms.

What specially drew my attention to this problem is the distribution of plasma cells in relation to other lymphoid cells in the infiltrates along arteries in the kidneys of old NZB and NZB × NZW $F_1$ mice (Hicks and Burnet, 1966). Plasma cells are always peripheral to the central core of lymphocytes, and as one moves peripherally the infiltrating cells around the afferent arterioles or infiltrating between tubules and around glomeruli are almost *all* plasma cells. The kidney has an exceptional blood supply and a high degree of oxygenation, while one can presume that the centre of a mass of infiltrating lymphocytes will be relatively anaerobic. The assumption that the presence of plasma cells indicates a certain level of oxygen tension would give just such a distribution as is observed. In the lung infiltrates an almost wholly similar picture is seen.

Plasma cells are rare in normal thymuses. When they occur it is usually in the vicinity of blood vessels entering through the cortex. The pallor of the thymus, taken with its high metabolic activity, would indicate that oxygen tension in the tissues, particularly of the cortex, is low. In a casual study of the thymus of the mono-treme *Tachyglossus* I was struck by the fact that the organ appeared to be well supplied with blood and that mature plasma cells were relatively common (*a*) immediately adjacent to blood vessels in the cortex, (*b*) at the cortico-medullary junction and (*c*) in some situations immediately under the outer membrane delimiting the cortex. In the spleen of all animals, plasma cells are mainly to be seen in the well-oxygenated red pulp or peripherally on the Malpighian bodies of the white pulp.

This is a hypothesis which should be amenable to a variety of experimental approaches. Do animals kept under hyperbaric oxygen produce more antibody than controls? What would be the effect of maintaining an oxygen pneumoperitoneum in an immunized animal? I have not been able to find any record of such experiments.

## Plasma cell production in birds

It has been shown by Jaffe and Fechheimer (1966) that when chick embryos of opposite sex are joined in parabiosis the bursa of Fabricius at hatching contains a proportion of mitoses of the opposite sex. If this is confirmed it suggests immediately that the bursa has a thymus-like function to take in circulating stem cells and allow their differentiation to fully functional immunocytes. It is not necessary that the stem cells should be wholly undifferentiated. It seems more likely that they are in fact recently differentiated immunocytes from the thymus which, if they are to become plasma cells still require to undergo proliferation and further differentiation in the bursa. We assume that there is a high concentration in the organ of what we may call 'bursal hormone' or, to leave the possibility open for future mammalian applications, 'plasmapoietic hormone'. This will have to be endowed with the following characteristics:

(*a*) active stimulation of cell proliferation;

(*b*) differentiation of immunocytes within the bursa to potential plasma cell and antibody producer but inhibition of actual development in this direction in the organ itself.

We have to remember that the bursa has a special relation to the

407

cloaca with its concentration of foreign antigens. There is a freely patent canal and an arrangement of 'leaves' of lymphoid tissue so disposed as to present a very large surface to the contents of the bursal cavity. There is smooth muscle in the duct and in the bursal capsule and it is likely that there is some sort of ebb and flow of fluid between cloaca and bursal cavity. It seems almost certain that the bursa has evolved as part of the birds' immunological mechanism in such a way as to 'sense' the dominant foreign antigens in the bowel and initiate appropriate immune response. There is a little evidence (Burnet and Lind, 1961) that a soluble antigen (BSA) introduced into the lumen of the bursa provokes a small but significant antibody production. An emulsion of alum-precipitated BSA which is a much better antigen when given intramuscularly had no effect whatever.

This allows a still speculative addition to the function of the bursa. We assume that when a differentiated cell of immune pattern $A$ present in a bursa follicle meets the corresponding AD $a$ entering from the bursal lumen, it is specifically stimulated. The effect is a delayed one, being manifested only when the cell reaches an appropriate nidus in lymphoid tissue. There it takes on the plasma cell character and initiates antibody production. In other words, for a cell differentiated to an immunocyte in the thymus to give rise to a plasma cell clone it must (*a*) undergo secondary differentiation with limited proliferation in the bursa and (*b*) make contact with the corresponding AD either in the bursa or in the circulation soon after it has left the bursa.

This hypothesis lends special interest to recent work from Good's laboratory (Cooper *et al.* 1967). They find that the disability produced by neonatal bursectomy followed by sublethal X-irradiation can to some extent be remedied by returning the chick's own bursal lymphocytes immediately after the irradiation. The chick, unlike equivalent controls not receiving bursal cells, develops germinal centres and shows numerous plasma cells in spleen and caecal lymphoid tissues. Both $19S(M)$ and $7S(G)$ immunoglobulins are present at 7 weeks of age. But in sharp contrast no antibody is produced to BSA or a *Brucella* vaccine.

In trying to interpret this result we have first to recognize that the only cells to have experienced or which can experience a sojourn in the bursa are those present at the time of bursectomy, i.e. only a few days since the bursa became lymphoid in structure, and over those 3 or 4 days the numbers of blood lymphocytes were very low. Quite small numbers of clones can have been established compared with the flush of activity in the following week or two. On the antigenic side we can probably assume that by the time bursectomy was accomplished bacterial invasion of the cloaca had been present for, say, 24 hours. This would make it likely that the emulsion of bursal cells injected into the bursectomized and irradiated chicken that had provided them, would include some cells that had been activated by the combined action of the bursal hormone (or plasmapoietic hormone, PL.H) and a few (bacterial) antigens. If only these clones are able to give rise to germinal centres, plasma cells and the limited number of corresponding antibodies, we have an adequate interpretation of the results of Cooper and his colleagues.

## A second differentiation hormone in mammals?

From this discussion of the avian position it appears that a further process of differentiation in an environment dominated by a humoral factor (PL.H) plus subsequent antigenic stimulation is necessary before an immunocyte differentiated in the thymus can become an antibody-producing plasma cell. The evidence, however, provided recently by Mitchell and Miller (1968) indicates that plasma cells are in some cases, and perhaps in all, not derived from cells which have been differentiated in the thymus. If this conclusion can be generalized, one must assume extensive differentiation of mammalian stem cells to immunocytes in other regions of primary differentiation. These can be conveniently referred to as bursa-like or gut-associated lymphoid tissue although there is little direct evidence for such a location of the site of differentiation.

There is, however, indirect evidence in favour of such an interpretation. I am highly sympathetic to the suggestion (Peterson *et al.* 1965) that many serious disorders of immunological function in man can be understood in terms of two complementary systems, one thymus-dependent and the other 'gut-associated lymphoid tissue'-dependent.

Since the function of the second system is by hypothesis to allow plasma cell development we can conveniently label any postulated hormone PL.H. In considering the aetiology of the human conditions I shall use Good's general approach but put more importance on PL.H than he does.

It is assumed that in man and other mammals, PL.H is produced in such sites as Peyer's patches, appendix, and tonsil, that it is in all probability a specific protein potentially antigenic, and that in its absence plasma cells do not develop and neither immunoglobulins nor antibodies can be produced in more than minimal amounts. Since writing this chapter, I have come to favour the idea that the hormone is mainly involved in the actual process of primary differentiation of the second (antibody-producing) system of immunocytes. If the hormone is a specific protein, its absence in pathological conditions can be due only to two causes:

(*a*) because of congenital failure of production associated with a genetic anomaly which could well be of more than one type. If, as we are assuming, congenital agammaglobulinaemia is basically an example of this, the gene(s) concerned are carried on the X-chromosome;

(*b*) because of an autoimmune process in which the auto-antigen is a determinant characteristic either of the PL.H itself or of some substance needed for functioning of the cells which produce it.

The clinical and immunological characteristics of sex-linked congenital agammaglobulinaemia are fully in accord with interpretation (*a*). In brief there are no plasma cells, very low levels of all three types of immunoglobulin, and no antibody production. On the other hand the thymus is histologically normal, homograft rejection takes place almost normally and measles infection runs a normal course. According to Good there is almost complete absence of tonsils and of appendical lymphoid tissue.

The clinical pattern of acquired agammaglobulinaemia is more variable but it has many aspects suggestive of an autoimmune condition. It involves mostly females aged from thirty to fifty, and relatives have been reported to show undue incidence of other autoimmune diseases. The association with thymoma and variably with aplastic anaemia is more appropriately discussed in chapter 12.

These clinical phenomena may have some quite different explanation but the postulation of two hormones, lymphopoietin and plasmapoietin, would bring the process of differentiation of stem cell haematocytes to immunocytes very closely into line with the known function of erythropoietin in red cell production.

If we accept this general approach, plasma cells could be produced from stem cells in two possible fashions. If Mitchell and Miller's work is of general application, the sequence will be:

(*a*) differentiation to immunocyte in gut-associated lymphoid tissue under the influence of PL.H;

(*b*) nonspecific stimulation to full competence by products from antigen-stimulated thymus-dependent immunocytes;

(*c*) antigenic stimulation to initiate a plasma cell line which in appropriate physiological situations proliferates and produces antibody.

At this stage it is probably necessary still to retain the alternative that the sequence may be:

(*a*) differentiation in the thymus under the influence of thymic hormone (lymphopoietin);

(*b*) antigenic stimulation to become a committed immunocyte;

(*c*) renewed antigenic stimulation in the presence of an adequate oxygen tension and an adequate concentration of PL.H (plasmapoietin).

# 5  The nature of antibody

In the last year or two there have been numerous reviews of the chemical, physical and antigenic structure of immunoglobulins and antibodies (Cohen and Porter, 1964a; Fahey, 1962; Kabat, 1966). In this chapter, therefore, no attempt is made to cover the development of the modern approach. The whole emphasis is on those aspects which are relevant to the basic problem with which we are concerned, the origin and deployment of immune pattern. A first section is designed to collect together information that is readily available but is convenient to have at hand. Most of the other sections are concerned essentially with the interpretation of what has been found.

*Nomenclature*

There has been virtually complete adoption by immunologists of the nomenclature proposed by a World Health Organization committee in 1964 (Memorandum, 1965). Immunoglobulins are defined as 'proteins of animal origin endowed with known antibody activity and certain proteins related to them by chemical structure and hence antigenic specificity'.

The three types commonly present in mammals are:

Ig G: classical precipitating antibody of molecular weight around 150,000, formerly $\gamma_2$.

Ig M: macroglobulin of molecular weight around 1,000,000, formerly $\beta_{2M}$.

Ig A: non-precipitating antibody of specific antigenic quality, of molecular weight 160,000, formerly $\gamma_{1A}$ or $\beta_{2M}$.

There have already been elaborations of this basic division and more will undoubtedly follow (see table 2 in Book 1).

All immunoglobulins are composed of light (or L) and heavy (or H) chains. The light chains of mol. wt 22,000 are of two serological types ($\kappa$ and $\lambda$ in man). They are common to all immunoglobulins and both antigenic types are present in all individuals.

The heavy chains of G, M and A immunoglobulins differ specifically by virtue of antigens which are called $\gamma$, $\mu$ and $\alpha$ respectively.

Fragments produced from Ig G by the use of papain are termed:

Fab—Porter I and II, the antibody-binding fragment.

Fc—Porter III, crystallizable portion of heavy chain.

Fd—'A piece', the portion of the heavy chain which can be separated from Fab (Porter, 1963).

## Species differences and allotypes

*Human.* An additional immunoglobulin, D, has been described by Rowe and Fahey (1965) which is present in at least 80 per cent of people in small amount. Its sedimentation value is 7 S, it has a specific heavy chain antigen, $\delta$, and can carry either $\kappa$ or $\lambda$ light chains.

The ADs of allotypes Gm(a+ b+) are on the Fc portion of the heavy chain of Ig G, while Gm(f) is on Fab and requires both light- and heavy-chain portions to be present for its activity (Gold *et al.* 1965).

The Inva and Invb determinants are on the stable halves of the light chains.

*Rabbit.* The allotypes A1, A2, A3 are on the heavy chain but it is very difficult to prepare light chain free from these determinants (Feinstein *et al.* 1963). Determinants for A3, A4, A5 are exclusively on the light chain. Both groups are present in Fab.

*Guinea-pig.* There are two 7 S antibodies of differing mobility; $\gamma_1$ and $\gamma_2$ (slower). The faster $\gamma_1$ resembles Ig A of other mammals in mediating passive systemic or cutaneous anaphylaxis but differs in that both $\gamma_1$ and $\gamma_2$ have the same low carbohydrate content, 0·74 per cent. They also have a common AD on Fc (Benacerraf *et al.* 1963 *b*).

*Mouse.* The conditions in mice seem to be highly complex. According to Fahey *et al.* (1964) there are four main immunoglobulins, Ig G1, Ig G2 (with subclasses 2a and 2b), Ig A (with sedimentation variable 7 S–19 S) and Ig M. Allotypes seem likely to be as numerous and as complexly interrelated as the histocompatibility factors and red cell antigens (Warner *et al.* 1966).

## The carbohydrate content of immunoglobulins

All immunoglobulins contain a carbohydrate moiety which has been studied in detail only in Ig G. For various species the carbohydrate content is given as 2·4 to 2·9 per cent and is almost wholly confined to the heavy chains. The results on enzymatic splitting are less definite and in both human and rabbit Ig G about one-third is associated with Porter's I and two-thirds more firmly bound to piece III. The carbohydrate contains hexose, fucose, hexosamine and sialic acid; according to Rosevear and Smith (1961) its composition in the rabbit is 3 galactose, 5 mannose, 2 fucose, 8 or 10 glucosamine and 1 sialic acid units in a single oligosaccharide attached to the polypeptide by aspartic acid (Fahey, 1962). There are unresolved differences of opinion on the last point.

Both Ig M and Ig A contain much more carbohydrate, 12·2 per cent and 10·5 per cent respectively (Cohen and Porter, 1964a). Since it is known that the capacity of antibody to bind complement and attach to cells is associated with the Fc piece, to which at least most of the carbohydrate is attached, there is a distinct possibility that many of the qualities perhaps particularly of denatured globulin will prove to be dependent on the polysaccharide. Human Ig A contains four times as much carbohydrate as Ig G and has a very much higher avidity for cells.

Some direct support for this suggestion comes from the result of mild periodate oxidation of antibody. Andersen *et al.* (1966) found it possible to oxidize particularly the fucose and neuraminic acid components of the carbohydrate without evidence of damage to protein. The treated antibody showed greatly diminished capacity to give passive cutaneous anaphylaxis (PCA) with very little change in precipitating ability. Since PCA depends on the fixation of antibody to tissue this points to a function of carbohydrate in this respect.

## The amino acid sequence of the light chains

The light chains of immunoglobulins have a molecular weight of the order 20–25,000 and are single polypeptide chains containing 214 residues. Adequate amounts of material for analysis can be obtained either as Bence Jones protein or as light chain chemically separated from myeloma protein. Several groups of workers are now actively engaged in determining the full amino acid sequence by various modifications of the methods developed by Sanger, Edman and others. It is quite certain that progressively more accurate results will appear between the writing and the publication of this chapter, but already the general outline seems to be clear.

It has been known for some time (Schwartz and Edelman, 1963) that the amino acid constitution of K- and L-type light chains was different and that while the C terminal groups of light chains were constant, those of the N terminal varied from one protein to another. Milstein (1965) found the C terminal sequences for K and L chains were

$$\text{K–ARG–GLY–GLU–CYS}$$
$$\text{L–THR–GLY–CYS–SER.}$$

The first virtually complete studies of the whole sequence were published by Hilschmann and Craig (1965) and by Titani *et al.* (1965). Both groups worked on the same material and obtained equivalent results. Three Bence Jones proteins of type K were shown to be almost identical from residue 108 to the C terminal end (214) with single amino acid differences at 175, 178 and 191. Subsequent work by Milstein (1966) and Baglioni *et al.* (1966) has shown that the change at no. 191 from leucine to valine is associated in change of allotype from Inva to Invb. The N terminal halves (nos 1–107) varied greatly but there were regions where two of the three were virtually identical, e.g. the residues at the N terminal (1–10) were the same in 'Roy' and 'Ag'.

Particularly in view of Smithies's (1964) hypothesis that antibody pattern may depend on the occurrence of intrachromatid inversions, all those concerned have sought for regularities that might throw light on the genetic processes involved. No regularities have been detected. Putnam's group (Putnam *et al.* 1966*b*) consider the results on the whole speak against Smithies's hypothesis and Milstein, who has made a limited examination of eight Bence Jones proteins, can also find no systematic interpretation of the differences. It may be highly significant that even the vital cysteine residues are out of place in one of his samples.

Although there is a natural tendency amongst immunochemists

to look for at least a clue to the nature of immune specificity in the variable segment of the light chain, there are important biological differences between antibody production and myeloma globulin production. On the form of clonal selection theory we are adopting, any immune pattern may emerge so long as it does not react significantly with self-constituents that are available in the circulation or in the thymus. There may even be a majority of the random patterns which are not suited to react with any existent or potential AD. The clone that gives rise to an antibody has established its right to survive by its ability to react with a real AD. If a myeloma arises from a mutation or sequence of mutations in an immunocyte, these will be totally independent of the immune pattern it carries. We may well have to consider the possibility that some immunoglobulin patterns may emerge which even lack combining sites.

Preliminary studies of mouse myeloma proteins have been made along similar lines by Dreyer and Bennett (1965) and Bennett *et al.* (1965). They find that mice myelomas with L chain proteins show, on 'fingerprinting', tryptic peptides which include a group of nineteen common peptides and in each case four to five individual ones. There are also some major differences in amino acid composition between different proteins, e.g. threonine shows 19 residues in one, 25 in another, serine 32 and 42, tyrosine 11 and 7. The common peptides of these myeloma proteins are also present in the L chains of normal mouse immunoglobulins. Relatively complete sequence studies are now available for mouse light chains (Grey *et al.* 1967; Kabat, 1967). It is evident that there is the same division of the light chains into variable and constant regions and there are many sequences in common in the variable segments of human and murine K light chains.

## Physical differences in light chains

When guinea-pig antibodies are purified and separated into heavy (A) and light (B) chains, electrophoresis in 8 M urea on starch gel at pH 3·5 showed a pattern of relatively sharp lines corresponding to the B fraction (Edelman *et al.* 1961, 1963 *a*). The pattern differed for different antibodies. Other mammalian sera gave broad indefinite bands only. Cohen and Porter (1964 *b*) re-examined the position

by carrying out starch electrophoresis at pH 7·8 with very interesting results. Total light chain from normal sera of guinea-pig,
bovine, horse, baboon, and man showed numerous well-spaced
lines of differing intensity; the rabbit fraction was unsatisfactory.
Each of the lines corresponds to one of ten possible positions, and
the difference between species depends only on which positions
are occupied and the relative distribution of material amongst
them. The bands are not artefacts of technique since myeloma
proteins give a very different pattern for their B chains. In each
instance there was a single heavy band and immediately beyond
it a weak band in the next position with some intermediate trailing
(table 11).

TABLE 11. *Starch electrophoresis of human light chains*

| Band | Normal immunoglobin | Myeloma | | | | |
|------|---------------------|---------|---|---|---|---|
|      |                     | 2 | 3 | 4 | 5 | 6 |
| 2 | + | − | − | − | − | − |
| 3 | + + + | + + + | − | − | − | − |
| 4 | + + + | ± | + + + | + + + | − | − |
| 5 | + + + | − | ± | + | + + + | − |
| 6 | + + | − | − | − | + | + + + |
| 7 | + + | − | − | − | − | + |
| 8 | ± | − | − | − | − | − |
| 9 | ± | − | − | − | − | − |

Data from Cohen and Porter (1964 *b*).

In the light of the foregoing discussion of the nature of the light
chains, these bands can probably be ascribed to the differences in
amino acid constitution of the variable segment of the light chains
present in any normal population of immunoglobulin molecules.
The regularity of spacing points strongly toward a single unit of
charge being responsible. As in the abnormal haemoglobins, a
change from a basic amino acid to a neutral one would cause a
one-step shift. In all probability the differences in charge are
essentially accidental to the replacement of one amino acid by
another. It may well be, however, that a certain average charge has
some selective advantage in the sense of favouring proliferation of

the cell line which produces it. This is presumably why human light chains are concentrated in bands 3, 4 and 5; guinea-pig in 4, 5, 6, 7; and bovine in 2, 3, 4.

At this point the amino acid analyses of purified antibody from rabbits by Koshland and Englberger (1963) and Koshland *et al.* (1964) should be considered. They used rabbits immunized with three types of hapten, based on arsonic acid (ARS) (acidic); trimethyl ammonium (AMM) (basic); and $\beta$ phenyl lactoside (LAC) (neutral). Significant amino acid differences in whole antibody were found, including a comparison of two antibodies produced in the same rabbit. What is of special interest is that the differences were such as would be expected to facilitate adsorption of the haptens (table 12). One would expect, too, that if half the value is ascribed to each chain the light chains would show charges as follows: ARS, $+1$; LAC, $-2$; AMM, $-5$, when tested by Cohen and Porter's procedure.

TABLE 12. *Amino acid differences in purified antibodies*
(*Koshland* et al. *1964*)

| Amino acids and charge | | ARS (acidic) | LAC (neutral) | AMM (basic) |
|---|---|---|---|---|
| | | Immunizing hapten | | |
| Arginine | + | 3 | 3 | o |
| Tyrosine | o | 5 | o | 5 |
| Serine | o | 5 | o | 5 |
| Leucine | o | o | o | 2 |
| Glutamic acid | — | 1 | o | 5 |
| Aspartic acid | — | o | 7 | 5 |

Figures show the number of amino acids per 1,600 residues in terms of excess over the lowest value found.

It is relevant to these results that Sela and Mozes (1966) have shown that the proportion of a given rabbit antibody which is found in the two peaks of a DEAE–Sephadex separation procedure depends on the net electrical charge of the antigen. In their opinion the peaks correspond to Ig G having Porter's piece I in the Fab fragment and those molecules with piece II, respectively.

Basic antigens such as lysozyme and DNP-polylysine gave from 73 per cent to nearly 100 per cent of antibody in peak I while acidic antigens, diphtheria toxoid and an acidic DNP synthetic antigen show from 9 to 15 per cent in peak I. It is perhaps equally significant (*a*) that there is this overall difference between antibody against the two types of antigen and (*b*) that with one exception every antiserum shows some specific antibody of both types.

Both sets of findings are precisely what would have been expected on the view that combining site patterns arise by a random process.

## Structure of the heavy chains

For obvious reasons, much less effective work has been done on the structure of the heavy chains. It is, however, certain that there is considerable heterogeneity. Heavy chains from Ig G globulin or myeloma protein show at least four serological differences when examined with antisera made in rhesus monkeys (Terry and Fahey, 1964) or rabbits (Grey and Kunkel, 1964).

Peptide maps of the Fc fragment of normal G immunoglobulins and of eleven myeloma G proteins showed them to be closely similar except for differences associated with the Gm type. Gm(a + ) has a specific peptide absent in Gm(a−), while Gm(b+) has another specific peptide absent in Gm(b −) (Meltzer *et al.* 1964). On the other hand the Fd fragment must be highly variable since heavy-chain peptide maps from myeloma proteins are widely different. In each of the myeloma proteins studied by Frangione and Franklin (1965) there were well-defined spots not present in normal Ig G heavy chains and *more* of such strong spots than in the normal maps. In addition there was the significant finding that normal heavy chains showed a greater amount of faint background staining on the peptide maps and fewer distinct well-defined peptides than would be expected. This, of course, indicates that the normal H chains are highly heterogeneous and that the heterogeneity resides in the Fd fragment (Frangione *et al.* 1966).

Direct evidence to this effect has been obtained by Koshland (1966) in detailed studies of fragments of the same rabbit antibodies against ARS and LAC haptens which have already been described. The differences involve the Fd region and particularly one active fragment obtained by cyanogen bromide treatment. The differences between the two light chains and the two active fragments of the heavy chains have only a partial resemblance. She found that treatment with cyanogen bromide split the molecule into an active fragment of mol. wt 45,000 carrying all the binding power, and therefore approximately equivalent to Porter's piece I and an inactive group of peptides. The variation between the two antibodies was wholly ascribable to the active fragment, that is, light chain plus Fd. The other peptides corresponding to mol. wt 30,000 were of the same amino acid composition for each antibody.

The significance of these findings will be discussed in the next chapter but table 13 shows how the differences in composition were distributed.

TABLE 13. *Differences in amino acid composition between ARS and LAC antibodies*

| Fraction mol. wt | Anti-body 160,000 | 1 Light chain 23,000 | 2 Fd fragment 23,000 | 3 Fc fragment 30,000 | 2(1+2+3) 162,000 |
|---|---|---|---|---|---|
| Asp | +5 | +2 | o | o | +4 |
| Ser | −6 | o | −2 | o | −4 |
| Tyr | −6 | −1 | −2 | o | −6 |
| Pro | o | +1 | −1 | o | o |
| Val | o | +2 | (−1) | o | (2–4) |

Data from Koshland (1966).

It can hardly be doubted, therefore, that the heavy chain resembles the light chain in being composed of a variable segment and a stable one. Although the evidence is good that synthesis of Ig G in the cell is mediated by m-RNA for only two chains corresponding to complete light and heavy chains (see p. 422) there is a growing tendency to think of the light chain as having two distinct genetic components corresponding to Lv and Ls. On the same reasoning the heavy chain of Ig G will have four equivalent components: a variable and a stable component in Fd, and two stable components (to one of which a polysaccharide moiety is attached) in Fc. Each Ig G molecule therefore, although composed of only four polypeptide chains, two light and two heavy, can be regarded as two half-molecules, each of which is coded for by 2 + 4 genetic units. It is not excluded that in some molecules the Lv and Fd (variable) units may be identical: in fact there are hints from recent work by Doolittle and Singer (1965) that they may be. This is discussed at a later stage (p. 427).

*Functional activity of Fc*

The antigenic specificities of the H chain of Ig G which include in man the sub-types We, Vi, Gc and Ne (Grey and Kunkel, 1964) and the allotypic Gm specificities (a), (b) and (f) are significantly interrelated (see Mårtensson, 1966*a*, *b*), since Gm(a) and (b) antigens are known to be located on the Fc chain (Harboe *et al.* 1962; Franklin *et al.* 1962) it also probably carries the sub-type antigens as well. On the other hand the antigens Gm(y) and (z) are found on the Fd fragment by Litwin and Kunkel (1966).

Attachment of antibody to various cells (other than in relation to immune reaction) is ascribed to the Fc fragment (Ovary and Karush, 1961; Terry, 1966) making it available for reverse passive cutaneous anaphylactic reactions. Nothing appears to be known in regard to the nature of the groups on immunoglobulin and on the absorbing cell which are responsible for the interaction. It is known, however, that there is a significant degree of species specificity.

From work by Brambell and collaborators (Brambell, 1966) the Fc fragment is intimately concerned with passage of Ig G across the foetal yolk-sac of the rabbit or the neonatal intestinal wall of rat or mouse. Brambell suggests that an active group on Fc is responsible for attachment to receptors in phagosomes of epithelial cells (and perhaps other types of cell) by which the whole molecule is protected from enzymic destruction and at an appropriate time can be released again from the cell that had taken it up. This is based on the evidence by Morris (1964) that Fc but not Fab interferes with the transmission of intact Ig G mixed with them. In Brambell's view it is also supported by work on the catabolism of Ig G in mice (Fahey and Robinson, 1963). This showed that the half-life of each of the three 7S immunoglobulins of mice,

Ig G 1, Ig G 2a, Ig G 2b, was influenced by their concentration and that the infusion of Fc piece was as effective on a molar basis as intact Ig G in shortening the half-life, while Fab was inactive or had a reverse effect. Spiegelberg and Weigle (1965) found that half-lives of homologous and heterologous G immunoglobulins in the rabbit showed values depending on the species providing the immunoglobulins. These values were approximately the same for the corresponding Fc fragments, but light chains or Fab fragments were all rapidly eliminated.

In the condition known as 'heavy chain disease' (Franklin *et al.* 1964) the abnormal protein has a molecular weight of about 51,000, lacks several peptides present in normal heavy chain and appears to be essentially Fc in character (Franklin, 1964*b*).

### THE POLYRIBOSOMAL SYNTHESIS OF LIGHT AND HEAVY CHAINS?

The evidence in regard to the synthesis of immunoglobulin on polyribosomes is summarized by Askonas and Williamson (1966), whose own experimental work was compatible with the earlier study of Shapiro *et al.* (1966*a*, *b*). The evidence is that mouse myeloma heavy chain (identified by Fc antigens) is synthesized on polyribosomes averaging 300S and light chains (also identified immunologically) on smaller 200S polyribosomes. A pool of light chain develops shortly before the appearance of immunoglobulin and it is suggested that the light chains release the heavy chains from the polyribosomes.

There seem to be a good many soft edges to the experimental demonstration that whole heavy chain and whole light chain are each produced by a single species of m-RNA coded by a single cistron. In particular the origin of the Fd piece is left nebulous, and it seems unlikely that heavy chain disease, in which part or all of the Fd segment is absent, could arise by the partial expression of a single gene to produce only the carboxy terminal half or three-quarters of the full chain. It seems equally difficult to believe that the genetic origin of the relatively small portion of the heavy chain which carries the antibody specificity (expressed almost certainly in either identical or analogous fashion to the variable portion of the light chain) should be the same as the highly uniform standardized Fc portion. The possibility of a more complex process of synthesis and assembly does not seem to be excluded.

## THE NATURE OF THE COMBINING SITE IN STANDARD IG G ANTIBODY

There is unanimity amongst immunochemists that in a standard Ig G antibody there are two limited areas of the macromolecule which are the specific combining sites to which the corresponding AD or hapten can attach. There is direct evidence from electron microscopy that these sites are at the two ends of the elongate molecule (Almeida *et al.* 1963).

Evidence for the bivalence of antibody can be obtained at the physical level by equilibrium dialysis (Karush, 1957 a; Klinman *et al.* 1964), and more directly by the study of antibody fragments. Nissonoff (1963) described experiments showing that by appropriate treatment to break disulphide bonds, univalent fragments with binding capacity for antigen but no precipitating power could be produced. When antibodies of two different specificities were used, hybrid antibodies could be produced by recombining a mixture of the two types of half-molecule. These can react with a mixture of the two antigens with precipitation of a type not seen by either alone. Hong *et al.* (1965), from Nissonoff's laboratory, used similar half-molecules of antibody and of normal globulin. In a recombination mixture of antibody fragments, more than 80 per cent of precipitating ability was recovered, but when recombination in the ratio of 1 antibody to 11 normal globulin was attempted, no precipitation resulted, indicating again the necessity for bivalence if lattice formation is to occur.

An elegant demonstration of bivalence is due to Fudenberg *et al.* (1962), using rabbit haemagglutinating antibody. When this was converted to half-molecules (3·5 S) by Nissonoff's method it failed to agglutinate red cells. The same result was obtained with 'incomplete' human Rh antibody. When split, this failed completely to agglutinate enzyme-treated cells fully susceptible to the original serum. The logical extension of this work by the same group, Fudenberg *et al.* (1964), was to study the action of hybrid

423

antibodies on red cells of human and chicken which had been coated with the respective antigens. The hybrid antibody gave visible chicken-to-human unions not seen with either pure antibody acting on the same mixture.

Evidence for the equivalence of combining sites at both ends of the molecule in specificity and avidity is very limited but equally there is no evidence to suggest that any antibody molecule naturally produced possesses two different specificities. An experiment for which Miss Freeman and I were responsible thirty years ago (Burnet, 1941) still seems relevant. A doubly antigenic particle was produced by coating bacteria with an actively antigenic phage and inactivating with formalin. Immunized rabbits produced high-titre antibody to both. Serum was absorbed with bacteria with removal of all bacterial agglutinin and no reduction of antiphage titre. The absorbing bacteria were thoroughly washed and the antibody eluted by alkali. The eluate contained no antiphage. Since univalent antibody is capable of inactivating phage, this is a fairly critical experiment.

Work in Kabat's laboratory (Gelzer and Kabat, 1964*b*) on anti-dextran allowed the fractionation of a serum into subpopulations of antibody more reactive respectively with trisaccharide and hexa-saccharide. This was related to differences in the nature of the combining site in the two populations and to some extent at least implies that in individual molecules both combining sites were similar.

The heterogeneity of all antisera will prevent a full answer to this problem which provides a further incentive to obtain a myeloma protein with a demonstrable combining site. With a homogeneous population the assessment of similarity or difference between the combining sites should be straightforward (see p. 439).

Possible approaches to the nature of the combining site are largely based on the use of reagents which will react with specific amino acids, applied both to antibody alone and to antibody in which the combining sites are blocked with hapten. Pressman (1963) observed that an antibody against a positively charged group like *p*-azophenyltrimethylammonium was inactivated by carboxyl combining groups like diazoacetamide but was protected if hapten was in position during the esterification. On the other hand, antibody to the negatively charged azobenzene haptens was not affected by such reagents. This is direct evidence for a difference in the combining site between two antibodies.

Iodination procedures in which tyrosine is involved are liable to inactivate antibodies. According to Grossberger *et al.* (1962) all antibody binding sites show some inactivation by iodination which is prevented by the attachment of hapten. The extent to which such combination with tyrosine reduces binding power for hapten, however, varies greatly from one example to another. Another approach is by affinity-labelling (Metzger *et al.* 1964), in which a reactive group with an appropriate radioisotope is attached to the hapten. On reaction the label is transferred to an amino acid residue in the combining site. Subsequent fractionation of the antibody in various ways can then give valuable information as to the site of the active area in the molecule.

A great deal of interest has centred on the question as to whether the combining site was located on the light or the heavy chain. A majority of such studies have shown that the heavy (A) chain carries most or all of the combining activity. This seems to be specially clear cut for horse antibodies (Fleischman *et al.* 1963; Franěk and Nezlin, 1963). In guinea-pig antibodies, Edelman *et al.* (1963*b*) found the heavy chains more active than the light, but to a variable extent. In each instance, recombination of homologous L and H chains gave a considerably higher activity than H chain alone. This was not seen with heterologous pairs.

With rabbit antibodies there is a bias toward the light chain. Roholt *et al.* (1963) carried out 'pair iodination' using the two isotopes $^{131}$I and $^{125}$I. Part of the antibody was iodinated with $^{125}$I in the absence of hapten, the other half in the presence of hapten with $^{131}$I. The mixed antibody was then fractionated for L and H chains and these were subjected to tryptic digestion and the peptides chromatographed. The relative concentration of $^{125}$I and $^{131}$I was determined for each spot. Where the ratio $^{125}$I:$^{131}$I was unduly high, it was assumed that the peptide concerned had formed part of the combining site. Three such peptides were from the light chain, one, with a marginal excess, from the heavy chain.

Metzger *et al.* (1964) also showed by affinity-labelling of rabbit antibodies that both light and heavy chains were involved. Doolittle and Singer (1965), from the same group, have studied the nature of the peptides that are labelled on each chain. They conclude that a very similar group of peptides is obtained from each. The resemblance involves the average size of the peptides, their predominantly hydrophobic character and the degree of heterogeneity.

It has already been noted that with guinea-pig antibody (Edelman *et al.* 1963 *b*), reconstitution of antibody activity by combination of L and H chains requires that they should be homologous. This was also clearly demonstrated in Franěk's work with horse antitoxins. More detailed studies by Roholt *et al.* (1965 *a, b*) show that this holds even when the animals concerned are all immunized against the same hapten. It was of special interest that where two *pools*, A and B, of rabbit sera were used, reconstitution of L and H from pool A gave 62 per cent of the original binding power; similarly, both chains from B gave 59 per cent, but with reciprocal reconstitutions the values were only from 5 to 9 per cent. This indicates a capacity of selective union, even in a heterogeneous pool. Results obtained by Metzger and Mannik (1964) differ in that to obtain fully specific reconstitution from a mixture of antibody chains and normal Ig G chains, a small amount of hapten was needed. This increased both the extent and the specificity of the reconstitution.

One final aspect of the combining site that should be mentioned is that its specificity can be restored after complete denaturation with loss of all tertiary structure of the fragment, by reduction and solution in strong urea or guanidine solutions. By careful renaturation of the fragment, both Buckley *et al.* (1963) and Haber (1964) regained most of the specific binding capacity.

## The duplex chain hypothesis of the combining site

The molecule of Ig G has dimensions, according to Edelman and Gally (1964), of 240 × 57 × 19 Å and contains 1,495 amino acid residues. The approximate length of a single polypeptide chain of that extent is 8,000 Å, which means that to pack all the chains into the space available there will have to be such coiling and recoiling that any cross-section of the molecule would cut through at least 30 chains, of which more than half will abut on the surface. Apart from the presence of a small oligosaccharide which is known to play no part in the combining site, we have the surface of the molecule made up of a mosaic of amino acid residues arranged in four major chains. The general structure of the macromolecule is determined probably largely by intramolecular disulphide bonds, but with its known flexibility most of the surface must represent an essentially random mosaic.

Such a surface would undoubtedly have many regions with special adsorptive quality for a variety of small molecules, but most of them would be essentially temporary and almost certainly not subject to genetic determination. In thinking over possible ways in which a combining site could develop, I have been greatly influenced by Doolittle and Singer's finding that affinity-labelling studies point to the similar character of the components from both light and heavy chains (Burnet, 1966).

Any hypothesis must be presented in a flexible and generalized form to allow for the present stage of rapid development in the knowledge of immunoglobulin structure. Several earlier sketches have already been outdated. The current suggestion can be based on the diagram (fig. 32) which represents a possible configuration based on the published sequence of the Bence Jones protein Ag (Putnam *et al.* 1966*a*) on the assumption that this would be equivalent to the light chain of an antibody. It is assumed that the tyrosine residues are vital to the combining site and that the heavy chain component of the duplex region has a broadly similar— possibly identical—structural sequence forming part of the 'variable' Fd segment.

The combining region may well be large enough to carry, as indicated in the figure, more than one combining site of the conventionally accepted size. A duplex coil of this type seems to have the qualities necessary to account for the findings in regard to the combining site described in the last section. Specific points of agreement with the hypothesis depicted in fig. 32 are as follows:

(*a*) The special partially congruent duplex region will provide a persisting feature of the surface differentiable from the general mosaic.

(*b*) A stabilized duplex region will provide a patterned field of potential reactivity with AD, which will be of more regular quality than a pattern dependent on a single chain sequence with random neighbour groups.

(*c*) Both components are derived by hypothesis from 'variable' segments and this, plus any random choice amongst alternative variable segments that may exist, is appropriate for the provision of the wide range of specificities which is needed.

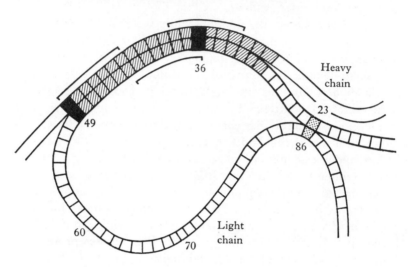

Fig. 32. Hypothesis of the nature of the combining site as an association of Lv with a portion of heavy chain of the same general quality. Cysteine (dotted) and tyrosine (black) residues shown as for Bence Jones protein Ag by Putnam *et al.* (1966*a*).

(*d*) It is clear that there must be a specific steric relationship of some sort between the regions of L and H chains which form the combining site, in order to account for the symmetry and equivalence of the two combining sites and for the need for homologous components to be present if separated L and H chains are to reconstitute a combining site. These are the qualities which make me reluctant to discard completely the possibility that both the light chain and the heavy chain contributions to the combining site are defined by a single region of the genome.

## THE STRUCTURE OF IGM AND IGA ANTIBODIES

There has been much less work done on Ig M and Ig A antibodies than with Ig G, but the outstanding feature is that the light chains are apparently of precisely similar character. There is also some evidence that as an animal being immunized moves over from Ig M-production to Ig G-production, the specificity and avidity of the antibody show no evidence of any associated change (Grey, 1964).

The main characteristics of Ig M antibody at the physico-chemical level are:

(*a*) A molecular weight of the order of one million, i.e. 5–6 times that of Ig G.

(*b*) A much higher content of carbohydrate, 12·2 per cent as against 2·8 per cent for Ig G in man; with considerable differences in composition (Cohen and Porter, 1964*a*).

(*c*) Significant differences in amino acid content in the heavy chains (Cohen and Porter, 1964*a*).

(*d*) A definite antigenic character ($\mu$) distinguishing M immuno-globulins from G or A.

(*e*) By treatment with mercaptoethanol the haemagglutinating activity of Ig M antibodies is destroyed (Deutsch and Morton, 1957) and 6·5 S units liberated. These can unite with antigen but do not precipitate (Onoue *et al.* 1964). They can be caused to recombine to give large molecules with most of the physical and biological properties of the original Ig M (Schroenloher *et al.* 1964).

(*f*) Electron micrographs (Humphrey and Dourmashkin, 1965) of Ig M haemagglutinating antibody on red cell surface membrane show appearances compatible with a pentamer or hexamer structure.

It seems probable that compared with an Ig G antibody of the same antibody pattern, the Ig M molecule has the same Lv and Ls components of the light chain and the same heavy chain unit—by hypothesis Fd (variable)—that is concerned with immune pattern. The Fc units are probably distinct and there is much more carbohydrate involved. The macromolecule, according to Cohen (1966), probably contains five 4-chain units of molecular weight 180,000 covalently linked to one another by disulphide bonds between heavy chains.

429

The work of Rockey *et al.* (1964 *b*) is almost the only satisfactory study of Ig A at the physical level but there is perhaps some doubt whether this fraction of equine serum, also called T and comprising the major part of a standard diphtheria antitoxin, is fully equivalent to human Ig A. They used an artificial *p*-azophenyl-*β*-lactoside antigen which allowed the use of hapten-binding studies. The antibody was non-precipitating but bound the hapten just as firmly as precipitating antibody and in amount that indicated that the antibody was bivalent. They deduced that the two binding sites might be close together on the molecular surface—so close that occupation of one site by antigen would sterically hinder occupation of the second but far enough apart to allow independent binding of hapten.

*The relationship of Ig A and Ig E to human reagin*

Heremans and Vaerman (1962) were apparently the first to suggest that $\beta_{2A}$ globulin (Ig A) was a possible carrier of allergic reaginic activity, on the ground that the peculiar tendency of Ig A to form complexes with other proteins might account for its strong affinity for tissue cells. More direct evidence by Ishizaka *et al.* (1963) was to show that, of the immunoglobulins, Ig A was the only one which, when infiltrated into the test site of a Prausnitz–Küstner (P–K) skin reaction, was effective in blocking the reaction. The same workers, however, now believe that the reagin is found in an antigenically distinct subfraction wrongly included in Ig A.

Ishizaka and Ishizaka (1966) agree that, when Ig A is physically isolated by a zinc precipitation method from serum of ragweed-sensitive patients, the fraction contains high reagin activity as tested by the P–K method. However, in their hands, when this fraction is precipitated by an anti-A serum prepared in a rabbit the reagin remains in the supernatant almost completely. Later, Ishizaka *et al.* (1966) find that a specific serum can be prepared which will precipitate all P–K reactivity from patients' sera. This is not done by authentic antisera against the other heavy chain antigens $\alpha$, $\gamma$, $\mu$, and $\delta$. The new immunoglobulin is therefore called Ig E, with its characteristic antigen, $\epsilon$. Evidence has been adduced for the existence of light chain antigens similar to those of Ig G but no statement has yet been made as to whether the light chains are of L or K type. No clear demonstration of Ig E in the serum of non-allergic normal human beings has yet been given and the possibility must remain that this is not a physiological type of immunoglobulin but one arising by a fairly common mutation in an Ig A immunocyte line, by which a modification of the $\alpha$ antigen and increased potential 'pathogenicity' go together. The association of hay fever with mucous membranes makes one reluctant to separate the peccant antibodies wholly from Ig A.

## The relationship of Ig A to glandular secretions

Much work on the economically important problem of the transfer of antibody to colostrum in the cow has made it clear that the colostral antibody is predominantly Ig A with some Ig M (Smith, 1946; Blakemore and Garner, 1956; Larson and Gillespie, 1957; Lovell and Rees, 1961; Pierce and Feinstein, 1965). There is very much less Ig A in the circulating plasma and if there is a direct transfer, as most of these authors believed, then there must be a very selective uptake, concentration and transfer of Ig A. It may be noted that the selectivity is wholly in the secretion of antibody into the colostrum. The young calf only absorbs antibody from the intestine for the first 36 hours of life but it will absorb Ig G from serum just as effectively as Ig A from colostrum (Brambell, 1958).

In other animals, including man, the same predilection for the presence of Ig A antibody in milk and other secretions such as parotid saliva can be shown. In man, Chodirker and Tomasi (1963) found that in colostrum, parotid saliva, nasal fluid and tears there were no more than traces of Ig M, with Ig A dominant. In a variety of other fluids including bile, cerebrospinal fluids, and intestinal, prostatic, vaginal and amniotic fluids, the ratio of Ig G to Ig A was not significantly different from that in the serum. All are in agreement that the specificity, i.e. the immune patterns of the antibodies in colostrum, are qualitatively equivalent to and at much the same titre as those in the blood. Adinolfi *et al.* (1966) found that the Ig A antibodies against *Escherichia coli* in colostrum were virtually absent in the blood where a similar antibody titre was carried almost wholly by Ig M. An interesting point was that

431

lysis of the bacteria was obtainable by Ig M plus complement or by Ig A plus lysozyme, but not vice versa.

The consensus of opinion points toward a simple selective trapping of Ig A from serum, concentration in the gland cells and secretion in combination with another protein, 'transport piece' of Pollara *et al.* (1966). Perhaps the most convincing evidence for this comes from two healthy men described by Rockey *et al.* (1964*a*) who lacked any Ig A in their serum although Ig M and Ig G were present, Ig M being in abnormally high concentration. Neither had Ig A in their saliva. This is strong but perhaps not decisive evidence that Ig M cannot be taken up by gland cells and converted into Ig A. Other evidence for this point of view also comes from genetically anomalous subjects. South *et al.* (1966) gave normal plasma globulins to agammaglobulinaemic children and found selective transport of Ig A to saliva combined with 'transport piece'. Immunological evidence was obtained of the pre-existence of free transport piece in saliva of these children.

On the other hand, Tomasi *et al.* (1965) report findings which suggest a more complex process in normal subjects. When $^{131}$I-labelled Ig A was administered intravenously it was not excreted in the saliva, all the label being associated with small molecules. They also found that an Ig A myeloma patient did not secrete the protein as such in parotid saliva.

Another finding which may become relevant in the interpretation of Ig A function is that of Heremans *et al.* (1966) that plasma cells in close relation to secreting cells and mucous surfaces, as in the duodenum of the rat, are in large numbers and 80 per cent of them produce Ig A. Crabbé (1967), working in Heremans's laboratory, has shown that this holds also for human beings. In fact 80–90 per cent of the numerous plasma cells in the submucosal areas from stomach to rectum were Ig A producers. An interesting sideline of this work was the recognition that certain cases of sprue and steatorrhoea showed a gross deficiency of Ig A in serum and of Ig A-producing cells in biopsies of the gut.

There is a persisting suggestion that while a proportion of Ig A passes unchanged with transport piece, much of the Ig A which is found in the secretion may be derived from depolymerized Ig M more or less basically reconstructed in the secreting cells. The existence of healthy individuals who have no Ig A suggests strongly the absence of an enzyme needed in the process of converting Ig M to Ig A which is presumably present in a proportion of immunocytes (subject to the usual degree of phenotypic restriction) and in secreting cells of the mammary, salivary, upper respiratory and lacrimal glands. It is obviously desirable that an experimental approach to disprove this hypothesis should be got under way.

An isolated observation which could lead to interesting developments is the report by Ainbender *et al.* (1966), who studied the antibody responses of a small series of adults who had been immunized by Salk vaccine, to determine the distribution of polio antibody in the three main immunoglobulins. They found that nearly half the women showed 30–67 per cent of the antibody in the Ig A fraction, while in none of seventeen men did the proportion exceed 20 per cent. This, taken with the early work of Kuhns and Pappenheimer (1952) in which non-precipitating antitoxin produced in allergic subjects was presumably almost all Ig A, indicates that there is a genetically controlled process (? an enzyme synthesis) which can rather readily swing the M:A ratio over a very wide range. This topic will need further discussion in relation to antibody production.

Accepting for the time being the simplest assumption that Ig A has been evolved to facilitate the defence of mucous surfaces and to provide in some mammalian species a means of conveying maternal immunity to the offspring, it becomes of interest to enquire why antibodies of this or closely related type are involved in hay fever. In hay fever the antigenic pollen coming into contact with nasal, conjunctival and respiratory mucous membranes produces a liberation of histamine. These are just the areas which are likely to be bathed with Ig A-containing fluids and one can reasonably assume that the epithelial cells are themselves excreting Ig A of the types circulating in the blood.

The hay fever patient may differ from the normal in that there is a greater frequency of pollen AD and Ig A-type receptor reaction on the surface of mucosal cells or on some other type of adjacent and accessible cells. The genetic disability could be either on the side of antibody and immunoglobulin structure or result

433

from some anomaly of cellular function. The appearance of an abnormal immunoglobulin, Ig E (Ishizaka *et al.* 1966), in allergic individuals, points toward the first alternative. If, instead of the normal process of concentration and secretion, an abnormal amount of 'combining site' either as Ig A(E) or in some other form, is retained in an accessible form on the cell surface and particularly if the antibody were of high avidity for the allergen, this would give a reasonable interpretation of the facts.

# 6 · The origin of immune pattern

It is accepted as virtually axiomatic that all specific immune reactions are mediated by steric association and binding (usually reversible) of AD and combining site (CS) on antibody or immunocyte receptor. The problem of the origin of immune pattern is then to establish reasons for the diversity of patterns of amino acid residues on the chains making up the CS. The exact relationship of the combining sites to the accepted structure of Ig G is not known but it is at least clear that portions of light and heavy chains are concerned and that the Fc portion of the heavy chain is not. I have suggested (Burnet, 1966) the possibility that a CS in the antibody molecule is produced by the apposition of two identical short sequences, one on the light chain, the other in the Fd fraction of the heavy chain, and forming part of the variable segments, Lv and Hv, of light and heavy chain. These sequences must therefore be present four times in each Ig G molecule. The more orthodox opinion is that pattern is generated by the interaction of two different variable segments, one on the light and one on the heavy chain. In principle, however, the problem is the same.

There is no suggestion that the CS is anything other than a configuration of peptide chains and it is only to be expected that the type of union observed at the CS may also be represented in the adsorptive power of soluble proteins other than antibody. Eisen (1959) in reviewing work in which the binding power of antibody for small molecular haptens was under study, pointed out that to obtain significant results the immunoglobulins must be separated from serum albumin. For some haptens there may be as many as 1,000 binding sites per serum albumin molecule as compared with two per antibody molecule, and the binding affinity per molecule of albumin may be as high as that of antibody. Different haptens vary greatly in their combining affinity for serum albumin.

Using purified Ig G antibody, equilibrium dialysis experiments with soluble haptens have consistently shown two binding sites per molecule and a wide range of binding avidity. According to Karush (1956) the heterogeneity of binding sites can be described as a Gaussian error-function in the free energy of binding.

In seeking the origin of immune pattern it is necessary to bear the heterogeneity of the antibody binding sites very much in mind. An antibody is merely an immunoglobulin with two CSs which for any reason have significant binding power for the ADs carried by the antigen under study. Once one has discarded the idea that a CS is made to order to fit the pattern of an AD, a high degree of randomness in many immune reactions becomes admissible.

## THE HETEROGENEITY OF ANTIBODY FROM A SINGLE INDIVIDUAL

If meaningful results are to be obtained from chemical studies of purified antibodies, e.g. in regard to the significance of amino acid differences between antibodies, antibodies from a single individual must be used. The best available human material is serum from a man studied by Kabat's group (Bassett *et al.* 1965).

From his serum Ig G (6·5 S) fractions were specifically isolated corresponding to anti-levan, anti-dextran, anti-teichoic acid, anti-A and anti-tetanus toxoid. The relevant genetic information is that the individual was Gm (a+ b+) as judged by the reactivity of his whole Ig G. Both K and L types of light chain were present and starch electrophoresis by Cohen and Porter's (1964b) method showed more than one light chain band for all the fractions tested. Amino acid analyses showed quite striking differences between antibodies. If one confines attention to three antibodies, all of which were Gm (a− b−), differences in the number of amino acid residues, calculated to a molecular weight of 160,000, are reported in table 14.

In reviewing this material, Kabat (1966) rightly says that it gives no information directly relevant to the structure of the CS, but it at least makes it extremely unlikely that the nature of the CS is *not* influenced by the amino acid composition of the immunoglobulin. There is plenty of evidence also to establish that in any large population of antibody molecules purified to contain only molecules whose CSs are reactive with a particular AD, there are differences among them in regard to the size or avidity of the CS. This has been demonstrated particularly with anti-dextran human serum and oligosaccharides of the isomaltose series by Gelzer and Kabat (1964*a*, *b*).

TABLE 14. *Content of certain amino acids in purified antibodies, each Gm(a− b−), from a single individual (modified from Bassett et al. 1965)*

| Amino acids | Antibody against | | | |
| | Levan | Dextran | Teichoic acid | Normal |
|---|---|---|---|---|
| Glycine | 101 | 100 | *120* | 107 |
| Valine | 132 | 134 | *102* | 139 |
| Lysine | 94 | 88 | *82* | 93 |
| Hydroxylysine | 0 | *3* | *4* | 0 |
| Proline | *101* | *101* | 108 | 107 |

The numbers correspond to the number of amino acid residues per Ig G molecule: significant differences from average normal globulin are italicized.

In rather similar fashion, Mäkelä (1964*a*, *b*) has shown that the antibody produced from a single cell may differ, in its cross-reactivity with related antigens, from that produced by another single cell, both being derived from the same immunized animal.

The essential features that seem to emerge from work of this sort on the heterogeneity of antibodies are:

(*a*) When an antibody is purified it frequently shows a gross deviation from the character of average immunoglobulin of the same type in regard to proportion of allotype, presence of hydroxylysine, or distribution of light chains in the electrophoretic series. None of these differences is decisive but all point toward the probability that a given antibody population is likely to be derived from a relatively small population of clones though only very rarely from a single clone.

(*b*) The differences among antibody molecules reacting with the same AD point strongly against any view that the CS is 'designed' to fit the AD and in favour of the alternative that AD

437

reacts with every CS which 'happens to correspond' sufficiently closely for adsorption to occur. Of special relevance in this connection is the effect described by Krueger (1965, 1966) of the action of streptomycin on the *in vitro* production of antibody by cells from a rabbit immunized with a standard RNA phage. The natural antibody in the rabbit's serum showed a certain range of cross-neutralization with a group of other RNA phages and the same pattern of cross-reactivity was given by the antibody produced in the control *in vitro* cultures. In the presence of streptomycin (which is currently assumed to act by interfering with the messenger RNA–ribosomal RNA action in the translation of m-RNA pattern to protein), the reactivity with the homologous phage is reduced and variable changes appear in the capacity to neutralize other phages. Sometimes the antibody is much more reactive against a heterologous phage than against the homologous antigen.

All this evidence of heterogeneity among antibody populations is consistent with the clonal selection point of view but it does not formally exclude other possibilities. One is struck, however, by the extraordinary complexity that would have to be built into any other type of theory to account for facts which fit comfortably in the clonal selection framework. The essential feature of the CS is its infinite variability, not only, in all probability, dependent on the amino acid residues actually composing it but also highly susceptible to modification of the heavy–light chain relationship by changes in almost any other part of the molecule.

### MONOCLONAL ANTIBODY

The chemistry of antibody as such is based almost wholly on what has been learnt about the monoclonal myeloma proteins as described in chapter 5. In discussion one finds a uniform agreement to regard a myeloma protein as in fact representing a monoclonal antibody. The justification for this does, however, require some critical consideration if use is to be made of information from myeloma proteins in any attempt to interpret the origin of immune pattern.

If, as is necessary in the present state of knowledge, we concentrate on Ig G myeloma proteins, normal immunoglobulin and

438

antibodies, the aspects in which myeloma protein resembles the (heterogeneous) normal immunoglobulin or partially purified antibody from a typical antiserum are as follows:

(*a*) All can be shown to have the classical Porter structure with two light and two heavy chains.

(*b*) The antigens κ, λ and γ are present in the appropriate chains and subject to the phenotypic restrictions seen in myeloma proteins, and all the allotypic antigens can be demonstrated in all three types of material.

(*c*) In mice (Bennett *et al.* 1965) the L chains of three mouse myeloma proteins showed ± 19 common peptides and 4–5 variable ones. The common peptides were also seen in normal immunoglobulin L chains.

(*d*) All Fc chains in man show similar peptide maps except for differences associated with Gm(a) and Gm(b) allotypes (Frangione and Franklin, 1965).

(*e*) In heavy chain from normal immunoglobulin or antibody there is evidence of great variability in the same Fd segment where differences between different myeloma proteins are shown (Frangione *et al.* 1966).

For general references and reviews on human myeloma proteins, see Fahey (1962), Franklin (1964*a*), Migita and Putnam (1962), Waldenström (1962), and for murine myeloma, Fahey (1962) and Potter and McCardle (1964).

The vexed question of whether a myeloma protein ever possesses a typical antibody CS seems now to have been answered affirmatively. Some years ago it was shown that a number of pathological Ig M proteins had the reactivity of rheumatoid factor (Kritzman *et al.* 1961; Franklin and Fudenberg, 1964). Metzger (1967) has described another such case and applied a complete range of physical and immunological investigations to the paraprotein. He has shown that the protein has the antigenic and structural qualities of Ig M, that all molecules are reactive with an AD on Fc of human Ig G but not with Ig G of other species, and that the valency is 5 as in Ig M antibodies. There can be no doubt, therefore, that this is an antibody.

Recently, workers in Waldenström's group (Zettervall *et al.* 1966) have found two sera in which the abnormal Ig G protein has the specificity of anti-streptolysin O (AST). One was from a typical case of multiple myelomatosis who showed an AST titre of 500,000 in the absence of any evidence of recent streptococcal infection. There was a typical electropherogram of monoclonal type and a limitation of the AST activity to the band. The second was from a case of monoclonal Ig G gammopathy without clinical characterization. In this instance the AST titre was considerably lower per gram of abnormal protein but it was limited to the abnormal band in appropriate experiments.

An even more striking example has been reported by Eisen *et al.* (1968) in the form of an Ig G myeloma protein which has specific reactivity for DNP (dinitrophenyl hapten). The protein is homogeneous in this respect; its avidity is relatively low but unlike a normal antibody population there is no significant scatter in the degree of affinity for the hapten. Like any Ig G antibody the myeloma protein is bivalent. The behaviour of this example is of special interest in view of the frequent argument against selectivist views that one could not picture the random production in the body of an antibody reacting with non-biological material.

Further extension of those studies should soon allow a much more fruitful approach to the chemistry of the CS.

There is, in addition, a possible approach from a different direction. Osterland *et al.* (1966) have described a rabbit which, after a short intensive course of immunization with a streptococcal antigen, gave an antiserum containing 55 mg/ml of Ig G, most of which was homogeneous electrophoretically and, when isolated, had the quality of specific antibody. Limited immunochemical work was consistent with a monoclonal origin of the protein. The antigen was essentially a polysaccharide composed almost wholly of rhamnose units, and there was some evidence that this type of response could only be obtained with one breed of rabbit. The antibody titre fell rapidly after its peak and there was nothing to suggest that a plasmacytoma had been induced.

Although the evidence is incomplete it is wholly consistent with the hypothesis that in this rabbit a single anomalous cell was responsible for the clone giving the dominant antibody globulin. One presumes that, as in the other rabbits of the group immunized, small amounts of other antibody populations were produced, but that these were insignificant in relation to the main clone. Preliminary work was said to indicate that one rabbit in 'five or ten' New Zealand red rabbits but none so far tested of New Zealand whites gave antibody of this type. From our point of view this is of special interest as indicating (*a*) that the capacity to produce an anomalous cell of the required type has some genetic component and (*b*) that there is also an individual quality, presumably the emergence of the appropriate somatic mutation.

### Differences in amino acid constitution of purified antibodies

No amino acid study has yet been reported on any 'monoclonal' antibody produced in an experimental animal but Koshland's review (1966) of her work on purified rabbit antibodies indicates that there are real differences between two antibodies. Recent studies have all been done on rabbits of a single ($a^1b^4/a^1b^4$) allotype constitution and two widely different haptens, ARS and LAC, on suitable carriers used for immunization. Antibody was 'harvested' after a brief course only of immunizing injections to minimize heterogeneity.

The new findings concern essentially the amino acid content of the light chains and of a fragment obtained from heavy chains by the use of cyanogen bromide, which breaks the chain at a methionine residue close to the papain cleavage point and gives as a major fragment what appears to be essentially the whole Fd portion of the heavy chain. In table 13 (p. 420) the significant findings were summarized. They indicate that the whole of the variation in amino acid content is located in the light chain and the Fd segment of the heavy chain. There is, however, no correspondence between the difference detected in the light chain and that in the corresponding Fd fragment. This excludes the possibility that the variable parts of light and heavy chain are in fact two identical polypeptide chains. Singer and Doolittle's (1966) evidence for virtual identity of the light and heavy chain constituents of the CS still holds but it is evident that this can involve only a small proportion of the variable area.

GENETIC ASPECTS OF IMMUNE REACTIVITY

It is well known that individual animals or human beings may show sharp differences in their ability to produce certain anti-bodies. Detailed studies with pure line strains of animals have rarely been published. Recent papers using pure line guinea-pig strains 2 and 13 have shown interesting individuality between the two lines, using on the one hand beef insulin and, on the other, synthetic polypeptides as antigens. Finn and Arquila (1965) found that strain 2 pigs could produce antibodies reacting with ADs of beef insulin with which strain 13 antibodies failed to react. Hybrid $F_1$ 2 × 13 guinea-pigs did not show reactivity to any other regions of the molecule than those reacting with one or other of the parent strains. Ben-Efraim and Maurer (1965) used wholly synthetic polypeptides and compared the same inbred strains 2 and 13 with outbred Hartley guinea-pigs. The capacity to produce antibody against seven polypeptides each derived from either two or three amino acid species is shown in a simplified table. The essential features are as seen with insulin as antigen. Each pure line has a limited range of patterns, the heterozygous animals being more versatile.

TABLE 15. *Reactions of pure line (2 and 13) and random bred Hartley guinea-pigs after sensitization with synthetic polypeptides*

|  | Polypeptides | | | | | | |
| --- | --- | --- | --- | --- | --- | --- | --- |
| Guinea-pigs | 1 | 2 | 3 | 4 | 5 | 6 | 7 |
| Strain 2 | + | − | − | + | − | − | − |
| Strain 13 | − | − | + | − | − | − | − |
| Hartley | + | + | + | + | + | − | + |

Data from Ben-Efraim and Maurer (1965).

Earlier, Levine *et al.* (1963) had shown that using both DNP and benzylpenicilloyl as haptens on poly-*n*-lysine carrier, all strain 2 guinea-pigs reacted while none of strain 13 did. In mice, some-what similar differences have been observed by McDevitt and Sela (1965). They used CBA and C57BL strain mice and a synthetic multichain polymer as antigen in Freund's complete adjuvant. The antibody response in CBA was very poor but detectable while in C57BL the response was more than ten times as great. In $F_1$ mice there was a wide range of response intermediate between those of the parent strains.

Pinchuck and Maurer (1965) found sharper 'all or nothing' differences in different pure strains of mice using a glu–lys–ala polymer GLA 5. Of six mouse strains, three showed antibody in all mice tested, while three others failed completely to respond. About half of a group of outbred Swiss mice failed to produce antibody.

The most important feature of these findings is that there are some animals which cannot produce antibody against an AD to which other animals of the same species can actively respond. This is something to be kept very much in mind in seeking a genetic origin of immune pattern. At the same time it is necessary to remember that the capacity to produce antibody corresponding to a given AD also depends on the nature of the protein or other carrier with which the AD is associated. This is discussed elsewhere (p. 144) in relation to the finding of Green *et al.* (1967 *a*) that a polypeptide–DNP complex might be antigenic in line 2 guinea-pigs but not in line 13, yet an equivalent anti-DNP-polylysine antibody be produced in the non-responders when BSA was complexed to the hapten.

## THE PROS AND CONS OF INSTRUCTIVE THEORIES OF ANTIBODY SYNTHESIS

In the last presentation of an instructive theory of antibody formation, Karush (1958) argued that the newly produced polypeptide chain was moulded into shape, i.e. into a definitive three-dimensional configuration, against a template of antigenic determinant. The configuration induced by the template determined how intramolecular disulphide and hydrogen bonds should be formed to stabilize the configuration. Once this was completed the antibody in some way liberated itself from the template.

Recent work on a number of well-defined proteins, including particularly ribonuclease, has led Anfinsen and his group (Anfinsen *et al.* 1961; Goldberger and Anfinsen, 1962; Epstein *et al.* 1964) to conclude that the tertiary structure of a protein is determined uniquely by the amino acid sequence of the primary polypeptide chain. The chain will take the three-dimensional configuration (including disulphide bonds) that is associated with the lowest configurational free energy. Ribonuclease has four intramolecular S—S bonds and, following reduction and unfolding in 8M urea, when no disulphide bonds persist, it can be re-oxidized to give enzymatically active protein with correct physical qualities (Anfinsen and Haber, 1961). Essentially similar results have been obtained for lysozyme, chymotrypsinogen, alkaline phosphatase and swine pepsinogen (Epstein *et al.* 1964).

Direct study of immunoglobulins and fragments have shown that in all probability this holds for antibody. Buckley *et al.* (1963) used Porter fragment I from a rabbit immune serum which showed univalent attachment to the antigen BSA. This was unfolded completely in 6M guanidine and, on controlled dialysis, complete reversal occurred with between 66 and 83 per cent retention of specific binding capacity in different experiments. In similar experiments Haber (1964) used fragment I of a rabbit anti-ribonuclease with six disulphide bonds. Complete reduction by mercaptoethanol and 6M guanidine gave full uncoiling as checked by ultraviolet rotatory dispersion. On dialysis and oxidation, up to 56 per cent recovery of binding capacity was obtained, indicating that 'the unique arrangement of disulphide must be determined by sequential information'. More recently, Freedman and Sela (1966) have found that by conjugating poly-DL-alanine to antibody, complete denaturation can be followed by reversal to more than 25 per cent binding power for specific antigen.

This evidence can be accepted as eliminating any possibility that antibody diversity arises by any instructive modification of a 'basic' immunoglobulin pattern common to all antibodies. The origin of that diversity must therefore be sought at the genetic level.

## THE EVOLUTION OF SPECIFIC PATTERN IN PROTEINS

Of the proteins which are readily amenable to amino acid sequence analysis, several can be obtained from a wide range of organisms so that some conclusions can be attempted in regard to how protein patterns developed in the course of evolutionary history.

Haemoglobin and myoglobin have been studied in greatest detail, particularly since the nature of sickle-cell anaemia and the other haemoglobinopathies have been recognized (see review by Zuckerkandl and Pauling, 1965). Normal human and other mammalian haemoglobins contain four chains in man, two $\alpha$ and two $\beta$, with foetal haemoglobin having two $\alpha$ and two $\gamma$. Myoglobin with 153 residues is clearly homologous to the single chains, $\alpha$ with 141 residues, $\beta$ and $\gamma$ with 146. The complex molecules have arisen by duplication of a gene and, once this has occurred, each of the genes becomes subject to independent mutation and evolution (Granick, 1965). Myoglobin now has a very different function from the $\alpha$-globin chain. Duplication of genes is probably one of the most important processes concerned with the refinement of functional adaptation during evolution, and Zuckerkandl (1965) points out that one of the advantages may be implied from the fact that in the earliest foetal haemoglobin in man there are two $\epsilon$ chains which are replaced first by two $\gamma$ and finally by two $\beta$ chains. There are strong possibilities, as discussed on p. 447, that the process of gene duplication has been important in the evolution of the immunoglobulins.

In all the globin chains there are a number of invariable key residues but as more and more haemoglobins and myoglobins are analysed the number has been steadily falling and, in 1965, according to Zuckerkandl and Pauling, stood at only 11 residues out of 140–155. This has some interesting implications when one remembers that in evolutionary changes the important matter is not the fact of mutation but the ability of a given mutation to allow adequate function of the cells concerned and eventually to replace the previously existent sequence as that characteristic of the species. What is specially surprising is that the number of alternative residues known for each site gives a distribution very close to a Poisson distribution with a mean of 2·1 alternatives per site.

445

Despite the requirements as to how viable sequences must be reached, the end result seems to be essentially a random one.

It is probable that only in fairly closely related forms can any information as to the sequence of change be established. One example is the work of Doolittle and Blomback (1964) on the fibrinopeptides split off from fibrinogen by the proteolytic action of thrombin which attacks only arginine–glycine bonds giving peptides A and B of 13–21 residues. In examining various ungulate mammals of reasonably well-known evolutionary history, Doolittle and Blomback calculate that an effective change in any given position may have occurred once in $3 \times 10^6$ generations. When one considers the large number (at least twenty-two) of abnormal human haemoglobins now known and the possibilities of genetic drift, it is obvious that attempts to lay down an evolutionary time scale for changes in amino acid sequences are hardly likely to be useful.

## GENETIC POSSIBILITIES FOR THE ORIGIN OF IMMUNE PATTERN

The following possibilities have been discussed for a genetic origin of diversity in both amino acid sequence and immune pattern.

(*a*) That the diversity results from the existence of a very large number of genetic loci that have arisen (perhaps by repeated duplication during evolution) and are transmitted to succeeding generations by the ordinary rules of genetics. The diversity of pattern within a given individual is generally considered to result from some somatic interaction within the genome by which a wide range of patterns can be generated through recombination between the duplicated cistrons.

(*b*) That diversity arises by accelerated somatic mutation in the cistron or cistrons concerned in the control of the structure of the CS.

(*c*) That diversity arises by some process which modifies the translation rules so that a certain pattern of m-RNA can give rise to many alternative patterns of polypeptide.

There are some requirements common to any approach. The

most important is the existence of phenotypic restriction by which at least most cells (and possibly all) produce only one type of immunoglobulin or antibody. The second is that there are inheritable differences in the capacity to produce antibody against specific antigens. The third is that, particularly in relation to avidity, there may be extremely large numbers of fine differences within one general type of antibody. Each of these three aspects justifies separate discussion in order to clarify the situation before discussing possible genetic mechanisms. It will become progressively more evident that none of the possibilities are mutually exclusive.

## Direct genetic interpretations of immune pattern

There can be no doubt that the antigenic qualities of immunoglobulin, as far as they have been investigated, conform to the normal rules of inheritance with the single important difficulty of the sharp phenotypic restriction seen in isolated mature plasma cells, myeloma protein and, with less complete evidence, in antibodies.

There is therefore good reason to look for as normal a genetic origin for immune pattern diversity as the facts will allow. Putnam *et al.* (1966b), for instance, present one of their alternative interpretations of the variable segments as 'resulting through the presence of many modified antibody genes resulting from successive duplication and mutation over many generations'. Dreyer and Bennett (1965) ascribed the variability to the incorporation into the light chain locus of any one of a large number of DNA rings that have arisen by gene duplication from a common gene ancestor.

The considerable number of invariant positions in the variable segment of the light chain (Lv) almost limits the possibilities to somatic mutation within a cistron or multiple duplications of the Lv locus with some mechanism for selection of one of the alternative rings or segments of DNA (see fig. 21). The popularity of the duplication approach almost certainly springs from the success of such an interpretation for the evolutionary history of the globins (Zuckerkandl and Pauling, 1965; Ingram, 1961). Zuckerkandl and Pauling refer to two types of duplication: (*a*) intralocus multiplication of some or all genes in the course of cell differentiation and

447

(*b*) interlocus duplication of individual genes or duplication of whole individual chromosomes in the course of evolution. Process (*a*) was assumed by Itano (1957) to account for quantitative differences in the proportion of normal and pathological haemoglobins in heterozygotes. It could be thought of as a means for speeding protein synthesis involving different genes in different tissues. Some support for the idea is obtainable from the phenomena of 'puffs' in defined regions of polytene chromosomes in cells engaged in protein synthesis. In these there is active DNA synthesis going on (Pavan, quoted in Sager and Ryan, 1961). No mention of somatic mutation differentiating one of such duplicated units from another is made and there seems to be no clear relevance to immunological processes.

The long-term evolutionary process (*b*) is regarded as the only way in which to account for isogenes, i.e. non-allelic, structurally different, yet homologous structural genes. To account for the origin of the $\beta$, $\delta$ and $\gamma$ haemoglobin chains in man and cattle it is necessary to postulate three sets of gene duplication.

In the Ig G molecule there are insistent suggestions of evolutionary duplication. The Lv and Ls segments of the light chain divide it accurately into two 107-residue sequences. There are common sequences in the K light chains and the Fc portion of the Ig G heavy chains. This has led to the suggestion that a series of duplications of a single gene could provide an indefinite number of cistrons coding for a length of 100–110 amino acid residues (Hill *et al.* 1966; Singer and Doolittle, 1966; Edelman and Gally, 1967). From this array, perhaps in most cases after somatic recombination between cistrons concerned with the Lv segment (Smithies, 1967; Edelman and Gally, 1967) two units are selected for expression in the light chain, four for heavy chain in Ig G. It is implicit that the other types of heavy chain are built up from similar units. This general approach can be speculatively developed in a number of different ways. The crux of the problem is, of course, to account for the diversity of immune pattern in a single individual.

From the discussions at the 1967 Cold Spring Harbor Symposium, 'Antibodies', one gathered that there was no longer significant support for the view that all possible patterns were each specifically present in the genome from conception. The view was

448

rather that from a structure of duplicated cistrons somatic recombination in one form or another could provide the necessary heterogeneity. This recombination was pictured as occurring during the embryonic development of the lymphoid cell series.

## The possibility of anomalous translation

There have been suggestions (Weisblum *et al.* 1965; Potter *et al.* 1965) that the variable sections of light and heavy chains arise by an anomaly of translation rather than by somatic mutation. If the cistron concerned has, say, 30 of its codons (out of 107) capable of producing one or other of two different amino acids, but at some stage is stabilized to produce one consistently, there would undoubtedly be scope for the appearance of many different patterns. The suggestion presumably is that when a cell differentiates, protein production falls into one pattern which may be any one of $2^{30}$ possible patterns (i.e. $10^9$).

The difficulty of this hypothesis is to devise a process by which systematic error in translation can be perpetuated through successive cell generations. In fact we should almost certainly have to postulate a continuing function of antigen within the cell. Nossal in conversation has suggested to me that one could consider the possibility that if an AD were present in the cell, contact with the 'right' pattern would act as a signal to fix the genetic mechanisms in the phase currently working. This would give a pseudo-instructive mechanism which may well appeal to those who are impressed, for example, with the necessity for antigen to be present if tolerance is to be maintained. An extremely elaborate control mechanism of a type for which there are no biological analogies would be needed and, from the point of view of simplicity alone, one feels that the hypothesis should be disregarded unless no other alternative is effective or until positive evidence in its favour is obtained.

## Somatic mutation

Since the first publication on the clonal selection hypothesis, I have intuitively favoured somatic mutation as the most likely process to generate diversity of immune pattern in the individual, and this preference has never weakened. There is, however, virtually no discussion of somatic mutation in texts on genetics although somatic crossing over and a variety of other rearrangements of genetic material, almost always in heterozygotes, are discussed. This has led nearly all who have considered the matter of antibody heterogeneity from a genetic standpoint to by-pass somatic mutation in favour of some type of rearrangement of genetic material evolved over geological periods.

To justify the stress placed on somatic mutation and specifically single-nucleotide change-point mutation in giving rise to immunological diversity, it seems necessary therefore to recapitulate the evidence for the widespread existence of somatic mutation in mammals. This was discussed with special reference to medicine in Burnet (1965a), in relation to neoplastic disease in Burnet (1957b), and in relation to autoimmune disease on several occasions.

There seems to be no reason why the basic incidence of somatic mutation should differ significantly from the frequency of mutation in germ cell lines. Since, however, there is much to suggest an elaborate process by which the germ cells are protected against mutagenic agents plus the automatic elimination, as lethals, of mutations which would not be lethal in a somatic cell, one could reasonably expect that the number of mutations per cell per replication which could, in principle, be detected might be much larger than any figures obtainable from genetic and evolutionary studies. Even if we take Orgel's (1963) figure of a frequency of error of $\pm 10^{-8}$ per nucleotide per replication there are likely to be $10^3$–$10^5$ point mutations involving each nucleotide pair of the human genome every day. The vast majority of such mutations can have no effect either because they are lethal or because the associated functional change is trivial when it involves only one cell in a million or more in a functioning tissue, or because the mutation involves a region of the genome inoperative in the particular differentiated cell concerned.

The whole key to the understanding of somatic mutation is the necessity that the initial mutation must in some way be 'magnified' if it is to be of any biological significance.

*The process of magnification.* Following the argument in an earlier paper (Burnet, 1965a), the standard way in which somatic mutation, occurring rarely and involving only a minute fraction of any cell population, can have physiological or pathological effects is by proliferation of the descendants of the mutant cell to produce a recognizably large clone of cells. There are several ways in which this can happen. If, for instance, a significant mutation takes place in a cell at an early stage in the primary segmentation of the ovum, then all descendant cells will be mutant, and amongst those cells in which the mutation can have a phenotypic effect a certain proportion, depending on the relationship of the mutant cell to morphogenetic processes, will show the effect in the adult animal.

The classic example is the description by Fraser and Short (1958) of fleece mosaics in sheep. Their survey of some twenty million Australian sheep for the occurrence of patches of long straight wool in animals of breeds on which the wool is normally short and strongly crimped, gave some 30 examples showing a continuous range of involvement from nearly 50 per cent of the skin area to scattered wisps of long wool. In the more heavily involved animals there was a clear indication that about a half, a quarter or an eighth of the wool-bearing area was involved and that the frequency of such examples was inversely proportional to the fraction involved. The simplest interpretation was that the likelihood of this particular form of somatic mutation was about $10^{-7\cdot3}$ per cell per fertilized ovum and that the rate was essentially the same over at least the first 4 or 6 segmentation divisions.

This example at least indicates that somatic mutation can occur from almost the earliest stage of differentiation and adds significant support to the hypothesis that mutational changes involving immune pattern could occur at any stage of development.

Passing to the more general question of somatic mutation in grown animals, a mutant cell will produce a recognizable clone of descendants if the mutant has a clear proliferative advantage over its unchanged congeners. This may be (*a*) because of ability to override local morphological controls, (*b*) by greater metabolic effectiveness in a competitive situation even though still responsive to local controls or (*c*) by its capacity to be stimulated to proliferation by a substance or condition naturally present in the internal environment.

In immunological fields we are primarily interested only in (*c*), but other proliferative processes may have great indirect importance in relation to the immune mechanism. The significance of the production by somatic mutation of new antigenic patterns within the body is, however, best left for chapter 13.

There are, however, at least two other ways by which somatic mutation may be demonstrable. The first is when, by the accumulation of more or less deleterious somatic mutations in a large proportion of the cells in vital tissues, there is a progressive failure of function. This is in fact a summary statement of the somatic mutation hypothesis of ageing. Its experimental support comes largely from the work of Curtis (1963), who showed a progressive increase with age in the proportion of liver cells which, on being stimulated to divide, showed aberrant mitotic figures.

The second is even more indirect. It depends on the reasonable assumption that when a tissue cell has undergone a somatic mutation, e.g. to produce an abnormal metabolite, and subsequently for unrelated reasons takes on neoplastic growth, a mass of mutant cells will arise. Any functional effect of the initial mutation may then become apparent. One can mention the fairly considerable number of hepatomas which have shown some unique metabolic anomaly (Dent and Walshe, 1953; Schapira *et al.* 1963). In all instances of tumours which show metabolic abnormalities there must always be some doubt as to whether the mutation arose before the malignant change or was in fact a secondary mutation or chromosomal rearrangement associated with the malignant change.

Another field in which evidence for somatic mutation has been claimed is in regard to blood groups. In a number of instances, large populations of human red cells have shown a proportion of inagglutinable cells after treatment with antibody or lectin which should agglutinate all cells of standard phenotype. Atwood (1958) working with blood from an $A_1B$ donor obtained exceptional cells with reactions corresponding to $A_2B$ and B. Hraba and Májský (1963) found that such findings seemed to be more frequent in patients with blood diseases, myeloid leukaemia and Hodgkin's disease. Scheinberg and Reckel (1960) caused an increase of inagglutinable erythrocytes by exposing pigeons to whole-body X-irradiation. In all these instances the changes are interpreted as somatic mutation or chromosomal lesion in erythropoietic stem cells. Either such cells produce larger numbers of erythrocytes or the mutation has involved a significant fraction of the effective stem cells.

*The frequency of somatic mutation.* Any attempt to elaborate a hypothesis that immune pattern arises by somatic mutation, mostly in the form of single nucleotide replacement, soon comes up against important gaps in knowledge. Two questions of special urgency are: (*a*) What is the approximate frequency of somatic mutation per nucleotide per cell generation? (*b*) Is mutation more frequent in the cistron(s) concerned with immune pattern than elsewhere in the genome? If those could be answered, a third might arise: (*c*) Once a clone of immunocytes has been 'committed', does the frequency of significant mutation decrease? In the present state of knowledge one can probably do no more than assume that the mutation rate is of the same order as is known to occur in organisms such as bacteria multiplying in asexual fashion. Orgel's figure of $10^{-8}$ per nucleotide per replication may be of the right order of magnitude.

It will be evident that although it has been implicitly accepted in this discussion that point mutation is the significant cause of somatic mutation, as it appears to be in those germinal mutations which modify protein pattern, the results we are concerned with could also arise by other somatic genetic processes which could give random heritable changes in a cell line. Somatic mutation must undoubtedly occur but its effectiveness in creating diversity could be greatly enhanced if associated with somatic recombination processes. As yet there is far too little evidence from sequence studies on Bence Jones proteins—there is none on classical antibody—to allow any definitive opinion on the nature of the somatic genetic processes concerned.

In the absence of that evidence, one must recognize the impossibility of deciding what somatic genetic process is involved. As has already been pointed out, this somatic process must be of random quality and capable of producing a very large diversity of patterns in the short period of embryonic development. It must work within the restriction that the variable segments remain of the same length and retain a substantial proportion of constant sequences. On the other hand, since the 'function' of the variable segment is almost wholly concerned with its diversity, many mutations or other genetic changes may be acceptable which on any other cistron would be lethal to the gene or the cell. It must also be continually reiterated that there is always a high degree of phenotypic restriction in every immunocyte line. However many duplicated cistrons there are, once a cell has differentiated to an immunocyte only one set is expressed.

*Brenner and Milstein's hypothesis.* The only genetic hypothesis which attempts to detail the process of somatic mutation responsible for the variability of the N terminal end of the light chain and, by implication, for the corresponding portion of the heavy chain is that of Brenner and Milstein (1966). They accept as obvious that some process involving a portion only of the gene concerned in programming the light chain is necessary to account for its hypermutability. They envisage the process as (*a*) the attachment of a cleaving enzyme capable of cutting a DNA chain at a 'recognizable' point or sequence at about the centre of the 'light chain' nucleotide sequence, (*b*) one of the strands of DNA is cut exposing a 3' end, (*c*) an exonuclease then degrades a length of the cut chain from the 3' end, (*d*) the gap is repaired with the possibility of error by a DNA polymerase and (*e*) on replication there will be conservation both of the original chain sequence and the altered sequence. The process will then go on presumably until a stage is reached at which the clone is genetically stable.

There are adequate analogies in bacterial and bacteriophage systems for the processes postulated by Brenner and Milstein. Excision of thymic-dimers and subsequent repair has been demonstrated in irradiated bacteria by Setlow (1964) and there are precedents for specific recognition of a DNA site for cleavage and for mutant DNA polymerases to produce a high frequency of errors. There is, however, no conclusive evidence for or against the hypothesis and no reason why it should not, if necessary, be associated with somatic recombination.

*Possible implications of the small size of the combining site in antibody*
Most of the information on the size of the CS in an antibody molecule is based on the contribution of Kabat and his collaborators using particularly human antisera to dextran (Kabat, 1966). Since this is a simple polymer of glucosyl units it is possible to use a

series of 2-, 3-, 4- ..., unit molecules, of which isomaltose is the dimer, to test their blocking ability or (when appropriately linked to a neutral carrier) their capacity to precipitate with antibody.

The conclusion was that the 6-mer was held as firmly as the 7-mer or higher and better than any compound lower in the series. Each of the 6 glucose units could therefore be concerned in determining avidity of union thus giving a maximum size of the CS of approximately $30 \times 12 \times 7 \text{ Å}^3$. This is about the size of two parallel peptide chains each of 5 or 6 residues.

Very little attention has been given to the relationship between the small CS and the much larger polypeptide sequence of the variable chains. On several occasions since 1964 I have tried to work out the implications of a hypothesis that the CS of an antibody is in some sense a genetic unit independent of the rest of the immunoglobulin molecule. One of the main incentives was to leave open a possibility of interpreting transfer of antibody pattern from one cell to another at a subcellular level. Although I at present prefer a different combination of hypotheses, I do not feel that the possibility of a specific genetic origin of the CS has been wholly eliminated. To bring such a view in relation to the variable segments of light and heavy chains with their presumptive influence on the CS, a fairly elaborate hypothesis is necessary including a picture of the small CS being a duplex structure in which the light and heavy chain contributions are identical or at least approximately equivalent. We can start with four premises.

(*a*) Amino acid sequence variation is the source of diversity in the CS.

(*b*) The CS is included in both light- and heavy-chain variable sequences.

(*c*) The CS is a duplex structure with 7–10 amino acid residues in congruent sequence on both chains. It must therefore arise by duplication of a linear sequence, presumably of DNA.

(*d*) Other portions of the light and heavy variable chains show no such correspondence.

In so actively developing a field, any hypothesis is at most only going to have provisional usefulness and the following is presented with that qualification.

The variable segments Lv and Hv are both derived from a single DNA sequence $3^1$–$5^1$ in which it is assumed (*a*) that the sequence is hypermutable and has been accumulating diversity through its somatic history, (*b*) that a short sequence of 20–30 nucleotides near the centre of the chain is duplicated, presumably at the time of primary differentiation to the immunocyte and (*c*) that the complex is then broken to give two DNA chains, $3^1$–A–B and A–B–$5^1$, which code for Lv and Hv respectively. The light and heavy chains are constructed in association with DNA coding for the stable portions of the two chains. The complex of polypeptide chains is now produced and the immunoglobulin molecule constructed. In the process of developing tertiary structure it is assumed that the two sets of congruent sequences A–B come to lie in two duplexes at the actual or potential ends of a symmetrical macromolecule (see fig. 19).

In place of the hypothesis that A–B is part of both the light chain and heavy chain cistrons, we could assume that there are three functionally distinct regions of the genome which are subject to the hypermutability process—these are A–B of 30 nucleotide pairs, Lv of 300 and Hv also of 300. In addition, of course, we have Ls and the Fds + Fc segment of heavy chain which between them contain all the allotypic markers and which can be neglected in the present context. Confining ourselves to the 'variable' components, all three must be used to code for the corresponding polypeptide chains and some mechanism at present unknown will be needed to allow their co-ordinated incorporation in the final macromolecule. The evidence is compatible with the simplest assumption that separate m-RNA molecules are involved and that as the chains are produced their intrinsic qualities plus the availability of appropriate enzymes result in the almost immediate construction of the immunoglobulin, with any chains in excess being discarded.

There is uniformity in the statement that in subcellular studies of the process there is an excess of light chains as recognized either antigenically or by physical means (Voss and Bauer, 1966; Shapiro *et al.* 1966*a*, *b*; Askonas and Williamson, 1966). The simplest explanation is that presented by Shapiro *et al.*, that if there were equal numbers of 'light' and 'heavy' chain cistrons which were transcribed and translated at the same rate, about twice as many light chains would be produced. To this we could add, 'and much more than the needed number of short A–B chains', should there be such a genetically distinct entity.

The problem of incorporating such peptides in the appropriate position within both light and heavy chains is perhaps no more intrinsically difficult than to put the oligosaccharide units on to the right place in the Fc segment. Undoubtedly there are more attractive hypotheses and the present one is only likely to come up for reconsideration if unequivocal subcellular transfer of immunological pattern can be established.

## THE SIGNIFICANCE OF AVIDITY IN RELATION TO IMMUNE PATTERN

Strictly speaking, the relationship between an AD and the CS on an antibody molecule or a cell receptor is expressible in terms of an association constant for the reaction $[CS] + [AD] \rightleftharpoons [CS.AD]$ which may be referred to as the avidity of the reaction. Only when the value exceeds a certain threshold will the interaction be recognizable as a specific immune reaction.

In discussing avidity, we accept the contention of earlier chapters that the differentiation of stem cell to immunocyte allows the synthesis of at least those immunoglobulin chains needed to produce a CS, and that the configuration of the CS is determined by a cistron (or by more than one) within which there has been an accumulation of random somatic mutations (nucleotide replacements). As will be more fully discussed in chapter 8, if the pattern of CS as it emerges is wholly random, it must necessarily follow that for any given AD there will be a range of cells with avidity from nil or negligible to the maximum that is physically possible within the limitations of the peptide CS. Their distribution will probably be as shown in fig. 6 (p. 35, Book I).

It must be recognized that while the shape of the curve in fig. 6 will be approximately the same for any AD tested against the whole range of immune pattern on newly differentiated immunocytes, there may well be important irregularities in the 'tail' of the distribution curve. Figure 33 is designed to show two aspects of this irregularity. Among the rare immunocytes with significant avidity, i.e. those to the right of point $X$, there may well be very rare patterns of uniquely high avidity for AD $A$ shown by the discontinuous termination of the tail. If we also have in mind a distantly related AD, $B$, this tested similarly would give a similarly shaped distribution curve but the detailed distribution of immunocytes (or immune patterns) would be quite different. If we could detect immunocytes which would be reactive with $B$, i.e. would lie to the right of $X$ on a distribution curve of avidity for $B$, they might well be scattered over the $A$ distribution as shown. If the cell shown at $D$ carried a pattern which 'by pure chance' was approximately equally avid for $A$ and $B$ it could well be scored as a 'double producer' in experiments with doubly immunized animals.

So far we have been considering distribution of the relevant immunocytes in an unimmunized animal. After immunization with $A$ there will be proliferation of cells to the right of $X$ and we can assume that a certain number of representatives of all the resulting clones will persist as memory cells. If now the animal is immunized

457

with antigen *B* it is obvious that the clone from cell *D* would provide a high proportion of the resulting antibody and a correspondingly high level of cross-reactivity with *A*. An example from Fazekas and Webster's (1964) work is described on p. 463.

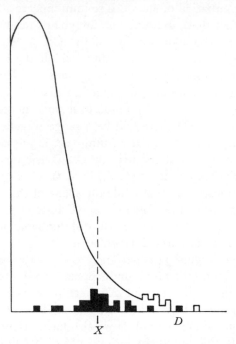

Fig. 33. Diagram based on fig. 6 to illustrate the behaviour of cross-reacting antibodies.

The main curve suggests the distribution of individual antibody molecules or immunocytes in terms of their affinity for antigenic determinant *A*. Only those to the right of line *X* are of sufficient avidity to react immunologically at a demonstrable level.

The distribution of reactive units for a related antigenic determinant *B* might be as shown by the black squares. The single immunocyte *D* highly avid for both will dominate the situation after double immunization with *A* and *B*.

## GENETIC INTERPRETATION OF IMMUNOGLOBULIN STRUCTURE

For fairly obvious reasons, almost all genetic study of the immunoglobulins has made use of antigenic characters. In man, study of the Gm allotypes has played a major part but, in addition, there are the K and L types of light chain and the specific differences between the $\gamma$, $\mu$, $\alpha$ and $\delta$ antigens of the heavy chains. As has been mentioned earlier (p. 419) there is a complex accumulation of Ig G molecules defined at the antigenic level in all human individuals. There are four primary antigenic types, all based on qualities of the heavy chains. These are We, Ig G1; Vi, Ig G3; Ne, Ig G2 and Ge, Ig G4 (Litwin and Kunkel, 1967). The group Ig G1 is responsible for 60–70 per cent of Ig G and for most of the Ig G myeloma proteins. The Gm specificities associated with this are Gm(a), (z), (f) and (y) while Ig G3, the next most frequent, has Gm(g) and a range of (b) specificities. Only one specific antigen, n, allows differentiation of Ig G2 subtypes. Of the G1 ADs, (a) and (y) are located on Fc, (z) and (f) on the Fd segment.

In the paper by Litwin and Kunkel (1967) a tabulation is made of the various common combinations of antigens found in Caucasoid and Mongoloid individuals as well as some rare anomalous combinations. Table 16 is an abbreviated version of this in which b.. is used in place of the complex set of subtypes that are on record for Ig G3. Although Litwin and Kunkel interpret their results according to normal genetic rules, there is no clear way of associating their results with the various possibilities which have been mooted to account for specific immune pattern.

Since all normal human sera contain K and L light chains and seven heavy chain types, Ig G (We, Vi, Ne, Ge) as well as those characteristic of Ig A, Ig M and Ig D, an individual heterozygous at all the loci concerned should have the potentiality of producing any one of 56 different immunoglobulin molecules (Mårtensson, 1966 a).

Perhaps the most significant feature in the genetics of immunoglobulins is the intense phenotypic restriction shown in all myeloma proteins and in all probability equally evident in pure clones of plasma cells and the antibodies they produce. A given plasma cell produces as a rule only one type—A, G or M—although there is a possibility that a single cell may change from M to G and at an intermediate stage show evidence of both (Nossal *et al.* 1964 c; Curtain and Baumgarten, 1966). It produces only Gm(a), Gm(b) or neither on heavy chain and only $\kappa$ or $\lambda$ antigens on the light chain (Pernis and Chiappino, 1964). The general opinion now is that the great majority of antibody-producing plasma cells produce

459

only one type of antibody, although it has proved difficult to exclude the possibility that equal amounts of antibody and of unidentifiable immunoglobulin of the same type might be produced (see Askonas and Humphrey, 1958).

TABLE 16. *The main Gm specificities seen in Caucasoid and Mongoloid individuals*

| Ig | Caucasoid | | | Mongoloid | | |
|----|-----------|-----|-----|-----------|-----|-----|
| | | | | | | (a) |
| G1 | (a) | (y) | (y) | (a) | (a) | (y) |
| | (z) | (f) | (f) | (z) | (z) | (f) |
| G3 | (g) | (b..) | (b..) | (b..) | (g) | (b..) |
| G2 | — | (n) | — | — | — | (n) |

Vertical rows show the common combinations found in individuals (from Litwin and Kunkel, 1966).

SOMATIC SEGREGATION AND THE LYON PHENOMENON

As has already been discussed, one of the most interesting genetic features of the mammalian immunoglobulins is the high degree of phenotypic restriction shown in immunoglobulins and in mature plasma cells. Irrespective of whether immune pattern arises by genetic factors undergoing some type of random selection for expression or as the result of somatic mutation during ontogeny, all immunocytes in humans can be expected to be heterozygous for a large number of the possible alleles. In some way, only one allele at each locus comes to phenotypic expression. In a real sense this is part of the problem of differentiation within any line of somatic cells, and at present there is little to be done beyond accepting the intense phenotypic restriction as a necessary aspect of immunocyte behaviour. Brief mention should, however, be made of two phenomena which may have some relevance. These are random allelic inactivation (the Lyon phenomenon) (Lyon, 1961) and somatic segregation.

*The Lyon phenomenon*

In mammalian females, only one X chromosome is functionally active in any somatic cell; the other is, as it were, segregated and appears in the interphase as the chromatin mass known as the Barr body. The 'choice' of which chromosome is inactivated is made apparently wholly at random early in embryonic life and each cell line thereafter has the P or M X chromosome selected for the rest of life. The female body is therefore a 50:50 mosaic of the two cell types. Very direct evidence for this concept can be obtained when individual cells can be examined for a sex-linked quality such as the presence or absence of glucose 6-phosphate dehydrogenase.

## Somatic segregation

This is postulated to involve the fusion of two cells and subsequent separation with a rearrangement of an autosome pair from MP MP to MM PP, if it takes place in uniformly heterozygous material. If it were to occur between cells which had undergone somatic mutation or some equivalent genetic process, as would be possible or probable in an immunological context, the results could be more complex:

$$M^1P^1\ M^2P^2 \rightarrow M^1P^1\ M^2P^2,\ M^1P^2\ M^2P^1,\ M^1M^2\ P^1P^2$$

and would not serve effectively to account for phenotypic restriction in immunoglobulins.

The process of somatic segregation can be directly demonstrated in animals with heteromorphic chromosomes. This is specially characteristic of the deer mouse (*Petromyscus*), and Ohno *et al.* (1966) have recently described a situation in which one autosomal chromosome pair may have either acrocentric (A) or subterminal (S) elements. When an AA male was mated with an SS female the offspring showed AS. In the mitosing cells in the spleen, however, both paternal and maternal homomorphic chromosome pairs were found in a ratio 19 AS:3 AA:1 SS. Essentially similar findings were obtained in nine $F_1$s of this cross, the paternal homomorphic pairs being regularly more frequent than the maternal ones. Ohno (1966) sums up the situation in the subspecies of *Petromyscus* studied as indicating:

(*a*) Somatic segregation occurs in mammals, perhaps particularly in immunoglobulin-producing cells.

(*b*) Somatic segregation involves some autosomes only, not all of them.

461

(*c*) Complementary products of segregation do not occur with equal frequency; one often has a selective advantage over the other.

More extensive evidence for somatic segregation is available in fish, but this is hardly relevant in the present context.

## CHANGES IN IMMUNE PATTERN ON SECONDARY STIMULATION

It has often been noted that when an animal which has developed antibody after natural or experimental immunization is exposed to a related antigen, the secondary response is distinct from what would be obtained from either antigen alone. Much of such information has been gained from field studies of virus disease. There is an important group of arboviruses, Type B, which includes yellow fever, Japanese and Murray Valley encephalitis viruses, as well as others of lesser virulence for man. In the course of studies, particularly of yellow fever, much information has been accumulated on cross-reactions by which serum from an individual infected with one virus will neutralize not only the homologous virus but usually to a less extent a number of other viruses which for this reason are regarded as antigenically related. In this discussion we can disregard the complications introduced by the existence of haemagglutinin inhibition tests and complement-fixation tests which give different indications of the degrees of relationships amongst viruses and concentrate on the capacity of antibody to destroy the infectivity of living virus, as the most direct measure. Omitting all technical details we may say that a series of six viruses, known to be related because of cross-reactions in anti-haemagglutinin tests, may show no cross-neutralization when tested by monovalent sera from animals that have been infected by only one virus. When, however, an animal once infected with virus $A$ is now infected with virus $B$, the serum taken afterwards will neutralize not only $A$ and $B$ but also, to a significant extent, $C$, $D$, $E$ and $F$ as well (see Theiler and Casals, 1958). Another example from viral immunology is presented in studies by Fazekas and Webster (1964) on cross-reaction between influenza virus strains.

Fazekas's findings are, essentially, that when a rabbit is immunized with SW virus and subsequently with FM1, it shows high titre both with SW and FM1. Single immunization with either shows relatively specific response with moderate cross-reaction. If such a serum from a doubly immunized animal is absorbed either with SW or FM1, the absorbed serum and the eluate have the same SW:FM1 ratio and have a much less wide range of avidity than normal antisera. He considers that only a small sector of antibody-producing cells can be concerned in producing the response.

On the interpretation being examined we must assume that on first exposure of the animal to FM1 there are small numbers of cells, each carrying one of a wide range of patterns capable of reacting significantly with one of the FM1 ADs. Among those which can react positively and produce memory cells will be some rare variants with high avidity and a wide range of reactivity with related, but not identical, ADs. In general terms, if we have a series of ADs, *A–E*, with difference from *A* increasing successively, we might expect a thousand antibody-producing clones stimulated by *A* to include numbers capable of reacting with related ADs which became smaller the more distant the AD structure, for example:

$$A \ 1{,}000 \quad B \ 900 \quad C \ 200 \quad D \ 10 \quad E \ 1.$$

The requirements for $A + D$ or $A + E$ would be much stricter than for $A$ alone and $A + B$, and it is quite possible that *only one* pattern is suitable to give an equal avidity for both patterns. There may thus be only one clone available for secondary-type stimulation when the animal is immunized with $D$ or $E$ (see fig. 33). The antibody population then will be as homogeneous as a myeloma protein unless the antigenic stimulus is adequate to produce a primary response as well. These findings, however, differ sharply from those of Jensen *et al.* (1956), who immunized ferrets successively with WS, PR8 and FM1. At each stage, WS absorbed out all reacting antibody but the other results were as shown in table 17.

On such findings we must assume that with the antigen dosage used, effectively all the clones stimulated by $B$ and $C$ had been committed to $A$ but, as in the previous example, relatively small numbers of the $A$-produced clones and their selected descendants were $A + B$, $A + C$, or $A + C + B$. If Fazekas's pair SW and FM 1 were at the virtual limit of cross-reaction so that only one possible avid SW + FM 1 combination was possible, all doubly avid clones would be identical and give his results.

TABLE 17. *Schematic representation of findings by Jensen et al. (1956) in ferrets sequentially immunized with three influenza A strains, A, B and C*

| | Unabsorbed | | | Abs. A | | | Abs. B | | | Abs. C | | |
| --- | --- | --- | --- | --- | --- | --- | --- | --- | --- | --- | --- | --- |
| | A | B | C | A | B | C | A | B | C | A | B | C |
| Primary | 100 | 5 | 0 | 0 | 0 | 0 | 100 | 0 | 0 | 100 | 0 | 0 |
| Secondary | 150 | 150 | 5 | 0 | 0 | 0 | 100 | 0 | 0 | 150 | 100 | 0 |
| Tertiary | 200 | 150 | 200 | 0 | 0 | 0 | 150 | 0 | 50 | 200 | 100 | 0 |

Anti-haemagglutinin titre of sera before and after absorption are shown.

In a different field we have the well-known experiments of Weigle (1961) in which animals formerly immunized with BSA were now injected with the rather distantly related human serum albumin (HSA). Analysis of the response showed that in addition to monovalent anti-BSA and anti-HSA there was antibody which could be adsorbed by *either* BSA or HSA, thus differing from antibody present in singly immunized animals.

As has been pointed out in discussion of avidity differences, in all such instances it is very difficult either to demonstrate or exclude the possibility of secondary mutation within established clones of immunocytes.

## FISHMAN'S APPROACH TO THE TRANSFER OF IMMUNOLOGICAL INFORMATION

In a series of papers, Fishman and his collaborators (Fishman, 1961; Fishman and Adler, 1963; Adler *et al.* 1966) have shown that when a bacterial virus as antigen is incubated with peritoneal exudate (PE) cells, such cells can give rise to RNA (in the form of the conventional phenol extract) with immunological properties. By adding such an extract to lymph node cells from a normal animal, small amounts of specific antibody as judged by low-titre phage inactivation are produced.

The results can best be discussed in terms of the most recent paper from the group, by Adler *et al.* (1966). In this, rabbit cells were used instead of rat cells and evidence is given that small amounts of Ig M antibody produced early by the lymph node cells have the allotypic specificity of the PE cells while the later Ig G antibody has the specificity of the lymph node cells.

On the surface the paper suggests that when phage is incubated in a mixed population of rabbit macrophages and lymphocytes, a proportion of immunocytes are stimulated to produce m-RNA corresponding to L chain and combining site. This when it is taken, presumably by pinocytosis, into a normal cell synthesizing Ig M-type globulin allows the production of a proportion of hybrid molecules with PE donor allotype and antiphage specificity. Simultaneously there is evidence of the production of a potent immunogen with phage specificity which stimulates appropriate immunocytes to specific activity. The second phenomenon is generally accepted. The first presents major difficulties if it is to be fitted into the pattern of immunoglobulin production.

The only comparable results are those of Mannick and Egdahl (1962). They used the transfer (erythematous) reaction seen in donor rabbits when injected with lymph node cells from a homologous animal draining a homograft of the donor's skin. Such cells gave twelve out of thirteen positive reactions in the donor and six out of thirteen in neutral rabbits. Neutral cells gave two out of twenty positive, but after these had been treated with RNA from immunized animals, seventeen out of twenty positive results were obtained. This on face value suggests that m-RNA for immune pattern was transferred and caused the formation of specific combining sites. Alternatively in immune lymphocytes the receptors are (or readily become) intimately associated with RNA and in the absence of that RNA are immunologically inert, as suggested by Janković and Dvorak (1962). It has also been found by Jones and Lafferty (1966) that the normal lymphocyte transfer reaction in sheep can be shown equally well by extracting crude RNA from homologous lymphocytes and inoculating it with bentonite. Neither cells nor the RNA extract produce a reaction in the donor.

A highly speculative possibility is that peptides carrying combining site activity and peptides or peptide–hapten units with AD potentiality are both liable to form complexes with RNA which survive conventional phenol extraction and retain a variety of immunological activity. This, however, is not adequate to account for Adler and Fishman's results. If these are confirmed we must either look for some unrecognized artefact or assume that there is a means by which information adequate to produce a specific combining site can be transferred, presumably in m-RNA. This virtually demands that the information is carried in part of the genome coding for only one of the chains or for a hypothetical genetically independent specific chain coding for the CS only (see p. 455).

As a pure exercise in speculation one might extend the idea to assume that when an immunocyte was stimulated to antibody production it was able to confer by liberation of excess m-RNA for the specific chain a capacity for any other neighbouring cell to initiate a short-lived production of Ig M antibody, all the other qualities of the antibody being specified by the cell's own genome. There are claims to increase the numbers of plaque-forming cells against sheep red cells by this type of procedure (Cohen *et al.*, 1965; Friedman, 1965 *b*), but in private discussion Jerne has stated that he has not been able to confirm this despite strenuous efforts to do so.

It is hard to avoid the feeling that in all these alleged or established instances of transfer of information by subcellular processes we are dealing with something not part of normal physiological processes. There are undoubtedly minor accidents—inevitable overproduction of components, partial breakdown of parts, piling up of unused raw materials—in any cell as in any human factory. In experimental situations devised to detect minimal significant events, genuine phenomena may be brought to light which with fuller understanding may prove to have no bearing whatever on the genetically programmed process which has evolved to meet the needs of the organism. The trimmings of cloth from a garment factory may have many virtues for making patchwork quilts or for paper-making but neither of these is relevant to the social function of the factory.

# 7 The role of macrophages, eosinophils and other auxiliary agents in relation to adaptive immunity

## MONOCYTES AND MACROPHAGES

In this chapter no attempt whatever will be made to go beyond those aspects of the subject which are immediately relevant to the nature and significance of adaptive immunity. The important topics seem to be as follows:

The concept of the macrophage as a preparer of antigen for antibody production in other cells.

The significance of the dendritic phagocytic cells in lymph follicles and germinal centres.

Specific enhancement of phagocytosis in immunized animals.

*The possible role of the macrophage in preparing antigen for antibody production in other cells*

In the 1930s the general opinion was that since macrophages—the reticulo-endothelial system of Aschoff—took up any identifiable antigen passing to lymph nodes and spleen, they were the obvious candidates for the antibody production known to occur in these organs. When detailed study of the process was initiated, Ehrich and Harris (1942) found that after antigen injection in the rabbit foot, the efferent lymph from the popliteal lymph node had a higher content of antibody than afferent lymph or serum. Later, Harris *et al.* (1945) found more antibody in the cells then assumed to be lymphocytes than in the lymph fluid of the efferent lymph. Subsequently it was recognized that a large proportion of the cells coming from an active lymph node are immature plasma cells but that there are very few, if any, macrophages or monocytes. In 1946 Harris and Ehrich suggested that the role of the macrophage was to break down foreign cellular antigens and release immunologically active soluble materials. This attitude has persisted, although from an early stage there were two distinct ways of looking at the macro-

467

phage–lymphoid cell relationship. In the first it was believed that the macrophage broke down the antigenic particle to the 'true' antigen which was then passed to lymphocyte and plasma cell, stimulating antibody production according to the direct template hypothesis. A recent modification is the assumption that the antigenic fragment becomes attached to soluble RNA, and this complex is the effective antigen which is transferred to the lymphoid cell.

The second suggestion is that the antibody-producing mechanism is fabricated in the macrophage and transferred to a nurse cell within which it is multiplied and activated. Both approaches are implicitly instructive and neither is compatible, except in tortuous fashion, with the clonal selection approach that I have adopted. The logical requirement is therefore to seek an interpretation within the bounds of clonal selection theory for the phenomena assumed to support the two-stage process.

Before doing this it is important to reiterate the conclusion of an earlier discussion (p. 316) that because under certain laboratory conditions a phenomenon occurs reproducibly, this does not necessarily prove that it occurs under natural conditions.

The results of Fishman's group have already been discussed in another context. They have shown that when peritoneal exudate cells from rats containing large numbers of macrophages are incubated for 30 minutes at $37°$ with relatively large amounts of bacteriophage as antigen, a cell-free extract can be obtained with immunologically active properties (Fishman, 1961). It was also found that extraction of the macrophage RNA by the conventional phenol method gave similarly active material (Fishman and Adler, 1963). Activity was initially demonstrated with cultures of rat lymph node cells which produced very small amounts of antiphage in a proportion of cultures. This was confirmed by Friedman *et al.* (1965), who added the significant finding that antigen disappeared if the time of exposure of antigen and macrophages was increased to 90 minutes. They found that active RNA preparations contained low-molecular-weight ribonucleic acid and phage antigens which could be detected by a micro-complement-fixation method. In later experiments Fishman and Adler (1963) used the introduction of RNA extracts with or without normal lymph node cells into X-irradiated rats to induce antibody production. There is

468

general agreement that limited exposure of certain antigens to macrophages allows retention of, and sometimes enhances, the antigenicity of phage, bacterial antigens (Galley and Feldman, 1966) and haemocyanin (Askonas and Rhodes, 1965). The commonest interpretation is that partial digestion by intra-lysosomal enzymes will often leave smaller fragments including ADs which, complexed with low molecular weight s-RNA, are antigenic—that is, on selection theory capable of stimulating a specific receptor on an appropriate immunocyte.

More recent work by Adler *et al.* (1966) makes the specific claim that antibody produced by the recipient cells has the allotypic specificity of the donor of the peritoneal exudate, not that of the donor of the antibody-producing cells. In view of the low molecular weight of the RNA units associated with Fishman's results, which are too small to code for any immunoglobulin, I do not feel that any weight should be given to this finding until it has been adequately confirmed and developed. One must be quite sceptical as to whether accurate allotyping of minute amounts of antibody is yet feasible. Should it be established eventually that precise immunological information can be transferred from a macrophage of one race of rabbit to a lymphocyte or plasmablast of another, fascinating possibilities of analysing the essential genetic representation of the combining site would be opened up. For the present it seems wiser to disregard the report.

Nossal and Abbot (1966) have made a number of observations on the fate of antigen (flagellin) labelled with $^{125}I$ and observed by electron microscope autoradiography. This shows clearly that antigen is taken up in the typical macrophages of the lymph node medulla by pinocytosis. The material is rapidly concentrated in phagolysosomes which, with time, tend to fuse. There is no accumulation of antigen on the surface of these cells.

An indirect indication of the probable unimportance of phagocytosis and intracellular digestion in preparing antigens for their immunogenic function is contained in the work of Lapresle and Webb (1960). They showed that by limited enzymic action new ADs could be exposed in serum albumin and that by immunization with the fractions new antibodies appear. These antibodies were not produced by immunization with the intact albumin.

Since the use of peritoneal macrophages is unrelated to the normal way in which foreign micro-organismal antigens are dealt with, no special significance need be placed on their behaviour unless it appears to be a legitimate model of what happens in lymph node or spleen.

One might conclude provisionally that the unspecialized macrophages of the tissues, including the medullary macrophages of lymph nodes, those of the splenic pulp and bone-marrow and the Kupffer's cells of the liver play no direct part in the normal process of antibody production. Indirectly they may have considerable importance.

*The immunologically significant function of the macrophages*
There is no doubt whatever about the physiological importance of the macrophages (the reticulo-endothelial system). They must represent the phagocytic wandering cells present in all invertebrates from the sponges upward, as well as in lower vertebrates, and in all probability the system can be regarded as the basic defence of the body against micro-organismal invasion. It is equally important that the function of macrophages in taking up effete or damaged body cells should be recognized. Vaughan (1965 b) points out that effete autologous red cells and lymphocytes are not phagocytosed by polymorphonuclears and their uptake by macrophages is independent of any form of opsonization.

We have only a peripheral interest in macrophage activity except in so far as it is involved with the general system of adaptive immunity. There are three significant areas.

(*a*) All types of foreign antigenic material are directly or indirectly taken up by macrophages and if we limit specific function relative to immunogenicity to the dendritic phagocytes of the lymph follicles the most important function of the general reticuloendothelial system is to prevent any possibility of paralytic flooding of the lymphoid tissues with antigen (see chapter 10).

(*b*) As Boyden (1963) has pointed out, macrophages must have some capacity for recognizing foreignness so as to be able to decide that a bacterium or other foreign material should be segregated in a phagosome and digested. It has proved, however, extremely difficult to exclude the activity of 'opsonins'—including, under this term, circulating antibody or bound cytophilic antibody and either natural or immune antibody. There is no doubt that phagocytosis is greatly enhanced when antibody is present, and the effect of normal serum is presumably due to specific antibodies. In both situations the presence of complement is either helpful or necessary. The situation varies according to the particle used and is discussed by Boyden *et al.* (1965). Vaughan (1965 *a*), using rabbit macrophages, found that in the presence of normal rabbit serum, foreign red cells from human, horse, cow, pig, sheep and kangaroo, as well as polystyrene and zymosan particles, could be phagocytosed. Appropriate absorptions showed that the immunoglobulins concerned were specific for each of the eight cells or particles being studied. Both Ig M and Ig G antibody can be effective and in view of the importance of natural antibody, Ig M, it is of interest that Rowley and Turner (1966) calculated from experimental data that 8 molecules of Ig M antibody or 2,200 of Ig G were needed for the phagocytosis of a single bacillus. The special virtue of Ig M here, as in red cell lysis (Humphrey and Dourmashkin, 1965), may be in its much greater capacity, molecule for molecule, to bind complement.

(*c*) There are situations where macrophages appear to achieve a great increase in killing power for a micro-organism used for immunization.

### Enhancement of phagocytosis in immunized animals

There is an enormous volume of literature concerned with the nature of enhanced phagocytosis and intracellular destruction of pathogens in immunized animals. After going through recent reviews (Elberg, 1960; Rowley, 1962; Suter and Ramseier, 1964) one finds it very difficult to obtain any general picture of what is taking place in the various experimental situations. Most discussion starts with the work of Lurie (1942), which he interpreted as indicating a specifically increased ability of macrophages from tubercle-immune rabbits to destroy tubercle bacilli. With a different technical approach this was not confirmed by Mackaness (1954). Several authors found that *Brucella* were more actively killed in peritoneal macrophages from immune than from normal guinea-pigs. Mackaness (1962) described extensive studies of

471

*Listeria* infection in the mouse. In this infection the lethal effect is associated with multiplication of the organism in the macrophages of the liver, spleen and peritoneal cavity. After recovery from infection by a sublethal dose there was substantial immunity and a striking increase in the ability of macrophages to proliferate in the peritoneal cavity when a culture was injected and to resist destruction by the organism *in vitro*. This change appeared about the fourth day and was associated with the appearance of DH to *Listeria* but no bactericidal antibodies were demonstrable. The enhanced killing capacity of the macrophages was nonspecific and was shown equally against *Brucella*. In other systems, for example sheep peritoneal macrophages (Njoku-Obi and Osebold, 1962), the killing of *Listeria in vitro* required both immune cells and immune serum.

In all probability we are dealing with a complex situation of which the important components are (*a*) a metabolic activation of macrophages associated with phagocytosis that may go on to active proliferation, (*b*) opsonization with perhaps a direct effect of antibody enhancing killing of the ingested bacterium and (*c*) the effect of cytophilic antibody transferred to macrophages (Jonas *et al.* 1965).

In addition there is the associated effect of DH, to be discussed more fully in chapter 11, as the so-called 'innocent bystander' effect. When sensitized lymphocytes meet the corresponding antigen, probably carried in the surface of other cells, the interaction gives rise to effects on adjacent cells, including macrophages. There is considerable scope here for a variety of nonspecific, specific or semi-specific results and corresponding difficulty in sorting out the mechanism of any given experimental situation.

Nonspecific effects of phagocytosis or exposure to endotoxins on the metabolic activity of polymorphonuclears and macrophages have been described by Cohn and Morse (1960). These may require a specific opsonin for the bacterium in question to initiate what then becomes a nonspecific activation. The work of David *et al.* (1964*a*) on loss of mobility of peritoneal macrophages when concomitant sensitive lymphocytes react with PPD, and the finding of Nelson and Boyden (1963) that PPD injected intraperitoneally after a mononuclear exudate had been induced with glycogen caused a sharp disappearance of macrophages, both point to a stimulus emanating from immunocyte-antigen contact (see chapter 8). Cytophilic antibody is still not well understood but Nelson and Mildenhall (1967) have shown that mice immunized with sheep red cells produce cytophilic antibody readily attached to and demonstrable by Boyden's method on mouse macrophages. In their opinion its titre was more closely correlated with the development of DH than with the titre of haemagglutinating or complement-fixing antibody.

It is of interest that macrophages have not been shown to have any acquired protective quality in virus infections. This is most evident in Mims's (1964) account of ectromelia infection in hyperimmune mice. On the other hand the capacity of a virus to multiply in macrophages seems to be a very important determinant of virulence or otherwise. They may be evidenced either by genetic differences between mouse strains in susceptibility to arboviruses also being manifested by the capacity of the virus to grow in macrophage tissue cultures (Goodman and Koprowski, 1962), or by differences between virulent and avirulent strains of ectromelia virus in the same respect (Mims, 1964).

The point which seems to emerge clearly from the discussions is that the macrophages defined as phagocytic mobile cells readily attaching to glass have no *active* immunological function. When active processes are demonstrated with peritoneal cells they are probably the responsibility of the lymphocytes which are always present in greater or smaller numbers. Weiler and Weiler (1965) showed this very clearly by separating 'macrophages' from 'lymphocytes' in exudates of immunized mice, on the basis essentially of the lymphocytes' failure to adhere to glass and the tendency of the macrophages to aggregate. When such cells in various mixtures were injected into irradiated mice along with bone-marrow cells, the antibody produced corresponded to that used to immunize the donor of the lymphocytes.

On the whole there seems to be little tendency at the present time to accept interconversion between lymphocyte and macrophage or to admit the existence of functional intermediates. As always there is the difficulty that many functionally distinct cells can look like small lymphocytes. It is quite probable that the bone-marrow stem cell from which monocytes and macrophages derive has the morphology of a small lymphocyte. It would, however, have none of the qualities of a committed immunocyte. What does appear to be clear is that the various types of macrophage show no evidence of differentiated immune character. They are not part of the immunocyte series.

*The phagocytic reticular cells of lymph follicles*

Although most studies of the significance of macrophages in relation to antibody production have made use of peritoneal cells, the attitude we shall adopt is that this is a highly artificial approach and that the results are only significant in so far as they are relevant to the behaviour of antigen in the phagocytic reticular cells of the lymph follicles. Following the work of Ada, Nossal and collaborators (Ada *et al.* 1964*a*; Nossal *et al.* 1964*a, b*; Ada and Lang, 1966) I shall adopt the hypothesis that these cells are the important means by which antigen is presented to the potentially reactive immunocytes involved in antibody production.

The Melbourne group's work was done with flagellar antigen from *Salmonella adelaide*, either as flagella or as semi-purified flagellin. The distribution of the antigen was followed either by autoradiography using small doses of antigen heavily labelled with $^{125}I$ or $^{131}I$ or with much larger doses of antigen by the use of fluorescent antibody (Miller and Nossal, 1964). The test object was the popliteal lymph node of the rat examined in relation to injection of antigen in the corresponding foot-pad.

A considerable amount of antigen was trapped in the draining popliteal node which, per mg of wet weight, showed 20–50 times as much radioactivity as the spleen (Ada *et al.* 1964*b*). In the lymph node itself the first appearance of antigen is in the cells forming the internal wall of the peripheral sinus and in the macrophages of the medulla. Initially there is no labelling in the lymph follicles but by 24 hours the dendritic phagocytic cells of the follicles are labelled, the diffuse cortex remaining clear. The label in the follicles persists for at least 28 days and is characteristically related to germinal centre production. Germinal centres become definite at about 5 days and show a characteristic cap of labelled dendritic cells peripherally (Nossal *et al.* 1964*b*). More detailed study using electron microscopy in combination with autoradiography showed that in the reticular phagocytes of the lymph follicle the antigen was held, apparently, on the surface of the dendritic processes— in contrast to its behaviour in medullary macrophages which is of classical character, involving ingestion and segregation in phago- somes with progressive digestion (Mitchell and Abbot, 1966).

An important feature of the germinal centre is the presence of a lacy network of material between the lymphoid cells which shows immunoglobulin staining (Ortega and Mellors, 1957) as well as dendritic concentration of antigen (White, 1963; Nossal *et al.* 1964*b*; Miller and Nossal, 1964; Hanna *et al.* 1966). In line with these results are the findings (*a*) of Ada and Lang (1966) that the labelled material concentrated in germinal centres is antigen com- plexed with antibody and (*b*) of Pernis (1966) that the immuno- globulins of germinal centres are not synthesized locally but are bound there secondarily. As has been well known (see, e.g., Curtain and Baumgarten, 1966), immunoglobulins of the germinal centres are uniform mixtures of molecules of different class and type.

Although most of these studies have been carried out with the popliteal lymph node in the rat the results are almost certainly also applicable to the spleen (Hanna *et al.* 1966) and to other species of animal.

The findings provide the justification for regarding one of the important features of antibody production, the initiation of activity in an immunocyte, as taking place within the lymph follicle. In summary, an antigenic particle opsonized by 'normal' or 'immune' antibody is taken up by the dendritic phagocytic cells. With or without partial digestion the antigen is held on the surface of the processes of these cells in such a way that it is accessible to adjacent cells. It is a necessary implication that these processes will also carry a wide variety of other antigens in similar accessible state.

By hypothesis, any accumulation of lymphocytes is composed of a more or less random selection of all the immunocyte clones— defined by the immune patterns of their combining sites—in the body. On the cytoplasmic processes of the dendritic phagocytic cells disposed along the reticulum fibres of the lymphoid tissue are presented all the immediately relevant foreign ADs. Having regard both to the local mobility of the lymphocyte and the high level of the circulation of lymphocytes from blood to lymph follicle to efferent lymphatic to blood, we can regard this situation as ideal from the point of view of giving every lymphocyte a relatively easy opportunity to test itself against every relevant AD and, if stimulated, to find a wide variety of niches in which to settle down and proliferate. In view of the rapid concentration of active plasma cells in the medulla it is simplest to derive them from a fraction of the cells stimulated in the follicle by the appropriate antigen. In line with the discussion in other sections we would regard the cells proliferating in the germinal centre as immunocytes —in part specifically stimulated by the antigen being studied but probably also including some nonspecifically activated cells of other clones. Cells produced in the centre may (*a*) be liberated as 'activated' large or medium lymphocytes via the efferent lymphatics into the blood (Morris, 1966), (*b*) move toward the medullary cords and develop as plasma cells probably influenced in part by the higher oxygen tension in this situation or (*c*) undergo immediate postmitotic necrosis possibly because of the co-existence of high vulnerability and contact with a destructive concentration AD which might be either autologous or foreign.

The fate and potentialities of the activated cells passing to the blood and becoming plasma cells or lymphocytes with memory cell functions and with special liability to take part in DH or homograft rejection reactions is discussed elsewhere (see p. 579). The possibility that some of these cells become mast cells is raised by the observation of Dr J. J. Miller (personal communication) that a number of labelled mast cells appear and persist in lymph nodes receiving tritiated thymidine at the time of an immune response. Little is known of the fate of germinal centres as such. When large amounts of foreign material reach the mouse spleen there is a rapid disorganization and disappearance of existent germinal centres (Congdon and Goodman, 1962). In other circumstances they may well fade into the associated small lymphocyte accumulations.

Since the earliest work on opsonin there have been suggestions that normal antibody may play a part in facilitating the processes leading to antibody production. Recent work by Williams (1966*a*, *b*) adds an additional support to this. He showed that lymphocytic depletion, either by thoracic duct drainage or by whole-body X-irradiation, greatly reduced the primary uptake of labelled antigen (flagellin) in the follicles of the draining popliteal lymph node but had no influence on the uptake by the macrophages of the medullary areas of the node.

The deficiency of the dendritic phagocytic cells could be remedied by the local injection of small amounts of specific antibody shortly before injecting the antigen. This fits in with a number of other observations on the ability of antibody to inhibit the recruitment of progenitor immunocytes to 19 S antibody production (Möller, 1964; Möller and Wigzell, 1965; Wigzell, 1966) but not (or much less actively) the production of Ig G antibody by committed cells.

The inhibition by antibody is specific. This holds also according to Wigzell (1966) for the minor but definite inhibition of 7 S antibody production up to 40 days after immunization. Specificity must mean that ADs are playing a significant part in stimulating antibody production for a long period. In macrophages, as ordinarily obtained from peritoneal exudate, antigen in the form of phage is fairly rapidly inactivated but immunogenicity survives at least 48 hours (Uhr and Weissmann, 1965). The evidence from labelled antigen studies with flagellin is that most of the antigen held in the early stages by littoral cells and medullary macrophages is gone while the follicular accumulations are still persisting (Ada *et al.* 1964*b*). Further, Uhr and Weissmann (1965) found, as would be expected, that most of the antigen was taken into phagolysomes where it would be subject to intense enzymatic action. The important feature of antigen in the dendritic phagocytic cells is that it is on the surface of the cytoplasmic processes of the cells (Mitchell and Abbot, 1965).

The tentative deduction would be that for some structurally and functionally defined reasons the surface of the cells has a special capacity for holding antigen and retaining it in effectively antigenic form. In addition, all the indirect evidence suggests that the associated presence of antibody does not destroy the capacity of antigen so held to stimulate specifically any committed immunocytes of appropriate pattern which make contact.

477

### THE EOSINOPHIL LEUCOCYTE

In view of the obvious association of eosinophil leucocytes with immune reactions it is disappointing to find that there is still no clear interpretation of this cell's functional role in immunity. Recent reviews on the subject include those of Archer (1966), Hirsch (1965) and Rytömaa (1960) while the series of papers by Litt (summarized in Litt, 1964) provides the most extensive experimental work on the association of eosinophils with lymph nodes in the course of response to antigen.

Eosinophils arise from stem cells in the bone-marrow, spend only an hour or two in the blood and seem to be end-cells with a total life of the order of 8–12 days. In the tissues they are particularly characteristic of loose connective tissue beneath skin, intestinal epithelium or bronchiolar epithelium (Rytömaa, 1960), that is, in situations where antigen–antibody interactions are liable to occur.

The characteristic intracellular granules contain a variety of enzymes similar in general to those of polymorphonuclears, with the addition of a stable peroxidase of different character. Chemically the crystalline material is largely protein (Archer, 1963). The granules can be seen to be discharged when antigen–antibody complexes are phagocytosed by eosinophils (Archer and Hirsch, 1963).

Although it is certainly not the only function of the eosinophil, the most important positive finding in the immunological field is its capacity to be attracted to and to phagocytose particles, such as red cells, coated with antibody. The nature of the product of immunological reaction which has a chemotactic effect on eosinophils is unknown. It was suggested by R. K. Archer (1959) that histamine was responsible for attracting eosinophils to a tissue region in the horse but this is not found in other species. Archer (1966) suggests that complement is almost certainly involved in the production of the chemotactic substance.

Phagocytosis of antibody-coated red cells or other antigen–antibody–complement complexes is followed by discharge of the granules. According to Archer (1966) this discharge may have important secondary results. The peroxidase has an active capacity

478

to discharge mast cells with local liberation of histamine. The basic protein is readily adsorbed to a variety of foreign particles and may act as a nonspecific opsonin. Some product of eosinophil breakdown has a stimulatory effect on bronchiolar mucus-secreting epithelium and it is now known that the Charcot–Leyden crystals of asthmatic sputum are derived from the cytoplasm of disintegrated eosinophils (Archer and Blackwood, 1965). There is a strong possibility in Archer's view that the local eosinophil response may play a part in producing the symptoms of asthma and other allergic conditions.

Litt (1964) has confirmed the importance of the antigen–antibody complex by showing that neither antigen nor antibody separately is taken up and that there are no differences in the behaviour of eosinophils from normal or immunized guinea-pigs. The most interesting feature of his studies was the appearance of diapedesis of eosinophils in the draining lymph node 5–10 minutes after injection of antigen (Litt, 1963) and continuing accumulation in the perifollicular region for at least 12 hours. At the earliest stage, eosinophils are the only cells migrating through the venular walls in the stimulated lymph node and Litt interprets this as implying the appearance, presumably as a result of contact between antigen and appropriate immunocyte, of an agent specifically adapted to facilitate this attraction of the circulating eosinophil, first to the local endothelial cells of capillary or venule and then to the lymphoid tissue. Since local eosinophilia is conspicuous in human urticarial lesions (Kline *et al.* 1932) the power of antigen–antibody reactions to produce specific diapedesis of eosinophils presumably holds also in other species and tissues.

The characteristic local accumulation of eosinophils in the neighbourhood of helminthic parasites in the tissues (Archer, 1966) is also associated with an increased blood eosinophilia beyond that seen with other types of infection. It seems likely that both phenomena are immunological in character, with antigen–antibody union as the initiating stimulus, although it would probably be wise to include other ways in which autologous $\gamma$-globulin could be denatured *in vivo*. We can postulate that the local presence of denatured $\gamma$-globulin in appropriate physical state and concentration liberates a pharmacologically active agent which makes the

adjacent capillary endothelium sticky and permeable for eosinophils and is locally chemotactic. The discharge and lysis of granules
may (*a*) act as a nonspecific opsonizing agent and (*b*) attract an increasing number of macrophages to the region. The special virtue
of worm infestations may be due to the absence of chitinases and
other enzymes capable of dealing with insoluble worm structures.
Progressive deposition of denatured γ-globulin, specific or nonspecific but ineffective on the worm surface, might provide the
stimulus necessary.

What may be of special interest is the nature of the feedback
which brings eosinophils out of the bone-marrow on demand or,
alternatively, how ACTH or cortisone (Thorn *et al.* 1948) turns
off the supply, presumably at the bone-marrow source. It must be a
positive stimulus and apparently a specific one; to be effective there
must be a rather large deposit of antigen, cf. asthma and helmin-
thiasis, and a developed immunological response. A possibility that
may be worth experimental study is that the local discharge of
eosinophil granules and their associated enzymes which follows
phagocytosis of antigen–antibody complexes may produce a peptide or other diffusible agent which acts both to attract circulating
eosinophils to the local site and to stimulate their liberation from
the bone-marrow to the circulation.

On this view the normal entry of eosinophils into the circulation
would depend on a small but detectable level of this agent being
maintained in the circulation, presumably mostly derived from
interaction of antigen, antibody and eosinophils in the submucosa
of the gut. The action of ACTH and any rise in circulating corticosteroid for other reasons in reducing the count of blood eosinophils
can be ascribed to the general corticosteroid effect of damping
down the pharmacological effects of antigen–antibody reactions.

### COMPLEMENT

The complement system as studied *in vitro* is a highly artificial set
of phenomena that has become a traditional object of laboratory
investigation because of the striking and easily measurable phenomenon of haemolysis.

Recent reviews of the functional structure of complement in

relation to immune haemolysis can be found in a Ciba Foundation Symposium (Wolstenholme and Knight, 1965) and in R. A. Nelson (1965). Based largely on Mayer (1965) the process can be formulated approximately as follows:

Antigen–antibody union initiates a process by which a precursor, $C^1$1 p, in association with components 1 q, 1 r, 1 s becomes an active esterase, $C^1$1 a. This has as substrates $C^1$2 and, to a lesser extent, $C^1$4. The unstable complex, which may be expressed as Ag–Ab–$C^1$1 a, 4, 2 a, is then stabilized by $C^1$3 components, 3 a, b, c, d, e, f. Within the complex a few molecules are generated of a haemolytic substance—according to Fischer and Haupt (1961), lysolecithin—and when the red cell surface membrane is the antigen this results in the drilling of a small hole visible in electron micrographs through which haemoglobin can be lost (Borsos *et al.* 1964). According to Green *et al.* (1959) the holes initially produced interfere with osmotic integrity by allowing passage of ions but not of larger molecules. This leaves the protein osmotic pressure unbalanced until with swelling of the cell and enlargement of the holes the haemoglobin is able to escape. Humphrey and Dourmashkin (1965) have made a quantitative study of hole production in sheep red cell membranes which has demonstrated the special facility with which 19 S antibody allows attachment of complement and hole formation. Their conclusions are that for each red cell there is space for fixation of approximately 90,000 M or 600,000 G antibody molecules, and that for a hole to be produced by complement, one M antibody is sufficient while, with G, two antibody molecules must be attached to immediately adjacent sites to allow the attachment of complement and hole production.

There seems to be no reason in the present context for any discussion of the basis of complement fixation in relation to conventional methods of titrating antibodies and antigens. This is available in all standard textbooks on immunology.

Recently immunofluorescent methods have been developed to detect complement bound to tissues. This may be done by treating a human tissue section, for example a kidney biopsy, with guinea-pig complement and, after washing, locating the areas of fixation with a fluorescent anti-guinea-pig globulin serum (Burkholder, 1961). More satisfactorily, complement bound *in vivo* can be

481

detected by an anti-human $\beta_{1C}$ globulin appropriately absorbed (Lachmann *et al.* 1962). This is an elegant approach to locating areas where antigen–antibody reaction has occurred.

Brief mention may be made of bovine conglutinin, summarized by Lachmann and Coombs (1965) as a rod-shaped protein of high molecular weight, not an immunoglobulin, which has capacity to bind with some component of aggregated complement. Immuno-conglutinin is found in a variety of immune sera and is an immuno-globulin which reacts with normally hidden determinants of complement which are exposed in antigen–antibody–complement aggregates. It is probable that this hidden determinant becomes immunogenic in the course of artificial immunization; for example, with bacterial vaccines when *in vivo* complement fixation probably occurs.

Immune-adherence is another rather artificial laboratory test reviewed by Nelson (1963*b*, 1965). When an antigen–antibody–complement union takes place in the presence of primate red cells, but not those from other mammals or birds, the complex is adsorbed to the red cell surface with agglutination of the cells. It is probable (Dacie *et al.* 1957) that the 'non-γ' antiglobulin tests given in many cases of cold-type acquired haemolytic anaemia are due to complement components attached in some such fashion to the red cells.

In autoimmune diseases of several types there is an abnormally low level of complement. This has been noted particularly in various forms of nephritis including systemic lupus erythematosus (Lange *et al.* 1960) and in myasthenia gravis (Nastuk *et al.* 1960).

The only condition in which there is some evidence of a clinical influence of imbalance amongst complement components is hereditary angioneurotic oedema. This is a disease in which patients suffer localized attacks of oedema in various parts of the body, and may be fatal when the larynx is involved. Donaldson and Evans (1963) showed that in this condition there is a congenital lack of an inhibitor of $C^1$1 esterase. During an attack there is a high level of circulating esterase and a reduction in the natural substrates $C^1$2 and $C^1$4 (Austen and Beer, 1964) and there is some evidence that there is greater reduction of $C^1$2 in blood coming from an oedematous area than elsewhere.

Perhaps symptomatic of an as yet hardly formulated relationship of the complement system with the complex 'enzyme cascades' that are initiated by trauma to the vascular system, it is known that $C^1$i esterase inhibitor is also an inhibitor of kallikrein (Kagen and Becker, 1963) which is a widely distributed enzyme, or group of proteolytic enzymes, acting on $\alpha$2-globulin to produce kinin polypeptides with an active effect on vascular permeability. On the whole the evidence (Becker, 1965) suggests that the clinical effect is mediated through the kallikrein system and that complement changes are more or less casual indications of the primary genetic anomaly.

# 8 The immunocyte and its response to specific stimulation

It is a necessary consequence of selective theories of antibody production that normal animals should possess small numbers of immunocytes of many different patterns, and traces of many types of Ig M antibody.

One of the outstanding results of the introduction of Jerne's antibody plaque technique has been the demonstration that small numbers of cells are present in mouse or other spleens which can produce plaques lysing several different foreign species of red cell and that the cells were specific. This was noted for sheep and rabbit red cells in the mouse spleen by Jerne and Nordin (1963) and for sheep cells by many other investigators including Friedman (1965 a) and Wigzell et al. (1965). Double producers, when looked or in test plates with mixed erythrocytes, were not found (Jerne, personal communication, 1964) and on stimulation with both types of red cell it was equally evident that two distinct populations of plaque-forming cells were concerned.

Hege and Cole (1966) found a background of 10–100 plaque-forming cells (against sheep red cells) per spleen in normal mice. The number increased to 150–400 three days after intravenous injection of phytohaemagglutinin (PHA), presumably representing a nonspecific stimulation of these cells to proliferate. Sublethal whole-body irradiation (to 500 r) produced no significant reduction in the number of plaque-forming cells and it was deduced that the generation time for these 'normal producers' was appreciably more than three days.

In a related type of experiment, Playfair et al. (1965) injected lethally irradiated mice with known numbers of normal spleen cells plus sheep or pig erythrocytes as antigen either singly or together. Four days later the spleen was removed and cut into a

484

large number of small volumes of known spatial relationship. Jerne-type tests were made with each piece. The results showed that there were discrete areas each containing 100–1,000 active cells and presumably each derived from a single precursor cell. When mixed antigen was used, all well-separated active areas contained cells producing antibody against sheep *or* pig, but never against both. Approximately 1 out of 20,000 spleen cells was effective, on the assumption that around 10 per cent of cells injected localized in the spleen. Kennedy *et al.* (1966), using a basically similar method but identifying foci of antibody-producing cells by laying thin sections of spleen on red cell containing plates, reached essentially the same conclusion. They calculate that the normal mouse spleen contains about 1,000 cells capable of responding to sheep red cell antigen.

## STIMULATION OF LYMPHOID CELLS TO PROLIFERATION AND ANTIBODY PRODUCTION

Everything that is known about antibody production is in accord with the view that ADs can, under appropriate conditions, stimulate immunocytes to proliferation and antibody production. It has, however, proved extraordinarily difficult to provide clear experiments in which can be shown the nature of the receptor by which the cell is stimulated and the form of the antigen when it acts as stimulus.

Small pieces of splenic tissue from an animal primed weeks or months previously produce relatively large amounts of antibody *in vitro* following antigenic stimulation (Ambrose, 1964; Michaelides and Coons, 1963). Fully dispersed cells produce either no response or a minimal one, and no one has so far claimed an unequivocal production of antibody by a continuing cell line of normal lymphoid cells.

Stimulation *in vivo* or, in what is equivalent to organ culture, *in vitro*, takes place in a complex environment where the antigen may be acting directly on the cell or after partial breakdown and linking to low molecular weight RNA in a phagocytic cell. Similarly, the susceptible cell which multiplies and/or produces antibody clearly carries the genetic mechanism which allows it to synthesize immunoglobulin, but we do not know whether the stimulus resulted

485

from contact of the AD with a cell receptor carrying antibody specificity, with adsorbed cytophilic antibody, or as a result of stimulation by pharmacologically active material liberated from some adjacent cell.

From the theoretical point of view the most direct action of antigen on the cells of immunized animals is the stimulation of pre-mitotic activity as shown by the uptake of tritiated thymidine *in vitro*. Dutton and Eady (1964) find that spleen cells from a rabbit spleen, primed some months previously with antigen, respond specifically to that antigen by a rapid uptake of thymidine. The antigenic effect is maximal with the shortest possible contact and is not shown by unrelated antigens. Active nonspecific stimulation, however, can be produced by PHA, zymosan and staphylococcal extracts.

In view of Haurowitz's contention that the secondary response is in some way due to the persistence of circulating antibody, Dutton and Eady's (1962) experiment is important. They showed that cells from rabbits immunized with human serum albumin or $\gamma$-globulin could be treated with rabbit antisera against egg albumin and retain only their specific reactivity. Egg albumin showed no increase in DNA synthesis while there was an active response to the specific antigen.

In Dutton's opinion, the kinetics of the response were such that it is most unlikely that the effect on the cells was mediated through another cell type. Experience from parallel experiments with the same general system in which antibody production was measured makes it virtually certain that the cells stimulated are those which proliferate to antibody-producing plasma cells.

Further evidence that the cell receptors concerned are the same as those which mediate stimulation to antibody production was obtained by the use of dinitrophenyl (DNP) as hapten. Cells from rabbits immunized with DNP-bovine $\gamma$-globulin (BGG) responded by DNA uptake to either DNP-BGG or BGG but not to DNP-lysine. This was, however, capable of blocking the action of DNP-BGG in stimulating DNA synthesis by about 50 per cent (Dutton and Bulman, 1964).

It is of interest that in later work Richardson and Dutton (1964), combining Jerne's technique with the thymidine uptake studies,

486

found that about 2 per cent of the cells showed early uptake of tritiated thymidine but plaque-forming cells appeared only on the fifth day in numbers about one-fortieth of those that had been initially stimulated. They concluded that antigen-dependent DNA synthesis represented the proliferative phase of antibody synthesis but that a process of maturation of the dividing cells must occur before substantial amounts of antibody are produced.

### THE 'IN VITRO' STIMULATION OF LYMPHOCYTES

The capacity for small lymphocytes to be stimulated to become large pyroninophilic cells (blast transformation) and undergo mitosis by PHA and other agents has already been discussed from the point of view of the potentiality of the lymphocyte (see chapter 4). In the present context we are concerned with the same phenomenon from the point of view of its implications for the nature of receptors with immune pattern and the process of stimulation. It is inevitable that points relevant to the production of antibodies will be introduced.

The immunological significance of the reaction was first indicated by Pearmain *et al.* (1963) who found that tuberculin would induce mitosis in lymphocytes taken from persons with positive Mantoux reactions but not from normal nonreactive individuals. This finding has been frequently repeated although it has also been stated (Gump, 1965) that with the use of graded doses of PPD and taking the uptake of tritiated thymidine as a measure of stimulation there was a very poor correlation with skin reactivity of the donor. The technique is a notoriously difficult one for quantitative work and the early statements that specific responses were obtained with any antigen that had been used to immunize the donor have not been confirmed. Perhaps the most persuasive finding that the effect on leucocytes from sensitized individuals is a true one comes from Heilman and McFarland's (1965) finding that mitotic reactions were greatly reduced, or absent, with cells from cases of active tuberculosis. The inhibitory effect was shown to be due to the serum.

The other presumptively immunological reaction is that seen when human lymphocytes from unrelated donors are mixed (Bach and Hirschhorn, 1964). Here there are strong indications that the intensity of the reaction is a measure of the 'distance' of the histocompatibility antigens concerned and it has been suggested as part of the procedure for seeking acceptable donors for kidney transplantation.

Neither of these approaches is particularly suitable for detailed analysis of the process. A few points of interest have, however, been recorded. Two groups have shown that non-cellular material from mixed lymphocytic cultures can stimulate other lymphocytes to blast transformation. Similar but less active material is produced by unmixed cell cultures. This is active on homologous but not on autologous lymphocytes (Kasakura and Lowenstein, 1965; Gordon and MacLean, 1965).

There is now considerable evidence that lymphocytes from tuberculin-sensitive persons or animals carry or can produce antibody reactive with tuberculin. Witten *et al.* (1963) used PPD conjugated with fluorescein to test small lymphocytes from patients of known status. Of fifteen positive Mantoux reactors the percentage of lymphocytes stained ranged from 1·5 to 8·8 per cent while eleven non-reactors showed an average of 0·36 per cent of reactive cells (0–1·3 per cent). Martins *et al.* (1964) used a sandwich-type fluorescence test to detect antibody against tuberculoprotein in sensitized guinea-pigs. Different animals showed 6–20 per cent positive. From a quite different angle, Rothman and Liden (1965), examining the proteins present in lymph node extracts from guinea-pigs by electrophoresis on acrylamide gels, found an extra band in material from sensitized animals. This was identified as Ig A but its immune specificity was not examined. Mills and Harden (1966), using sensitized guinea-pigs, found that the specific morphological changes and increased thymidine uptake could be obtained by contact *in vitro*. The reaction was specific and when a hapten–carrier sensitization was under study the reaction was not induced by a conjugate of the hapten with a different carrier, despite the fact that much anti-hapten antibody was circulating.

Although the evidence is not wholly satisfactory, it points firmly to the conclusion that blast formation and mitosis of lymphocytes is correlated with the presence of DH and is not directly related to the presence of circulating antibody.

488

In 1963 Gräsbeck *et al.* showed that an immune serum produced from rabbits immunized with human leucocytes produced blast transformation and mitosis in human lymphocytes analogous to what was observed with PHA or, in reactive subjects, tuberculin. This indicated that an antigen–antibody reaction at the surface of the lymphocyte could be effective irrespective of on 'which side' the reagents were situated. In view of the widespread evidence that lymphocytes frequently produce immunoglobulin (see chapter 9) it became of much interest to examine the activity of anti-globulin (anti-allotype) sera in this respect. This has not yet been reported for human beings but it has been extensively studied by Sell and Gell (1965) in rabbits.

They found that rabbit lymphocytes can be stimulated to blast transformation and thymidine uptake by PHA, rabbit anti-allotype serum and a sheep serum against rabbit immunoglobulin. The results of interest from the anti-allotype sera were as follows:

(*a*) Specific stimulation of lymphocytes corresponded to the allotype of the donor Ig G and the anti-allotype used, and was not due to a surface-coating from hetero-allotypic serum.

(*b*) Heterozygous neonatal animals which are producing no Ig G give specific lymphocytic reactions equivalent to their eventual allotype and uninfluenced by the allotype of any temporary maternal globulin.

(*c*) The results indicated that each lymphocyte from a hetero-zygote carried only one of the allotypes 4, 5 and 6 of the b locus which is located on the light chain. An anti-4 serum which produced 38 per cent of blast transformation with A4-4 animals averaged 18 per cent with A4-5 or A4-6 rabbit cells, and when anti-5 and anti-6 are used alone or together against an A5-6 heterozygote, the percentage transformation by the mixed sera is a simple summation of the results with the single sera (table 18).

The conclusions to be drawn from these results are that this particular type of antigen–antibody contact is an effective stimulus to blast transformation and, with some reservations that adsorbed immunoglobulin may be concerned, we can accept the likelihood that up to 40 per cent of the lymphocytes are immunocytes in the sense that they show the presence of light chain. Further, it seems that there is strict phenotypic restriction to one or other light chain allotype. The evidence for the heavy chain allotypes is inadequate for any conclusions.

An additional observation recorded in this paper is that cells from a rabbit immunized to produce an anti-allotype serum showed a proportion reacting by blast transformation when suspended in serum of the allotype used as antigen. This is, of course, the reverse situation from that operative in the main investigation.

TABLE 18. *Summation of anti-allotype action*

| | Uptake of $^{14}$C thymidine by donor cells | |
| --- | --- | --- |
| Antisera used | A2356 | A156 |
| Anti-A5 | 35,900 | 55,800 |
| Anti-A6 | 31,800 | 37,100 |
| Anti-A5 and anti-A6 | 80,200 | 88,200 |

The sum of counts in seven experiments with mixed antisera and the corresponding controls with single antisera are shown. Calculated from Sell and Gell (1965).

## REACTIONS OF MACROPHAGES AND MAST CELLS WITH ANTIGEN

There is nothing to suggest that macrophages have any intrinsic capacity to recognize or react to the pattern of an AD. It is, however, certain that antigen–antibody complexes of any sort are avidly taken up by macrophages, as is denatured globulin. The part played by macrophages in antibody production as possible carriers and modifiers of antigen for presentation to the antibody-producing lymphoid cells will be dealt with later. Here, all that need be said is that the enhanced activity with which an antigen is phagocytosed in an immune or even a tolerant animal (Ada *et al.* 1965) can best be ascribed to the opsonizing effect of even minute amounts of antibody.

Another important aspect is the readiness with which some macrophages at least take up cytophilic antibody. This was first demonstrated in rabbits by Boyden *et al.* (1960), who were examining the capacity of cells from the spleens of normal and immunized animals to take up $^{131}$I labelled antigen. They found that this capacity could be conferred on normal spleen cells by treatment with immune serum and that the cytophilic antibody so demonstrated was not necessarily proportional in amount to the titre of precipitating antibody in the serum.

In some systems, at least, cytophilic antibody is readily attached to macrophages, and Nelson and Mildenhall (1967) have used the capacity of peritoneal macrophages from normal mice to take up antibody from anti-sheep red cell serum as a measure of cytophilic antibody titre. Using Boyden's method, adherence of specific red cells to the macrophage is the index that antibody has been taken up. In mice there is a correlation between the existence of delayed hypersensitivity and the cytophilic antibody content of the serum at the time.

This work makes it likely that the findings of a number of workers (Nelson and Boyden, 1963; Perkins and Leonard, 1963; Vaughan, 1965 a, b) that there are distinct differences in the ability of mouse macrophages to take up red cells of various species, depend on the existence of natural antibody present as adsorbed cytophilic antibody on the cell surface.

The part played by mast cells in immune responses is still problematical but there is general agreement that in those reactions due to the liberation of histamine the histamine comes from mast cells, basophil leucocytes or platelets. It now appears to be clear that two different mechanisms exist by which passive cutaneous anaphylaxis can be produced in the rat (Movat *et al.* 1966). If hyperimmune rabbit serum is used for the local intradermal injection followed by antigen and dye intravenously, the reaction is not prevented by the administration of antihistamine plus anti-serotonin agents and appears to be essentially an Arthus reaction in which the toxic action of antigen–antibody complex is exerted on polymorphonuclear leucocytes, with release of lysosome constituents and local capillary and tissue damage. The second is seen when the antiserum is produced in rats using pertussis vaccine as adjuvant and foot-pad injection. Here, there is direct evidence that mast cells in rat omentum of immunized animals discharge their granules on contact with antigen and that passive cutaneous anaphylaxis reactors in the rat can be inhibited by the joint use of antihistamine and anti-serotonin reagents (Mota, 1964).

It is not clear whether mast cells are ever immunocytes in the sense that they either produce antibody or carry specific receptors. Humphrey *et al.* (1963) have shown that although sensitized rats and guinea-pigs do not in general show any reactivity of their peritoneal mast cells to antigen, rats immunized with antigen and pertussis do show granule discharge. The likelihood is that this merely represents an increased uptake of cytophilic antibody. This may also be the explanation for Lichtenstein and Osler's (1964) finding that in patients with ragweed allergy the antigen could liberate histamine from blood leucocytes, presumably from basophils.

### THE PATHOGENIC ACTIVITIES OF ANTIGEN–ANTIBODY UNION

The basic approach in this chapter is that contact of an AD with a CS on an immunocyte can act as a signal initiating a variety of processes ranging from lysis and death to proliferation and antibody production. For many reasons we can be certain that the heterogeneity of the populations concerned and the variability of microenvironments in the body will ensure that in any nominally uniform set of conditions there will, in fact, be not a uniform response but a distribution of various types of response.

In experimental work, however, it is almost always necessary to confine quantitative studies to one particular type of response. For the present we are concerned with the damaging potentialities of antigen–antibody complexes. The presumption that most cell-surface damage to an immunocyte arises from the process initiated by antigen–antibody union (or more strictly, AD–CS union) makes it desirable to start with the toxicity of antigen–antibody complexes as such.

### *The toxicity of antigen–antibody complexes*

With standard soluble antigens such as serum albumin, Campbell (1962) points out that soluble complexes are always in the region of antigen excess with a maximal antigen–antibody ratio of 2:1. Such complexes, with full occupation of both antibody binding sites, are non-toxic and do not fix complement. These are seen when the ratio of antigen to antibody is more than 1 and less than 2. That toxicity is more related to the antibody than the antigen since, as Ishizaka and Ishizaka (1962) have shown, nonspecifically aggregated immunoglobulin, human or rabbit, can fix complement and produce skin reactions comparable on a weight basis with the activity of antigen–antibody complexes. Similarly, suitably aggregated human $\gamma$-globulin, in the presence of complement, will agglutinate human erythrocytes in a reaction formally equivalent to 'immune-adherence' (see Nelson, 1963 $b$) in which antigen–antibody–$C^1$ complexes are adsorbed to red cells. In a later section (p. 611) the related phenomenon observed in some human 'drug reactions' is discussed. Here, a combination of drug (acting as hapten) with antibody and complement can become adsorbed to red cells, causing haemolysis.

The capacity of either antigen–antibody complex or aggregated normal $\gamma$-globulin to become attached to red cells or platelets and to release histamine from rabbit blood, presumably by its action on platelets, is ascribed by Ishizaka (1963) to the presence of the Fc (Porter III piece) segment of the heavy chain of Ig G. This also seems to be involved in reactions such as passive cutaneous anaphylaxis in which antibody is bound to tissue cells, and in fixation of complement (Benacerraf and McCluskey, 1963).

The toxicity of antigen–antibody complexes *in vivo* is related mainly to kidney damage. Germuth (1953) was the first to show that after large intravenous injections of bovine serum albumin in the rabbit, the danger period for kidney damage was when soluble antigen–antibody complexes were likely to be present. Dixon *et al.* (1961) showed that when rabbits were given repeated injections of a soluble antigen such as bovine serum albumin or human $\gamma$-globulin, the effect on the kidneys depended on the extent of the antibody response. Rabbits producing no antibody had no lesions; those producing much more antibody than was equivalent to the amounts of antigen injected showed a temporary proteinuria with recovery. In animals producing more or less equivalent amounts of antibody in which temporary solubilization of antigen–antibody complexes in antigen excess could occur, there was severe glomerulonephritis with proteinuria and azotaemia. Histologically the glomeruli showed diffuse hyaline thickening of the basement

493

membrane with deposits of foreign material mainly on the epithelial side of the membrane.

McCluskey *et al.* (1962) produced severe glomerulonephritis in mice injected intravenously three times in 24 hours with immune precipitate solubilized in hapten excess. Rabbit antibody was used and could be shown to be present and to persist in the glomerular basement membrane.

In similar experiments using rabbit antibody against soluble antigens, Mellors and Brzosko (1962) found that on injection of antigen–antibody mixtures into mice, most material was taken up by the fixed phagocytic cells of the liver and spleen but some lodged in the glomerulus. Again, definite membranous glomerulonephritis was produced by antigen–antibody complex but not by antibody or antigen alone. In another similar study, Weiser and Laxson (1962) showed that fluorescence-labelled antigen–antibody complex persisted longer in the kidney than elsewhere in the body.

Damage *in vivo* by antigen–antibody complexes is not confined to the kidney. To produce arteritis in rabbits, intermittent injection of soluble antigen is necessary (Heptinstall and Germuth, 1957). From work on autoimmune disease it seems very likely that many of the lesions *in vivo* are associated with adhesion of antigen–antibody–complement aggregates to platelets and the initiation of fibrin deposits (Vassalli and McCluskey, 1964; Hicks and Burnet, 1966).

### THE NATURE OF THE ANTIGENIC STIMULUS

The general picture emerging is that every immunocyte carries a series of accessible receptors of which the only theoretical essential is a CS equivalent to that characteristic of antibody produced by the clone. The receptor may be antibody or any portion of antibody carrying the CS. The other basic postulate is the relationship expressed by the statement that with various qualifications the product of avidity of receptor–AD union and effective concentration of AD is related directly to the likelihood that the immunocyte will suffer damage (or death) and inversely to the likelihood of proliferation and antibody production.

I know of no experimental or observational evidence against any of these contentions, but from the nature of immunological phenomena, in particular the heterogeneity of both antigen and immunoglobulins and the large populations of heterogeneous cells always involved, it is of extreme difficulty to provide decisive evidence for each point. The strength of any such formulation can only be gradually assessed by regarding every recorded phenomenon from the angle of whether it fits conformably to the current hypothesis. An excellent example of the limitations imposed by available experimental methods is in regard to the different specificities associated with stimulation to delayed hypersensitivity and to antibody production.

*Differences in the antigenic determinants involved in delayed hypersensitivity and antibody production*
The direction in which the study of DH has developed derives from the discovery of the tuberculin reaction by Koch and the clinical observation that many laboratory and industrial chemicals of low molecular weight, initially without irritating action on the skin, could, after a period of exposure, give rise to a specific hypersensitivity with oedema and erythema.

The tuberculin reaction is very susceptible to detailed study in man or animals, notably the guinea-pig, and with the semi-purified reagents, PPD or tuberculoprotein, now available it still remains the classical type of DH with which reactions with other reagents must be compared. It was found by Landsteiner and Chase (1937) that contact dermatitis could readily be induced and studied in the guinea-pig. Both approaches have come together in the numerous studies that have been made with small-molecular-weight haptens combined with protein or synthetic polypeptide carriers.

Hypersensitivity as manifested in the experimental guinea-pig skin may manifest itself in various forms and it is important to distinguish the rapidly appearing and persisting Arthus-type or immediate reaction from DH which is best seen at 24–48 hours and never before 12. In general, the Arthus-type reaction is parallel to the production of antibody. Delayed reactions may vary in intensity and persistence, and Raffel and Newel (1958) considered that the reactions seen with, e.g., bovine serum albumin in guinea-pigs sensitized by the method of Uhr *et al.* (1957) were not analo-

495

gous to the tuberculin reaction and should be differentiated as
'Jones–Mote' reactions. Martins *et al.* (1964) would define the
reactions to tuberculoprotein in the guinea-pig as DH—increasing
up to 48 hours, where sensitization was by killed BCG in water and
oil; or as a Jones–Mote reaction—liable to fade after 24 hours, in
guinea-pigs sensitized by tuberculoprotein without bacterial
bodies. Both arise within 7 days of the sensitizing injection. An
Arthus reaction to the same test reagent is obtained when there
has been prolonged immunization with tuberculoprotein for
several weeks.

The fact that both immediate and delayed-type reactions can
be induced by hapten carrier conjugates has obvious advantages to
investigators of the basis of immune specificity and the results can
readily be extended to the agents concerned in contact dermatitis
by making the assumption, for which there is experimental basis
(Salvin and Smith, 1961), that the actual antigenic agent is a natural
conjugate of the chemical with a skin protein.

The standard finding can be exemplified in the work of Gell
and Benacerraf (1961), who studied the reactions and antibodies
produced with 2,4,dinitrophenyl antigens. A guinea-pig sensitized
with the conjugate DNP-*A* will give a skin reaction of delayed
type with DNP-*A* but not with DNP-*B*, where *A* and *B* are unrelated carrier proteins. However, when immunization with DNP-
*A* results in the formation of antibodies these will also react with
DNP-*B*, and immediate skin reactions of Arthus type can be
elicited with either antigen.

In similar fashion, guinea-pigs sensitized by skin contact with
unconjugated picryl, that is, by picryl skin protein conjugate produced *in situ*, react poorly with picryl conjugated with a foreign
protein and are poor antibody producers. Guinea-pigs sensitized
by picryl-BGG are poor reactors when tested for contact sensitivity
(Gell and Benacerraf, 1961).

The implication that both carrier and hapten are concerned in
the DH reaction is strengthened by the finding that both are also
necessary if unresponsiveness is to be produced in animals under
cyclophosphamide treatment (Salvin and Smith, 1964).

A more detailed approach to the nature of the AD in DH reactions can be obtained from the experiments of Schlossman and

Levine (1967) in which haptens of known structure and varying length were used. These were composed of $\alpha$-DNP united to a peptide chain of 3 to 10 L-lysine residues. Guinea-pigs were sensitized by DNP-lys 9 or an unresolved mixture containing long chain peptides. It had previously been shown that only peptides with 7 or more lysine residues were immunogenic. The sensitized guinea-pigs were then tested with the various hapten-peptides and with DNP-(lys)5 and DNP-(lys)9 conjugated to a protein carrier. Results showed consistently that where the number of lysine residues was 6 or less, an Arthus (3–6 hour onset) reaction only was produced, while with 7 (usually) and with 8 or more a regular DH reaction was produced. Experiments on the blocking effect of hapten on precipitation of a DNP-(lys)11-protein by antiserum showed that maximal blocking was reached by the heptamer, that is, at the same point where immunogenicity appeared.

The interpretation by the authors is that the DH reaction depends on the local biosynthesis of sufficient antibody by sensitized lymphocytes to form antigen–antibody complexes capable of producing tissue damage. In our view it can be equally well interpreted on the basis that only those cells with which AD and receptor can make a sufficiently precise (avid) union are effectively stimulated either to initiate a primary antibody response or to show the degree of damage and liberation of pharmacologically active products characteristic of DH. This precise union necessarily involves a larger segment of the receptor (or CS on antibody) than is involved in stimulation to secondary antibody production or than is needed to block antigen–antibody union. This is a widely accepted point of view but it is usually stated with an implicit or explicit assumption that the antigen in some way causes the construction or appearance of the larger CS. This is inadmissible on any theory of the genetic origin of immune pattern so that it is necessary to provide an acceptable interpretation of the different specificity in terms of a strictly selectionist approach. As in so many other contexts, a full discussion would require material more appropriate to other chapters. In this instance there will need to be consideration of the range of specificity for the same group of antigens of (*a*) DH, (*b*) antibody production, (*c*) inhibition of (*a*) or (*b*) by *in vitro* action and (*d*) induction of tolerance. The

497

fourth, which we interpret as lethal stimulation of immunocyte by antigen, will be treated in chapter 10.

In concentrating on the difference in the specificity of DH on the one hand and immediate hypersensitivity with the correlated production of antibody on the other, we should consider two sets of facts. There is first the well-known finding that special conditions are needed to obtain pure DH without antibody production, such as to present the animal with material in low concentration or of low intrinsic antigenicity. The work of Uhr *et al.* (1957) making use of small doses of antigen–antibody precipitate has already been referred to. Using a different approach, Leskowitz (1967) showed that a very small antigen, arsanilate as hapten conjugated to N-acetyltyrosine, can evoke DH which is hapten-specific (but which must be tested for with a hapten–protein carrier solution) and gives no production of antibody. With more conventional types of antigenic material in larger doses, antibody production and de-sensitization is likely to occur.

The second point to be underlined is that, subject to the inevitable limitations arising from the heterogeneity of the experimental material, the evidence favours the view that a single type of immune receptor pattern is concerned in both DH and antibody production. Martins *et al.* (1964) demonstrated fluorescence by the sandwich technique in lymphocytes implicated in DH reactions; this indicates at least close correspondence of the two CSs concerned. Uhr and Pappenheimer's (1958) work on desensitization and its relation to antibody formation points in the same direction.

If we accept both these points, it becomes necessary to postulate that there are conditions, other than the fact of specific contact between AD and CS, which determine the nature of the response. The simplest approach which seems adequate to cover the facts is based on the avidity–effective antigen concentration relationship (fig. 34).

We assume that for an uncommitted progenitor immunocyte to become a committed immunocyte, it must be stimulated by an avid union of minimal degree, that is, involving one receptor only. If more than a certain minimum were concerned, death of the cell would presumably result, hence the need for a very low effective concentration of antigen. In the committed immunocyte stage we have more receptors (or fixed antibody) present and more easy of access for ADs with less than the full specificity–avidity needed for primary stimulation. Any stimulation of an immunocyte which is capable of producing a DH reaction is, by definition, one which will provoke cellular damage. Combining our present point of view with that of Karush and Eisen, we assume that only immunocytes with a closely complementary steric relationship to the AD will react in this fashion. Such contact will be lethal for the cell if the effective concentration of the antigen rises beyond a certain level.

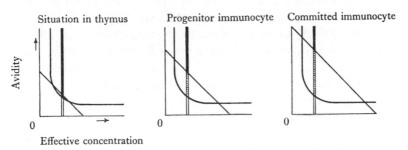

Fig. 34. To show the use of fig. 22 under three circumstances: 1. On differentiation in the thymus; 2. Progenitor (uncommitted) immunocyte; 3. Committed immunocyte. Open = no action; hatched = proliferation and/or antibody production; black = destruction of stimulated cell.

A less avid contact involving a smaller proportion of the AD will be less damaging and therefore capable of allowing antibody production or memory cell formation.

This is one point of view, but the trend of opinion is that there may be qualitative factors, only indirectly or not at all concerned with relative avidity for antigen, which determine whether immunocytes can function as agents of DH. For reasons developed in other sections my current predilection is to regard cells subserving DH as being equivalent to one of the defined types of immunoglobulin producer, perhaps the most primitive of them all. Immunocytes then would be classed as those potentially or actually concerned with DH, Ig M, Ig G, Ig A or one of the rarer types. The possibility that DH cells might be members of a clone capable also of giving rise to immunoglobulin producers must still remain but, as in so many other fields of immunology, an experimental

499

decision may have to await refined tissue culture methods which can handle pure clones of functional immunocytes. It also seems highly probable that the number and interconvertibility of the basic types of immunocyte will differ from one species to another.

Other aspects of DH, and in particular the process by which suitably reactive immunocytes accumulate in the body, are better discussed at a later stage (chapter 11).

### The signal produced by antigenic determinant–receptor union

With most writers still reluctant to rule out the possibility that specific stimulation of immunocytes may be due to external antigen–antibody union secondarily involving the cell, and an even greater majority considering that a two-cell situation—macrophage or dendritic reticulum cell on the one hand and lymphoid cell on the other—is involved, no investigator seems to have approached the problem from the angle of Paton's (1961) and Shulman's (1964) theories of drug action. The following are among the factors which may be concerned.

(*a*) The value of association and dissociation co-efficients with the important qualification that in view of the normal heterogeneity of dispersal of ADs on the antigenic particle or in relation to the surface of another cell, it is most improbable that model situations in which these values can be determined will ever be found.

(*b*) The number of receptors accessible on the cell and if an equilibrium situation is attainable the proportion occupied by the AD at any given moment. If this situation exists it would correspond with the 'occupation' hypothesis of drug action theory.

(*c*) According to the 'rate' concept of drug theory, if the rate of association of drug with receptor reaches a threshold value, stimulation occurs. In the immunological situation it becomes very difficult to know what 'rate' means. Conceivably it could refer to the speed with which, once receptor and AD have been manoeuvred (as it were) into position, the complementary steric patterns can reach a provisional stabilization. The intensity of response would then be a direct function of avidity as so defined with the corollary that multiple or rapid recurrent stimulation of this type would be destructive.

The nature of the process set going by the AD–receptor contact is unknown. According to Cooper and Rubin (1966) the change induced in human lymphocytes by streptolysin O, presumably acting as an antigenic stimulus, has the general character of RNA mobilization for synthetic activity, microsomal RNA being conspicuous. Stimulation by PHA has a quite different result, with much increased production of rapidly labelled RNA of messenger type without the normal production of ribosome precursor and ribosomal RNA. This points strongly toward the illegitimacy of regarding PHA stimulation as equivalent to specific antigenic stimulation.

### The release of pharmacologically active material from specifically stimulated cells

Almost any form of cellular damage will produce one or more of a large number of pharmacologically active materials, some, such as histamine and serotonin, being well-defined small molecules. There is a relatively large group of peptides of kinin character and a number of proteins with enzymatic activity (see Cameron and Spector, 1961). All of these have been shown to be liberated in some of the direct or indirect results of antigen–antibody reaction *in vivo*.

Histamine is the most important agent concerned in the early stages of most vascular reactions following cellular injury. It is probably mostly derived from mast cells being liberated (with serotonin in mice and rats) when discharge of granules occurs. Spector and Willoughby (1964) believe that an SH-containing enzyme may be concerned in its release. There appears to be no evidence whether histamine is or is not liberated on immunological or other damage to lymphocytes. The fact that in the mouse thymus massive conversion of thymocytes to mast cells (Burnet, 1965 b) can occur would perhaps make it likely that refined methods might well show liberation of small amounts of histamine. In the rat, serotonin (5-hydroxytryptamine) is liberated in association with histamine and, if passive cutaneous anaphylaxis is to be prevented, an anti-serotonin drug as well as antihistamine must be used (Humphrey *et al.* 1963).

The later phases of vascular responses associated with immunological or other injuries to cells are probably largely due to kinins, peptides produced by cell proteases. The globulin permeability factor of Miles and Wilhelm (1955) may be a precursor of one or more such enzymes.

In a direct attempt to study the pharmacological aspects of DH reactions, Willoughby *et al.* (1964) extracted from guinea-pig lymph node cells a 'lymph node permeability factor' (LNPF) with a powerful action in increasing vascular permeability. Extracts from the tissue involved in a tuberculin reaction in guinea-pigs taken at different stages of the response show an effect on permeability, which rises and falls with the intensity of the tuberculin reaction (Spector and Willoughby, 1964). Recently, Willoughby *et al.* (1965) have found that LNPF is antigenic and that treatment of an animal with antibody against LNPF of its own species greatly reduces the extent of typical DH reactions. Neonatal treatment of rats with rat LNPF results in an unresponsive state later in life, shown by failure to react to LNPF and to certain hypersensitive reactions (Willoughby, 1966). The recently introduced drug 'Indomethacin' appears to be a strong inhibitor of LNPF action but also has an action against the enzyme kallikrein, perhaps indicating an enzyme action of LNPF (Walters and Willoughby, 1965).

The significance of these findings for the interpretation of DH reactions in general can be left to chapter 11. In the present context the most important aspect is in regard to the circumstances in which stimulation of lymphocytes by antigen leads to reaction in adjacent cells. The experiments most directly relevant are those of Dutton and Harris (1963) which arose from previous work by Dutton and Eady (1962) already referred to, in which it was shown that antigenic stimulation of spleen cells from a previously primed rabbit caused a sharp uptake of thymidine. The later experiments showed that cells from an immunized rabbit lost nearly all their capacity to take up thymidine after brief contact with antigen if they had previously been irradiated (300 r) or frozen and thawed. Cells which were not 'primed' by addition of antigen also took up little thymidine. If, however, irradiated cells were primed with antigen, washed, and then added to intact but unprimed cells, there was a very active uptake. It appears that a process was initiated in the irradiated cells which gave rise to something which, passing to non-primed cells, could set off the synthetic process in them. The most likely interpretation is that antigen–receptor contact allowed the formation and liberation of a pharmacologically active substance which could stimulate any cell of the appropriate type and physiological state to proliferation. The alternative of a transfer of specific immunological pattern seems vanishingly unlikely, particularly in view of the nonspecific action of PHA.

This type of process becomes of special significance in instances such as those described by Mitchell and Miller (1968) where the interaction of thymus-dependent cells is necessary to allow the appearance of antibody-producing cells which can be identified as of non-thymic origin. The simplest hypothesis, also relevant to Nossal and Mäkelä's (1962a) finding that plasma cell precursors are activated *before* contact with antigen, is that antigen contact of a thymus-dependent immunocyte capable of reacting specifically with antigenic determinant A will release products stimulating any adjacent immunocytes of non-thymic origin to the first stages of mitosis including any of pattern reactive with A. Such a mechanism obviates the undesirable necessity of postulating any transfer of genetic information.

The second example is taken from the work of David *et al.* (1964a) who were interested primarily in studying DH *in vitro*. They used peritoneal exudate cells from normal, sensitized or immunized guinea-pigs and tested their ability to migrate from a packed deposit in a capillary tube into fluids with and without added antigen. Without going into any details about the technique it can be said that cells migrated freely in the absence of antigen but when the antigen used for sensitization or immunization— PPD, diphtheria toxoid or ovalbumin—was added, no migration occurred. The inhibition was specific. It seemed evident that contact with antigen led to some surface change 'stickiness' which prevented normal separation and movement.

Three aspects of this work were of special interest:

(*a*) It was found impossible to confer this reactivity on normal peritoneal cells by any treatment with serum from sensitized or immune animals.

(*b*) If sensitive cells were mixed with normal cells and tested against the antigen used for sensitization, the mixture behaved like cells from sensitized guinea-pigs even when the proportion of sensitive cells to normal cells was as low as 2·5 to 5 per cent. This, of course, makes it quite possible that the 'real' number of reactive cells might have been very much less than 100 per cent of those obtained from the peritoneal cavity of the sensitized animal.

(*c*) Treatment of cells by mild trypsinization completely removed the capacity of the cells to be inhibited by antigen (David *et al.* 1964*b*).

Subsequent work by Bloom and Bennett (1966) was concerned with the respective roles in this phenomenon of the macrophages and the lymphocytes present in guinea-pig peritoneal fluid. They separated the two types primarily by using adhesion to glass as the differentiating factor and showed that the observed migration was a function of the macrophages and that these, in the absence of lymphocytes, were not inhibited by PPD. The inhibition of a mixed cell suspension was due to the release of an agent produced by contact of sensitized lymphocytes with the corresponding antigen which in some way paralysed macrophage migration. As few as 0·6 per cent of sensitive lymphocytes with normal macrophages could be effective in inhibiting migration in the presence of PPD. The agent was not produced by the stimulated lymphocytes in the presence of mitomycin and is presumably protein in character.

# 9 Antibody production

There has been an intimidatingly large amount of work on various aspects of antibody production and there are many recent seminars and reviews (Šterzl, 1965; Wolstenholme and Knight, 1963; Haurowitz, 1965; Gowans and McGregor, 1965; Nossal, 1962).

It will be necessary, therefore, to confine the various sections of this chapter to those topics which bear directly on the interpretations introduced by the clonal selection approach. Nothing will be included, therefore, in regard to the general character of primary and secondary responses or the influence of different routes of inoculation or infection. No mention will be made of the influence of antibody production on the course of infection.

From a clonal selection point of view the topics which seem to require experimentally based justification are:

(a) Phenotypic restriction of cells to production of one antibody and one immunoglobulin type. This is the general rule but its qualifications, notably the existence of double producers and modification of antibody after secondary stimulation with a related antigen, also need consideration.

(b) The concept of the phagocytic reticulum of lymphoid tissue as the dominant route by which antigen is presented to immunocyte.

(c) The significance of nonspecific immunoglobulin production in the light of Nossal and Mäkelä's (1962a) findings.

(d) The nature and mechanism of the Ig M → Ig G transition and the possibly related problems of uncommitted and committed immunocytes.

(e) The ontogeny of the antibody response.

## NONSPECIFIC GLOBULIN PRODUCTION

### The control of immunoglobulin levels

If we are correct in assuming that all immunoglobulin molecules are *ipso facto* antibody molecules, there are some important questions to be answered.

In the absence of severe infection the normal human individual has a fairly constant level of about 0·7–1·5 g per cent of immunoglobulins in his serum. In adults the ratio of Ig G:Ig A:Ig M is approximately 10:3:1 (Fahey, 1962). Since the half-life of Ig G and antibody in man is about 14 days, there must be a constant replacement of immunoglobulin and presumably some type of homeostatic control. In other words, we must have a constant replacement of Ig M-producing cells and the conversion of an appropriate proportion of these to proliferating plasma cells producing Ig G and Ig A. There is, however, very little direct evidence bearing on the nature of the homeostatic mechanism if it exists. It is possible, indeed, that the best approach to understanding may be to look to morphogenetic controls determining the overall size and distribution of lymphoid tissue in the body as indirectly determining the dynamics of the immunocyte system and hence immunoglobulin levels. This could well be enough to account for the rather wide range of standard levels of plasma immunoglobulins without the necessity of a feedback based on plasma concentration with some appropriate sensor–effector circuit. The weakness of such an approach is, of course, the absence of any substantial knowledge on the mechanism of morphological control. At present the only practical approach is to look at the conditions, clinical or experimental, associated with abnormally low or high levels of circulating immunoglobulins.

Abnormally low levels of immunoglobulin are characteristic of sex-linked congenital agammaglobulinaemia. There are traces only of Ig M, and antibody is not produced with any of the standard stimuli. Plasma cells are absent and the basic lesion seems to be an inability of the immunocyte to differentiate to the plasma cell series. Since the abnormality is based on a single gene it is reasonable to assume that there is a failure to produce some key protein hormone or enzyme in functional form. I have suggested that acquired agammaglobulinaemia may well represent an autoimmune condition in which the target auto-antigen is this same key protein.

In these conditions we have the elimination of a major function rather than a deviation of a homeostatic mechanism. In the absence of any way of identifying or assaying what was elsewhere called a hormone from 'gut-associated lymphoid tissue', it is impossible to know whether changes in the circulating level of that hormone are related to the ease with which plasma cells develop and immunoglobulins are produced. This could certainly represent an important aspect of a homeostatic control.

There is evidence that abnormally high levels of immuno-globulin, resulting from intensive immunization, fall at a rate considerably greater than that corresponding to the normal half-life. This was shown for rabbits given polyvalent pneumococcal antigen by Catsoulis *et al.* (1964). This is perhaps the best evidence for the existence of a homeostatic mechanism controlled by the actual level of circulating immunoglobulin.

Hypergammaglobulinaemia of a polyclonal character is seen especially in persons with autoimmune disease or in close relatives of such patients (Leonhardt, 1957). Mackay and Burnet (1963) have regarded hypergammaglobulinaemia as one of the 'markers' of autoimmune disease. Particularly high levels are observed in lupoid hepatitis.

The other medical field in which high levels of immunoglobulins are constantly being observed is in relation to tropical disease. It is almost regular to find abnormally high values in adult natives of regions where malaria is hyperendemic, such as West Africa (Cohen and McGregor, 1963) and New Guinea (Curtain, 1966). Although the possibility of other types of infection also being concerned is not wholly excluded, the work of Rowe and McGregor (1967) makes it likely that malarial infection is directly responsible. In Gambia, children show low immunoglobulin levels when maternal Ig G has disappeared but show a rapid development of Ig M associated with experience of malaria and later Ig G. The high immunoglobulin level in West African adults is associated with a rate of synthesis of Ig G three to five times that of Europeans.

Although there are some discrepancies in the literature, the weight of opinion (Schofield, 1957) is that expatriate West Africans free from malaria show γ-globulin levels of the normal European range.

## Globulin increases associated with antigenic stimulation

In experimental animals any intensive course of immunization is likely to produce an easily measurable increase in 'γ-globulin' or in one or more immunoglobulin types. In general, the increase in the amount of globulin is greater, sometimes much greater, than the amount accountable for as specifically adsorbable antibody.

Humphrey (1963) provided a striking instance from his analysis of the Ig G rise observed in rabbits injected with Freund's complete adjuvant alone, that is, with only killed tubercle bacilli as antigen. The animals showed a large increase in Ig G, far more than could be accounted for by antibody against tubercle bacilli. Examination for antibodies that the animals were known to have been producing and for standard 'natural' antibodies showed trivial increases at most. In the discussion of this paper, Kabat confirmed the variability of the relationship between immunoglobulin increase and antibody rise in titre. In one immunized rabbit, almost all the immunoglobulin increase can be accounted for by antibody, in another less than 10 per cent. There are two reports which suggest that in many instances approximately equal amounts of specific antibody and of nonspecific or random immunoglobulin are produced. Such experiments, requiring estimation of the amount of isotopically labelled amino acid incorporated into antibody and immunoglobulin, have been reported by Askonas and Humphrey (1958) using isolated perfused lungs of immunized rabbits and by Valette-Robin (1964) who used the intact rabbit.

Another experimental approach comes from the finding by Rittenberg and Nelson (1960) that when rabbits were X-irradiated (400 r) 24 hours before primary immunization with bovine serum albumin, there was a sharp rise in both $\beta$- and $\gamma$-globulin levels but no demonstrable production of antibody. There is evidence of nonspecific stimulation of immunoglobulin (19S) and natural antibody production following injection of bacterial lipopolysaccharide (endotoxin). Rowley (1964) found '$\gamma$-globulin' to be increased by up to 50 per cent over 1–4 days afterwards.

On the axiom that all immunoglobulin is antibody, all these findings must mean that under certain conditions a random sample of immunocytes must be stimulated nonspecifically to antibody production. There are several possible interpretations. For some years I regarded Askonas and Humphrey's (1958) results as probably indicating that each immunocyte had the capacity to produce two (allelic) types of CS. If, for example, a dozen clones of cells were combining to produce antibody *a* under stimulation by antigen *A*, each cell would also produce an equal amount of some unrelated antibody, *b, d, f, x*, etc. The mixture of these would not be recognizable as antibody. Such a suggestion only makes sense if immunocytes, except under special circumstances, are regarded as heterozygous and the two regions concerned in determining immune pattern are co-dominant. The apparent universality of phenotypic restriction in immunocyte product, however, makes this a very unlikely answer. It will, however, not be formally excluded until a full study of monoclonal antibody (Osterland *et al.* 1966) has been carried out from this point of view.

A second possibility is that wherever cells are being stimulated immunologically by contact with their specific antigen, there is a liberation of pharmacological substances from the stimulated cells which, in their turn, can provoke antibody production by adjacent cells of unrelated immune pattern.

A third related hypothesis is that, with the local liberation of antibody, some cytophilic antibody will attach to unrelated immunocytes which will then be liable to stimulation with the antigen present.

There are analogies for both these processes, while the striking degree of phenotypic restriction in all aspects of immunoglobulin production points against the general assumption of two immune patterns per cell.

A fourth possibility may also need consideration. This is simply that, for one reason or another, an immunocyte which can be stimulated by antigen *A*, that is, having a CS receptor that can react with *A*, produces antibody of such low avidity that it is not demonstrable by the assay being used.

It is unlikely that any final conclusion can be arrived at on the available experimental evidence but the following points can be referred to.

In discussing the immunocyte response, Dutton and Harris's (1963) work was described in which incorporation of tritiated thymidine was induced in non-sensitized cells in association with X-irradiated immunologically primed cells, in the presence of antigen. Other effects of the same general character are described on p. 513.

The production and distribution of cytophilic antibody has hardly been studied, but Nelson and Mildenhall (1967), using the capacity of mouse immune serum to transfer reactivity to peritoneal macrophages, found no correlation with other measurable antibodies against the same antigen (sheep red cells).

It is a universal experience to find when assaying what is assumed initially to be the same antibody by two or more different methods, that extensive differences in the ratio of titres are apt to occur in different animals and in the same animal at different stages of immunization. Blumer *et al.* (1962) noted that precipitin titres fell rapidly in immunized rabbits, but if the titre was measured against the same antigen bound to benzidine-treated red cells, it persisted much longer and might even rise when the precipitins were disappearing. It is not necessary to refer to the extensive literature on 'incomplete' antibody of low avidity which can be shown to be bound to particulate antigens, red cells or bacteria, by using an antiglobulin serum. A discussion of the mechanism of action of these low-avidity antibodies will be found in Long *et al.* (1963).

*The production of immunoglobulin and antibody in germ-free animals*
Germ-free animals are not necessarily free from foreign antigens but they at least lack the persisting and heterogeneous source of antigens in the micro-organismal population of the bowel. All workers find an abnormally low level of immunoglobulins. According to Gustafsson and Laurell (1959), germ-free rats show only 10–15 per cent of the normal $\gamma$-globulin level, while Ikari (1964) found an even lower level (2·8 per cent) in germ-free NIH strain mice.

When the rats used by Gustafsson and Laurell (1959) were exposed to normal rat intestinal flora there was a lag of 4 weeks before the immunoglobulins started to rise. With a 'mono-contamination' of *Staphylococcus albus* only, there was no significant increase. The implication is clearly in line with the view that before active specific antibody formation against many ADs can take place, there will have to be a gradual accumulation of cells nonspecifically activated by adjacent specific reaction.

'Normal antibodies' may be found in the serum of germ-free animals. For obvious reasons most work has been done with bactericidal antibodies to common gram-negative bacteria using foetal bovine serum as a source of complement free of natural antibody. In general, the results such as those reported and summarized by Ikari (1964) show that although antibody appears, it does so at a slower rate than in conventional animals and never reaches such a high level, and the antibody is more readily inactivated by heat. The amount is related to the level of $\gamma$-globulin and probably depends on the production of heat-labile 19 S antibody independently of specific stimulation by antigen. In the conventional mice there is a rapid re-stimulation from bowel antigens giving rise to heat-stable antibody; in the germ-free animals the minimal stimulation from small amounts of antigen in the sterile diet gives only a small late response.

In mice, Swiss–Webster type, germ-free animals show many fewer potential antibody-producing cells ($\pm 8$ per cent) than in conventional animals. However, when immunized with human $\gamma$-globulin (Ig G) or *Salmonella typhimurium* vaccine, the response was very little different from that of conventional mice and the proportion of potentially antibody-producing cells rose from 8 to 33 per cent (Olson and Wostmann, 1966 *a, b*). Once large amounts of antigen are used the build-up to almost full capacity takes place much more rapidly.

## Persistence of antiviral immunity

Another important series of phenomena which points strongly toward nonspecific stimulation of antibody production is found simply in the persistence of antibody after infection when there is no evidence of persistence of the infectious antigen (virus) in the body. It is a truism that immunity, usually with demonstrable antibody, may last for many years after a single infectious exposure to a virus such as measles or yellow fever. The classical study of measles in the Faroe Islands (Panum, 1847) showed that following an absence of measles for 60 years the only persons who escaped infection were those over 60 years old. This at least establishes that after recovery the patient is completely incapable of liberating virus in infectious form.

More recent studies of measles in which antibody titrations were used are reported by Black and Rosen (1962) for Tahiti. They found hardly any reduction in titre over 8 years and there were numerous individuals, whose only attack had been 30 years previously (with one possible re-exposure) whose sera showed the full immune titre. Sawyer (1931) described yellow fever antibody in a man whose only exposure must have been 60 years previously.

There is no basis in what is known of the behaviour of viruses *in vivo* to assume in such cases that there is a steady trickle of infectious virus from some cellular reservoir to spleen or lymph nodes. Wherever such persistence does occur it is readily recognizable from epidemiological and clinical data, as in the examples of herpes simplex, Brill's disease (chronic typhus infection) and transfusion infections with malaria, serum hepatitis or syphilis. Many years ago Jensen (1933) studied the rate of fall of antitoxin titres in children given a single injection of purified diphtheria toxoid. Using an empirically derived formula which, with appropriate constants, could be fitted to the curves obtained, he worked out that the time required for antitoxin titre to fall below the Schick level (0·03 AU) would vary in different children from a few days to more than 65 years.

Prolonged maintenance of a slowly falling antibody level in the absence of antigen can only be interpreted within the pattern of our theoretical approach if there are nonspecific ways by which cells can be stimulated to form plasma cell clones and produce antibody. The actual method is immaterial provided that it involves a random sample of the populations of immunocytes present in the body. Whenever a severe infection results in a massive proliferation of immunocytes, including necessarily large numbers of memory cells, the immunocyte population of the body will remain proportionately weighted in that direction for many years.

### A mechanism for nonspecific antibody production

The approach that will be adopted is that the stimulation of an immunocyte by the corresponding AD has the effect of stimulating adjacent immunocytes, particularly committed memory cells, to initiate DNA synthesis which is associated with an increased mobility and the possibility of a variety of immunological functions. As indicated earlier, it is immaterial whether the stimulus is transmitted via the liberation of pharmacologically active agents or by the joint action of antibody liberated by the primarily stimulated cell and antigen.

There are several sets of findings other than those already referred to which point in this direction. Recently activated mononuclear cells play an improbably large part (*a*) in being stimulated to proliferate to plasma cells in response to subsequent injection of antigen and (*b*) to accumulate at sites of subsequent challenge for delayed hypersensitivity.

The experiments of Nossal and Mäkelä (1962*a*) and Mäkelä and Nossal (1962) gave an unequivocal indication that the plasmablasts and plasma cells producing antibody in either primary or secondary responses were derived from cells which were incorporating thymidine into DNA *before* the antigenic stimulus. This has neither been confirmed nor refuted, although the possibility that the results depend in part on re-utilization of label at later stages has been raised by Nossal himself. Gowans and McGregor (1965), in reviewing the work, incline strongly to the view that at least in the primary response the small lymphocyte is the cell stimulated by antigen, but agree that this is not compatible with Nossal and Mäkelä's results. It seems to be quite out of line with general immunological thought to accept Nossal and Mäkelä's (1962*b*) picture of a series of proliferating lymphoblasts persisting indefinitely and sufficient in number to cover all immune patterns.

A simple hypothesis will bring Gowans and Nossal's points of view into concordance and make the results compatible with a standard selective approach. This is to accept the ability of antigen–immunocyte contact to activate and mobilize adjacent immunocytes from the small lymphocyte to a DNA-synthesizing blast form. In the normal animal, the commonest site for multiple contacts of antigen with specific immunocyte will be the mesenteric lymph nodes. We can adopt the evidence already discussed that there is a highly dynamic mixing of lymphocyte clones in the body so that in every lymphoid follicle there is likely to be a representative of every significant immune pattern in the body (see p. 399).

In the mesenteric lymph nodes and elsewhere, a random population of activated lymphocytes will then be passing into the efferent lymph vessels and lymph trunks. In their turn these lymphocytes will pass to other lymphoid accumulations via the bloodstream and postcapillary venules (Marchesi and Gowans, 1964). In every lymph node, then, one can reasonably expect to find DNA-synthesizing cells of all available clones, ready to accelerate their proliferation under the impact of *specific* stimulation and, in part at least, take on the plasma cell form. I have already noted how this approach is highly relevant to the recent findings of Mitchell and Miller (1968) which indicate the interaction of two sharply differentiated immunocyte populations, thymus-dependent (mediating delayed hypersensitivity) and non-thymus-dependent (concerned with plasma cell formation and antibody production).

Work on the cellular response during homograft rejection, or in the tissue where a delayed hypersensitivity reaction is developing, points strongly to the accumulation of recently proliferating cells not necessarily carrying specific immune patterns. A number of examples are given in Gowans and McGregor's (1965) review. McCluskey *et al.* (1963) showed that in sensitized guinea-pigs tested 1–2 days after an injection of tritiated thymidine and 9 days after the sensitizing dose, 90 per cent of the mononuclear cells in the infiltrate were labelled. In the same sense is the result of experiments in which sensitivity was conferred passively by adoptive transfer of lymphocytes from a sensitized donor. If tritiated thymidine was given to the recipient on the two days before adoptive transfer and skin test, 70–90 per cent of the infiltrate cells were labelled, i.e. were of recipient origin. All who have investigated the origin of cells involved in delayed hypersensitivity reactions are agreed that only small numbers of specifically sensitized cells are concerned (Turk, 1962; Najarian and Feldman, 1961). The only point to be made in the present context is the regular availability of large numbers of activated lymphocytes and their capacity to pass into tissues, both general (Gowans, 1962) and lymphoid (Hall and Morris, 1965). From the point of view of immunoglobulin production it must be assumed that a proportion of such nonspecifically activated cells become plasma cells but either do not proliferate or show a much lower level of proliferation than specifically stimulated cells. It will be recalled that most of the large cells in rat thoracic lymph which took up tritiated thymidine *in vitro* settled in the gut of recipient rats and took the morphological appearance of typical plasma cells.

## The significance of lymphocytic mobility

It is appropriate to re-emphasize in the present context the immense mobility of the immunocyte system. There is a constant discharge of lymphocytes from thoracic duct and other lymph trunks into the bloodstream; most of these are small but there is also a significant proportion in the DNA-synthesizing phase. These circulating lymphocytes leave the blood preferentially through postcapillary venules into lymphoid tissue, but others, especially at any site where capillary permeability has been increased by inflammation or otherwise, pass into the general tissues. From the lymph nodes there is a steady release of lymphocytes including about 4 per cent of activated cells (Hall and Morris, 1965) from a quiescent node. With local antigenic stimulation, very much larger numbers of DNA-synthesizing cells, both blasts and immature plasma cells, are found in the efferent lymph. These are the findings from the popliteal lymph node of the sheep. In all probability the behaviour of mesenteric lymph nodes, with their constant antigenic stimulation from the gut, lies in between the quiescent and the artificially stimulated popliteal nodes.

Again recalling the likelihood that every primary lymphoid follicle is in life a writhing mass of mobile lymphocytes in a sponge-like framework of reticulum fibres and dendritic phagocytic cells distributed along the fibres, we can see a highly efficient process of uniform mixing. It is a process by which every immunocyte in the body—progenitor or committed, quiescent or activated—has an equal chance to make effective contact with any antigenic unit held on the surface of the phagocytic reticulum cells. It is precisely the mechanism needed to make a selective mechanism for antibody formation 'work'. Probably the only speculative component in the picture is the assumption that indirect nonspecific stimulation of an immunocyte may (*a*) stimulate it to limited proliferation and production of the antibody proper to its genetic character and (*b*) render it highly susceptible to react with its corresponding AD should opportunity for contact occur. It is such contact that gives rise to the actively proliferating and maturing clones of plasma cells which dominate the histological picture of antibody production.

## THE RELATIONSHIP OF IGM TO IGG ANTIBODY

Since Bauer and Stavitsky (1961) showed that in rabbits the first antibody produced following immunization with protein antigens was of Ig M character with, after a few days, change to Ig G, there has been much study of this phenomenon. For technical convenience the differentiation is usually made by comparing the antibody titre before and after treatment with 2-mercaptoethanol (2-ME); the antibody resistant to 2-ME being regarded as Ig G. Where comparative assessments by sedimentation in the ultracentrifuge are made there is good agreement between 19 S and 2-ME-sensitive antibody.

From all the work that has been reported, the impression emerges that the basic response is in Ig M antibody and that ontogenetically, and perhaps phylogenetically, Ig G and Ig A are secondary developments. The natural antibodies found in human serum with bactericidal activity against Gram-negative bacteria are almost wholly Ig M. These are also present in those cases of dysgammaglobulinaemia in which essentially only Ig M is produced (Michael and Rosen, 1963).

### The early immune responses of human foetus and infant
The human foetus is immunologically almost inert under normal circumstances. Maternal Ig G antibodies pass relatively freely into the foetal circulation but not Ig M. Minute amounts of Ig M, presumably actively produced, are present in the serum of newborn infants (West *et al.* 1962).

In man and primates generally, Ig G appears early in foetal life (16–20 weeks in monkeys, Bangham, 1961). Appropriate tests with labelled proteins in monkeys indicated that γ-globulin was selectively transferred across the placenta from mother to foetus and not vice versa. The other plasma proteins with the exception of a small movement of albumin from mother to foetus did not cross the barrier in either direction. Human cord blood usually has a slightly higher γ-globulin content than the mother's blood and it is not uncommon to find a higher antibody titre (Bryce and Burnet, 1932), suggesting a positive transfer across the placenta.

In man there is some evidence that antibody production by the foetus can result from antenatal infection with rubella virus. Human foetuses 24–38 weeks old suffering from congenital syphilis or toxoplasmosis were examined for plasma cells by Silverstein and Lukes (1962), and ten out of sixteen syphilitic foetuses and one out of three from toxoplasma infections contained very large numbers of plasma cells. None was found in control embryos of the same age range. Premature infants respond significantly to some antigens. Smith (1960) injected premature infants on the first day of life with TAB vaccine. Most of them gave a good agglutinin titre to one of the six potential antigens present, *Salmonella typhi*, H antigen; no antibody against typhi O or against any of the paratyphoid antigens was produced. Physical study of the H agglutinin produced by these children showed that it was sensitive to 2-ME and sedimented as 19 S, that is, it was composed of Ig M. Fowler *et al.* (1960) found that premature infants readily reject skin homografts.

The sequence of γ-globulin levels in the serum of children after birth has been extensively studied. The clearest picture is probably seen in the studies of infants in families subject to agammaglobulin-aemia. Children born of mothers who are agammaglobulinaemic have values under 10 mg per 100 ml and maintain those low levels for some time, 3 to 6 weeks in the two instances examined by Bridges *et al.* (1959). Thereafter there is a steady rise which attains the normal level at around six months of age. Three mothers who had previously borne sons with congenital agammaglobulinaemia produced male infants subsequently shown to be agammaglobu-linaemic. They were born with amounts of γ-globulin equal to or slightly above that of the mothers. The levels fell steadily at first, quite parallel with those of normal children but, when after three to six months the normal values rose, these continued their logar-ithmic fall to a level about 20 mg (Good *et al.* 1962).

Passing to the behaviour of normal neonates it is clear that, in man, antibody responses are poor in the neonatal phase but the response varies with the antigen and the degree of maternal immunity. Relatively good responses to living polio-virus were obtained in infants under one month by Koprowski *et al.* (1956) while a good response to diphtheria toxoid is obtained from neonates provided their maternal antitoxin is absent or very low (Osborne *et al.* 1952). Using a small particle phage, Uhr *et al.* (1962*a*) found a rapid production of antibody in premature infants to the same level as in older children although in the same infants there was no response to diphtheria toxoid. All the early antibody was of 19S type.

Taking these results with other investigations, the human position can probably be summarized as follows:

Maternal 7S globulin and antibody fall logarithmically with a half-life of the order of 20–30 days so that there are only traces remaining after six months.

Actively produced γ-globulin production is initiated between three and six weeks with a steady following rise to reach full adult levels around three years.

Trace amounts of 19S antibody (Ig M) may be produced in foetal and neonatal life, but insufficient to influence γ-globulin levels unless very delicate assays are used (Roth, 1962).

*Foetal and neonatal responses in experimental animals*

An excellent review of the relevant aspects of foetal and neonatal immunoglobulin levels is given by Good and Papermaster (1964). The main patterns seen in mammals can probably be summarized in three groups—the ungulates, with pig, lamb and calf as examples; the rabbit; and man.

The absence of transfer of antibody across the multi-layered ungulate placenta and the consequent importance of the colostrum for the transfer of maternal immunity has long been known.

Pre-colostral calf or pig serum contains only very small amounts of immunoglobulin and there is some evidence that natural antibodies are present (Weidanz and Landy, 1963; Muschel and Jackson, 1963) though this was not found by Šterzl *et al.* (1961). Where the point has been examined the globulin was of Ig M type.

The foetal lamb is a relatively active potential producer of antibody from an early stage (Silverstein, 1964). The gestation period is 150 days and antibody was produced by the following antigens:

$\Phi$X 174 from 41 days; ferritin from 66 days; egg albumin from 125 days. Two antigens, *Salmonella typhi* vaccine and diphtheria toxoid, were ineffective at any stage. As previously shown by Schinckel and Ferguson (1953) homograft rejection was seen at 85 days but sensitization by BCG was never achieved. First production was always of 19 S antibody but 7 S appeared in older foetuses (Silverstein *et al.* 1963). Plasma cells are found in the course of these reactions, though not in association with homograft rejection.

Similar studies in cattle using *Leptospira* as antigen have been reported with production both of plasma cells and antibody at less than 164 days of gestation (Fennestad and Borg-Petersen, 1962).

In rabbits, Brambell and his group (Brambell, 1958) have shown that transfer of maternal immunoglobulin in the uterus is a complex process via yolk sac and amniotic fluid. Extensive studies of antibody production in the rabbit show in general that during the first four days of life there is liable to be no response to a protein antigen (Dixon and Weigle, 1957) and a 'good' response is not obtained till four weeks of age. By giving large doses of particulate antigens such as red cells and bacteria, an earlier response can be obtained (Deichmiller and Dixon, 1960). Newborn rabbits show no plasma cells and only Ig M is produced during *in vitro* culture of spleen and lymph node tissue though some Ig G may be produced by lymphoid tissue of the intestinal tract (Thorbecke, 1960).

Active production of $\gamma$-globulin is first observed about the fourteenth day of life, and it is of interest that the first appearance of plasma cells (Thorbecke, 1960; Thorbecke *et al.* 1962) is in the appendix at 8 to 10 days of age. Essentially similar results are seen with other animals; guinea-pigs (Bishop and Gump, 1961), pigs (Šterzl *et al.* 1961) and chickens (Wolfe *et al.* 1957).

*The Ig M–Ig G sequence in experimental animals*
The most detailed study available is that of Uhr *et al.* (1962*b*) and Uhr and Finkelstein (1963) using $\Phi$X 174 as antigen and guinea-pigs as experimental animal. In this system, production of 19 S antibody begins on day 3 after a single injection of antigen,

reaches a peak at 6–7 days and is falling when Ig G antibody first appears at day 9. Thereafter there is an exponential rise for 4–5 days and persistence of 7S antibody production for at least a year. If a suitable dose is chosen, some animals will produce only 19S antibody, while others give in addition a small amount of 7S. If such animals are given a second injection of the same antigen in a larger dose some weeks later, the first group producing initially only 19S now again gives a primary-type response with 19S followed by small amounts of 7S. The second group gives a true secondary response with a rapid rise of 7S antibody obscuring any Ig M which might have been produced. The implication is that immunological memory is established concomitantly with the ability to produce Ig G antibody. Another important feature observed in this system is the short duration of the primary 19S response. If, on the other hand, the first injection of antigen is large enough to provoke a substantial 7S response as well, this persists for at least a year.

In rabbits, Svehag and Mandel (1964*a*, *b*) used polio-virus as antigen to show almost precisely similar phenomena, but with some additional points. They found that most normal rabbits have minimal amounts of 19S-type neutralizing antibody..With a small dose of antigen this starts to rise in 8–12 hours and stops being produced in 4 or 5 days; no 7S is produced. With a larger dose, all rabbits make both 19S and 7S, the Ig G response being longer lasting. By choosing an intermediate dose only some rabbits give an enduring Ig G response. This is an 'all or nothing' phenomenon and presumably means that one progenitor cell is adequate to allow emergence of an enduring reaction. The secondary response is seen only when a primary 7S production has occurred. By direct assay only 7S is present, but by the use of zone ultracentrifugation it can be shown that a proportion of new 19S antibody is also produced.

In rats, using *Salmonella* flagella as antigen, Nossal *et al.* (1964*a*) found that with doses of 1 *μ*g or less the first response was 2-ME-sensitive initially, being largely replaced by 7S after 7 days. The size of the 19S response was clearly dose-related but there was virtually no difference in the persistence or height of the primary 7S response for doses from 0·1 to 10·0 *μ*g. The response to soluble flagellin was quite different. No 19S antibody was produced and with doses from 0·1 to 10·0 μg there was no significant difference in the course of the slowly initiated but long-lasting response.

Finally we can mention a somewhat similar set of studies in rabbits by Fazekas and his group using purified influenza virus and, in some experiments, the solubilized antigenic units of the virus coat. According to Fazekas and Webster (1964) the significant AD is carried by a protein of around 60,000 mol. wt of which approximately 2,000 molecules are present on each virus particle, one at the outer tip of each of the surface spikes seen in electron micrographs of virus. Immunization of a rabbit (and human results are comparable) with virus particles by the intravenous route gives a rapid appearance of antibody in 2–4 days which is initially wholly 19S in character. Around 7–8 days there is often an indication of a second rise in titre which corresponds to the appearance of 7S antibody. As in Nossal's experiments, a very small dose of antigen is sufficient to produce a maximal response (0·3 μg). With smaller doses the results become more irregular and the response is greatly delayed, and 100- to 1,000-fold below maximal.

Less extensive studies have been made with the antigenic sub-units, but with a maximal dose the same antibody level is reached but the rise begins about 7 days after the single intravenous injection and is wholly of 7S-type antibody.

## Cellular relationships of Ig M and Ig G

The only specific tests of single cells for the type of globulin they are producing are reported by Nossal et al. (1946c), who examined cells from rat lymph node around the time at which change in the character of antibody from Ig M to Ig G was taking place, that is, around the sixth and seventh day. Before this time all positive cells were 19S. In all, they noted 42 Ig M producers, 64 cells producing Ig G only and 17 producers of both Ig M and Ig G. Mellors and Korngold (1963), in studying the distribution of cells in human lymphoid tissue for the antigens μ, γ and α, found most cells produced only one type but noted that an occasional plasma cell showed the presence of both Ig M and Ig G. In rabbits, Cebra et al. (1966) using similar fluorescent antibody methods found all reactive cells to contain one type of immunoglobulin only.

The overall impression is that a cell when first stimulated by the corresponding AD produces 19 S antibody but at a certain point the cells currently producing 19 S antibody switch·rather rapidly to 7 S production. This, however, is not universally accepted and certainly requires qualification to account for the fact that with polysaccharide antigens Ig M antibody may go on being produced

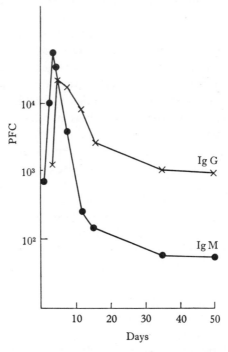

Fig. 35. The response to immunization with sheep red cells in mice as shown by the numbers of plaque-forming cells producing Ig M and Ig G antibody. (Redrawn from Wigzell, 1966.)

for long periods. There is, however, no doubt that in the best-studied system concerned with the production of Ig M antibody—the Jerne antibody plaque system in the mouse spleen—the number of Ig M-producing cells reaches a peak about 4 days after immunization, and thereafter cells producing the much more weakly haemolytic Ig G antibody rapidly increase in number (Wigzell *et al.* 1965) (fig. 35).

There is evidence from several sources that the change may be associated with the action of 7S (Ig G) antibody on the 19S producers. It has been clear since the work of Uhr and Baumann (1961) that there is an inhibitory effect of antibody (Ig G) on the production of antibody of the same specificity. Further work shows that the main effect is the inhibition of the 19S response by relatively small amounts of specific 7S antibody. The effect is clearly seen in the experiments of Möller and Wigzell (1965) where the administration of Ig G antibody stopped any increase in the number of plaque-forming cells in the mouse spleen. Sahiar and Schwartz (1965) used rabbits in which the Ig M response to BGG was abnormally prolonged by the administration of sub-suppressive doses of 6-mercaptopurine. By infusing Ig G antibody of the same specificity the production of Ig M was sharply inhibited. Unrelated antibody was inert.

The specificity of the effect must involve antigen at some point and it is tempting to think of Ig M production being necessarily associated with continued attachment of AD to cell receptor— an attachment that can be released by AD–Ig G antibody union. This is virtually equivalent at a dynamic level to the alternative hypothesis that, in the presence of Ig G antibody, any circulating antigen is unable to make effective contact with the receptors of progenitor immunocytes. Rowley and Fitch (1968) would in fact almost define the progenitor immunocyte as a cell which can be prevented from reaction with antigen by the presence of specific antibody.

Once any Ig G antibody is produced the induction of new immunocytes into Ig M activity ceases and it would be in accord with the observations if the presence of 7S antibody, in addition, accelerated a switch of 19S producers to 7S producers. This, however, has not been established.

Other immunosuppressive agents such as X-irradiation (Svehag and Mandel, 1964b) or 6-mercaptopurine (Sahiar and Schwartz, 1965) can also prolong the period of Ig M antibody persistence and this may mean that conversion to Ig G production is an intrinsic quality of the cell, or of the sequence of its early descendants.

Santos (1967), reviewing the relative effect of immunosuppressive drugs on Ig M and Ig G responses, emphasizes that under appropriately controlled conditions the appearance of Ig G in the later stages of the primary Ig M response can be eliminated by the administration of 6-mercaptopurine, cyclophosphamide or methotrexate as well as by X-irradiation. There is, however, consistent evidence that immunocytes engaged in the primary response, that is, producing mainly Ig M, and those in the later stages producing Ig G are equally susceptible to destruction by X-rays (Makinodan *et al.* 1962) and by cyclophosphamide (Santos, 1967). The differential effect cannot therefore be explained as resulting from the elimination of a more sensitive population of Ig G cells, and so favours the concept that cells initially producing Ig M either switch to become Ig G producers or have descendants with that capacity.

### The significance of the response to 'good' antigens

It is obvious that in rabbits, guinea-pigs and rats, given a single dose of a small amount of a potent microbial antigen, there is a consistent pattern of response. Using the common features of this response and bearing in mind the related findings that Ig G antibody has a capacity to swing Ig M-producing cells to Ig G production, it is possible to provide a relatively satisfactory interpretation of the findings in regard to the sequence of Ig M–Ig G antibody production to first and subsequent injections of a good antigen.

The interpretation which emerges is based on the postulate that there is a difference in quality between progenitor cells (X cells of Coons (1965), potential antibody-forming cells of Rowley and Fitch (1968), PC1 of Albright and Makinodan (1965)) and committed immunocytes (Y, antibody-forming cells or PC2 cells, respectively, of the above authors).

In any normal animal are small numbers of progenitor immunocytes capable of reacting with the relevant AD of a good antigen previously unmet. Nonspecific stimulation may result in the liberation of sufficient Ig M antibody to allow its experimental detection as normal antibody. Minimal effective contact with such a cell by antigen stimulates an immediate increase in antibody production and sets a process of commitment in action. Multiple or repeated contact with antigen may be destructive (see chapter 10) and the capacity of antigen to stimulate such cells is lost in the presence of Ig G antibody (Rowley and Fitch, 1968). Commitment for most cells is probably limited to (*a*) increased capacity to produce Ig M antibody and (*b*) limited proliferation with or without change to plasma cell morphology but with retention of Ig M-producing quality.

A small proportion of cells (which determine the 'all or nothing' effect of Svehag's experiments) take on under antigenic stimulation the character of the committed immunocyte. This can be stimu-

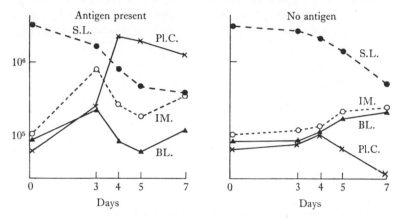

Fig. 36. Behaviour of cells from a primed mouse in diffusion chambers with and without antigen. Redrawn from Nettesheim and Makinodan (1965). S.L. = small lymphocytes; Pl.C. = plasma cells; IM. = intermediate cells; BL. = blasts.

lated by antigen in the presence of antibody, is inhibited only by large concentrations of antigen and can be readily provoked by renewed antigenic stimulation to produce either a subclone of typical plasma cells producing Ig G antibody (see fig. 36) or a population of memory cells.

In view of the evidence by Ada and Lang (1966) and others that antibodies are present in relation to the dendritic phagocytic cells of lymph follicles, it is probably necessary to assume that primary stimulation of progenitor immunocytes takes place elsewhere. Any accumulation of antigen in the dendritic phagocytic cells following the first injection will, however, be available for recurrent stimulation of memory cells once there is an initial population of committed immunocytes. The probability that 7 S antibody can specifically swing an Ig M-producing immunocyte to Ig G antibody production has been discussed earlier (p. 523).

The failure of 19 S antibody production with soluble antigens of flagella or influenza virus presumably depends on the failure of the signal resulting from contact with a progenitor cell to initiate production of antibody, although it can facilitate or at least not hinder the change to committed immunocyte. Once these are in existence, an Ig G response is possible.

It has been shown recently that a fragment of the flagellin molecule has virtually no antibody-producing power but is a potent agent in producing specific tolerance (Ada *et al.* 1968). Again, the phenomenon presumably depends on modification of the signal produced in the progenitor immunocyte according to the concentration and mode of presentation of the AD.

## *The meaning of the secondary response*

Many years ago I showed (Burnet, 1941) that when a rabbit was given a single intravenous inoculation of staphylococcal toxoid only a minimal production of antitoxin appeared 13–18 days later. A second intravenous injection of toxoid gave a very brisk response rising rapidly between the second and fifth days (see fig. 23, p. 195). This is something very different in character from what is observed with viral or flagellar antigens. It is therefore of much interest that although these experiments were done nearly 30 years ago we did show that whereas bacteriophages, herpes virus and bacterial vaccines used as antigens provoked local production of antibody in the draining popliteal lymph node, no such effect was seen even in second injections of staphylococcal toxoid. At that time, of course, nothing was known of the dendritic phagocytic cells but in retrospect a simple explanation of the difference is that the soluble toxoid antigen was readily broken down by body proteases and did not survive to persist on the dendritic phagocytic cells of lymphoid tissue. Only the initial 'pulse' of antigenic stimulation occurred with the production of a population of committed immunocytes but no opportunity for their stimulation until the second injection. In some ways this was similar to Nossal *et al.*'s

(1964a) finding with monomer flagellin, but here the antigen persisted and the first inoculation could serve as a source of continuing antigenic stimulus.

If this is true the conventional differentiation between primary and secondary responses is almost meaningless. The term 'primary response' should be limited to situations in which it is impossible for immunocytes and their descendants to receive more than a single 'pulse' of stimulation, either because the AD is rapidly destroyed in the body or when the first dose of antigen is so low that it is statistically impossible for any given immunocyte to be influenced by more than one antigenic contact.

With standard antigens the so-called primary response as observed is in fact made up, for the most part, of a continuing and slowly decreasing 'secondary' response mediated by the stimulation of committed immunocytes by antigen on dendritic phagocytic cells. Similarly, the short-lasting 'observed secondary' response is merely an accentuation of the process due to the large numbers of memory cells available plus a 'feedback' inhibition of the process by the high concentration of specific Ig G antibody produced. This inhibition by antibody presumably might act at more than one level—by union with antigen in body fluids, by release of antigen from immunocyte receptors or by partial blocking of antigen at the dendritic phagocytic cell–immunocyte contact interface.

## PHENOTYPIC RESTRICTION OF IMMUNOCYTES AND ANTIBODIES

One of the most striking findings in regard to immunocytes and antibodies that has been made in recent years is the restriction of a cell to the production of one form of immunoglobulin and antibody of one specificity. This has already been frequently mentioned and its significance in relation to the origin of immune pattern is discussed in chapter 6 (Book II) where the relevant information in regard to the antigenic qualities of the various subtypes of human immunoglobulins is also given.

527

*Myeloma proteins*

The interest in phenotypic restriction in immunoglobulins springs primarily from the character of the abnormal paraproteins present in the blood and urine of patients with multiple myelomatosis or symptomless 'monoclonal gammopathy' (Waldenström, 1960, 1962). All the productive chemical work on myeloma proteins discussed in chapter 5 is in fact wholly dependent on the finding that in nearly all instances the excess immunoglobulin is in the form of a homogeneous molecular population. The abnormal plasmocytes responsible for its production are accepted by virtually all workers as representing a pure clone derived from a single initiating cell. Each cell is clearly secreting only one type of protein; there is no allelic product and there is none of the heterogeneity of the normal immunoglobulin of the same heavy chain type.

Confining ourselves to the antigenic qualities of immunoglobulins of the human Ig G groups, a given individual will have both K and L light chains and at least three of the antigenic subtypes of heavy chain. His Gm and Inv status will depend upon his inheritance but it does not necessarily follow that a person Gm(a+b+) will produce immunoglobulin uniformly of that character. Since from standard teaching, all potentialities to produce anything that can be produced by one cell of the body are present in every cell, the uniformity of each myeloma protein shows a very striking degree of phenotypic restriction. It also provides a strong *a priori* case for believing that it is a general property of *normal* immunoglobulin-producing cells to show a similar degree of phenotypic restriction.

It seems unnecessary to present any detailed evidence for the uniformity of myeloma proteins. Virtually all the criteria by which one immunoglobulin can be distinguished from another have been devised by the experimental comparison of one myeloma protein with another. It was only their homogeneity that made this possible. The other point of equal importance is the immense heterogeneity of myeloma proteins as a class. Each individual patient has a single type of abnormal protein apart from a very small proportion, about 2 per cent (Imhof *et al.* 1966), which show a diclonal character. If, however, one took an equal amount of, for example, each of the 222 myeloma proteins in the series of

Imhof *et al.*, the resulting mixture would be very close indeed to the heterogeneous mixture of immunoglobulins present in the serum of any normal human adult, except perhaps for a relative excess of Ig M. The ratio of K to L in this series of myeloma proteins is 61:39, which is close to the proportion of the two antigens in the immunoglobulin from any individual. A detailed antigenic study of any group of Ig G myeloma proteins will show that each possesses ADs not wholly identical with those of any of the others. Korngold wrote in 1961: 'Every multiple myeloma globulin contains unique antigenic determinants. If these represent normal constituents [there are] as many gamma globulins as there are patients with multiple myeloma.' Grey *et al.* (1965) found the same individuality and showed that the individual specific antisera could usually be exhausted by absorption with *normal* globulin. Each antigenic pattern of myeloma protein therefore seems to correspond to a proportion of the globulin molecules in normal individuals.

The point is perhaps being over-emphasized because of its outstanding support for a clonal interpretation of the whole process of immunoglobulin and antibody production. The whole corpus of work on myeloma proteins becomes completely intelligible on a single hypothesis—that any cell in the process of antibody production as an immature plasma cell is liable to a somatic mutation whose only effect is to prevent the normal maturation of the cell. Instead of having as descendants only mature non-proliferating plasma cells after 6–8 generations (Nossal and Mäkelä, 1962a), the clone continues indefinitely in the immature phase in which both proliferation and antibody production go on simultaneously. I know of no facts which refute or even cast serious doubt on that hypothesis. Its implication, that all immunocytes producing antibody produce a single genetically determined immunoglobulin, is equally in accord with the facts as far as they can be established.

The existence noted elsewhere of myeloma proteins or Waldenström macroglobulins with antibody character involving the whole population of paraprotein molecules is of course the most cogent confirmation of this point of view. The finding by Eisen *et al.* (1968) of an Ig G myeloma protein with antibody activity against an artificial hapten, DNP, probably clinches the identification of all myeloma proteins as monoclonal antibodies and strongly supports the view that in fact all immunoglobulins are antibodies in a perfectly legitimate sense.

*Restriction in antibodies*

The best example showing significant differences amongst the antibodies produced by a single individual comes from studies of a human serum described by Allen *et al.* (1964), Bassett *et al.* (1965) and discussed extensively by Kabat (1966). The subject was Gm(a+b+) and had been extensively immunized, antibodies against dextran, levan, teichoic acid and tetanus toxoid being present in sufficient concentration to allow isolation from specific precipitates. Isohaemagglutinin anti-A was also present. The Gm status of the purified antibodies can be tabulated (table 19).

TABLE 19. *The allotypic constitution of human antibodies in a single Gm(a+ b+) individual (Kabat, 1966)*

| Antibody against | Gm type | Probable components |
|---|---|---|
| Levan ⎫<br>Dextran ⎬<br>Teichoic acid ⎭ | (a−b−) | All (a−b−) |
| Tetanus toxoid | (a+b+) | (a+b−), (a−b+), (a−b−) |
| A | (a+b−) | (a+b−), (a−b−) |

There are clearly differences amongst the antibodies pointing toward the origin of each type from a limited number of initiating cells. The possibility that for a given AD an immunocyte of a particular Gm type is more likely to be stimulated than one of a different Gm type may be important. It seems more likely, however, that initially there are only a very small number of progenitor immunocytes of appropriate pattern which are available to make primary contact with the AD. By a process not dissimilar to genetic drift, the eventual population of immunocytes and antibodies will have an individual character dependent on the small number of initial cells from which it was derived.

Studies of the amino acid content of the three purified antibodies with common (a−b−) allotype showed quite sharp differences in composition amongst them (see p. 437). In the absence of adequate evidence of homogeneity, however, such differences cannot be given any simple interpretation, and in the case of the anti-dextran it could in fact be shown that both K and L light chains were present (Bassett *et al.* 1965).

Somewhat similar findings have been reported in rabbits. Gell and Kelus (1962) immunized a heterozygous rabbit with a hapten–ovalbumin conjugate and found that while antibody precipitating with ovalbumin contained all the allotypes, that specific for the hapten had undergone 'deletion' showing only one allotype at *a* and one at *b* locus. In Rieder and Oudin's (1963) experiments, rabbits of known allotype constitution were immunized, each with three unrelated antigens. The isolated antibodies showed quite wide differences in the amount of each allotype antigen present. As an example, rabbit B126 A4,5/A1 gave results which required, in addition to 4,5/1 immunoglobulin, the following combinations as a minimum for the three antibodies tested: pneumococcus II, $-5/- -5/1$; DNP, $4-/1$ $4-/-$; ovalbumin, $-5/-$.

### Antigen restriction in immunocytes

The standard approach to determining the type of immunoglobulin being produced by a single cell (usually, but not necessarily, a plasma cell) is by the use of fluorescent antibody (Coons *et al.* 1941; Nairn, 1964). When human material is being studied the reagents are usually prepared by immunizing rabbits with purified myeloma protein or Bence Jones protein of the required type. These will allow the recognition of the heavy chain antigens $\gamma$, $\mu$, $\alpha$, and $\delta$, and the light chain $\kappa$ and $\lambda$. Similar methods can be applied to the detection of cells producing specific Gm and Inv allotypes and using appropriate species for antibody production for similar studies in other mammals than man.

Results with human material show consistently that any given plasma cell produces only one type of heavy chain antigen; $\mu$, $\gamma$ or $\alpha$; and one or the other of the light chain antigens, $\kappa$ or $\lambda$. Bernier and Cebra (1965) found the proportion of cells G, A and M to be 45, 39 and 16 per cent respectively, with the K:L ratio approximately 3:2. Mellors and Korngold (1963) found occasionally plasma cells staining for both Ig G and Ig M and noted similar

531

appearances in germinal centres. Pernis *et al.* (1966) found a small proportion of mature plasma cells (0·06–0·2 per cent) showing type D immunoglobulin in human spleen. According to van Furth *et al.* (1966*a*), small lymphocytes show either no evidence of immunoglobulin or small amounts of Ig M only. Cells of lymphocyte morphology in bone-marrow or thymic cortex show no immunoglobulin.

Immunoglobulins in plasma cells of Gm(a+b+) individuals were found by Curtain and Baumgarten (1966) to be either Gm(a+) or Gm(b+), or neither, in the ratio 45:25:30 respectively. The marker Inva was not segregated and 25 per cent of the previously unlabelled cells, that is, negative for Gm(a+) or (b+), carried Inva. They found that in germinal centres, multiple reagents appeared to be present in the same cells. This has been observed by others for K and L (Pernis and Chiappino, 1964; Pernis, 1966), and may well represent an artefact related to the persistence of many antigens on dendritic processes of the reticular cells (see p. 475).

In the hyperimmunized rabbit, Cebra *et al.* (1966) found that plasma cells in the spleen showed only one type of immunoglobulin in the proportions: G 71–81 per cent, M 14–21 per cent and A 5–8 per cent. The heavy chain allotype markers A1 and A2 in 1,2 heterozygotes were also separate, 53–88 per cent A1, 12–47 per cent A2.

The use of anti-allotype sera to stimulate blast transformation and mitosis by Sell and Gell (1965) to show phenotypic restriction in rabbit lymphocytes has been discussed in chapter 8.

Results reported by Sell (1967) are to some degree at variance with the accepted picture of phenotypic restriction to one type of globulin apart from cells in transition from Ig M to Ig G producers. Using rabbit lymphocytes he finds that blast transformation of small lymphocytes can be induced by sheep antisera specific for the heavy chain antigens $\gamma$, $\mu$ and $\alpha$ of rabbit immunoglobulins. The results are interpreted as showing that 25 per cent of rabbit lymphocytes carry all three antigens while the remaining 75 per cent have $\gamma$ and $\mu$ only. This cannot be accepted as proof that the corresponding immunoglobulins are being synthesized by the cell. The technique is a very sensitive one and the immunoglobulins detected may be adsorbed from the plasma. Virtually none of these cells showed evidence of immunoglobulins by immunofluorescent techniques.

*Restriction of antibody pattern in immunocytes*

From what has already been said about the homogeneity—including every detail of amino acid sequence—of the myeloma proteins and their provisional identification as the exact equivalent of *monoclonal antibodies*, we should expect to find an equally definite phenotypic restriction in each immunocyte and, apart from the possibility of secondary mutation, in the corresponding clone.

From the first discussions in Melbourne between Lederberg, Nossal and myself in November 1957 on the experimental implications of the clonal selection approach, it was clear that the first move must be to determine whether a single cell from a doubly immunized animal could produce more than one antibody. It was Lederberg's suggestion that a modification of micro-manipulative methods well known in microbial genetics could readily be applied and that *Salmonella* antigens could well be used as immunizing antigens. In this way there was initiated a long and fruitful series of investigations by Nossal and his collaborators beginning with a paper by Nossal and Lederberg (1958). This work was summarized by Nossal and Mäkelä (1962b) and Nossal (1962).

Probably the clearest picture of the results obtained is in the paper by Nossal and Mäkelä (1962b) which gives the results of an extensive series of experiments done with a background of several years' experience and in the knowledge of the conflicting results that had been obtained by Attardi *et al.* (1959). The experiments involved the immunization of rats with *Salmonella* antigens singly or in various combinations, the vaccine strains being chosen to provide wholly distinct H and O antigens. Single-cell suspensions were then prepared from lymph node or spleen and tested in microdroplets (*a*) for the liberation of immobilizing antibody anti-H (flagellar), (*b*) for O antibody shown by micro-agglutination of an appropriate strain with a different H antigen from the vaccine strain and (*c*) for adherence to cells usually by way of the O antigen. Method (*c*) has the advantage that relatively large numbers of cells can be scanned in a single microscopic field. In this way, when rats were immunized with two distinct strains of *Salmonella*, at least four different types of antibody could be sought. In summarizing the results, Nossal and Mäkelä (1962 *b*) state that of some 7,000 cells examined from multiply immunized animals, 38 per cent produced antibody (2,660). Of the active cells, 98·2 per cent produced one antibody only while 1·8 per cent appeared to produce two, but of these only two cells were found which released easily measurable amounts of two different antibodies. On occasion, therefore, double producers can be found but none of a variety of régimes of immunization, etc., allowed more than this almost insignificant minority to be found.

Another series of experiments of comparable scope and effectiveness made use of bacterial viruses as antigens (Attardi *et al.* 1959, 1964). Single cells isolated from spleen or lymph node of immunized animals and suspended in a small amount of fluid can provide enough antibody for standard tests to be made with two or more bacteriophages. As a typical example of their results from a rabbit immunized for over a year with phages T2 and T5 (which show no cross-reaction), 392 cells were tested of which 48 produced one antibody only; 29, T2; 19, T5; while 11 produced antibody to both phages. A variety of controls make it quite certain that the result is experimentally valid. There is, however, one essential control which is lacking. This is to show that the antibody from a single cell which neutralizes both A and B is in fact composed of two distinct antibodies in the sense that absorption with a suitable preparation of A removes anti-A but not anti-B and vice versa. The significance of this in relation to cross-reaction between virus antigens has been discussed earlier.

From limited work with sandwich-type fluorescent antibody techniques refined by using two different coloured fluorochromes, it appears that the plasma cells in a doubly immunized animal are producing one or other type of antibody but not both (Coons, 1958; White, 1958).

There is, however, a curious phenomenon reported by Hiramoto and Hamlin (1965) in regard to cells from guinea-pigs immunized with human γ-globulin. The antiserum from these guinea-pigs has two bands, I and II, when tested against a papain digest of the antigen. The components of the antigen responsible for the bands were separated by column chromatography and used to treat cells from lymph nodes. A section would be treated with I, washed and developed with rabbit anti-whole human γ-globulin conjugated with rhodamine. After reading and washing, the section was treated with II and developed with the same antiglobulin conjugated with fluorescein. Of the plasma cells checked, 30 per cent reacted with I only, 25 per cent with II only and 45 per cent with both I and II.

These findings, however, have not been confirmed by Benacerraf's group (Green *et al.* 1967*b*) using a strictly analogous system. They immunized rabbits with guinea-pig Ig G 2 and tested for antibody against Fab and Fc fragments using specific reagents and applying both immunofluorescent and autoradiographic techniques. Less than 4 per cent of the cells examined were apparent double producers, and this proportion was regarded by the authors as arising from technical inadequacies associated with purification of the antigen fractions.

When artificial antigens with DNP as hapten and a carrier protein, were used the results were unequivocal. Despite the fact that hapten and carrier were part of the same molecule, all the lymph node cells from immunized rabbits or guinea-pigs reacted either with the hapten or with carrier protein antigen, never with both. Amongst 1,043 reacting cells there were 622 reacting with the carrier and 321 with DNP.

A related set of experiments in 'non-responder' guinea-pigs gave a similar result, 526 single producers and no cell producing two antibodies (Green *et al.* 1967*a*). It is clear that there is no special virtue in having two different ADs on one molecule.

*The significance of double producers*
The existence of immunocytes with two wholly distinct types of specificity and capable of producing two unrelated antibodies is incompatible with the early form of clonal selection theory. Here, phenotypic restriction was implicitly assumed and the highly complex genetic relationships amongst the immunoglobulins was unknown. The present position can perhaps be expressed as follows:

(*a*) In technically satisfactory experiments the number of cells producing two unrelated antibodies is extremely small.

(*b*)  The high proportion of double producers in the experiments of Attardi *et al.* (1959) may represent the proliferation of very rare cells carrying an immune pattern reactive with both antigens. They would then not be double producers in the accepted sense.

(*c*) The small numbers of acceptable double producers will represent the result of anomalous genetic processes related to the still incompletely characterized mechanisms of diversification and phenotypic restriction of immune patterns.

If one takes the approach that biology is concerned with matters relevant to the evolution and survival of the organism under study, it is not usually needful or illuminating to probe deeply into rare and anomalous findings. Eventually they must find an explanation, but in the interim period undue concentration on anomalies may seriously impede an effective understanding of the normal process.

# 10　Immunological unresponsiveness

As in other chapters, the discussion of acquired immunological tolerance and paralysis will be centred on those aspects which are of special importance for the clonal selection approach. One of the most controversial aspects of this theory has been the contention that tolerance is essentially the absence of immunocytes capable of reacting with the AD concerned. Others, with Gorman and Chandler (1964) the most explicit, consider that 'tolerant cells' must be postulated to account for the phenomena, particularly those of partial tolerance. If the necessity for elimination of immunocytes by antigen under some circumstances is accepted there may still be serious resistance to the view foreshadowed in chapters 3 and 4 that the thymus is the primary site for 'censorship' and elimination.

As in every other field of immunology, the available data are very incomplete. Tolerance is normally assessed only on the basis of one or, at most, two ways of measuring the intensity of immune response. Where there is a diminution or disappearance of antibody recorded, the G, M or A character is rarely stated. It will be necessary usually to consider the data from the point of view: Having in mind the broad theoretical approach being supported, is there, in the information provided by a given series of papers, anything which requires significant modification of the approach or makes it desirable to clarify and interpret some special implication of clonal selection theory?

To facilitate such an approach the following paragraph summarizes the working interpretation of tolerance.

Tolerance to a given AD results when the tissues of an animal are exposed to the antigen concerned under conditions which allow destruction of all immunocytes carrying immune pattern receptors with more than a minor degree of avidity for the AD. This is much easier to achieve when all the potentially reactive immunocytes are progenitor cells than when there are substantial numbers of committed immunocytes.

THE THYMUS AS THE MAIN SITE FOR THE DEVELOPMENT
OF TOLERANCE .

The working hypothesis is that in man and the usual placental mammals used in laboratory work, the thymus is the primary location at which differentiation of stem cells to progenitor immunocytes occurs, and that with this differentiation they become immediately susceptible to destruction by any ADs present in the thymic environment and for which their receptors have a significantly avid capacity for union. On this view removal of the thymus neonatally or at any other time prevents both differentiation and censorship of any new stem cells circulating in the blood. It would be reasonable to believe that a limited amount of the same combined process of differentiation and censorship could and does take place in other sites of active lymphoid cell multiplication, including those associated with the gastro-intestinal tract from pharynx to rectum.

To support this hypothesis it is first necessary to show that antigens can readily be distributed throughout the thymus in the neonatal period. Mitchell and Nossal (1966) followed the distribution of heavily labelled flagellin given in a dose known to produce tolerance to newborn rats. There was extremely diffuse distribution of antigen through all lymphoid tissues, including the thymus, where the grain count indicated there were thousands of molecules of antigen per cell. In this paper the authors suggested that direct contact of antigen with a reactive lymphoid cell results in tolerance, while 'when a cell meets antigen appropriately fixed to membrane and processes of dendritic phagocytic cells it embarks on the proliferative events of immunity'.

The greater ease with which tolerance is produced at the neonatal stage is well documented. Eitzman and Smith (1959), for instance, gave 200 mg/Kg of bovine serum albumin to rabbits of graded ages and obtained the following percentages of tolerant animals: 2–8 days, 100; 14–15 days, 80; 21–22 days, 27; 28–30 days, 0. A contributory factor may be that catabolism of foreign protein is slow in newborn animals (Robbins *et al.* 1963) and, with labelled antigen, phagocytosis is very inconspicuous (Mitchell and Nossal, 1966).

There are major technical difficulties in direct experimentation on the thymus but Staples *et al.* (1966) have obtained some fairly direct evidence that the thymus is a major region concerned with tolerance induction. In rats given sublethal irradiation with thymus shielded and immediately thereafter inoculated with 20–40 mg of bovine gamma globulin (BGG) into the substance of the thymus or intraperitoneally, the first group subsequently showed no skin sensitization and reduced antibody production while those injected intraperitoneally showed standard responses to both tests.

Other aspects of thymic involvement in tolerance are not directly relevant to its postulated censorship function. There is now a general consensus that acquired tolerance to homografts in mice is always associated with persisting chimerism of lymphoid tissue. Galton *et al.* (1964) showed that in an appropriate system of co-isogenic mice, donor cells could be recognized in the thymus of tolerant mice for at least sixty days. The antigen concerned is clearly accessible in the thymus. When tolerant mice are thymectomized they retain their tolerance for considerably longer (Claman and Talmage, 1963; Taylor, 1964). This, however, is merely an indication of the failure of new immunocytes of appropriate pattern to be differentiated.

## THE SPONTANEOUS DISAPPEARANCE OR EXPERIMENTAL ABROGATION OF TOLERANCE

The essence of our approach to tolerance is that for effective failure of response there must be elimination of all immunocytes capable of effective reaction with each of the significant ADs of the antigen to which the animal is tolerant. Once antigen is reduced to a low enough level in the relevant regions of the body, newly differentiated immunocytes will find an opportunity to survive, and in due course there will appear the same range of progenitor immunocytes capable of reacting with the various ADs of the antigen as is found in an untreated animal. The experimental manifestation of this spontaneous return of responsiveness to the tolerant animal would be irregularity amongst individual animals and the initial appearance of incomplete antibody reactive against only a fraction of the ADs.

The available evidence is in line with this expectation. Humphrey (1964) studied the duration of unresponsiveness in rabbits given a single injection of human serum albumin (HSA) at birth. The time at which an immune elimination response developed varied greatly from rabbit to rabbit and some were unresponsive for periods well beyond the time when all antigen must have been eliminated from the body. On occasion, an animal showing a definite immune response became unresponsive again at a subsequent test. When a series of these rabbits rendered tolerant by neonatal injection were tested about a year later with alum-precipitated antigen, the response was 'poor'. Antibody was usually non-precipitating and small in quantity; failure to bind to various fragments of antigen was also observed. In commenting on these results, Humphrey concluded that the initiation of paralysis was an effect of whole molecules (that is, of all their ADs) and that the 'capacity to respond to antigen returns piecemeal in respect of different parts of the antigen mosaic'. Weigle (1964) described similar findings although in his experiments tolerance was usually terminated by administration of a distantly related antigen (see p. 464). He noted that a proportion of 'terminated' animals regained the tolerant state and that the small amount of antibody produced was related only to constituents shared by both the tolerance-producing and the challenge antigen. With prolonged immunization of animals tolerant to BSA with HSA, tolerance was terminated in the sense that a subsequent series of injections with BSA produced antibody. This antibody was, however, very largely anti-HSA and when absorbed with HSA the small amount of anti-BSA disappeared. In no sense was there a true secondary response to BSA.

Using adult rabbits rendered unresponsive by the administration of a large dose of BSA after sublethal irradiation, Linscott and Weigle (1965) found that reactivity returned between 2 and 4 months. It was characteristic that antibody, as judged by Farr's method, returned before any antibody that could be estimated by a quantitative precipitin method.

Less extensive studies in mice gave essentially similar results. Dietrich and Weigle (1963) found that tolerance induced by a variety of foreign serum albumins given to newborn mice in a single large dose was lost as judged by the occurrence of immune elimination in 5–10 weeks, yet these mice, when challenged with the homologous antigen in adjuvant, never produced antibody capable of giving bands with antigen in gel-precipitation tests. In extension of this work, Dietrich and Grey (1964) found that when tolerance to BSA was diminishing after 10 weeks, the antibody produced on challenge with antigen in adjuvant was in smaller amount and much less avid than in control mice.

Weigle's work on the abrogation of tolerance by the injection of a rather distantly cross-reacting antigen has been referred to already in regard to the nature of the antibody produced against the original antigen when tolerance is broken. The phenomenon by which tolerance to BSA in rabbits can be broken by the administration of a related antigen has been widely quoted as a model for the origin of autoimmune disease. The common assumption is that normal body antigen $A$ is recognized as 'self' and none of its ADs, $a$, $b$, $c$, $d$, etc., can function as immunogens. If, however, a change of, say, AD $c$ to $x$ results in the change of antigen $A$ to $A^1$, then $A^1$ is recognized as foreign and it is assumed that $a$, $b$, $d$, etc., as well as $x$ can thereby become immunogenic. The result is that such antibody can attack $A$ in virtue of the common components $a$, $b$, $d$, etc., and an auto-antibody is born. This is incompatible with clonal selection theory in sharply differentiating the recognition function from the signal function that leads to antibody production. It is therefore necessary to look at the facts in some detail.

The basic phenomenon can be shown in table 20. Rabbits were rendered tolerant to BSA by neonatal injection and subsequently tested with purified serum albumins of other mammalian species. If an antigen is very close to BSA, as sheep SA is, an injection merely serves to reinforce tolerance. If it is wholly unrelated like BGG it has no influence of any sort. HSA which cross-reacts, but only to a small extent, is the most potent reagent for breaking tolerance. In later work (Weigle, 1962), BSA conjugated with sulphanyl or sulphanyl + arsanyl groups was as effective as HSA for this purpose. Only in one of the groups of animals tested, however, was there production of precipitating antibody to BSA.

In many ways this is a test situation for selection theory. If a preceding process has eliminated all immunocyte lines capable of

541

reacting with the ADs of BSA, how can anti-BSA be produced by *any* manipulation? The answer, in our opinion, should not be sought by completely dissociating 'recognition' from antibody production, but by accepting the situation that all immunological phenomena are what we have called 'soft-edged'. Antibody pattern and AD pattern fit by accident and the fit can never be perfect. Poor fits allow some types of observable reaction; for other responses to be evident a highly avid relationship is required. Another factor, which may turn out to be very important if it is ever definitively established, is the difference between the stimuli mainly responsible for the destructive stimulation of an immunocyte and those mainly responsible for its stimulation to antibody production.

TABLE 20. *Breaking of tolerance by related antigens*

| Albumin used as desensitizer | Cross-reaction with BSA (per cent) | Rabbits losing tolerance to BSA |
|---|---|---|
| Human SA | 15 | 12/14 |
| Pig SA | 32 | 2/5 |
| Sheep SA | 75 | 0/7 |
| Bovine SA | 100 | 0/9 |
| BGG | 0 | 0/5 |

From Weigle (1961).

The essence of high-grade tolerance, for example, toward normal body components is that each potentially reactive immunocyte meets the ADs immediately it is differentiated. There is no possibility of immunocytes with more than minimal reactivity emerging and there is no production of reactive antibody. In particular, antigen is not opsonized and taken up by the dendritic phagocytic cell where it can find optimal conditions for stimulation of immunocytes to antibody production.

In an animal tolerant to BSA, administration of an antigen with related but not identical ADs will be capable of stimulating any immunocytes with the differential ability to react with the related antigen but not with BSA. This will lead eventually to the production of a population of reactive immunocytes and some circulating antibody. Such antibody will bring the modified antigen and its whole range of ADs, modified or unmodified, into the optimal situation on the dendritic phagocytic cells of the lymph follicles. This is a situation that would never be reached by an antigen to which tolerance was complete. We can therefore accept the likelihood that amongst the immunocyte population are cells with receptors capable of reacting positively with ADs optimally situated on the dendritic phagocytic cells but which did not react sufficiently intensely to be destroyed by any small concentrations of antigen they had encountered in the circulation.

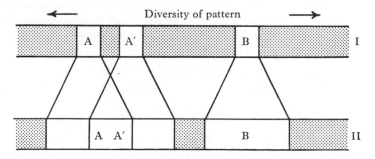

Fig. 37. Abrogation of tolerance by immunization with a related antigen.
Each bar is to be taken as a one-dimensional display of diversity of pattern on immunocyte receptors in terms of the antigenic determinants which will react with them. Tolerance means failure of primary stimulation and bar I shows the situation in regard to primary stimulation by three ADs; two related (A and A′) and one unrelated (B).
In bar II the situation concerns committed immunocytes. Those derived from A will be capable of reaction with a wider range of antigenic determinants including A′ but not with more distant determinants such as B.

Under these circumstances the result to be expected would be precisely what Weigle found, i.e. the production of poor-quality antibody to those ADs common to BSA and the modified antigen with an increasing likelihood of re-establishing tolerance with further injections of BSA (see fig. 37).

## IMMUNE PARALYSIS IN MICE TO PROTEIN ANTIGENS

It is now well known that soluble proteins such as serum $\gamma$-globulin or serum albumin, when cleared of aggregates by centrifugation and administered without adjuvant, are non-immunogenic in mice, when judged by the standard technique of the immune elimination of $^{131}$I-labelled antigen (Dresser, 1962; Celada, 1966). Immunization with either antigen emulsified in Freund's complete adjuvant will, however, give a typical response except in animals which have been treated previously with the soluble antigen. This is then an example of specific immunological paralysis. The other feature of importance is that a mouse primarily immunized with antigen plus adjuvant will respond to a secondary stimulus of soluble antigen (Celada, 1967).

Since BSA is perhaps the most popular 'pure' antigen used in experimental immunology, with bovine or human $\gamma$-globulin almost equally fashionable, this somewhat disconcerting behaviour in mice has provoked a good deal of study.

Mitchison's (1964) work on high-level and low-level paralysis provides a good example of the complexities that arise. The general form of his experiments was to 'immunize' CBA mice with a range of doses of BSA from 10 mg to 10 $\mu$g, giving three injections a week for varied times. Ten days after the last injection they were bled and immediately challenged with a standard dose of BSA (7 mg in Freund's complete adjuvant). From the response to this challenge, two zones of 'partial paralysis' could be identified (fig. 38). Mice given 10 $\mu$g doses of BSA give only a low-titre delayed response, yet from an early stage in the process these mice do not respond, or respond very poorly, to a standard challenge of antigen in adjuvant. At the other end of the scale, mice given a large dose of antigen show a quick rise and a quick fall in titre. Subsequently they develop a diminished reactivity to the standard challenge with a relatively large dose of antigen in adjuvant. Sercarz and Coons (1963), using the same antigen in mice, studied the cellular situation in animals immunologically paralysed by large doses. Using fluorescent sandwich techniques they found no antibody-producing cells in fully paralysed mice. When, with a borderline dose, small amounts of antibody were produced, they saw rare brightly fluorescent cells. Transfer of cells to normal syngeneic mice gave no evidence that they were capable of producing antibody.

Several of these studies of soluble antigens in mice have given evidence that there are small amounts of other antigens in the immunizing preparations. In Dresser's (1963) experiments, there was a minor component which immunized even in the absence of adjuvant. If, as appears to be legitimate, we make the simplifying assumptions that the immune elimination used as assay is an over-all measure of amount multiplied by avidity of antibody molecules directed against a single AD present only on the antigen, the situation can be visualized as follows.

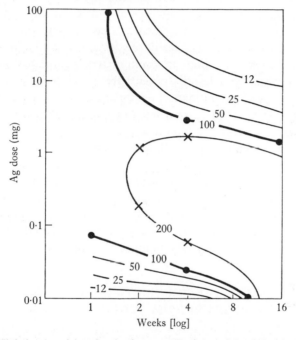

Fig. 38. High-level and low-level tolerance. (Redrawn from Mitchison, 1964.) Mice were given three 'immunizing' inoculations with the dose shown, followed by a standard challenge. The lines of equal antibody response show the antibody titre as percentage of that in control animals.

Soluble antigen *A* will be present in high molecular concentration ($10^{13}$ molecules per mouse in the lowest dose) and the great majority of contacts will tend to be destructive; either directly lethal or, as suggested by Šterzl (1966), by way of terminal differentiation to short-lived antibody-producing cells. In Mitchison's (1964) opinion the age-kinetics of paralysis loss and the effect of

545

thymectomy indicate that an individual 'lymphocyte cannot recover from paralysis (or alternatively, according to clonal selection, death)'. On the other hand there are many instances on record, including two already noted (Sercarz and Coons, 1963; Mitchison, 1964), in which paralysis can be imposed on an animal already producing specific antibody by a further large dose of antigen.

It follows, therefore, that the results of injecting soluble antigens into mice will be highly variable unless there is strict control of the strain and age uniformity of the mice and of the timing and number of injections. There seem to be two basic rules. First, that an immunocyte destroyed as a result of antigenic stimulation can play no further part, but that an immunocyte stimulated positively toward proliferation is still potentially liable to destruction by further contact with antigen. The second is that, for paralysis, all immunocytes above a certain level of avidity must be destroyed but, given the necessary circumstances, a single immunocyte can initiate a clone of antibody-producing cells.

With such a background there seems to be no necessity whatever for hypotheses such as that of Dresser (1963), in which both paralytic and immunogenic receptors in each cell are postulated, with the further implication that paralytic and immunogenic receptors for all possible antigens must be postulated. It is much more satisfactory to adopt the point of view presented in chapter 8.

### FELTON'S PHENOMENON: IMMUNOLOGICAL PARALYSIS BY PNEUMOCOCCAL POLYSACCHARIDE

In 1932, Felton, in studying the possibility of immunizing mice against pneumococcal infection with the specific soluble substance (SSS, polysaccharide antigen), found that while a minute dose of less than 10 $\mu$g will immunize, an injection of 100 $\mu$g upwards paralyses. If a dose of 1 mg is given to each of a group of mice they will remain susceptible to lethal infection usually for the whole of the mouse's lifetime. In addition, attempts to immunize, using the small dose of 1 or 2 $\mu$g which is regularly effective in previously untreated mice, have no influence whatever on the paralysed mice (Felton *et al.* 1955).

This is a clearly defined observation that has been readily confirmed by all subsequent investigators. In some ways it provides an important test case for any immunological theory. The significant findings when SSS III is used as antigen can be stated summarily.

1. The antigen is taken up by macrophages, is not susceptible to enzymatic breakdown, and persists for long periods in the body (Dixon *et al.* 1955).

2. When otherwise paralysing doses are given with water in oil adjuvant, immunization can result (Neeper and Seastone, 1963 *b*).

3. Mice in the early stages of immunization with a small dose can be paralysed by a subsequent large dose of antigen but hyperimmune mice do not lose their immunity (Brooke, 1966).

TABLE 21. *Adoptive immunity in SSS III paralysis in mice*

|  | Recipient | |
| --- | --- | --- |
| Donor | Normal | Paralysed |
| Immune | P | P |
| Paralysed | O | — |

P—protected; O—unprotected.

4. Adoptive immunity experiments, using lymph node suspensions from immune or paralysed mice and normal and paralysed mice as recipients, gave the tabulated results (table 21). The essential feature is that lymph node cells from an immune animal can protect a syngeneic paralysed mouse (Neeper and Seastone, 1963 *a*; Brooke and Karnovsky, 1961).

5. Paralysis is highly type-specific (Brooke, 1966).

6. There is some discrepancy in regard to the possibility of breaking paralysis. Neeper and Seastone (1963 *b*) found partial immunity after paralysed mice were given antigen in Freund's complete adjuvant. Brooke found that the use of neither adjuvants nor pneumococcal vaccines could protect paralysed mice.

7. Using SSS III and a specific depolymerase, Brooke (1964) found that paralysis could only be broken by administration of the depolymerase during the first week after antigen injection.

The interpretation of these results would be based on the

assumption that a newly emerged immunocyte reactive with the antigen is stimulated to destruction with great regularity by more than minimal contact. Once a population of committed immunocytes is in existence this is much more resistant, as always, to destructive stimulation. This makes it possible to transfer immunity to paralysed mice even though the latter have enough polysaccharide in circulation to effectively eliminate any progenitor immunocytes they are themselves producing. In agreement with Siskind and Howard (1966) there is probably a constant amount of circulating antigen maintained from the supply held in phagocytic cells and perhaps, in part, returning to that reservoir. There is no longer any basis for the earlier contention that antibody was being bound by fixed antigen as fast as it was produced. The importance of phagocytic processes is underlined by the findings of Kerman *et al.* (1967). They compared the doses of SSS needed to immunize and to paralyse two sets of newborn mice, one born of normal mothers, one from mothers paralysed with SSS. As might be expected, they found that the minimal dose required to immunize or to paralyse was much lower ( $\pm$ 10 per cent) in the offspring of paralysed mice than in mice of the same age born of normal mothers. The effect is clearly related by the authors to the absence of any natural antibody against SSS in the first group and hence absence of effective reduction of the antigen in the circulation.

This is perhaps a particularly striking example of two general aspects of phagocytosis, (*a*) that phagocytosis, particularly of soluble antigens, is very greatly facilitated by even minute amounts of antibody to act as opsonin and (*b*) that one of the major functions of the macrophage system is to reduce the concentration of antigens in the body to a level which will allow them to be handled effectively by immunological processes.

### TOLERANCE BY LYMPHOCYTE DEPLETION

The most physiological method of depleting the lymphocyte population of an animal is chronic drainage of the thoracic duct for several days. In addition to lymphopenia this causes a sharp reduction in the weight of lymph nodes associated with depletion of small lymphocytes from the cortex (McGregor and Gowans,

1963, 1964; Gowans and Knight, 1964). Such depleted rats showed prolonged survival of first-set homografts of skin, provided the difference in histocompatibility antigens was relatively minor, and severe depression of primary antibody response. On the other hand, depletion of a primarily immunized rat had no significant effect on the secondary immune response.

Two elaborations of this result are important. The depression of antibody response produced by X-irradiation can be remedied by injecting almost pure suspensions of small lymphocytes. The second point of particular relevance in the present context is that cells from *tolerant* donors are incapable of this effect (Gowans and McGregor, 1963).

These demonstrations of the importance of the size of the lymphocyte population strongly suggest that the effects of X-irradiation and cytotoxic drugs, allowing under appropriate conditions the development of tolerance, may, in most cases, be a simple result of the associated lymphocyte depletion. Perhaps the results are particularly important in underlining the fact that all tolerance is partial tolerance, and in showing once again that the population of small lymphocytes in the cortex of lymph nodes is as labile and mobile as the lymphocytes of the blood. All the findings are concordant with the view that these large labile populations of lymphocytes, mostly small lymphocytes, are very largely composed of immunocytes, both progenitor immunocytes whose ancestors have not been 'committed' by contact with specific antigen, and memory cells.

### THE INHIBITORY EFFECT OF ANTIBODY ON ANTIBODY FORMATION

This is a topic which has been discussed in part earlier but requires reconsideration in relation to tolerance, since there is important evidence that what can be called 'partial tolerance' is due largely to this action of antibody (Rowley and Fitch, 1965). If a rat is injected at birth and subsequently twice weekly with sheep red cells a high degree of tolerance develops; for example, after 15 injections there is no circulating antibody and the median number of plaque-forming cells is very low and almost within normal limits.

549

Analysis of the early stages shows that tolerance is not immediate. After 5 injections (that is, at two weeks of age) plaque-forming cells in the spleen were up to ±7 per cent of the response in a normal animal of the same age and there was detectable antibody. The interpretation by Rowley and Fitch (1968) is that primarily produced antibody, 7 S or 19 S, is effective in preventing initiation of proliferation and antibody production by what we have been calling 'progenitor immunocytes'. It does not, however, prevent the continuation of antibody production by committed immunocytes, nor does it block the capacity of the red cell antigen to invoke a secondary response in a primed animal. Under the conditions of repeated antigen injections in the rat the committed cells are apparently 'directed' almost wholly into the end-cell process of antibody production with absence of memory cell formation.

Rowley and Fitch (1968) believe that the antibody suppresses response either by acting directly on the antigen or on an early type of 'processed' antigen. An alternative, not necessarily excluding this, is that AD in any form, either reversibly or irreversibly blocked by specific antibody, is much less likely than unimpeded antigen to make effective contact with a progenitor immunocyte —the necessary preliminary to the production of a committed immunocyte. In the presence of adequate antibody, progenitor immunocytes of potential reactivity with the antigen are in essentially the same position as progenitor cells with immune patterns wholly unrelated to the antigen. This may be an important consequence of the accumulation of antibodies which is found in relation to dendritic phagocytic cell processes, especially in germinal centres.

Committed immunocytes (memory cells) are susceptible of stimulation in the presence of antibody and by antigen–antibody complexes (Rowley and Fitch, 1968). From other work they are also relatively or absolutely insensitive to the effect of excess antigen (Michaelides and Coons, 1963). In the opinion of Möller and Wigzell (1965), 7 S antibody-producing cells are end-plasma cells not capable of proliferation or of giving rise to other proliferating types. The antibodies they produce 'inhibit the initiation of new 19 S-producing cells by combining with antigen, thus suppressing the recruitment of their own precursors'. Both groups regard the phenomenon as an important feedback control. Its existence must clearly be kept in mind in considering all tolerance situations and related phenomena.

Some further points in regard to the development of this characteristic type of partial tolerance to sheep red cells can be drawn from two papers from Czechoslovakia. Šterzl (1966), using colostrum-free piglets, immunized them with graded doses of sheep red cells, testing after the primary and the secondary response. If 1 per cent of cells were used for each inoculum, the primary haemagglutinin titre was 22, while the secondary response had an Ig G titre of 22,000 and an Ig M titre of 2,600. If, however, a similar volume of packed red cells was used, the primary titre was 380 while the secondary response showed no Ig M and only 800 Ig G. This, in Šterzl's opinion, points to lethal differentiation of progenitor immunocytes to short-lived non-proliferating antibody-producing cells. This is very close to Rowley and Fitch's view. In a study of the rise in plaque-forming cells following a standard injection of sheep red cells in normal immune and tolerant rats, Hašek *et al.* (1965) make the important point that the small rise in plaque-forming cells seen in partially tolerant rats is due to cells of standard antibody-producing capacity. It is the number of active cells that is reduced, not their average quality.

### UNRESPONSIVENESS FOLLOWING THE USE OF IMMUNOSUPPRESSIVE DRUGS

In the first chapter the point was made (p. 337) that the most impressive evidence for the validity of the clonal selection theory was the way in which each of the major new immunological discoveries, since its enunciation, has fallen easily into place within the concept. Almost the first of these was the demonstration by Schwartz and Dameshek (1959) of specific tolerance following the immunosuppressive action of 6-mercaptopurine. This was a discovery of the greatest practical importance in providing very soon a practical means of facilitating organ transplantation in man, but it was of almost equal importance at the theoretical level.

The drug is an antimetabolite interfering with purine synthesis and acts by forming an abnormal ribonucleotide. Its effect, as shown in its cytotoxic and partially therapeutic action in leukaemia, is on the cell as a whole and presumably renders normal mitosis impossible, with subsequent necrosis. In this connection it is relevant that most of the pyknotic cells in the thymus and germinal centres can be shown to have died in mitosis. It seems almost certain, therefore, that 6-MP greatly increases the likelihood of a destructive contact of antigen with immunocyte even when the contact would normally result in proliferation and antibody production. If, in fact, the action of 6-MP is confined to the destruction of cells activated for proliferation, the occurrence of specific tolerance after its combined action with antigen is almost categorical proof of the *clonal* nature of tolerance. The surviving population of cells can produce other patterns of antibody but not the pattern that was carried by the eliminated clones. To interpret this in some alternative fashion, implying that only one of many thousand potentialities in each cell was eliminated by this cytotoxic drug, would seem to be mere sophistry.

The production of specific unresponsiveness by 6-MP in rabbits requires relatively large doses of the drug and daily injections for a week or more after the antigen injection (Schwartz, 1963). 6-MP is lethal to rabbits at about 15 mg/Kg which is only a little above the usual 10 mg/Kg daily used to demonstrate tolerance. Guinea-pigs are much more resistant, possibly because of their higher cell content of xanthine oxidase, and the lethal dose is of the order of 100 mg/Kg/day. Even with large doses there is no evidence that tolerance can be produced in guinea-pigs against either antibody production (Genghof and Battisto, 1961) or DH (Humphrey and Turk, 1961; Salvin and Smith, 1960). On the other hand, immuno-suppression of allergic encephalomyelitis can be demonstrated (Hoyer *et al.* 1960). It seems clear from the pattern of results that the immunosuppressive action is only marginally specific in relation to general cytotoxicity.

Although, in their first paper, Schwartz and Dameshek (1959) found only a minimal effect on the secondary response in rabbits, others have found a well-marked production of suppression and subsequent tolerance in rabbits (Forsen and Condie, 1963).

Studies in mice have shown that where there is not a full H2 compatibility difference tolerance to a skin homograft can be induced in the adult by the administration of a large dose of spleen cells after some days' treatment with 6-MP (McLaren, 1961). The experiment did not succeed when there was a major H2 difference.

Rabbits rendered tolerant behave like normal rabbits in showing no acceleration of removal of the antigen from circulation (Schwartz and André, 1960) and when, after several months, tolerance is lost, the response to antigen injection is of the primary type. When

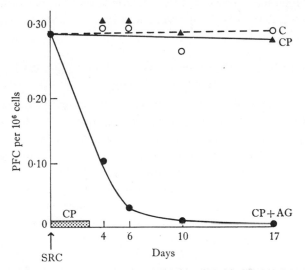

Fig. 39. The effect of antigen plus cyclophosphamide in reducing the number of plaque-forming cells in immunized rats. (Redrawn from Aisenberg and Wilkes, 1967.) C = control rats; CP = cyclophosphamide alone; CP+AG = cyclophosphamide + antigen. Times of administration of antigen (SRC) and a drug (CP) are shown.

tolerance was induced in rabbits with 6-MP in solution, Feldman *et al.* (1963) found it impossible to break tolerance with preparations of DNA or spleen cell homogenates that had been successful with animals made tolerant with insoluble 6-MP. This last discrepancy is another example of the marginal quality of the processes concerned. It is not a legitimate argument against a clonal interpretation.

Cyclophosphamide is also capable of producing specific unresponsiveness to foreign red cell antigens. As with 6-MP, the joint action of drug and antigen is required. Figure 39 (taken from

553

Aisenberg and Wilkes, 1967) shows the influence of cyclophosphamide alone, and cyclophosphamide plus antigen, on the number of plaque-forming cells per million spleen cells. The situation is complicated by the general cytotoxic effect of the drug which sharply reduces the number of cells per spleen but, when this is allowed for, it becomes evident that the disappearance of plaque-forming cells under the influence of cyclophosphamide is antigen-dependent. The disappearance of Ig M plaque producers was not associated with any appearance of Ig G antibody producers.

## THE BEHAVIOUR OF SYNTHETIC ANTIGENS WITH D-AMINO ACIDS

It has been generally found that synthetic polypeptides of D-amino acids are ineffective as antigens (immunogens) when compared with similar L-polypeptides. Benacerraf *et al.* (1963*a*), however, found that antibody to the azobenzarsonate (ABA) group produced by immunization with the hapten combined to poly-L-GAT would react in passive cutaneous anaphylaxis tests with ABA-poly-D-GAT. Leskowitz *et al.* (1966), however, find that when guinea-pigs are sensitized with ABA-acetyltyrosine they react well when challenged with ABA on a variety of natural protein or poly-L synthetic peptides as carriers, but not with ABA on a poly-D-peptide carrier. Mixing an excess of the latter with a small dose of active conjugate has no inhibitory effect.

Leskowitz feels that this excludes the possibility that any sort of preformed antibody, whether circulating or fixed on lymphoid cells, plays a significant part in the DH response. In sorting out possible interpretations we must also bear in mind Janeway and Sela's (1967) finding that in fact a poly-D-peptide is antigenic in the mouse when given in extremely small dose. Its reputed non-immunogenicity is due to its paralysing ability.

A substance carrying an active hapten can be non-immunogenic because it cannot make effective contact with cell receptors or because all effective contacts result either in destruction of the immunocyte or in blocking receptors without stimulation. The first possible interpretation of Leskowitz's experiment is that union of receptor with ABA hapten can act as a positive stimulus only when an L carrier is present. With a D carrier the union is inert. This is difficult to fit in with the absence of inhibition of the L activity by excess of D.

The following is an attempt to bring these findings into line with Mitchison's (1964) findings on high- and low-level tolerance. For any AD at a certain concentration and in a defined mode of presentation, contact with a potentially reactive immunocyte may be (*a*) ineffective and, if effective, induce (*b*) a lethal process rapidly becoming irreversible and probably associated with the release of pharmacologically reactive agents or (*c*) initiate a process of mitosis leading to antibody production. If (*b*) is, say, ten times as likely to happen as (*c*), then since, for a period at least, a cell in which (*c*) has been initiated can still respond in (*b*) fashion to further contacts, antibody formation can only occur with very low concentrations of antigen. This is the simplest explanation of the findings in mice.

The inability of ABA-D-GAT to provoke a delayed reaction in guinea-pigs, while ABA-L-GAT is active, may not be too closely related to the phenomena seen in mice. In the first place, on any reasonable hypothesis we are dealing not with progenitor immunocytes but with committed immunocytes, perhaps of a rather special type, since DH can rather easily be dissociated from antibody production in the guinea-pig. Leskowitz's method of sensitization with a small hapten-acctyltyrosine conjugate in fact sensitizes but does not produce circulating antibody. It is widely believed (see p. 582) that the cells mediating DH are those small lymphocytes which respond to antigen stimulation *in vitro* by blast transformation. In the early stage of such change *in vitro* they presumably influence associated cells, including vascular endothelium and mononuclear cells of various types, by the release of a variety of soluble agents which induce the progressive development of the cellular and vascular aspects of the DH reaction. Clearly the D antigen fails to set this action in train, but one's impression is that the difference is probably to be considered a pharmacological one with no very close relationship to the immunogenic qualities shown in mice. It is probably relevant that neither L nor D ABA-TYR, which are effective as sensitizing immunogens, can provoke a delayed reaction (Leskowitz *et al.* 1966).

555

## TOLERANCE IN RELATION TO DELAYED
## HYPERSENSITIVITY

One of the major difficulties in developing a general approach to immunological phenomena is that most experimental work on antibodies and antibody production has been done in rabbits, on the cellular concomitants of antibody production in rats, on quantitative studies of antibody-producing (plaque-forming) cells in mice and on DH in guinea-pigs. There are, of course, excellent reasons for choosing the most convenient species for the type of work in which one is interested, but it makes for great difficulty in obtaining for instance a clear picture of the interrelationship of DH and antibody production.

With the development of interest in synthetic polypeptide antigens and carriers for haptens it has often been noted that an artificial antigen will sensitize a guinea-pig to give a typical DH reaction but no detectable antibody, while the same antigen is a relatively good antibody producer in the rabbit (Ben-Efraim *et al.* 1963). Leskowitz *et al.* (1966) used low-molecular-weight antigens by conjugating a hapten (ABA) to acetyltyrosine and produced sensitization of guinea-pigs to the hapten conjugated with carrier protein but no antibody.

On the other hand, Uhr *et al.* (1957), by immunization with small amounts of antigen in the form of antigen-rabbit antibody complex, produced DH without antibody. A further injection produced 'desensitization' with active production of antibody and no skin reaction (Uhr and Pappenheimer, 1958).

Contact sensitization with simple haptens can provide a wide range of dissociated reactions. In the standard sequence of sensitization and test by skin contact, no antibody is demonstrable. When picryl or dinitrofluorobenzene in oil is fed to a guinea-pig repeatedly over three weeks (Chase, 1946) the animal cannot be sensitized by skin contact and has no antibody. It can, however, be sensitized to a conjugate of the homologous hapten and a carrier protein, the sensitivity being directed essentially toward the carrier protein (Coe and Salvin, 1963).

Still another variant can be seen when a non-immunogenic polypeptide, polyglutamic acid, given neonatally to rabbits, renders them tolerant to polyglutamic acid 60-alanine 40 which is normally an active antigen (Maurer *et al.* 1965).

Clearly, any interpretation of unresponsiveness must be based on an understanding of the nature of DH. A good deal has been mentioned in regard to lymphocytic function that is relevant to the nature of DH, but there are still many important difficulties in the way of any accepted interpretation. It has proved very difficult to find a suitable position in the argument for DH and associated phenomena and I have decided to include it in chapter 11. For the present, one can accept the general conclusion of Uhr (1966), modified only by using our term 'immune receptor' instead of 'cell bound antibody produced by the sensitive cell'. Uhr considers that such receptors have a high affinity for antigen, that the long duration of the hypersensitive state is mediated by sensitized small lymphocytes, that the interaction of sensitized cells and antigen can release a factor involving non-sensitized cells in the reaction and that the tissue damage that results is due mainly to the release of lysosomal enzymes.

These points will be discussed in more detail at a later stage.

*Unresponsiveness to delayed hypersensitivity*

With this background the various types of unresponsiveness that have been studied can be discussed in more detail.

The first clearly defined examples of desensitization were given by Uhr and Pappenheimer (1958), who showed that in guinea-pigs sensitized with antigen–antibody mixtures given intradermally, desensitization was easily accomplished by a moderate dose of antigen and the desensitization was specific. A point of interest is that if a challenge injection of a minute dose of antigen in a sensitized guinea-pig is followed 2 hours later by an intravenous injection of an adequate dose of antigen, the skin response fails to occur. Since the desensitizing action of antigen is specific it must act by union with specific receptors on the sensitized cells. The most likely result of that union is destruction of the fraction of the immunocyte population with high-avidity receptors. If what takes place in the site of a DH test over some 24 hours was initiated at once throughout the body there would be no local sign of what

557

was happening and, with the disappearance of reactive cells from the circulation, no local response.

During the process of desensitization there is often a nonspecific partial desensitization against other antigens. This is probably an effect of the widespread antigen–immunocyte reactions taking place throughout the body. It is very much in line with this explanation that a positive tuberculin reaction disappears in man around the time of appearance of a measles rash (von Pirquet, 1908).

The first demonstration of unresponsiveness to a simple chemical hapten was shown by Sulzberger (1929), who used neoarsphenamine in guinea-pigs, some stocks of which were readily sensitized by repeated intradermal injection. If a guinea-pig was sensitized by a single intradermal injection and next day given a single intravenous injection of the drug, sensitization failed to occur. Frey *et al.* (1964) found that with another hapten (DCB) they could temporarily desensitize a fully sensitive animal by an intravenous injection followed within 6–12 hours by a small intradermal test inoculation. If the latter were delayed beyond three days, unresponsiveness was not maintained. This would require some complex *ad hoc* explanations in detail but broadly is obviously similar to the desensitization against contact sensitization by feeding the active chemical already referred to. In many fields, intradermal injection favours sensitization; intravenous injection, unresponsiveness.

In a recent review of his experience in this field, Chase (1966) considers the feeding method to be the most effective way of producing unresponsiveness, but a substantial proportion can also be made unresponsive by a single intravenous injection (Chase *et al.* 1963). Most animals remain unresponsive indefinitely, but as old normal animals often fail to become sensitized by procedures which are effective in young guinea-pigs it is not possible to give a clear statement on this point. The failure to sensitize old animals is itself of interest, in suggesting that appropriate immunocytes become fewer with atrophy of the thymus and other ageing processes in the immune mechanism.

Interrelations between two contact haptens may differ according to whether the comparison is positive or negative. Coe and Salvin (1963) found that picryl chloride and haptens with dinitrophenyl groupings (for example, dinitrofluorobenzene) showed extensive cross-reaction in sensitized animals. On the other hand, when a guinea-pig rendered unresponsive to dinitrofluorobenzene by feeding was tested with picryl chloride it could be rendered specifically sensitive to picryl chloride while still nonreactive to dinitrofluorobenzene. Desensitization thus appears more specific than sensitization. It must be remembered, however, that with a clonal approach, unresponsiveness to *A* means removal of *all* immunocytes that can react with or produce antibody against *A* while a positive response, whether of sensitization or of antibody production, merely means that a proportion of any immunocytes that can react with *A*, with or without cross-reactivities, have been stimulated. This is the point to which Chase refers as not explicable on clonal selection theory but, with the present perhaps more sophisticated approach, this phenomenon is precisely what would have been expected. A similar finding by Salvin and Smith (1964) with unresponsiveness induced by hapten–protein conjugates probably has a similar interpretation. They used as haptens, *p*-aminobenzoic acid (B) and *p*-amino-arsonic acid (A) conjugated to bovine *γ*-globulin (Pg) or egg albumin (Pe). Guinea-pigs were rendered unresponsive to B–Pg by giving that conjugate with cyclophosphamide. Two months later they were tested for capacity to become sensitized to each of the four possible conjugates. The animals maintained their unresponsiveness to B–Pg but showed normal sensitization with the other three materials B–Pe, A–Pg and A–Pe.

### TOLERANCE BY PARABIOSIS

The qualities of transplantation phenomena and homograft immunity are of great practical importance but do not lend themselves easily to the elucidation of immunological principles. The whole development of the concept of tolerance was, however, in terms of tissue transplantation or red blood cell survival and some phases of this aspect of tolerance obviously call for discussion. These are tolerance by parabiosis, split tolerance and the apparent adoptive transfer of tolerance. In all cases the basic phenomenon appears to be the development of lymphoid tissue chimerism.

This is most clearly seen when animals are lethally irradiated and saved by the injection of cells from bone-marrow or neonatal spleen of an antigenically distinct donor. Ford *et al.* (1956) used a chromosomal marker, T6, to detect donor cells in various situations, and many variants of this approach have been subsequently used. In such chimeras, for a time at least, the whole lymphoid system is of donor type. If rat bone-marrow is successful in saving lethally irradiated mice it can be shown that the surviving mice may be producing serum albumin of mouse type but immunoglobulins of rat specificity (Grabar *et al.* 1957). There is also abundant evidence that animals rendered tolerant neonatally or, when adult, by large inocula intravenously, are also chimeras although in most cases only a small proportion of donor cells (1–5 per cent) is present (Brent and Gowland, 1963).

Most of the papers on tolerance induced by parabiosis in mice are based on the work of Martinez and his colleagues at Minneapolis, reviewed by Martinez and Good (1963). Most of the experiments took the form of joining in parabiosis $F_1$ A × Z to the parental Z strain. In such a combination there is a gradual development of tolerance by the Z partner for $F_1$ A × Z skin. Only two out of seven tests showed skin retention at the 23rd day, while twenty-three out of twenty-five were accepted at 42 days (Martinez *et al.* 1961). When unrelated strains are joined it is common for one partner to dominate in producing a graft-versus-host reaction in the other but even when mutual tolerance develops it may develop asymmetrically. In the same paper, Martinez *et al.* note that when Z–Ce parabioses were separated after 27 days' union and tested with the partner strain skin, the Z partners showed eleven out of twelve acceptances of Ce skin but none of fourteen Ce partners accepted Z skin. After 64 days, however, the corresponding figures were eleven out of thirteen and eight out of thirteen acceptances.

This progressive development of mutual tolerance is operationally equivalent to a continuing *intravenous* infusion of foreign cells, and parallels the results of Brent and Gowland (1962) who produced tolerance in adult mice by repeated intravenous administration of allogeneic lymphoid cells. When these were given intraperitoneally only sensitization with accelerated rejection of homograft resulted.

The interpretation of parabiotic tolerance may be along the following lines. We assume for simplicity that the difference between the strains depends on their respective possession of ADs $A$ and $B$. If the immunocytes active against histocompatibility determinants are represented by corresponding lower-case letters, the composition of the two strains will be $A-bcd$, etc., and $B-acd$, etc. With the development of parabiotic tolerance, a progressive elimination of clones $a$ and $b$ must be postulated so that while each partner retains its antigenic quality $A$ or $B$, the common population of immunocytes is $cde$, etc., with both $a$ and $b$ absent. Each partner will therefore tolerate a graft of the other's skin and when $cde$, etc. immunocytes are transferred to a neonatal or irradiated host which they can colonize, the host will tolerate, temporarily at least, either $A$ or $B$ skin.

### ADOPTIVE TRANSFER OF TOLERANCE AND LYMPHOID CHIMERISM

Many authors have declined to accept the simple view that tolerance equals an absence of cells with appropriate receptors and have preferred to retain at least the possibility that 'tolerant cells' can exist. These presumably are cells with the potentiality of producing many types of antibody but in which the possibility of producing the type of antibody being sought experimentally has been inhibited. To accept this suggestion is to scrap the whole concept of clonal selection.

Gorman and Chandler (1964) introduced a new way of looking at tolerant cells. They accept the general clonal approach but believe that in addition to clones of immunologically competent cells (immunocytes) we must also postulate another parallel system of 'immunologically incompetent' cells. Their concept is that for every antigen, self or not self, there are or may be two specific clones of cells (*a*) immunologically competent (IC) cells which, on contact with antigen, multiply and produce antibody and are morphologically of the plasma cell series and (*b*) immunologically incompetent (II) cells which, on stimulation with antigen, also multiply but whose only function is to 'occupy a niche' open to both II and IC cells. For every AD the body has encountered there is a niche for correspondingly reactive IC and II cells such that (II) + (IC) equals a constant. Whether II or IC cells are produced depends on the conditions under which the stimulus inducing proliferation takes place. The II cells are said to be small lymphocytes, most of which correspond to the very large numbers of clones of II cells which are responsible for the maintenance of tolerance to normal body components.

The phenomena which are claimed as calling for the existence of II cells are:

(*a*) partial tolerance.

(*b*) Dresser's results in mice with particle-free γ-globulin as antigen.

(*c*) the Battisto–Chase phenomenon in inhibiting contact sensitivity in guinea-pigs.

(*d*) the emergence of plasma cells into prominence when lymphocytes are eliminated by irradiation or other procedures.

(*e*) accentuation of autoimmune disease by irradiation.

(*f*) transfer of tolerance by cells (alleged).

Almost all these phenomena have been discussed in this or other chapters in terms of a positive clonal selection concerned with immunocytes only. In particular it has been emphasized that tolerance is always partial tolerance and that immunological phenomena are always soft-edged. Only a naïvely rigid concept of clonal specificity could allow Gorman and Chandler's statement that partial tolerance is impossible on any clonal selection theory. There is one point, however, which does require fuller discussion.

If it can be shown that an animal tolerant to a defined antigen can transfer that tolerance to another by cell transfer with regularity and with appropriate controls to ensure that the antigen is not significantly concerned in the phenomenon, then the existence of a 'tolerant' cell would be established and the concept would have to be implanted on to clonal selection theory to give a fuller picture of reality.

Most of the evidence for adoptive transfer of tolerance comes from transplantation experiments. It is now accepted that where substantial tolerance persists the animal in question is always a lymphoid cell chimera. Martinez *et al.* (1961) used parabiotic pairs of CC and C3H mice which developed mutual tolerance. Cells from lymphoid tissue of the C3H partner transferred to newborn C3H conferred full tolerance to CC skin. Argyris (1963) obtained C3H mice tolerant to CBA by neonatal injection. When these mice were 2 months old, their spleen lymph node and bone-marrow cells were transferred to lethally irradiated CBA or C3H mice. These were saved and showed tolerance to both C3H and CBA skin grafts. In a final example, Cole and Davis (1961) found that LAF$_1$ mice lethally irradiated and saved with C3H bone-marrow remained chimeric for at least 14 months. Their lymphoid cells were effective in blocking the saving effect of homologous bone-marrow from unrelated strains of mice on lethally irradiated LAF$_1$ hosts. They showed, however, no activity against either C3H or LAF$_1$ bone-marrow, which retained their activity.

In all these experiments we can be confident that mixtures of chimeric cells were being transferred to an immunologically ineffective host. The donor mixtures were necessarily mutually tolerant; that is, on our picture of tolerance they had been cleared of *A* immunocytes reactive against *B* antigens and of *B* immunocytes reactive against *A* antigens. In the incompetent host these mutually tolerant populations will colonize the lymphoid tissue and confer the same range of tolerance that was characteristic of the donor.

Transfer of tolerance to a non-living antigen is conceivable if the donor still contains relatively large amounts of antigen but, in fact, the experiments of Brooke and Karnovsky (1961) with mice paralysed by pneumococcal polysaccharide were negative. The only positive claim I have noted is from work by Isaković *et al.* (1965), who grafted thymus from rats made tolerant to BGG by neonatal exposure, to thymectomized and irradiated recipient rats. These showed some inhibition of DH and produced 19S antibody, in these respects resembling partial tolerance. The results, though significant, were partial and irregular and could not be claimed as evidence for the existence of positively tolerant cells. The most likely explanation is that the elimination of all early differentiated immunocytes of appropriate competence from the thymus might be adequate to account for the short-term results reported.

## TOLERANCE FOLLOWING X-IRRADIATION:
## RADIATION CHIMERAS

There is a very extensive literature reviewed by Koller *et al.* (1961), on the properties of radiation chimeras produced by lethal X-irradiation of the host followed by injection of an adequate dose of normal bone-marrow cells of a different strain or even a different species. Such animals are initially completely recolonized by donor-type lymphocytes, granulocytes and erythrocytes, the immunoglobulins are of donor type (Grabar *et al.* 1962) and the animal will accept a homograft of donor-type skin. The whole immune mechanism has been replaced and it is even found that sometimes a syngeneic skin graft will be rejected (Doak and Koller, 1961). Such mice are completely tolerant but the effect is so gross that there is not much that bears on the finer points of immunological theory.

Perhaps the following are the most important results with relevance to the nature of immune mechanisms.

1. The ability of bone-marrow cells to provide a full complement of immunocytes. Bone-marrow is known to contain few immunocytes (Chin and Silverman, 1960) and long-lived radiation chimeras producing donor-type immunoglobulins only gradually develop the capacity to give a normal response. The use of syngeneic chimeras avoids many complications and, in these, capacity to reject an alien tumour (Koller and Doak, 1960) or to produce antibody to foreign red cells (Makinodan *et al.* 1956) approaches the normal some 50 to 60 days after irradiation.

2. The persistence of cells carrying immunological information in the lethally irradiated host has been claimed but, in at least one well-controlled investigation (Garver *et al.* 1959) no retention of host antibody-producing capacity was present in rat → mouse chimeras.

3. Ford *et al.* (1959) described some instances in which rat → mouse chimeras gradually reverted to mouse immunological quality. Some showed one or a small number of karyotypic marker patterns arising from radiation damage which would indicate that the repopulation was derived from a very small number of mouse cell clones. There is no statement as to the immune status of these

mice apart from the fact that they were apparently healthy. This gives an *a priori* probability (but no more) that they possessed a substantial inventory of immune patterns among the immunocytes. The same points arise here as were discussed in regard to the experiments of Trentin *et al.* (1967) earlier (p. 335). Both sets of observations seem to indicate that the variety of immune patterns that arises from the zygote in 30 or 40 days (in mice) can also arise from a single stem cell in circumstances where it can proliferate at a comparable intensity to embryonic cells. This in itself points rather strongly against the process of diversification of pattern being a controlled process of differentiation and favours the idea that the diversity arises by wholly random processes, the result being trimmed to the requirements of the body by the various processes of selective destruction and proliferation.

# 11 The integration and deployment of immune responses

In dealing with the somewhat miscellaneous topics covered in this chapter, it does not seem necessary to do more than provide a very brief summary with current references for sections such as the nature of viral immunity which have no special bearing on those aspects of immunological theory with which we are concerned.

There are, however, three topics which have important theoretical implications. These are the immunological relationship in placental mammals of mother and foetus, the blood groups and their association with haemolytic disease of the newborn, and finally, the nature of DH and the conditions which may be related to it.

## VIRAL IMMUNITY

Modern discussion of the serological response to viral antigens has been largely the responsibility of Fazekas's group at Canberra. They have been principally concerned with the response of experimental animals to infection with influenza viruses or to immunization with the surface antigens. Their particular interest in 'original antigenic sin' has been referred to earlier (p. 463).

A recent summary at the biological level will be found in Fazekas and Webster (1964) and a discussion of physico-chemical and mathematical aspects in Fazekas and Webster (1963).

Perhaps the most important technical aspect of the use of bacterial or animal viruses as antigens is the sensitivity of the means of detecting specific antibody. In a standard test in which the percentage neutralization of 100 phage particles after exposure to serum for 1 hour is examined, the actual weight of antigen involved in the reaction is very much smaller than in any other practical approach. There is still some uncertainty in regard to the mechanics of phage neutralization, but it is clear that one of the tail antigens is concerned (Lanni and Lanni, 1954) and although

566

it has been shown that neutralized phage can still be adsorbed to the bacterial surface, the most likely interpretation is that attachment of antibody interferes with the appropriate functioning of the complex attachment and injection of DNA which initiates infection of the host (Stent, 1963).

There are no features of the immune response to viruses that are not observed with 'good' antigens of other origins. Reference to the use of viral antigens has been made earlier at various points but it may be worth referring again to the study by Svehag and Mandel (1964*a*, *b*) of the production of antibody to polio-virus in the rabbit. This showed clearly (*a*) that most normal rabbits have minimal amounts of 19 S neutralizing antibody in their sera, (*b*) with a very small dose of antigen there is a rapid response confined to 19 S and (*c*) with larger doses of antigen 7 S antibody appears and persists longer than 19 S. This is wholly in line with work with the more commonly used antigens.

## IMMUNOLOGICAL FEATURES OF THE MATERNAL REALATIONSHIP TO FOETUS AND NEONATAL OFFSPRING

From a purely immunological point of view, the foetus in a placental mammal is more or less equivalent to a homograft of an $F_1$ animal or a parent. It should therefore be rejected in virtue of various paternal antigens not present in the mother. This apparent paradox has provoked much experimentation, reviewed by Billingham (1964), with the net result that the only well-established effect on record is that repeated pregnancies by a male of different histocompatibility type may introduce a minor degree of tolerance in the female (Breyere and Barrett, 1960, 1963). There is clearly a well-organized means of rendering any physiological association of maternal and foetal cells during pregnancy innocuous to both individuals.

### The function of the trophoblast

In discussing the possibilities, Billingham (1964) concludes that, in man, the effective segregation of the two systems is due to the existence of a specialized layer of foetal cells, the trophoblast, which in species (including man) with a haemochorial placenta

567

constitutes a continuous layer in direct contact with the maternal blood. The work of Simmons and Russell (1962) showed that the virtue of the trophoblast in mice, where the conditions are broadly similar to those in man, was its absence of antigenicity. Grafts of $F_1$ embryo tissues on the maternal strain were rapidly rejected with the single exception of trophoblast. This showed proliferation and formation of villi as effectively as in a compatible host. This finding is in itself of considerable interest in showing that it is possible to have actively functioning cells without apparently any histocompatibility antigens.

The non-antigenic character of the foetal trophoblast interposed at every point between maternal and foetal blood circulations is presumably the most important of the physiological defences against sensitization of the mother, but minor breaches of the barrier undoubtedly represent an occurrence common enough to have some evolutionary significance.

### Accidental cell interchange

Leakage of foreign cells from mother to foetus, and vice versa, probably occurs to some extent in every pregnancy but normally has no harmful significance. Movement of soluble proteins of maternal origin into the foetal circulation is, of course, normal and presumably invokes a temporary tolerance to any maternal antigens foreign to the foetus. Studies of appropriate Gm genotypes in mother and child indicate that a small proportion of children capable of producing antibody against a maternal antigen do so spontaneously (Steinberg and Wilson, 1963) indicating that any tolerance must have disappeared before the maternal globulin had been completely broken down.

*Foetus to mother.* Entry of foetal red cells into the maternal circulation is most frequent during actual delivery, and this, of course, is of primary importance in the aetiology of haemolytic disease of the newborn.

The possible significance of nucleated cells which must enter along with red cells is also of interest. As mentioned already, there is no evidence that the foetus can sensitize the mother but a considerable amount of evidence that when only minor histocompatibility differences are concerned, repeated pregnancies by a male of different strain produce a certain degree of tolerance to skin grafts of paternal type. According to Billingham *et al.* (1965) the histocompatibility antigen determined by the Y chromosome appears to be specially prone to produce tolerance in female mice after several pregnancies, always, of course, including some male foetuses.

Entry of trophoblast into the maternal circulation appears to be an extremely common event (Douglas *et al.* 1959) during human pregnancy usually without any untoward effect, or any evidence of sensitization. The malignant tumour, chorionepithelioma, is, however, a derivative of foetal cells of very similar character and its behaviour has interesting immunological aspects. The high degree of virulence with rapid metastasis indicates a failure of any effective immune response in the patient, and Hackett and Beech (1961) considered that this was due to the same absence of antigenic activity that is evident in normal trophoblast. It has been reported, however, that in two cases the husband's skin was accepted for much longer periods than skin from a random donor (Robinson *et al.* 1963). This suggests that histocompatibility antigens are present but produce tolerance rather than sensitization. As has been noted in other contexts, exposure to an antigen solely by the intravenous route has a general tendency to induce tolerance while dermal or local tissue entry favours sensitization.

There is also important evidence pointing in the same direction from the occasional occurrence of spontaneous regression and from the abnormally favourable responses to treatment with amethopterin (Hertz *et al.* 1961, 1964). Both suggest that there is a certain degree of immunological constraint which is normally insufficient to hinder the high intrinsic proliferative energy of the tumour cells, and that the malignant foetal cells carry significant amounts of foreign (that is, paternal) histocompatibility antigens.

*Mother to foetus.* Entry of maternal blood cells into the foetal circulation also occurs, and again there is very little evidence of immunological effect, either tolerance or sensitization. Another possibility that has often been considered is the entry of cells which could provoke a graft-versus-host reaction in the foetus. There have been suggestions that a runt disease syndrome can very rarely be produced in human infants. At least one case has been described of a chronically ill male infant with XX–XY lymphocytic mosaicism which did not involve the bone-marrow (Kadowaki *et al.* 1955). This is the pattern one would expect from implantation and proliferation of maternal immunocytes during pregnancy. Abstracts of the American Pediatric Society meeting in 1966 mention infants with severe immunological deficiencies who suffered fatal reactions in attempts to reconstitute the lymphoid system with donor cells, in one case maternal bone-marrow. Obviously, gross runt disease will only be seen in immunologically deficient infants, but it may be much more difficult to rule out the possibility that some cases of delayed lymphoreticular proliferative disease in childhood have been initiated in this fashion.

The conditions which come to mind are acute leukaemia, Hodgkin's disease (a suggestion made by Green *et al.* 1960) and the Burkitt lymphoma. The last-named condition is generally regarded as of viral origin but the cells are immunocytes and two independent groups, Wakefield *et al.* (1966*b*) and Tanigaki *et al.* (1966), have found that established lines of Burkitt cells in tissue culture produce Ig M. The quite exceptionally favourable response to chemotherapy strongly suggests that the tumour is under some degree of immunological constraint. Although it seems probable that the popular hypothesis will eventually be substantiated, the evidence pointing to a viral aetiology is all circumstantial and what may be direct evidence of viral presence or action in the tumours or cells derived from them (Klein *et al.* 1966; Old *et al.* 1966) seems to be rather more readily explicable as colonization of an established tumour by virus rather than by assuming an aetiological role. It would seem, however, worth investigating whether the karyotypic sex of Burkitt tumour cells always corresponds to that of the patient and whether any other evidence can be obtained of a common origin of maternal lymphocytes and the cells of the corresponding Burkitt lymphoma. One possibility would be to use Hellström's approach and test lymphocytes from a range of donors including the mother of the patient for evidence of cytotoxic effects against the lymphoma cells.

THE INTERPRETATION OF BLOOD GROUPS

One of the important obligations of any theoretical treatment of immunological phenomena is to provide a satisfactory interpretation of the existence of blood groups in man and of their clinical and evolutionary effects. The discovery of the ABO blood groups by Landsteiner depended on the existence of the natural isoagglutinins anti-A and anti-B in all individuals lacking the corresponding red cell antigen. This in itself has provided a major problem for immunological theory. From 1941 onwards it has been recognized that haemolytic disease of the newborn depends on differences in the antigens carried by the red cells of mother and foetus, and is mediated by the passage of maternal antibody into the foetus in the late stages of pregnancy. This, with the demand for compatible blood in transfusion, has led to an unprecedentedly detailed study of the genetic, chemical and immunological aspects of the antigens concerned. This has been largely but not exclusively guided by the use as reagents of antisera of human origin that have developed as a result of casual immunization by repeated pregnancy or transfusion. Blood-group work is a highly specialized branch of immunology that has necessarily to deal with practical matters and with minutiae of detail in individual clinical episodes, but there is much that is highly relevant to our general theme. It is inevitable that any comprehensive work like Race and Sanger's (1962) *Blood Groups in Man* will be largely concerned with fine differences between antigens and their genetic significance. Virtually all the factual material for this section has been taken from this classic text, but selection has been for points of theoretical significance and genetic aspects will, for the most part, be left out of consideration.

*Normal isoagglutinins*

The haematologist uses both naturally occurring antibodies and those derived from accidental immunization by pregnancy or transfusion in man. Less frequently, antisera produced by experimental immunization of animals or human volunteers are used. Of the normal isoagglutinins, those of the ABO groups are the most important, and will be treated separately in the next section. Among the numerous other blood group systems that have been uncovered since 1940, natural isoagglutinins are much less common. As in the ABO groups they are found, if at all, only in the serum of individuals lacking the antigen toward which the antibody is directed.

With only very rare exceptions the blood group systems—MNS, Rh-, Lutheran and Kell—are not associated with natural antibodies, either complete or incomplete. Even when there has been extensive deletion of the gene complex in the Rh system, as in the genotypes -D-/-D- or ---/---, no natural antibodies against Rh antigens are found. Several -D-/-D- individuals have been artificially immunized with cells of the common Rh phenotypes and have produced active and complex antisera.

In the P system with the types $P_1$, $P_2$, $P^k$ and, extremely rarely, pp, anti-$P_1$ in rather weak activity is found in most $P_2$ individuals while the two known pp individuals both showed active antibody against $P_1$, $P_2$ and $P^k$ cells.

The recently discovered Ii system behaves quite differently from any other blood-group system. Cord blood from normal infants has an antigen i which is progressively converted into or replaced by I as the infant grows, the adult status being reached at about 18 months (Marsh, 1961). Here, cold agglutinins only are observed. A proportion of cord bloods contain anti-I of Ig M character (Adinolfi, 1965) and low-titre cold anti-I may be found in some normal adults. It is almost always the dominant component of cold-type acquired haemolytic anaemia (Wiener *et al.* 1956). High titre anti-I is regularly found in the very rare adults with i red cells. Anti-i is not found in normal individuals but has been observed in patients with some form of reticulosis.

## The ABO group antibodies

Omitting rare and anomalous conditions, the ABO blood group system shows six phenotypes $A_1$, $A_2$, B, $A_1$B, $A_2$B and O. The antigens concerned are A, $A_1$, B and H, all of which are manifestations of a parent mucopolysaccharide H (Morgan, 1960) as modified genetically. The antigens for the $A_1$ and $A_2$ phenotypes are A + $A_1$ and A, respectively. According to Morgan, A specificity is associated with $\alpha$-D-N-acetylgalactosamine, B with $\alpha$-D-galactose and H with L-fucose.

572

The mucopolysaccharides are present as AH, BH, ABH and H in cells of the standard four major groups and the only anomalous one which need be considered is the 'Bombay' group Oh in which none of the three antigens is present. With only quantitative deviations, all individuals show in their sera antibodies to each of the components A, B or H not present in the red cell surface.

TABLE 22. *The ABO blood group system*

| Group | Antigens | Normal antibodies |
|-------|----------|-------------------|
| A | A, H | anti-B |
| B | B H | anti-A |
| A B | A B H | nil |
| O | H | anti-A    anti-B |
| Oh 'Bombay' | — | anti-A, anti-B, anti-H |

The newborn infant either shows no isoagglutinin or small amounts from the mother, but begins to produce the characteristic Ig M isoagglutinin at the same age (3–6 months) as other immunoglobulins and the peak titre is usually reached at 5–10 years (Thomsen and Kettel, 1929).

Antibody is never produced against any of the antigens present in the red cells and this is fully in accord with the general basis of tolerance in selective theories. It is less certain what accounts for the regularity of the appearance of the isoagglutinins in all individuals in whom they are not 'forbidden'. The most reasonable view is probably to recognize the immense variety of mucopolysaccharides and other substances presenting monosaccharide or oligosaccharide ADs amongst bacterial products. Many of these will correspond more or less closely to the characteristic A, B and H ADs and a regular, if minute, leakage of bacteria from the gut into the circulation could ensure proliferation of appropriate clones and production of isoantibody. A point of interest against any counter-suggestion that 'forbidden' antibody is liberated, but is mopped up by the circulating cells as soon as it is produced, is the absence of cold antibody against autologous cells in normal persons. Chickens produce anti-human blood group B agglutinin as a natural component of serum but fail to do so if they are reared in the absence of living bacteria (Springer *et al.* 1959). The chemical nature of the Rh antigens is unknown and the failure of normal Rh-negative individuals to produce anti-D indirectly supports the hypothesis that the ABO isoagglutinins are there because of casual correspondence between red cell and bacterial ADs.

From the beginning of work on blood groups the normal iso-agglutinins have been tacitly equated with antibodies but it has been necessary to demonstrate that they are, in fact, immuno-globulins. This has been fully established for both 'normal' and 'immune' anti-A and anti-B, most of the normal being Ig M (Kekwick and Mollison, 1961). Most interest attaches to the anti-I cold agglutinin of autoimmune haemolytic anaemia (cold type). This was shown by various authors to produce a 'non-$\gamma$' coating on red cells but it seems to have been effectively demonstrated by Harboe (1965) that what first occurs is union in the cold of Ig M antibody with adsorption of the complement components $C^1 4$ ($\beta_{IE}$) and $C^1 3$a ($\beta_{IC}$) followed by elution at 37 $^\circ$ of antibody, with persistence of complement components on the red cell surface. I have found no identification of natural anti-$P_1$ antibodies as immunoglobulins, but as they have similar specificity to antibodies produced by rabbit immunization there is no reason to exclude them. The other blood groups are defined and studied by anti-bodies produced by natural, accidental (via transfusion) or experi-mental immunization.

There is every justification, then, to accept the serological reagents used by haematologists for blood grouping as immuno-globulins produced by processes similar to those involved in producing other normal and immune antibodies.

### *'Private' and 'public' blood groups*

With the universal interest in blood transfusion, haemolytic dis-ease of the newborn and post-transfusion reactions, enormous numbers of cross-tests between individual cells and sera have been made. As an incidental result, individuals are uncovered with a 'private' blood group; that is, it is found that their cells are agglutinated by a certain serum but that cells from all other individuals remain unaffected. Conversely, there are also 'public' blood groups, in that a patient suffering from a post-transfusion reaction provides a serum which agglutinates cells from virtually every other individual; for example, the antigen Vel has been tested for in approximately 30,000 individuals of whom 8 only have been found to lack it.

It is of interest from our present point of view that when cells carrying private antigens are available for test, a proportion of normal sera agglutinate some of them while sera from cases of autoimmune haemolytic anaemia are much more active in this respect (Cleghorn, 1961). On the other hand, the very few negative individuals in the 'public' blood groups do not show a corresponding isoantibody.

A possibility that may need consideration here is that the private antigen presumably arises by mutation and could therefore in principle also arise as a result of somatic mutation in a proportion of stem cells of normal or pathological individuals. Once somatic mutation had resulted from the presence of the antigen in a very small proportion of red cells, an antibody could be produced. The fact that Cleghorn (1961) found at least six separable antibodies against private antigens in a single serum from a case of autoimmune haemolytic anaemia may have other explanations, but *a priori* suggests that the average individual may in the course of a lifetime produce by somatic mutation a large crop of private antigens below the level necessary to provoke antibody in normal individuals. The implications for the interpretation of autoimmune disease are discussed elsewhere (p. 608).

### The Rh groups and haemolytic disease of the newborn

The Rh blood-group system has grown more and more complex with time but for practical purposes we can consider that there are three common gene complexes which can be represented: CDe, cde, cDE. Combinations of these three patterns produce the genotypes of 78 per cent of English people, of whom 15 per cent are Rh-negative, having the constitution cde/cde. Another 1–2 per cent will also have no d gene and are hence susceptible to be immunized by the entry of Rh-positive (that is, containing D antigen) blood.

There are many rare deviations from the classical scheme which are reviewed by Race and Sanger (1962), but the only ones of general immunological interest are the deleted forms -D-/-D- and ---/---, in view of their application to the serology of acquired haemolytic anaemia. Nothing is known of the chemical nature of the Rh antigens but there are hints that they all represent enzymic modifications under genetic control of a basic substance. This would have the same sort of relationship to the observed ADs that H has to A and B. Indications in this direction include the behaviour of autoimmune antibodies and the fact that a guinea-pig immunized with semi-purified antigen from Rh-negative (cde/cde) human cells produced an anti-D serum (Murray and Clark, 1952).

The definition of the Rh system began with the experiment by Landsteiner and Wiener (1940) in which they immunized rabbits and guinea-pigs with rhesus monkey cells. Such antisera, now known to carry anti-D, agglutinated red cells from 85 per cent of human beings, who were therefore 'Rh-positive', and failed with those of the other 15 per cent 'Rh-negative' people.

Subsequently animal sera have been little used. Isoantibodies have been obtained from the standard sources. These are from patients given incompatible blood transfusions, from women immunized by paternal antigens via the foetus in pregnancy, and from patients with autoimmune disease especially if they had been transfused.

On ordinary principles a person receiving a transfusion from another individual can produce antibody only against antigens not carried in his own red cells. It does not by any means follow that he or she will produce every antibody which it is possible in theory to produce.

Haemolytic disease of the newborn was first recognized by Levine and Stetson (1939), and within a year or two was shown to represent the result of damage by the mother's antibodies on the foetal red cells. The only common situation is where a Rh-negative mother (d/d) carries a d/D child of an Rh-positive father and there is a significant leak of foetal cells into the mother's circulation. This occurs most frequently shortly before birth and can be recognized by examining the mother's blood immediately after delivery for small numbers of cells containing foetal haemoglobin (Kleihauer *et al.* 1957).

576

Immunization of the mother occurs in only about 5 per cent of the possible combinations. The requirements for immunization include the dosage of foetal cells reaching the maternal circulation and the number of pregnancies in which this happens. The existence of ABO compatibility of foetal cells toward the mother (Levine, 1943) increases the likelihood 2–2·5 times over that when there is incompatibility (Race and Sanger, 1962). For maternal antibody to affect the foetus, it must be Ig G, and in sufficient concentration.

There is now very active research going on as to how best to exploit the method initiated by Finn *et al.* (1961) to prevent maternal immunization by administration of high titre anti-D antibody to vulnerable mothers immediately after delivery. An interim summary of results (Combined Study, 1966) shows complete prevention of antibody production as against 12–25 per cent with antibody in the control series.

## Significance for immunological theory

Most of the findings which I have summarized are in line with other immune phenomena and have no special bearing on the validity or otherwise of the present theoretical approach. Only a few points call for special comment.

(*a*) Most antibodies concerned in haemolytic disease of the newborn are 'incomplete' in the sense that they do not agglutinate corresponding cells under ordinary conditions of suspension in saline. Such antibodies can become attached to red cells and block the effect of agglutinating anti-D as well as making the cells agglutinable by anti-human globulin—the direct Coombs test. Incomplete antibody will agglutinate cells suspended in 20 per cent bovine albumin or previously treated with trypsin or papain.

Incomplete antibody is not monovalent antibody (which probably does not exist) and is best interpreted as antibody of relatively low affinity for the AD. If the AD is not made easily accessible by removal of surface protein, a second bridging union by an antibody molecule already attached to one red cell is only possible to a significant effect when the affinity of the combining site is above a certain level. Degrees of incompleteness vary from case to case and the pages of Race and Sanger are full of accounts of individual

577

sera showing some exceptional quality or other. The picture is not the production of some definable substance, anti-D, but of the appearance of a variety of new immunoglobulins whose common feature is a capacity to react with D ADs to produce a demonstrable effect. The range of goodness-of-fit is obviously broad.

(*b*) The effect of high titre anti-D in preventing immunization of the mother by foetal cells recently received is in line with recent work on standard antigens (see p. 549). In part, the effect may be due to the resulting rapid elimination from the circulation of the foetal cells and their accelerated intracellular destruction. It is probably more important that any liberated antigen will be sufficiently associated with antibody to eliminate the possibility of effective stimulation of a progenitor cell.

(*c*) The 'compound antigens' of the Rh group represent one example of the complexities that have involved every one of the blood groups as investigation has been intensified. The case of antigen f, probably equivalent to ce, is instructive. It was found in the serum of a 'much-transfused' white man of genotype CDe/cDE and reacted only with the products of gene complexes cde, cDe and cD$^u$e, that is, where c and e were present together on the same chromosome. It is to be noted that this was not the case in the donor of the serum. Other examples require Ce or CE to be in the *cis* position for their appearance. Still others seem to be specific for the product of a whole complex (Cde$^s$, for example) (Race and Sanger, 1962, pp. 164–71).

What emerges is wholly in line with the interpretation that an antibody has no relation to an AD other than the casual one that it happens to fit the pattern presented. The symbols of the haematologist are in a continual process of becoming inadequate for the facts. One may contrast the beginnings of the Rh system—when it was covered by the statement: Rh-positive 85 per cent, Rh-negative 15 per cent—with the 1962 formulation in Race and Sanger and the further complexities that have accumulated since then. Clearly, there is a complex basic antigen macromolecule subject to strict genetic control by a pair of gene complexes subject to point mutation, deletion, etc. Given the commoner basic configurations of human immunoglobulin combining sites, certain antibody patterns will be more commonly produced allowing the

Fisher CcDdEe convention to be uncovered. It only needs, however, an exceptional individual to produce a pattern of antibody or receptor that has special affinity for some aspect of the basic molecule neglected by the great majority of patterns, for new serological concepts such as compound antigens to be introduced. There could hardly be a clearer indication of the virtually infinite potential complexities of macromolecular structure both in AD and in combining site. It is inconceivable that anything but a stochastic process could be capable of keeping up with the changes of antigenic pattern which are presented by pregnancy, by transfusion, by infection or by somatic mutation of the body's own tissues.

(*d*) The final feature of special interest is the number of antibodies against private antigens which may be found in normal sera and their much more frequent occurrence in sera from cases of autoimmune haemolytic anaemia. This has been touched on above, but the relation to autoimmune disease requires further consideration in the next chapter.

### DELAYED HYPERSENSITIVITY

Although tuberculin hypersensitivity must probably remain the prototype of DH reactions, much work has also been done in three closely related fields: (*a*) the production, usually in guinea-pigs, of skin reactivity to soluble protein antigens especially by the use of Uhr *et al.*'s (1957) method of sensitization by injection of antigen–antibody complexes in small dose; (*b*) sensitization of guinea-pigs (or human beings) to simple chemicals, often derivatives of dinitrophenol and (*c*) the rejection of skin homografts.

Several aspects of DH have been discussed in other contexts, notably in relation to immunological unresponsiveness. In this chapter the most important features requiring discussion are the concept of DH as mediated by sensitized lymphocytes, the special relationship of such cells to the thymus and the cellular mechanics of the classical DH reaction exemplified by the Mantoux reaction in man. In order to obviate an extensive review of earlier work the discussion will be initiated by using the conclusions reached by Uhr (1966) in his recent comprehensive review. The other current

579

review which has been utilized in developing a theoretical approach is the *British Medical Bulletin* dealing with delayed hypersensitivity. The concluding 'perspective' by Humphrey (1967) is probably closer to my own picture than that of Uhr, but the two reviews taken together provide a comprehensive picture, with full references, of those aspects of DH which seemed important in the 1965–7 period.

## Uhr's interpretation of delayed hypersensitivity

At the conclusion of a critical and authoritative survey of DH, Uhr (1966) presents a concept of its nature which, in his opinion, 'reflects more accurately current speculation in this field'. It is convenient to traverse his seven points for their concordance or otherwise with the general thesis of this book.

1. 'The antigen-specific factor is a gamma globulin.' This is only acceptable if the concept is broadened to cover any situation by which a cell carries those portions or attributes of a specific immunoglobulin which will allow interaction with the AD concerned.

2. 'Both 19 S and later 7 S gamma globulins are concerned.' This is based by Uhr on the early appearance and long persistence of sensitization. It is acceptable with the same type of qualification as 1.

3. 'The antibody molecules [in this and later sections we would replace this by 'The cell receptor molecules'] are produced by the sensitive cell, are located at the surface of the cell and can react at the cell surface with antigen outside the cell.'

4. 'The [cell receptor molecules] have a relatively high affinity for antigen which allows the induction, elicitation or de-sensitization in the presence of...serum antibody.'

5. 'The long duration of the delayed hypersensitivity state is mediated by sensitized long-lived small lymphocytes.' Here one would add, 'or small lymphocytes derived by descent from formerly sensitized cells (memory cells)'.

6. 'The interaction of sensitized cells and antigen results in the release of a factor that involves non-sensitive cells in the reaction.' Of the three candidates mentioned by Uhr, neither cytophilic antibody nor antigen–antibody complex seems appropriate and we

should prefer to speak at present of a product of cell damage of unspecified character.

7. 'The tissue damage of delayed hypersensitivity results from the release of lysosomal enzymes probably from macrophages but possibly also from lymphocytes.' Here agreement must be qualified by a realization of the complexity of the cell population, the degrees of damage and the interacting pharmacological reactions in any region of tissue damage.

Essentially there is nothing in this summary by Uhr which is inconsistent with the view we have developed in regard to immunocyte function in earlier chapters of this book. It is, however, a conservative approach which leaves many problems of current interest untouched. Uhr has fully recognized this and his seven points represent something like the commonly acceptable factors of many earlier theoretical formulations.

The major lack is any indication of the nature and significance of the dissociation of antibody production and DH. Why, for instance, can children with agammaglobulinaemia develop DH (Good *et al.* 1960) and reject homografts (Schubert *et al.* 1960) while unable to make antibody? Then there is the generally accepted finding (Benacerraf and Levine, 1962) that reactions mediated by serum antibodies require a smaller area of identity of the immunizing and challenging antigens than do DH reactions. Using various related haptens, cross-reactions in the DH system are appreciably more extensive than those reported in rabbit antibody systems employing identical haptens. In summing up, the same group of workers concluded that desensitization studies with cross-reacting antigens show that the DH response is characterized by the production of a heterogeneous population of cells, [the immune receptors of] which are all more or less closely adapted to the structure of the homologous protein conjugate.

### A modified approach

The following elaboration of Uhr's formulation would be in line with the approach I have adopted.

Delayed hypersensitivity is probably still an incompletely analysed concept which may cover phenomena with only a superficial resemblance. Keeping as far as possible to the prototype tuberculin reactivity as a model, it is clear that there are two aspects, the process of sensitization and the phenomena that follow challenge of the sensitized animal with antigen.

Most of the observable phenomena concern the second phase but even here there are major difficulties in, for instance, deciding such an apparently simple point as whether the mononuclear cells in the reaction are lymphocytes or monocytes. My own interpretation of the information in the literature is necessarily in the general terms of a clonal selection approach and is very similar to that adopted by Humphrey (1967) in summarizing the number of the *British Medical Bulletin* devoted to delayed hypersensitivity. The key cells are specifically sensitized lymphocytes which enter the tissue containing the antigen with a larger number of unsensitized lymphocytes and other leucocytes from the blood (Boughton and Spector, 1963). Contact with antigen induces changes within the sensitized cells which include blast transformation and proliferation, lysosome formation and discharge, and the liberation of a variety of soluble agents capable of 'activating' or damaging adjacent cells in a variety of ways.

The weight of the evidence suggests that the specific quality of a DH immunocyte is sufficiently different from the antibody-producing cells and their progenitors for it to be seen as representing a series equivalent in status to the Ig M-, Ig G- and Ig A-producing immunocytes—the DH immunocyte. The possibility is by no means excluded that such immunocytes are also actual or potential Ig M producers.

In line with the discussion in chapter 8 and with the Eisen–Karush theory, the specific quality of the immunocyte could be due simply to the configuration of the immune receptor, allowing a more avid association and hence a more destructive stimulation of the cell, but additional factors are probably also involved. The requirement for a more extensive association of the combining site with a larger AD involving both hapten and a portion of the carrier protein is probably almost equivalent to the demand for high avidity. The nature of the presentation of the primary antigenic stimulus that leads to sensitization may also be important.

Obviously, some potential antigens are much more likely to provoke DH than others; tissue homografts, tuberculoprotein and sensitizing chemicals, for instance. This probably will provide the key to the significance of the particular type of immunocyte which mediates DH. In the case of the skin-sensitizing chemical compounds it is clear that union of the hapten with a body protein is necessary for sensitization. For reasons to be discussed shortly, it seems probable that the complex may be in part or predominantly attached to mobile cells, lymphocytes or monocytes, and be carried in this fashion to the paracortical region of the draining lymph node (see fig. 25). The work of Oort and Turk (1965) points strongly to processes in this area being vital for the development of skin sensitivity.

TABLE 23. *Antibody production and delayed hypersensitivity*

| A. Plasma cell production | B. DH immunocyte production |
|---|---|
| 1 Antigen held on dendritic phagocytic cells of the lymph follicles in the form essentially of ADs held on the surface of the cell processes and usually partially blocked by antibody. | 1 Antigen held as a closely integrated part of the lipoprotein surface layer of mobile mononuclear cells which pass to the paracortical area of lymph nodes. |
| 2 Contact with AD made by the rapidly changing mobile population of committed and uncommitted immunocytes in the lymph follicle. Stimulation, particularly of committed cells, is not inhibited by the association of antibody. | 2 Contact with AD is made predominantly by cells of relatively recent origin from the thymus, that is, progenitor cells only. Only a high-avidity union is effectively stimulating, that is, an extensive area of the AD needs to have a steric relationship to the combining site. The presence of antibody inhibits stimulatory contact. |

The special character of what may be called DH immunocytes would then depend on the fact that for them to be effectively committed a special type of presentation of the antigen may be necessary.

A certain symmetry would emerge if the mechanisms for commitment (*a*) to plasmablast-plasma cell proliferation and (*b*) to proliferation as lymphocytes with capacity for DH were related as shown (table 23).

Such a formulation would leave it open whether the progenitors of (*a*) and (*b*) were already differentiated when they left the thymus or whether commitment to (*a*) or (*b*) is determined by subsequent circumstances. It is possible that no operational decision on the point can be devised at the experimental level. Phenomena that may be relevant include the inability of human agammaglobulin-aemics or bursectomized and irradiated chickens (Szenberg and Warner, 1964; Cooper *et al.* 1966*b*) to produce antibody, although they can develop DH and reject homografts. The special facility of immunization with Freund's complete adjuvant to give DH is obviously important. Although the results as reported are irregular, it seems clear that thymectomy has a more marked effect in diminishing reactions in the DH field than most types of antibody production (Parrott and de Sousa, 1966).

Without attempting to decide where (*a*) and (*b*) progenitors arise, we can make use of available knowledge to summarize the requirements for the development of committed clones of the two types as follows: The requirements for (*a*) seem to be (i) presence of bursal hormone or its mammalian equivalent, (ii) stimulation by antigen held in the dendritic phagocytic cells of lymphoid tissue, (iii) opportunity for plasmablast proliferation, including probably a high local $O_2$ tension. Of these, probably only (i) is mandatory. For (*b*) the governing conditions are (i) presence of thymic hormone, (ii) stimulation by AD intimately associated with 'self-components' and probably carried on a cell surface, and (iii) a high-affinity association of AD and combining site.

### Lawrence's transfer factor

The approach we have adopted may have some bearing on one of the most puzzling of immunological phenomena, the transfer factor described by Lawrence (1955) and subsequently intensively investigated (for review, see Lawrence, 1959).

The basic findings are that leucocytes taken from a human subject actively sensitive to tuberculin and injected subcutaneously in adequate dose into tuberculin-negative subjects will induce a positive skin reactivity in the recipient (Lawrence, 1949). This could also be observed with leucocytes disrupted osmotically or by repeated freezing and thawing, and later studies indicated that the agent concerned was highly resistant to physical and enzymatic agents. Essentially similar results were obtained with a number of other sensitizing antigens, of which the most important was probably the transfer of coccidioidin sensitivity to individuals who had never been in an endemic region (Rapaport *et al.* 1960*a*, *b*). In most experiments the transferred sensitivity lasted for months or years.

The results cannot be duplicated in experimental animals, and human leucocytes active in man are quite inert in guinea-pigs and rabbits.

In the light of studies on the physico-chemical character of the transfer agent and what seem to be fully adequate efforts to eliminate experimental artefacts such as sensitization by the preliminary tests for non-reactivity in recipients, Lawrence and his colleagues have sought an interpretation in terms of some information-carrying agent which, by its replication, can endow sensitivity on the lymphocytes of the recipient. If, as we have suggested above, the essential feature of antigenicity which induces DH is the integration of the AD into the cell surface of mobile cells, presumably monocytes or lymphocytes, a wholly different solution becomes possible.

The hypothesis is that the AD which confers specificity becomes incorporated in the lipoprotein on the cell surface in such a form that when the cell is autolysed in the body or disintegrated *in vitro* the complex AD plus cell-surface component, which we can call the immunogen, can be transferred to another cell. The immunogen incorporated in the new recipient cell surface can continue to react with immunocytes newly produced from the thymus and of appropriate avidity.

This suggestion has affinities with the hypothesis of Pappenheimer (1956) that transfer factor is essentially an antigen–antibody complex that can be transferred from one leucocyte to another. Lawrence's (1959) view that transfer factor is neither conventional antigen nor conventional antibody must be accepted, and his 'self + X' hypothesis (Lawrence, 1960) has very much in common with the present approach. In particular, his virtual identification of DH and homograft immunity represents an important advance. I find it impossible, however, to accept the concept of a self-replicating agent resistant to a battery of enzymes and to prolonged freezing and thawing.

585

The only experimental evidence directly opposed to the cell-surface hypothesis is the fact that with highly sensitive donors a positive reaction is present within 48 hours. This would almost demand a passive component but if, as others have suggested, the subjects had once been positive to the reagent used this is perhaps just possibly an active response. The further finding that a positive reaction can persist for a year or more in the recipient rules out all possibilities other than an antigen of special quality and persistence or a self-replicating agent which produces directly or indirectly the receptor of the specifically sensitive immunocytes.

Admittedly, the hypothesis does not completely cover the facts as reported but this holds even more for every other hypothesis which has been promulgated to interpret Lawrence's findings. The hypothesis of a transferable antigen in the form of AD built in like a histocompatibility antigen into the lipoprotein surface layers of mobile cells accounts for the following features.

1. Resistance to distilled water, lysis of cells, repeated freezing and thawing, prolonged storage in deep freeze, and the action of DNA-ase, RNA-ase and DNA-ase + trypsin.

2. The sensitivity is specific, involves the whole skin surface and is long lasting, that is, equivalent to an active response. It does not cross species barriers because of the necessity to be incorporated into the type of surface characteristic of the species. For the same reason, the interaction of leucocytes with antigen *in vitro* does not give them transfer capacity.

3. There is no detectable antibody in white blood cell extracts or in the serum of a positive recipient. In some recipients there is a sharp increase in the intensity of reactions after 6–8 weeks.

These points are taken from Lawrence (1960, table 1, p. 258), as are the following, where difficulties exist.

4. The early onset already discussed.

5. There is also evidence that effective leucocytes can be rendered inert by contact with tuberculin—which could have a number of explanations.

6. Finally, many of the donors in the tuberculin and coccidioidin experiments showed no sign of active disease and may well have been sensitized years previously. In both cases there would remain the possibility of a slow liberation of immunogen from quiescent

lesions containing no living pathogens. This might be sufficient to maintain a population of antigen-carrying cells for many years.

*Thymic associations with delayed hypersensitivity*

The findings in lymph nodes of mice and guinea-pigs draining skin areas treated with sensitizing chemicals such as oxazolones point toward the existence of a different population of lymphocytes from those involved in normal antibody responses. Both in the guinea-pig (Oort and Turk, 1965) and in mice (Parrott and de Sousa, 1966) there is a rapid appearance of pyroninophilic blast cells in the region which Parrott *et al.* (1966) called the 'thymus-dependent area' (fig. 25). This reaction is seen 3 and 4 days after painting but has disappeared by the 7th day in mice. It is wholly absent in mice neonatally thymectomized and sensitized with oxazolone at 5–7 weeks of age. By contrast, administration in the same skin area of polysaccharide pneumococcal antigen (SSS III) in a dose which produces antibody in both intact and neonatally thymectomized mice produced no significant change in the area. Small germinal centres in the cortical follicles and increased plasma cells in the medullary areas were seen in both thymectomized and normal mice.

The thymus-dependent area was originally recognized by the absence of its normal complement of small lymphocytes in neonatally thymectomized C3H mice. It is of special interest that such mice show inability to reject homografts and little or no DH but with some, though not all, standard antigens normal antibody production occurs. According to Turk and Willoughby (1967), treatment of guinea-pigs with an anti-lymphocytic serum produced by immunizing rabbits with guinea-pig thymus cells removed small lymphocytes from the paracortical areas of lymph nodes. Sensitization by oxazolone was virtually abolished and there was marked depression in the number of immunoblasts in the paracortical area.

There are several points for discussion in these findings. The striking appearance of *many* pyroninophil blasts in the area at 3 days makes it unlikely that these are all or mostly specifically stimulated immunocytes. The probable sequence of events, if one combines suggestions made by Parrott and de Sousa with our own hypothesis, may be as follows. The treated skin area will liberate into the afferent lymphatics, complexes of chemical with skin protein or lipoprotein, and mobile cells (either lymphocytes or monocytes) carrying the same or similar oxazolone-antigen on the surface. These cells will lodge primarily in the paracortical region where one assumes that there are large numbers of immunocytes of the type specially concerned with the surveillance function. Significant numbers will react with cells carrying the antigen and as a result a process stimulating adjacent cells will be set in train. The simplest explanation, for which precedents are discussed elsewhere (p. 513), is that pharmacologically active agents from the primary reacting cells are concerned. Another possibility which might repay experimental study is that the hapten-body protein acts directly as a phytohaemagglutinin-like stimulus to the cells to which it becomes attached.

Reverting to the simplest hypothesis, one would assume that stimulated immunocytes would proliferate either locally or elsewhere to produce descendant small lymphocytes much in the fashion of the pyroninophil cells which developed in Gowans's (1962) experiments (p. 342). These descendants would serve as memory cells to maintain the sensitization for a long period.

The 'dependence' of these areas on the thymus presumably indicates that many of the cells normally present are recently derived from the thymus and that this is for some reason the most likely region for such progenitor immunocytes to settle for a time before rejoining the circulating population. The proposition therefore becomes tenable that capacity to mediate DH responses is particularly characteristic of the newborn uncommitted immunocyte. This has some attractive features as many writers have felt that the cells responsible for homograft rejection and graft-versus-host reactions had a different quality from the antibody-producing series.

When one considers the existence of foetal haemoglobin, with its implication that erythroblasts can switch from one type of chain to another ($\gamma$ to $\beta$), it does not seem impossible for a 'surveillance-type' combining site to be replaced at a certain stage of maturity by an 'antibody A-, G- or M-type' combining site. One can speculate, rather irresponsibly, perhaps, about several possible forms that such a process might take. One would be the presence of a special type of 'variable' chain replacing the standard N terminal segments of light and heavy chains. This might have been specially evolved to recognize minor deviations from standard histocompatibility patterns amongst the individual's own cells. With our present state of ignorance, we might also consider the possibilities that the degree of exposure of the combining site of the cell receptor is responsible for the difference or even that there is a special sequence of the variable chain exposed in the immature cell and a different sequence at a later stage of development.

It may prove extremely difficult to obtain any experimental evidence directly relevant to such possibilities but one might hope to see progress develop from continued study of myeloma proteins and of the functional activity of the pathological cells in a variety of human lympho-proliferative disorders. Sooner or later we are almost certain to find the type of conditioned malignancy (? Hodgkin's disease) which has the corresponding relationship to the DH immunocyte that myelomatosis has to the antibody-producing plasmablast.

### Delayed hypersensitivity and cytophilic antibody

One experimental approach to DH is the peritoneal macrophage-disappearance reaction in guinea-pigs described by Nelson and Boyden (1963). In this, a specifically sensitized animal has a peritoneal exudate induced nonspecifically. If an injection of antigen is given, preferably but not necessarily into the peritoneal cavity, there is a disappearance of cells from the fluid as a result of their aggregation and adhesion to the peritoneal endothelium. In all probability the reaction is due to the presence of cytophilic antibody on the surface of the 'sensitized' macrophages. Boyden (1964) showed that in guinea-pigs immunized with antigen in Freund's complete adjuvant (F.C.A.), peritoneal macrophages were reactive with antigen in virtue of cytophilic antibody attached to their surface. This type of antibody did not appear unless F.C.A. were used. Immunization with red cells alone or with incomplete adjuvant produced haemagglutinin but not skin sensitization or cytophilic antibody. According to Jonas *et al.* (1965) only macrophages and blood monocytes take up this type of cytophilic antibody in demonstrable fashion.

There is as yet no adequate interpretation of the cytophilic quality of antibody. Since one of its qualities is to hold antigen either as a labelled molecule or a visible particle such as a red cell to a cell surface, one would expect that, other things being equal, Ig M antibody would be more effective than G or A and that a high avidity of combining site would be better than a low avidity. There have also been suggestions that the associated carbohydrate may play a part in attaching the antibody to the macrophage or monocyte. As in other connections, a strong suggestion arises that high-avidity antibody is associated both with the use of F.C.A. and with the development of DH.

It remains for the future to determine to what extent cytophilic antibody is directly concerned in the reactions included under DH. On present indications it seems likely to have only an indirect relationship to the main process.

# 12 Autoimmune disease as a breakdown in immunological homeostasis

No theoretical approach to immunity can be satisfactory unless it adequately covers the phenomena of immunopathology. It is in fact of special interest that many of the enlightening developments of modern immunology have come from observations in pathology. Congenital agammaglobulinaemia and multiple myelomatosis come immediately to mind: delayed hypersensitivity developed from Koch's studies of tuberculosis, and quantitative serology was a by-product of the need for a standard dose of a new therapeutic agent, antitoxin for diphtheria. Autoimmune disease will in due course be equally revealing. At the present time, however, there is no accepted interpretation of autoimmune disease, and in this field almost more than any other there is an unintelligent resistance to the concept that disease can arise by other means than the impact of something from the environment.

In regard to autoimmune disease there are three current attitudes all opposed to the selection approach. These are: (a) the completely negative approach, which puts 'autoimmune' in quotes and says that the cause of the clinical conditions so-called is unknown. (b) The concept that when a body antigen is modified by toxins, drugs or virus action (that is, by something environmental impinging on the organism) it becomes antigenic and provokes the appearance of antibodies not only against the modified antigen but also against its normal unmodified form. The response is therefore equivalent to the production of auto-antibodies in the strict sense. (c) The third approach is that among the myriad of ADs on bacteria and other micro-organisms which may enter the mammalian body, patterns may be found which correspond by simple accident to ADs carried by cells in the body. When an invading bacterium with one such AD is dealt with, it provokes antibody which can also react with the body component in question and under suitable conditions give rise to the corresponding auto-immune condition.

I believe that all the 'environmental' explanations of auto-immune disease can be included under (*b*) or (*c*). They are all implicitly based on an instructive view of antibody production. Something abnormal from without impresses a complementary pattern on a system of cells; the antibody is produced and, by the accident of its configuration, attacks cells which could have been expected to be exempt from immune attack. There are, however, well-established immunological phenomena which can be used to support both these points of view, and both will need detailed consideration.

The attitude that I have adopted since 1958 is summarized in the term 'forbidden clone', which I borrowed from the 'forbidden lines' of spectroscopy. Autoimmune disease arises, on this hypo-thesis, from the functioning of clones of immunocyte whose only necessary difference from a normal clone is that the cells resist destruction by AD–CS union under conditions which would be lethal to a normal immunocyte. The difference in behaviour of such cells is ascribed to a metabolic change resulting from somatic mutation, usually on the background of a genetic anomaly. In one instance there is evidence to suggest that the somagenetic metabolic change can be mimicked by a drug effect on the progenitor immunocytes.

The form to be taken by this chapter will be primarily to present the observational and experimental evidence in favour of the two instructive views of autoimmune disease and to use this to test any weaknesses of the forbidden clone approach. In addition, three classical groups of autoimmune disease will be looked at critically for any light they throw on the nature of forbidden clones, the processes that allow their emergence, and the part played by the thymus. The field of autoimmune disease is far too large to be covered in a general discussion and, apart from the three examples of autoimmune haemolytic anaemia, systemic lupus erythematosus and myasthenia gravis, no other specific disease will be more than incidentally mentioned.

THE EFFECT OF FREUND'S COMPLETE ADJUVANT

Modern experimental work on the production of models of auto-immune disease in laboratory animals is almost wholly the creation of Freund's complete adjuvant (F.C.A.). The antigen is mixed with killed acid-fast bacilli and homogenized as a water-in-paraffin oil emulsion by the use of a suitable emulsifying agent. The effectiveness of this manipulation and its probable mechanism when used with conventional antigens has already been discussed (p. 210).

By repeated injections of mammalian cell extracts in F.C.A., normal experimental animals will (*a*) produce antibodies which will react by complement fixation or agglutination of tanned and antigen-coated cells with tissue antigens, (*b*) induce localized auto-immune disease with lymphocytic infiltration of the target organ and antibody and (*c*) produce fatal disease with a general resemblance to systemic lupus erythematosus.

The readiness with which these results are obtained depends on the experimental factors, species, dose of immunizing extracts and site and sequence of injections. There are some points of interest. In rabbits, F.C.A. alone (but necessarily including antigenic bacterial substance) causes a sharp rise in immunoglobulin, most of which is not demonstrable antibody (Humphrey, 1963). This is also seen in guinea-pigs (Binaghi, 1966), while in these animals and also in mice (Barth *et al.* 1965) administration of antigen with F.C.A. induces an increase of Ig G1 incommensurately large compared with the antibody response.

Boughton and Spector (1963) injected turpentine into left guinea-pig testes and produced degenerative changes in the *right* by the injection of F.C.A. in the foot-pad. Either manipulation alone gave no significant lesion.

It is well known that simple injection of F.C.A. in rats can produce a characteristic arthritis (Pearson and Wood, 1959) which, in the opinion of Waksman *et al.* (1960), may be hypersensitization to the antigen present in the adjuvant.

Passing now to the production of auto-antibody by injection of tissue extracts in F.C.A., the first finding is that, in general, auto-antibodies are more readily produced if the extract is derived from

593

species other than the one being immunized. In most instances the antibody produced reacts with the tissue antigen of the injected species as well. The following are typical references: Asherson and Dumonde (1963)—rat liver, kidney, etc., into rabbits gave antibody reacting by complement fixation with rabbit extracts. Asherson (1964)—rat colon but not rabbit colon effective in rabbits. Weigle (1965)—heterologous but not homologous thyroglobulin produced antibody in rabbits. Barnett *et al.* (1963)—adrenal into rabbits gave antibody to homologous and heterologous inocula but evidence of damage to adrenal only with heterologous material.

If rabbits are immunized with mammalian ribosomes or RNA plus F.C.A. for prolonged periods they will eventually die, but in the process give rise to antibodies against rabbit ribosomes, leucocytes, erythrocytes and almost every soluble protein in the plasma (Dodd *et al.* 1962). Okabayashi (1964) reported similar lethal disease in a proportion of rabbits given prolonged injections of egg albumin.

Results of this type make it quite clear that auto-antibodies can be produced against the animal's own tissues by suitable manipulation of genetically normal animals. It has been claimed that this makes nonsense of any suggestion that there is a homeostatic mechanism which prevents the production of antibody against components of the animal's own tissues. This is approximately equivalent to saying that the body has no homeostatic control over pH because a big enough dose of ammonium chloride will have highly harmful effects when it overrides that control.

In the absence of enormous doses or the use of unphysiological manipulations like injection with F.C.A., body components are not *harmfully* antigenic and, if they are accessible, are not antigenic at all; in addition there are many substances which are poor or very poor antigens in that a small proportion only of rabbits inoculated give rise to demonstrable antibody. F.C.A. can reverse all these findings and in the process muddy the approach to any evolutionary interpretation of the processes of immunity.

The situation in the local Freund granuloma is highly complex and no serious attempt seems to have been made to analyse the total situation. It is known that an important feature is the very slow, long-continuing release of antigen (Herbert, 1966) but the complex cellular relationships are obviously also important. Spector and Lykke (1966) found that most of the initial cells come from blood monocytes and that many of these cells proliferate *in situ*. At a later stage, small lymphocytes accumulate, being probably derived *in situ* from larger precursor cells reaching the granuloma from the blood. One could also expect lodgment of stem cells from the bone-marrow and movement both of antigen and of mobile cells from the granuloma to the draining lymph nodes.

Any ascription of specific function to the Freund granuloma must be speculative, but it is in line with all the effects observed to regard the complex situation involving not only the local granulomas with their antigen depots but also the draining lymph nodes and possibly also more distant lymphoid tissues as having the following effects.

(*a*) By slow release of antigen to provide a continuing specific stimulus, involving both the nominal antigen and the antigens of the associated acid-fast bacilli, to immunocytes throughout the body.

(*b*) For reasons discussed earlier, this specific activity has a nonspecific stimulatory effect on adjacent immunocytes with increased production of immunoglobulins and an activation of immunocytes with patterns potentially reactive against any available ADs.

(*c*) To allow stem cells to differentiate to immunocytes but to provide a much less effective censorship of self-reactive and unduly avid immunocytes than is the case in the thymus and any other normal regions of differentiation and censorship.

(*d*) To provide increased likelihood for the incorporation of ADs into the surface layer of mobile cells as suggested in relation to DH in the previous chapter.

### TISSUE ANTIGENS IN RELATION TO EXPERIMENTAL AUTOIMMUNE DISEASE

The first experimental production of an autoimmune state—by Rivers and Schwentker (1935) who produced allergic encephalomyelitis in monkeys by repeated injections of autolysed brain

595

material—was long before F.C.A. was developed. The immensely greater ease with which allergic encephalitis could be produced by the use of adjuvant has resulted in its use in all subsequent work. The following sentence from Paterson's (1966) review of the field is significant. 'This conspicuous void [on how Freund's adjuvant functions] is the more disturbing when it is realized that without the introduction of adjuvant in the 1940s we might not be writing about experimental autoimmune disease in the 1960s.' All our discussion of accessible and inaccessible tissue antigens and of experimental demonstrations of their antigenicity is against the background that without F.C.A. they are virtually non-antigenic.

Dumonde (1966) has published a comprehensive review of the large recent literature on tissue antigens, and his conclusions provide a convenient basis for discussion of the points with which I am primarily concerned. The chemical nature of the antigens is unknown, but it is important that they tend to occur in corresponding organs of a wide range of species. They have some features which suggest hapten-carrier structure.

The inaccessibility of tissue-specific antigens from the immunological system 'implies that full tolerance of these determinants need not be attained even though the antigenic determinants may be present during embryonic life'. This is in line with current thought but probably requires some qualification to account, for example, for the age and sex incidence of Hashimoto's disease. There is certainly a genetic influence at work and it may be that the genetic differences are expressed in the relative ease with which destruction of immunocyte by antigen contact occurs. Wherever, for any reason, the critical level of the function (avidity multiplied by concentration) for cell destruction is raised, the likelihood of serological manifestations of tissue damage and symptoms of autoimmune disease will be increased.

### Abrogation of tolerance by related antigens

This leads to the most interesting aspect of tissue antigenicity which has already been mentioned: the greater likelihood that antibody and organ lesions will result if tissue from another species is used as immunizing agent. Dumonde (1966) considers three possibilities of interpretation which need not be mutually exclusive.

The first is based on Weigle's (1964) experiments discussed already in chapter 10 of Book II in which tolerance to BSA in rabbits was broken by the cross-reacting antigen HSA. Essentially the same interpretation can be offered when we consider a normal component of the body antigen A and the related but not identical antigen A' from another species. The experimental subject is tolerant to A, that is, there are no immunocytes carrying immune patterns (CS) which are sufficiently complementary to A to be stimulated to proliferate or even to take on the status of committed immunocytes. There will therefore be no circulating antibody against A for opsonization or deposition on dendritic phagocytic cell surfaces in lymphoid tissue. When A' is injected there will be a certain probability that immunocytes differentially reactive with A' and not A are present in the A-tolerant animal. These can therefore be stimulated to commitment and proliferation and so provide a new situation *vis-à-vis* A. Although the A'-reactive progenitor immunocytes show no significant reactivity to A, it is in line with the discussion in chapter 8 that committed A'-reactive cells, particularly when associated with A' antibody in dendritic phagocytic cells, should be capable of stimulation by A with production of antibody reactive with A, but of low avidity. Weigle (1965) has published an account of experiments with heterologous or modified homologous thyroglobulin in rabbits which apply this approach to an experimental autoimmune situation.

The second hypothesis suggested by Dumonde is that the tissue antigen can be regarded more or less as a hapten–carrier complex; the hapten being organ-specific and inaccessible, the carrier species-specific and fully tolerated. This does not seem to be a helpful approach, and most of the evidence quoted as providing model experiments is open to the alternative explanation that linkage of hapten to carrier protein does not necessarily leave the ADs of the carrier as they were in the native protein.

The third hypothesis is simply that there exists partial tolerance of the complex tissue antigens. This is the point of view which has been consistently adopted in this book. Complete tolerance can only be a limiting condition asymptotically approached in some natural or experimental situations. All natural situations are highly complex with multiple ADs, while for each AD there will be a unique distribution of avidity of potential union amongst the immunocyte populations of the body. Figure 6 could in principle be applied to each AD under consideration. Dumonde is essentially in agreement with this general approach in explaining the phenomena of experimental autoimmune disease in normal animals.

Because a satisfactory experimental and logical interpretation of broken tolerance by administration of distinct but related antigen is possible, it does not necessarily follow that all or any autoimmune disease in man is produced in analogous fashion. There is no well-established example and the possible instances are all capable of alternative interpretations. In a sense, the Wassermann reaction is an autoimmune reaction but is much more likely to be directed against an AD unmasked by the process of treponemal infection than toward an AD specifically modified.

None of the haemolytic syndromes associated with drugs shows any indication that the haemolytic effect results from immunization with a 'm⁻dified' red cell antigen. The only well-documented account of autoimmune-type haemolytic anaemia being 'caused' by a drug is worth close consideration in this regard. The responsible drug is $\alpha$-methyldopa and its effect is discussed in some detail in relation to autoimmune haemolytic anaemia later in this chapter.

## PRODUCTION OF AUTOIMMUNE DISEASE BY EXTRINSIC ANTIGENS

The second major alternative to the forbidden clone hypothesis of autoimmune disease is represented by Stevens's (1964) elaboration of the evidence from various sources that certain diseases of doubtful aetiology—for example, rheumatic fever, glomerulonephritis and multiple sclerosis—are associated with chronic micro-organismal infection. He then extends this to seek similar aetiological factors for all diseases currently classified as autoimmune. He interprets autoimmune diseases as being due essentially to the accident that certain common bacteria possess antigens which have ADs in common with some present in body tissues and particularly in certain membranes; for example, the basement membrane of the kidney, synovial membranes and the myelin sheath of nerve. It is assumed that, following infection, antibacterial antibodies are produced often in very large quantity (to account for the hypergammaglobulinaemia common in autoimmune disease) and by their activity produce the observed damage.

There is much to be said for welding the essential features of this approach into the 'Darwinian', 'forbidden clone' thesis which Mackay and I (1963) have supported. It is in fact self-evidently important in relation to rheumatic fever and only slightly less so in regard to acute scarlatinal glomerulonephritis. These are not normally included in the autoimmune diseases but there is no real reason why they should be separated, since there is general agreement that the lesions are produced by immunological mechanisms.

The important point is to assess the relative contribution of environmental factors and of genetic individuality (germinal and somatic) in each disease. To postulate an immune response to universally present saprophytes as the 'cause' of rare diseases like systemic lupus erythematosus and multiple sclerosis is as operationally useless as to assume a universally present cancer virus as the cause of malignant disease. The common saprophytes and semi-parasites represent a continuing aspect of the environment in which we have evolved and which is compatible with full health in 99·9 per cent of person-years. In the case of streptococcal infections of the throat, conditions are different. There is no doubt that rheumatic fever is associated with symptomatic throat infections and in environments such as military camps in winter, where such infections are frequent and severe, the incidence of rheumatic fever is higher than normal. There is, however, a well-established genetic factor, and one never hears of more than 3 per cent of individuals in a heavily infected environment developing diagnosable rheumatic fever. It is equally clear that there is a gradient of genetic and possibly somatic genetic susceptibility. Similar conditions will apply to acute glomerulonephritis and perhaps to erythema nodosum and other manifestations of tuberculosis.

One might in fact arrange immunologically mediated diseases in a gradient from those where the immunological effect is a necessary sequel to the environmental process to those in which there is no real evidence of any environmental 'cause' or 'trigger'.

599

A few examples at various levels in sequence may be given as follows:

(*a*) The rash in measles.

(*b*) Serum sickness.

(*c*) Rheumatic fever and acute glomerulonephritis.

(*d*) Late syphilis, assuming as is probable that the lesions are immunologically mediated.

(*e*) Drug purpuras and haemolytic anaemias.

(*f*) Rheumatoid arthritis.

(*g*) Acquired haemolytic anaemia.

(*h*) Systemic lupus erythematosus.

In the last three the intrinsic factor is so strong that all that can be looked for from the environment is a possible trigger. If we allow the validity of Burch's (1963) reasoning, this is equally illusory.

In rheumatic fever we can accept that the streptococcal infection is all-important but the individual factors responsible for the development of rheumatic fever in only about 3 per cent of those who suffer streptococcal pharyngitis call equally for elucidation. A detailed discussion of rheumatic disease and more cursory treatment of acute glomerulonephritis is therefore desirable.

## Rheumatic fever

It is well accepted that rheumatic fever, which may include poly-arthritis, carditis and a less conspicuous group of other mani-festations, is a sequel of pharyngeal infection by haemolytic streptococci of type A. In general populations, only a small propor-tion of streptococcal pharyngitis gives any rheumatic sequelae; 2·5–3 per cent in military populations is quoted by Stollerman (1964). The onset of rheumatic fever occurs a week or more after the peak of streptococcal pharyngitis and is regularly associated with high titres of anti-streptolysin O and other antibodies against soluble streptococcal products (Halbert *et al.* 1961). It has therefore become just as generally accepted that, in one way or another, immunological factors play a major part in the pathogenesis of rheumatic fever.

The characteristic Aschoff nodule of rheumatic carditis has counterparts in other parts of the body and these focal granulomas appear to be the only well-recognized histological lesions. There is no evidence that they contain streptococci and they have no special resemblance to any autoimmune manifestations. Much work has been done in an attempt to reproduce the rheumatic granuloma in experimental animals. The most interesting analogues are the recurrent skin nodules produced in rabbits by intradermal injection of sonically disrupted streptococci of types A and C (Schwab and Cromartie, 1957) and the cardiac lesions found in a small proportion of rabbits in which repeated streptococcal infections had been induced over several months (Murphy and Swift, 1950). These are focal lesions pointing to a focal rather than a generalized process.

Kaplan (1965) has reviewed his work on streptococcal antigens with cross-reactivity with cardiac muscle. The cross-reactivity is shown by the fact that when rabbits are immunized with cell wall material from an A (type 5) strain, approximately one-quarter produce serum reactive with human cardiac muscle by both complement fixation and immunofluorescence. Similar antibodies are found in human beings after streptococcal infections, about 24 per cent of those not having rheumatic fever and more than 50 per cent of rheumatic fever patients. In addition, Kaplan has described five cases of acute death with severe rheumatic carditis in children in which there was heavy accumulation of $\gamma$-globulin (Ig G + A) in the cardiac muscle. The cross-reactive antigen appears to be a protein related to but not identical with the M protein of the streptococcus.

There can be little doubt that some form of antigen–antibody interaction is taking place during acute rheumatic fever and in the tissues symptomatically involved. One of the most striking features differentiating non-rheumatic from rheumatic streptococcal infections is the much higher anti-streptolysin titre in the latter. In the rheumatic individual there is also an average of 5 bands against streptococcal products in agar-diffusion tests as against 1·2 in control individuals (Halbert *et al.* 1955). A highly active immune response to the streptococcal infection therefore seems to be a requirement if rheumatic fever is to follow.

The most straightforward hypothesis to cover the points which have been outlined depends on accepting the indications that streptococcal cell-wall materials are concerned and have ADs closely related to an AD in cardiac muscle. The assumption, then, is that in a proportion of children there is both an acute strepto-coccal infection of the tonsillar region and a highly active antibody response. As a result there is release into the bloodstream of either damaged streptococci or partially disintegrated streptococcal wall materials. In either case this will be associated with antibody. The complex will lodge in various regions of which the synovial mem-branes or subsynovial tissue is a likely one as judged, for example, by the predilection of serum sickness for the joints. At the height of an attack of rheumatic fever we can picture residual sources of streptococcal antigens but virtually no living streptococci widely distributed. The cardiac muscle is specially liable to be involved because of the existence of the cross-reacting antigen in the strepto-coccal wall. At an early stage in antibody production, occasional immunocytes with invasive quality could initiate minor myocardial lesions which, in turn, would lead to lodgment of streptococcal antigen–antibody complexes. The damaging effects will be essenti-ally due to cellular damage produced by direct antigen–CS effects either on immunocyte or adjacent tissue plus, as suggested by Halbert *et al.* (1961), possible liberation of toxic products such as streptolysin O from complexes with antibody.

On this interpretation there is relatively little relationship of rheumatic fever to autoimmune disease. It is very much a strepto-coccal process—an abortive invasion that, like a receding guerrilla war, gives great trouble during the mopping-up process. One might even use the metaphor that the enemy disappears but leaves behind a vast number of land mines and booby traps to be dealt with!

The features of relevance to the general topics of immuno-pathology on this interpretation include, first, the necessity to assume a certain invasiveness of immunocytes capable of initiating scattered cardiac lesions. This is simply part of the general problem of the production of organ-specific autoimmune disease. As in the general case this can be equated closely to a DH reaction requiring a sufficient population of immunocytes with CSs of high avidity for the tissue AD, and some region of increased capillary permea-bility to allow their adhesion and entry into the tissues. It is possible that the feature which differentiates the child susceptible to rheumatic fever from the insusceptible is his capacity to produce and allow to proliferate immunocyte clones of exceptional avidity for the cross-reacting cardiac and streptococcal ADs. If abnormal mutants can occur in immunocytes reactive against self-antigens, they can equally arise in cells with receptors directed against extrinsic antigens. One of the variables within any population of antibody molecules is their avidity for antigen and the extent of cross-reactivity where two or more related antigenic patterns are simultaneously present in the body. Both these qualities will have preferential survival value under the conditions of streptococcal infection. As has been discussed earlier, high avidity for the tissue antigen may be one of the main determinants of the 'aggressiveness' that allows an immunocyte to penetrate target tissues. It is not unreasonable in this connection to look on the subacutely infected tonsil with its heterogeneous cells and purulent crypts as the equivalent of a Freund's granuloma, and to endow it with the same immunological qualities.

Since cellular infiltration in the cardiac muscle is only moderately intense in rheumatic fever, any invasive effect of specific immunocytes is likely to be initiatory rather than progressive. Once a point of attack is available, antibody, especially avid antibody, may play a greater part. It is certainly involved in producing diffuse severe carditis.

The second feature of interest is the persistence of lesions in chronic rheumatic carditis which, according to Kaplan and Dallenbach (1961), are characteristically associated with immunoglobulin. The liability for recurrence of rheumatic fever is well known but the evidence is now very strong that this is always associated with renewed streptococcal infection. The continuing existence of points of minor damage in cardiac muscle would obviously increase the ease with which the process leading to carditis could develop.

The third feature is the appearance of antibody reacting by immunofluorescence with myocardium after surgical manipulation of the heart (Kaplan, 1964) and after some cases of cardiac infarction (van der Geld, 1964). These usually have no harmful sequelae but their existence confirms the presence of a cardiac tissue antigen normally inaccessible but capable of actively provoking antibody production when it is released.

*Acute glomerulonephritis*

The facts about acute glomerulonephritis following on scarlet fever and other streptococcal infections are clear cut, but very little light has yet been thrown on its pathogenesis. The position seems to be considerably more obscure than for rheumatic fever. Signs of glomerulonephritis, haematuria and proteinuria appear relatively soon after the fibrile streptococcal episode; the mean latent period, according to Rammelkamp (1964), being 10 days as against 19 days for rheumatic fever. The great majority of recent American cases have been associated with haemolytic A streptococci of type 12, though several other types may more rarely be involved. The indication is strong but not absolute that different strains of type 12 differ sharply in nephritogenic capacity. There have been epidemics in which up to 13 per cent of an exposed population developed signs of glomerulonephritis but a prospective study of 100 throat infections with type 12 streptococci in Chicago gave only one case of extremely mild glomerulonephritis.

None of the experimental models shows the characteristic histological signs of typical post-scarlatinal glomerulonephritis, which is a proliferative and exudative glomerulitis without any changes to be seen in glomerular basement membranes beyond scattered minor irregularities (Jennings and Earle, 1961). When nephrosis or chronic nephritis develops on the initial post-streptococcal lesion, crescents and basement membrane lesions develop, but this is not the immediate result of the streptococcal process with which we are concerned. The only experimental model which has a reasonably close resemblance is acute serum sickness in the rabbit (Feldman, 1958). Rabbits were examined after a single large dose of BSA and showed lesions at the time of early antibody formation when soluble antigen–antibody complexes, presumably with attached complement, were circulating. At this time Feldman found marked proliferation and swelling of the glomerular endothelium often with occlusion of the glomerular capillaries and trapping of polymorphonuclear or mononuclear leucocytes. The basement membrane could be shown to be holding antigen and γ-globulin by immunofluorescence but this had little effect on the electron microscopic appearance.

There are many other ways of producing experimental kidney damage but none has provided any closer model. Lesions have been produced in mice by the implantation of streptococci in diffusion chambers in the peritoneal cavity, but in Tan and Kaplan's (1962) experience the action was a wholly tubular one and in all probability due to the non-antigenic streptolysin S. Kantor (1961) produced renal lesions in mice by the injection of streptococcal M protein but the effect was apparently due to the precipitation of fibrinogen by the protein and had no significant resemblance to the histology of streptococcal glomerulonephritis in man. Methods not involving the use of streptococci include Masugi (nephrotoxic) nephritis produced by injection of an appropriate anti-kidney serum. The damaging effect is mediated in the first instance by the deposition of anti-host kidney antibodies, particularly on the basement membrane, followed by the development by the host of antibody against donor globulin which in turn increased the glomerular damage where donor antibody was fixed (Ortega and Mellors, 1956). Recent work indicates that much of the actual damage is due to fibrin deposition and can be prevented to a considerable degree by the administration of heparin.

In line with the general behaviour of tissue antigens, active immunization of sheep with heterologous glomerular basement membranes in F.C.A. produces a severe lethal nephritis with gross lesions. Finally, we may mention the spontaneous kidney lesions in NZB and particularly NZB × NZW mice (Helyer and Howie, 1963; Hicks and Burnet, 1966) which again show little resembling the human glomerulonephritis.

Markowitz and Lange (1964) report that human glomeruli contain an antigen which cross-reacts with an antigen probably derived from type 12 streptococcal cell membrane. Preliminary evidence suggests that only a minority of human glomeruli contain the cross-reacting antigen. Obviously, more work is needed before this finding can be accepted as relevant to the pathogenesis of the human disease, but it could have a bearing on the relatively rare susceptibility amongst children to nephritis.

The current consensus of opinion is that immunological factors are concerned in acute post-streptococcal nephritis, but there is no indication as to the antigens concerned. Perhaps the most likely interpretation is that the situation is related to mild capillary endothelial damage by erythrogenic toxin or some other streptococcal product, which leaves the glomerular capillaries unduly prone to damage by any antigen–antibody complexes circulating in the 7–14-day period after the primary episode. As in rheumatic fever there is no real resemblance to an autoimmune process in the typical case of post-streptococcal glomerulonephritis which heals normally.

There is no doubt that chronic damage following the acute phase may develop into clinical and histological forms much more closely resembling experimentally produced or clinical autoimmune disease.

## Bacterial antigens in relation to ulcerative colitis

In 1959 Broberger and Perlmann described auto-antibodies in the serum of children with ulcerative colitis. In order to eliminate possible bacterial antigens in any extract of adult human colon they used foetal colon as source of antigen. There were indications that this antigen was a polysaccharide. Subsequently (1963) they showed that immunofluorescence reactions with serum and cytotoxic effects with blood leucocytes could be demonstrated on tissue cultures of foetal colon. A similar antigen can be obtained from the colon of germ-free rats (Perlmann *et al.* 1965). In the course of these experiments it was observed that if non-sterile (conventional) rat colon was used the reaction was masked by bacterial antigen–antibody reactions. Analysis showed that a crude lipopolysaccharide antigen from *Escherichia coli* 014 could block the antibody against colonic tissue. It appears probable that the antigen concerned is the so-called 'Kunin-antigen' which is almost universally present in masked form in Gram-negative intestinal bacteria. *E. coli* 014 is unique only in not possessing an inhibitory lipopolysaccharide which in some quite unusual fashion renders the common (Kunin) antigen non-immunogenic in all other strains. According to Neter *et al.* (1966) this antigen, in the presence of endotoxin, can prime immunocytes for a secondary response to 014 antigen but does not provoke actual antibody production.

These results obviously justify more study, but there is no indication as yet that bacterial antigens which are universal and heterogeneous in the gut play any significant role in initiating ulcerative colitis in the small proportion of children and adults who suffer from the condition.

## AUTOIMMUNE HAEMOLYTIC ANAEMIA

The first disease to be recognized as resulting from the action of an antibody-like substance on body cells was paroxysmal cold haemoglobinuria (Donath and Landsteiner, 1904), and the recognition by Dameshek and Schwartz (1938) that haemolysins could be demonstrated in some acute haemolytic anaemias initiated the modern phase of immunopathology. Since the antigenic structure of human red cells has been more deeply studied than that of any other cell, it is of special importance that the behaviour of human autoimmune haemolytic anaemia (AHA) should be used as a test for the validity or otherwise of the clonal selection approach.

For obvious reasons, most of the material in this section will be drawn from Dacie's text (1962) and later material largely from the discussion at the New York Academy of Sciences in 1964. The general significance of human blood groups for immunology has already been discussed.

### *The specificity of auto-antibodies against human erythrocytes*

It is highly characteristic of auto-antibodies in haemolytic disease that they are in no way concerned with the ABO antigens. The specificity falls into three distinct classes.

Warm-type AHA: About a third of these can be shown to have a definite specificity for one of the Rh system antigens of which anti-e and anti-c are the most common. Most antibodies react with all unselected human red cells but these nonspecific reactions can be shown also to be related to the Rh system by the many fewer antibodies which react with red cells of the very rare genotypes -D-/-D- and ---/--- (Weiner and Vos, 1963). Even with completely deleted cells, however, about 40 per cent of AHA sera will react. Warm incomplete antibodies are always of Ig G type, and in a majority of patients the antibody is of monoclonal origin as judged by the restriction of light chains to either K or L types, not both.

Cold haemagglutinins: These react almost exclusively with the peculiar I antigen present in at least 99·9 per cent of adults but absent in foetal and early postnatal life, when it is replaced by i. There are very rare individuals whose blood retains the i antigen throughout adult life. These usually show relatively high titre anti-I isoagglutinins. Most cold agglutinins, including those of normal sera and the high-titre agglutinins of *Mycoplasma pneumoniae* infections as well as the cold type of haemolytic anaemia, are specific for I. It is of great interest, however, that those of infectious mononucleosis and many of those associated with lympho-proliferative diseases are anti-i. Cold agglutinins are Ig M and, except in cold agglutinin haemolytic anaemia, have both K and L light chains (Costea *et al.* 1966), but the latter are all of type K.

Paroxysmal cold haemoglobinuria: Donath–Landsteiner (D–L) disease is due to the interaction of a cold haemolysin on the antigen Tj$^a$ (Levine *et al.* 1965). The disease is often but not always associated with syphilis and it is probably significant that the antigen Tj$^a$ is also present on a wide variety of cells other than erythrocytes. The vast majority of persons are Tj$^a$ but there are occasional Tj(a−) persons whose cells are unagglutinable by D–L serum.

The most interesting feature of these results is the way in which auto-antibodies are directed—not against the ADs which differentiate one person from another and define the classical blood groups: instead they act on (*a*) aspects of the Rh complex common to the great majority which include e, present in 97 per cent of people; (*b*) the I antigen present in 99·9 per cent and (*c*) the Tj$^a$ antigen which in 1958 was present in all but eight individuals of many thousands tested (Race and Sanger, 1962).

Previous mention has been made of Cleghorn's (1961) finding that a high proportion of AHA sera (as well as a much smaller proportion of normal individuals) may show antibody against red cells carrying 'private' antigens. The possibility was raised that the possession of a private red cell antigen might result from a rare germinal mutation and that the same mutation occurring at the somatic level in an erythropoietic stem cell could become evident if a small clone of mutant red cells in an otherwise normal individual acted as an antigenic stimulus. Patients with AHA are *ipso facto* capable of producing antibodies against red cell antigens which escape the normal homeostatic controls. Without particularizing these 'fail-safe' controls, any weakness would be expected to increase the proportion of people who would produce antibody against a minor red cell antigen such as a private antigen arising by somatic mutation.

*Evidence for monoclonal auto-antibodies*

On the 'forbidden clone' hypothesis, one assumes that the relevant somatic mutation toward resistance to censorship is quite unrelated to the immune pattern. It will therefore be distributed according to the frequency natural to the individual of his immune patterns. Once an immunocyte with a forbidden pattern emerges it will have to overcome the other 'fail-safe' controls before it can emerge as an effectively pathogenic clone. If mutation to resistance is relatively rare and secondary controls not easy to overcome, then it would be expected that not uncommonly the pathogenic cells at least in the early stages of autoimmune disease should belong to a single clone.

The only practical approach to determining monoclonal character is by eluting specifically attached auto-antibody from its target cell or antigen and determining the antigenic character of the immunoglobulin concerned. The results of such studies are still scanty and have been briefly noted already. All human beings produce both K and L light chains and each committed immunocyte is either K or L in the proportion of approximately 60:40. If a given antibody is derived from more than three initiating cells it is very unlikely ($\pm$ 1:40) to be pure L, and if from more than four, pure K would be almost equally rare (about 1:20). Leddy and Bakemeier's (1965) findings of eleven out of twenty-one K, four out of twenty-one L with two unequivocal K + L suggests that the average number of clones is around 1·2. Since the proportion is compatible with random choice of K or L, one cannot assume that K clones have some special quality linked with the K character. Despite the qualifications raised by the authors, this points strongly toward the monoclonal origin of most warm-type cases of AHA (see fig. 26, p. 260). The position in regard to cold haemagglutinins is more interesting. All appear to be Ig M in type

(Mehrotra, 1960 $b$) and those found in normal serum at low titre, in *Mycoplasma pneumoniae* cases and in classical cold agglutinin disease, are directed against antigen I. In infectious mononucleosis (Costea *et al.* 1966) and in at least some haemolytic anaemias associated with chronic lymphatic leukaemia (Marsh and Jenkins, 1961) the antibody is in fact anti-i. It should be noted that cold agglutinin disease antibody may also show a lower titre against i as well as the dominant anti-I. All those tested show K and L light chains except the antibody of cold haemagglutinin disease, which in five cases (Costea *et al.* 1966) was consistently K. This has been confirmed (Franklin and Fudenberg, 1964; Harboe and Deverill, 1964) and, in addition, Angevine *et al.* (1966) have described a cold agglutinin of Ig A type which was also wholly K. In the absence of any L-type antibodies the result cannot be used as evidence of monoclonal character. Indirect evidence, however, favours the possibility, such as Mehrotra's (1960 $a$) finding that isolated cold agglutinins were antigenically individual as well as Schubothe's (1965) evidence for physical individuality and for the existence of very high titres of macroglobulin pointing to a monoclonal paraproteinaemia (Waldenström, 1960, 1962). The limitation to K presumably means that only on the K variable chain is the disposition of stable residues such that anti-I readily develops.

The D–L antibody in paroxysmal cold haemoglobinuria has been extensively studied in regard to the physical aspects of its characteristic cold–warm haemolysis but not apparently in regard to its Ig type and K or L antigenic quality. Dacie regards these antibodies as being less remote from those of cold agglutinin disease than is usually held. The features of major interest are that a high proportion of cases occurred in children with congenital syphilis and that in many instances the clinical and serological signs of the disease could last for many years, the patient remaining quite well unless he was exposed to cold. The disease was always rare and now is hardly ever seen. It is in no way a regular feature of either acquired or congenital syphilis but presumably represents the emergence of an anomalous clone of immunocyte, which is in some way influenced by the existence of treponemal infection. Its rarity might suggest that if investigation were practical, the antibody would in most cases be found to be monoclonal in character.

The possibility raised by Schubothe (1965) that some cases of cold agglutinin disease are essentially indistinguishable from Waldenström's macroglobulinaemia has already been mentioned. In line with our discussion of myelomatosis, one can regard the emergence of the autoimmune haemolytic anaemias as resulting from the random impact of mutation on an immunocyte irrespective of the immune pattern it carries. Only when the nature of that immune pattern and the type of mutation are appropriate and when there has been successful escape from all the relevant control mechanisms, can free proliferation of the mutant occur and autoimmune disease be initiated.

### Drug-induced haemolytic and thrombocytolytic disease

Shulman (1964), in discussing certain drug-induced haemolytic anaemias and thrombocytopenic purpuras, has suggested that if instead of the drug we postulate a metabolite or virus (neither detectable by current methods) which functions as the drug, an interpretation of 'autoimmune' diseases of similar symptomatology would be available. This would provide an extrensic 'cause' for the conditions and eliminate the need for forbidden clones, etc.

The essence of the drug action, for example, in stibophen haemolytic anaemia is that the drug in a very small proportion of patients who receive it acts as a hapten and provokes the appearance of an avid antibody. This antibody leads to haemolysis *in vivo* by a process which can be conveniently analysed *in vitro*. The reagents concerned are drug, antibody from the patient, complement, and any type of human red cells. The sequence is (*a*) binding of drug to the high-avidity antibody; (*b*) adsorption of the drug–antibody complex to the red cell surface; (*c*) attachment of complement components and (*d*) haemolysis. The adsorption to the red cell surface seems to be significant only as a means by which complement lysis can be initiated. This is *not* an autoimmune process but it has one important feature in common, i.e. the rarity of its occurrence. As always, rare responses to a common environmental stimulus usually mean a genetic anomaly which includes somagenetic anomalies. The rarity of stibophen anaemia or quinidine purpura is interpreted as being due to the requirement that a highly avid immune pattern must be carried by the clone(s) concerned. Highly avid immunocytes are, by hypothesis, prone to destruction by antigenic contact and those concerned in these reactions may have a certain degree of resistance analogous to that of a genuine autoimmune clone. Whatever the detailed requirements the emergence of such clones and antibody is unusual. In any such situation the criterion of autoimmunity is the specific relation of antibody or immunocyte to the normal ADs of the target cell. Since this is lacking in Shulman's model and is clearly demonstrable in the three classic types of autoimmune haemolytic anaemia there is no basis for seeking imaginary extrinsic factors.

The commonest drug causing thrombocytopenia of this same type is quinidine. Shulman (1964) suggests that drugs which provoke Ig M antibody affect red cells, while those provoking Ig G unite as a drug–antibody complex, preferentially with platelets. In more recent work, investigations by Shulman *et al.* (1965) point directly to the presence in patients with idiopathic thrombocytopenic purpura of a 7 S antibody reactive with both autologous and homologous platelets. In the idiopathic disease there is therefore no indication of any intermediary corresponding to the drug-hapten of quinidine purpura.

## The haematological effects of α-methyldopa

The drug 'Aldomet' (α-methyldopa) is an effective and popular antihypertensive drug which is given for prolonged periods. The recognition in Dacie's clinic of two cases of autoimmune-type haemolytic anaemia (AHA) in patients on 'Aldomet' treatment (Worlledge *et al.* 1966) stimulated investigation of a large series of patients on continuing therapy of this type (Carstairs *et al.* 1966). Of patients who had been taking the drug for more than 6 months, approximately 20 per cent showed a positive direct Coombs (DC) reaction of Ig G type. The first appearance of the positive DC reaction was usually between 6 and 12 months after starting the drug and there was little evidence of any increase in the percentage positive with further continuance of therapy. There was, however, a higher proportion of DC-reactors in those with higher maintenance doses. The DC reaction was not associated with any signs of haemolytic anaemia in the series studied. An active effort to identify cases of AHA in England led to the conclusion that perhaps one-third of the cases of AHA recognized in England in 1965–6 were patients taking α-methyldopa. It was evident, therefore, that the drug had a significant effect (Worlledge *et al.* 1966) but the incidence of clinically recognizable disease was of the order of only 0·02 per cent. The outstanding features of these cases of autoimmune haemolytic anaemia were:

(*a*) Serologically the condition was identical with the idiopathic disease, the incomplete antibody being largely nonspecific but in many showing anti-e specificity as well.

(*b*) There was no evidence that the drug played any part as hapten or influenced immunological specificity in any way.

(*c*) The cases could be brought to normality by cessation of the drug and the use of corticosteroids.

Two problems arise: (*a*) what is the process leading to the DC reaction in 20 per cent and (*b*) what is the (presumably) additional process which gives rise to symptomatic blood destruction in 0·02 per cent of persons on maintenance therapy with the drug.

In regard to (*a*) the first point to be noted is that the effect of the drug is not immediate. The DC reaction usually arises between 6 and 12 months but sometimes later. This is not an effect of the drug on a physiological process in the ordinary sense, but it is compatible with α-methyldopa allowing the initiation of disease by a process random with respect to time which would be ineffective in the absence of the drug. There are many random possibilities but in view of everything else known about immunity and immunopathology, somatic mutation seems by far the most likely type of random event to be responsible. To make such an hypothesis more specific, we assume that a minor mutation toward resistance occurs occasionally and will by chance involve progenitor immunocytes which can react with red cell antigen. Normally such a minor degree of resistance would not prevent the destruction of the newly differentiated immunocyte, but in the presence of an adequate concentration of α-methyldopa or some derivative metabolite such a cell becomes a phenocopy of the fully resistant mutant responsible for standard AHA. It emerges from the thymus and can be stimulated to produce 'DC' antibody and to proliferate under stimulus. The avidity of the antibody and the receptor-CS of the cell is low; the antibody is very incomplete, and the stimulus to proliferation a weak one.

For AHA to supervene in a minute proportion it seems necessary to assume a background genetic abnormality. The situation can be schematized as indicated in table 24 in terms of genetic changes from the normal in the immunocytes involved.

TABLE 24. *An interpretation of autoimmune haemolytic anaemia and the effect of α-methyldopa*

| Clinical states | |
| --- | --- |
| Idiopathic AHA | abc |
| α-MD AHA | ab+ |
| α-MD DC | b+ |
| Symptomless | ab |
| Symptomless | b |

Key to factors involved: a—genetic anomaly; b—somagenetic anomaly; c—somagenetic anomaly; +—phenotypic action of drug equivalent to c.

There are two aspects ·which should probably be elaborated slightly. The first is the apparent harmlessness of a well-marked typical Coombs test of true Ig G type. Traditionally this would be expected in itself to initiate accelerated destruction of the coated cells in the spleen. It is equally traditional, however, not to carry out antiglobulin tests on persons showing no signs of anaemia, and one wonders how often DC-positive reactions would be found if looked for in a large random population. Darnborough (1958) described an instance in a perfectly healthy blood donor who had been giving two or three blood donations per annum for several years. His blood gave a typical DC of warm type and the eluate showed anti-c specificity. There is not much to be said except perhaps to underline the multiple 'fail-safe' controls that must be involved.

The second point is to mention the possibility that a drug effect of this type could conceivably act as a trigger to break down an unstably balanced situation in a person genetically predisposed to autoimmune disease. A possible case of this sort reported by Walzer and Einbinder (1963) was in a girl of fifteen taking propylthiourea who developed a generalized illness resembling systemic lupus erythematosus. On discontinuing the drug she gradually recovered over some months. The serum contained an antibody directed toward polymorphonuclear leucocytes (autologous or homologous) and with no relation whatever to propylthiourea.

Such an interpretation of the $\alpha$-methyldopa situation is completely in line with the interpretation of AHA given in chapter 12 of Book 1. The wholly unexpected occurrence of this condition in 1965–6 has in fact provided another example of the way in which the clonal selection approach has been able to accept each new development in immunology with a minimum of *ad hoc* adjustment.

## SYSTEMIC LUPUS ERYTHEMATOSUS

SLE is the autoimmune disease *par excellence* and has provoked an enormous amount of study mostly since the discovery of the lupus erythematosus cell test by Hargraves *et al.* (1948) provided a satisfactory diagnostic criterion for what is liable to be a symptomatically heterogeneous group of cases. From the point of view of the forbidden clone hypothesis the important features of the disease are:

(*a*) The multiplicity of antibodies.

(*b*) The apparent and probably real concentration on ADs in nuclei.

(*c*) The characteristic age and sex incidence.

(*d*) The interactions with the thymus.

We may recapitulate the point of view built up from looking at the phenomena of SLE from the selectivist angle.

The disease has no extrinsic cause; it arises because of a genetic anomaly which increases the likelihood that censorship of immunocytes reactive against self-antigens will be ineffective either because of increased liability to somatic mutation alone or this plus a genetic inadequacy of the control mechanism (by hypothesis largely centred on the thymus).

### Auto-antibodies in systemic lupus erythematosus

There is now general agreement that the reactive materials in sera from SLE patients are antibodies (Kunkel and Tan, 1964) and all discussion will be based on this assumption. The feature that differentiates SLE from all other diseases is the presence of a wide spectrum of antibodies against nuclear constituents, notably DNA. To obtain the full spectrum of possibilities, many different patients must be studied, but large numbers of antibodies can be demonstrated in a single serum. In general, the more refined the approach the more complex the findings and the greater the individuality from one patient to another.

Despite certain technical difficulties of handling purified DNA as an antigen for *in vitro* tests it is fully established that DNA of any origin will react by precipitin (Deicher *et al.* 1959), or complement fixation (Robbins *et al.* 1957). The general interpretation from detailed work with single- and double-stranded DNA and with inhibition tests by oligonucleotides, purines and pyrimidines (Stollar *et al.* 1962) has led to the conclusion that the reaction is with a wide variety of sites on the DNA, and particularly the denatured DNA, molecule in which purine and pyrimidine bases play the main part. It is of interest that no antibody against DNA had been produced experimentally until very recently with the significant exception of bacteriophage T2 or T4 in the rabbit (Murakami *et al.* 1962) where the antigenicity was ascribed to the content of hydroxymethyl cytosine and glucosyl differentiating phage DNA sharply from any present in the rabbit. More recently it has been found possible, by combining denatured calf thymus DNA with methylated serum albumin and using F.C.A., to produce in rabbits antibody which reacts with denatured DNA but not with native DNA (Plescia *et al.* 1964). There is a general consensus that antibodies specifically reactive with DNA are found in no human condition other than SLE. The standard diagnostic approach to SLE since 1950 has been the lupus erythematosus cell test. This is essentially an appearance resulting from the phagocytosis of nuclei damaged by an antibody, the lupus erythematosus factor, which is probably directed against ADs characteristic of nucleoprotein as such rather than of either DNA or histone (Miescher, 1957; Goodman *et al.* 1960). In most sera giving the lupus erythematosus cell test there will usually be antibodies of other specificity. Most of the group are of 7S Ig G type and can pass the placenta (Beck and Rowell, 1963).

The antinuclear factor shown by immunofluorescence tests is regularly present in SLE and normally shows solid fluorescent staining of the test nuclei. There are other types of staining (Beck, 1961); speckled, with discrete flecks of fluorescence over the nucleus; and nucleolar, where fluorescence is limited to the nucleolus. According to Kunkel and Tan (1964) the speckled appearance may be related to their 'phosphate extract' antibody. Beck *et al.* (1962) found that nucleolar ANF reactions occur only rarely in SLE, only one instance in 67 sera being observed. Antinuclear factors from SLE sera may be of Ig M, Ig G or Ig A type and, according to Barnett *et al.* (1965), tests to determine the type of antibody absorbed on to reactive nuclei indicate that in all cases both K- and L-type light chains were concerned.

Amongst other antibodies found in SLE sera may be mentioned the soluble nuclear antigen obtained from calf thymus and other nuclei by Tan and Kunkel (1966). This appears to have a polysaccharide as AD, since its specificity is destroyed by periodate. Anti-histones and a variety of anticytoplasmic reagents have also been described. The possibility that the 'aggressive' immunocytes of SLE and other autoimmune diseases have a deviant (? mutant) metabolism is raised by Epstein and Tan's (1966) finding of an excess of light chains in SLE serum and in the synovial fluid of rheumatoid arthritis patients.

From the general point of view in which we are interested, this summary of serological studies leads to a simple conclusion: that there are present in SLE sera a wide variety of antibodies whose range of reactivity suggests that the stimulation for their activity arises in a region where nuclei are present in various stages of disintegration. The two sites which immediately come to mind are germinal centres and the thymic cortex.

The second point which seems important, when one is seeking an explanation of the pathogenesis of SLE in terms of a clonal selection approach to immunity, is the rather rapid transition from normality to a serious immunopathy. Unfortunately, it is not practical to do extensive serological studies on a young woman *before* she develops symptoms of SLE. From family studies, for example Leonhardt (1957), there is a suggestion that a high level of serum globulin may be present before symptoms appear. Until someone finds himself in a position to follow the successive development of the different types of auto-antibodies in SLE, a variety of possibilities will remain. As far as present information goes, however, everything suggests that a rather sudden and presumably autocatalytic breakdown in homeostasis takes place at some point. If the current interpretation is correct that when a lupus serum shows active interaction with DNA a very wide range of individual antibodies is present, it does not seem reasonable to take the attitude that each clone is appearing wholly independently.

*Possible role of the thymus in systemic lupus erythematosus*

In looking for the type of breakdown that would give the observed picture, we are necessarily limited to the consideration of hypotheses within the framework of the clonal selection interpretation adopted for the normal processes of immunity. The most obvious approach is to postulate a failure of the censorship process at the site of differentiation. The situation might well be such that immunocytes emerging with self-reactive receptors were stimulated to proliferation rather than being destroyed by the dominant auto-antigens available at this site—by hypothesis, the thymus. Since SLE is a wholly human disease and the thymus is almost the least accessible of organs for biopsy or functional study, the hypothesis can be tested only by its indirect implications. One postulates a basic genetic weakness making mutation toward resistance to censorship more likely but adds an additional genetic weakness on the positive side of censorship. If we return to the standard diagram, fig. 22, the effect of significant changes in either resistance of nascent immunocyte or effectiveness of thymic factors will be manifested by movement of the operative line along the diagonal O—D. The effect of the thymic environment is to move the operative line $x$ units *toward* the origin and to make the situation as simple as possible the resistance induced by somatic mutation moves it $x$ units *away* from the origin. On fig. 40 is also shown a vertical bar representing the concentration of a 'self' AD in the thymic environment, and two horizontal lines, the lower one at the level of avidity necessary for immunocytes to have a demonstrable effect by the production of low-avidity antibody. This will be very close to the null limit in the standard diagram. The upper one represents the level of avidity needed for an immunocyte to have potentially pathogenic action.

The diagram serves merely to show how a breakdown in thymic effectiveness could convert a balanced situation with low-grade antibody, anti-nuclear factor for example, resulting from a proportion of resistant cells to a disastrous situation with full symptomatology.

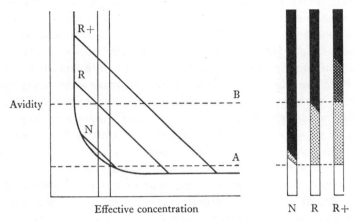

Fig. 40. Application of fig. 22 to conditions of autoimmunity with appearance of 'forbidden clones'.
The appearance of resistant (R) or very resistant (R+) cells capable of reacting with antigenic determinants present in the thymus at an effective concentration (represented by the open vertical bar) will be equivalent to raising the diagonal from the normal value (N) to R or R+.
If A is the level of avidity needed to allow the detection of auto-antibody and B the minimum level at which symptomatic damage to cells carrying the AD concerned, the nature of the immunocyte population which can develop is shown in the corresponding bars on the right.

The implications of such a process fit well with what has been observed in thymuses removed surgically from cases of SLE. The thymus is the most morphologically labile of all organs and it is easy to picture a stress situation which could temporarily lower the effectiveness of censorship. When high-avidity cells are capable of multiplying on antigenic contact with self-antigens in the thymus, one would expect a complex situation including (*a*) the appearance of germinal centres, (*b*) plasma cells in greatly increased numbers, (*c*) stress atrophy of the cortex and (*d*) proliferation of epithelial cells leading to a shrunken and functionless thymus. At such a stage there will be no differentiation of stem cells to immunocytes in the thymus and hence no need for censorship.

In the patient BE, described by Mackay and de Gail (1963), Burnet and Mackay (1965), the thymus was removed surgically from a typical case of early SLE in a girl of fourteen who had not been treated with corticosteroids. The thymus was atrophic with no definite cortex; there were numbers of active germinal centres in the remaining thymic tissue, and elsewhere a variety of cellular accumulations, some definitely epithelial in character, others spindle-celled; Hassal's corpuscles were numerous and often large and cystic in appearance; plasma cells were abundant at the periphery of the islands of thymic tissue and there were many mast cells in the same situations. Studies by Goldstein and Mackay (1965) on a series of thymuses from SLE patients obtained by autopsy or operation showed no germinal centres but otherwise the same general histological character.

Thymectomy in SLE had no significantly curative effect but in Mackay's (1965) opinion 'there may be a delayed ameliorative effect [from] a progressive diminution in the frequency of relapses'. On the present hypothesis no more than this could be expected.

The only animal model of SLE available for routine study is the disease observed in $F_1$ NZB × NZW mice (Helyer and Howie, 1963; Burnet and Holmes, 1965). The model is not exact but includes a high incidence of positive tests for anti-nuclear factor and lupus erythematosus cells, striking vascular lesions and regularly fatal renal disease. In the present connection it is interesting that these hybrids, particularly the females, show gross thymic lesions with masses of granulomatous tissue including germinal centres and pyroninophil cells, often macroscopically visible as a 'lumpy' thymus. The cortex was atrophic to variable degrees and often showed massive conversion of thymocytes into pyroninophil cells over considerable areas. The appearance of the medullary 'granuloma' was precisely similar to that of lesions often seen in lymph nodes in these mice. We have noticed the relative absence of epithelial cells in the thymus of NZB mice, and de Vries and Hijmans (1966) and East and Parrott (1965) have suggested that this may in some way be related to the poor control against autoimmune processes. In young NZB × NZW mouse thymuses there are more epithelial cells than in NZB but it is possible that there is a similar functional weakness. Certainly in our experience the thymic lesions are more conspicuous in NZB × NZW mice than in any other strain.

This analogy of SLE with conditions in autoimmune mice may turn out to be irrelevant, but as things stand at present it tends to support the importance that has been ascribed to the thymus in the pathogenesis of SLE. In both the murine and the human diseases there is evidence of a genetic weakness in the censorship role of the thymus, which may have more than one component. In both, this results in stimulation rather than destruction of relevant immunocytes by the antigens freely available in the thymus. Local formation of germinal centres is common but is followed by partial or complete destruction of thymic function as the organ becomes increasingly a target tissue for attack by the immunopathological process.

## MYASTHENIA GRAVIS

Myasthenia gravis is a disease, probably autoimmune in nature and certainly involving the thymus, which still presents a number of fascinating problems in neurophysiology as well as in immunology. There is little evidence as to the nature of the lesion responsible for the characteristic muscle weakness. It has never been shown to be directly due to circulating antibody, and Goldstein and Whittingham (1966) suggest that it is a hormonal effect. There is, however, still more than a possibility that an immunological basis for the functional lesion will eventually be found. Confining ourselves to established immunological aspects the following features are relevant.

(*a*) The presence in 20–30 per cent of sera from cases of myasthenia gravis of an antibody which will react with the A band of skeletal muscle (Strauss *et al.* 1965).

(*b*) About half the cases positive with muscle also react by immunofluorescence with epithelial (? myoid) cells in the thymus (van der Geld *et al.* 1963).

(*c*) Surgical specimens of thymus removed from patients show

621

in 60–70 per cent of cases numerous germinal centres in medullary tissue (Sloan, 1943). These contain immunoglobulin-producing cells (White and Marshall, 1962).

(*d*) About 15 per cent of cases are associated with a lympho-epitheliomatous tumour of the thymus (Castleman, 1960). Such tumours may be unassociated with general symptoms, but most are associated with myasthenia gravis and much less frequently with aregenerative pancytopenia or acquired agammaglobulin-aemia (Good *et al.* 1964).

Cases with tumour in older persons have much more consistent serological reactions and are not benefited clinically or serologically by thymectomy (Osserman and Weiner, 1964).

(*e*) Thymectomy in early cases with germinal centres only is beneficial both serologically and clinically but rarely curative.

(*f*) Neonatal myasthenia in children of myasthenic mothers is not unusual and is not necessarily associated with circulating maternal antibody. It may be seen in thymectomized mothers (Harvey and Johns, 1962).

(*g*) Biopsy of muscle may show small accumulations of lympho-cytes. On occasion a more definite myositis is found (Klein *et al.* 1964), almost always in cases with thymoma.

(*h*) Patients and their siblings show a higher incidence of rheu-matoid arthritis, thyrotoxicosis and other conditions in the 'auto-immune' category than a normal population (Simpson, 1960).

This summary of the immunological facets of myasthenia gravis justifies an attempt to visualize its pathogenesis in terms of an autoimmune disease whose target cells may include striated muscle and a proportion of the epithelial cells of the thymus. The most interesting theoretical requirement is to provide an understanding of how and why myasthenia gravis differs from SLE, which we have regarded as also a disease in which thymic cells are attacked. One of the important differences is that, according to Strauss *et al.* (1965), *no* immunofluorescent reactions with muscle were obtained amongst 67 cases of SLE and 30 of rheumatoid arthritis.

If myasthenia gravis is due to the appearance of forbidden clones of the general character we are postulating there are some important restrictions which must be incorporated in any hypothesis. There is the individualistic character of the target cells and, by implication, the ADs concerned. The absence of any reactive antibodies in SLE sera points strongly against a wholly random origin of the pathogenic patterns. The possibilities seem to be:

(*a*) There is a genetic factor influencing the abnormally frequent appearance of immune patterns capable of reacting with an interrelated group of ADs which, by hypothesis, are concerned with the neuromuscular junction and the A band of muscle and which are also present in thymic myoid epithelial cells.

(*b*) Only in a genetically susceptible group is there available during young adult life any accessible AD of the specific myasthenia gravis character which can stimulate any immunocytes of appropriate quality to proliferate.

The alternatives are not mutually exclusive and the second is chosen for first consideration mainly because it offers an important hint as to the significance of the thymic germinal centres. If, as suggested by Sainte-Marie and Leblond (1958), thymocytes produced in the cortex normally pass into the medulla, this may well provide a second set of ADs for 'censorship' testing. In view of the existence of a rather wide range of derivatives of branchial epithelium in the course of normal development and of the heterogeneous character of 'epithelium' in the thymus it seems reasonable to expect that there may be a similarly wide range of tissue determinants available, one of which is evidently related to skeletal muscle.

It would be reasonable, therefore, to picture each germinal centre arising in relation to a source of antigen acting as a stimulant to proliferation of an immunocyte capable of abnormal resistance to the censorship function in the thymic environment. Abnormal resistance, presumably derived by mutation, is the operative condition and since the restricted range of pathogenic action in myasthenia gravis is highly characteristic, one is driven back to the first hypothesis; i.e. that there is some inherent (genetic) quality in the basic immune pattern of individuals prone to myasthenia gravis which makes it probable that a resistant immunocyte will carry this particular pattern rather than any other.

There are various possibilities which could be imagined *ad hoc* to provide a special ability to proliferate in the thymic environment for cells carrying the combination of resistance with the specific myasthenia gravis pattern. For the present it seems adequate to rely on a combination of the two features listed above.

### The role of the thymus in autoimmune disease

It must not be forgotten that there is only one disease, myasthenia gravis, for which the thymus is frequently removed surgically and examined histologically. Thymuses removed at autopsy after more than the briefest illness are notoriously atrophic and unrevealing. There is a single paper by Gunn *et al.* (1964) describing surgical operations on the thyroid in which a commonly present fibrous cord that connects the apex of the thymus to the thyroid was used as a guide to allow a small piece of the thymus to be taken for histological examination. They found sixteen out of fifty cases of thyrotoxicosis with lymphoid medullary follicles in the thymus and a partial correlation of a positive finding in the thymus with similar follicles in the thyroid. This draws attention to Sloan's (1943) early statement that thymuses taken at autopsy from three conditions resembled those of myasthenia gravis in showing excessive lymphocytes in the medulla. They were: hyperthyroidism, five out of twenty; acromegaly, two out of five; Addison's disease, three out of seven.

There is more than a hint here that if it were possible to take a thymic biopsy at the early stage of development of a number of organ-specific autoimmune diseases, proliferative processes would be evident in the medulla. I have already mentioned the findings in the only case of early SLE in which the thymus was removed surgically, and to round off the findings I should mention two cases of severe AHA (Wilmers and Russell, 1963; Karaklis *et al.* 1964) in infants in which, after failure of splenectomy and corticosteroid therapy to modify the disease, the thymus was removed. *Post hoc* or *propter hoc*, in both cases a rapid recovery ensued. I have seen sections of both thymuses, one of which is briefly described in Burnet (1964). This showed a highly cellular medulla, the most conspicuous feature of which was the large number of plasmablasts, many showing mitosis in certain areas. The other specimen was also highly cellular but was poorly fixed, and no cellular detail could be made out.

In discussing the significance of the changes in a thymus

removed from a girl with SLE (Burnet and Mackay, 1965), we tabulated four ways in which the thymus might be involved in autoimmunity. Some modification, particularly of the first, is necessary in the light of further knowledge but the four headings still provide a satisfactory summary. With the appropriate modifications, they are:

(*a*) As the major site of differentiation of stem cells to immunocytes the thymus is the region in which first significant contact of normal or mutant immunocyte with 'self' ADs takes place. In this sense the thymus may be the *source of the pathogenic clones concerned*.

(*b*) The thymus does not normally allow the development of germinal centres and, except along vascular bundles, plasma cells are not normally seen. The presence of germinal centres in the thymus can therefore be taken as an *index of the presence in the body of forbidden clones*.

(*c*) If, for genetic reasons, the special quality of the thymus that governs its censorship function is inadequate, autoimmune disease may in part be related to this *ineffectiveness of homeostatic control*. Reasons have been given for thinking that this may play a part in human SLE and in the murine disease of NZB and NZB × NZW strains.

(*d*) The thymus may be a *target organ of autoimmune attack* either in a general sense, as in SLE, or in virtue of the varied samples of tissue antigens carried by the epithelial cells of the organ. It is probable that the latter is responsible for those conditions in which numerous germinal centres are present in the medulla.

### THE AGE AND SEX INCIDENCE OF AUTOIMMUNE DISEASE

All well-defined human diseases have a broadly reproducible age incidence, sometimes throughout the world, and always when a large reasonably homogeneous population is available for study. The age incidence of death from infectious disease can be a very important clue to epidemiology (see Burnet, 1952) and there has also been great interest in the specific age incidence of death from various types of malignant disease. To a close approximation, those forms of malignant disease not clearly influenced by hormonal changes in the body show a specific age incidence which, when plotted on a double logarithmic scale, gives a straight line (Armitage and Doll, 1957, and many others). This type of pattern is given by a stochastic process in which chance effects, random with respect of time from birth, are accumulated according to some appropriate rules. There have been suggestions, for example, that random events only became significant when they involved 5 or 6 contiguous cells simultaneously, when 5 or 6 sequential events involved the same cell line or when a lesser number of sequential events were connected by a proliferative phase of the cell line concerned.

In view of the essential similarity of the forbidden clone approach to the somatic mutational theory of cancer, Burch was led to apply a similar stochastic approach to a wide range of chronic illness. Originally, autoimmune processes were chosen and, using a fairly elaborate mathematical approach, Burch (1966a) found that there was in fact a reasonable fit of sex- and age-specific incidence curves to general stochastic equations of the form:

$$P_{n,t} = So(1 - e^{-kt})^n$$

where $P_{n,t}$ is the age-specific prevalence at time $t$ of a condition that is initiated by at least one of each of $n$ independent random events each of an average rate, $k$, per year. *So* is the proportion of the population genetically subject to risk. A different equation of the same general form applies to conditions in which the random events are dependently related.

It is outside my ability and interest to attempt to follow Burch's reasoning in detail. To one with a medico-biological experience with a minimum of mathematics, what seems to emerge is:

(*a*) a reasonable certainty that stochastic processes of the general quality of somatic mutation are playing a major part;

(*b*) that there are striking differences from one disease to another;

(*c*) that Burch's requirement that the *only* factors entering into the determination of age incidence are genetic or somagenetic does not ring true. One thinks of the changes by cohorts of the age-specific incidence of lung cancer over the last 50 years (Metcalf, 1955) without more than minor deviations in the shape of the curve.

(*d*) that many 'normal' processes, such as greying of the hair and loss of teeth (Burch, 1966*b*), fall into the same sort of pattern and that there may be an important nascent approach to the general problem of ageing implicit in Burch's work.

I find the theoretical interpretation given by Burch and Burwell (1965) too hypothetical for serious consideration. The lymphoid cells strike me as highly adequate for damaging roles but hardly an appropriate vehicle for morphogenetic and morphostatic activities. So far there seems to have been little work by any other group on the lines pioneered by Burch. The approach, like genetic and somagenetic concepts in general, offers no promise of new treatment of human disease and little that can be applied to bring prestige to a laboratory worker. Yet it is hard to avoid the conclusion that in the long run it may be the best approach to a real understanding of one of the many human problems for which there can be no practical solution because they are of the very nature of organic evolution.

# 13 Immunological surveillance and the evolution of adaptive immunity

In discussing the evolutionary significance of the immune mechanism in Book I, I adopted the hypothesis that variability in histocompatibility antigens and the associated capacity for immunological recognition of self from not-self has an important evolutionary significance which can be crudely expressed in the phrase: 'If it were not so, cancer would be an infectious disease.'

The experimental and observational justification for such a point of view is necessarily rather indirect. The evolution of the vertebrates covered 500 million years and, with only minor exceptions, immunological work has been confined to man and half a dozen mammalian species plus the domestic fowl. There is, however, a rapidly increasing interest in evolutionary aspects of immunity and in the last year or two such unorthodox experimental animals as hagfish, lampreys, dogfish, toads, the Australian monotreme the spiny ant-eater and the American opossum have been subject to at least preliminary investigation by modern immunological techniques.

Relevant studies may be grouped under two headings. The first concerns the phylogenetic aspects of the immune process and the justification for the accepted position that adaptive immunity is solely a vertebrate characteristic. In this section the natural division is in regard to (*a*) the processes of defence and repair in invertebrates and (*b*) the behaviour of the living forms most relevant to the early stages of vertebrate evolution.

The second section should concern those phenomena observed clinically or experimentally which bear on the origin, extent and significance of the heterogeneity of significant tissue antigens between individuals of all normal mammalian species and on the appearance of new antigens by somatic mutation within the body and their immunological significance. The evidence on somatic mutation will necessarily be largely drawn from phenomena involving malignant cells and will involve a discussion of the

628

'clonal selection theory of carcinogenesis' and the evidence that immune responses may be concerned in the survival or destruction of emergent malignant cells. This will necessarily require consideration of the alternative favoured by Hellström, Klein and others that 'allogeneic inhibition is of more importance than immune processes'.

PROCESSES OF DEFENCE AND REPAIR IN INVERTEBRATES

Most of the material to be used in this and the next section is reviewed by Good and Papermaster (1964), who reach the firm conclusion that immunity as observed in man and the classical mammals of the laboratory is essentially a vertebrate phenomenon. There are well-defined cellular defence processes in those invertebrates that have been studied, but adaptive immunological responses have never been demonstrated. Only a few of the more recent experimental studies need be cited to confirm that conclusion. Bernheimer *et al.* (1952) made comprehensive studies on caterpillars inoculated with a variety of standard antigens, phage, *Escherichia coli*, streptolysin O, human red cells and egg albumin. No antibodies could be detected in body fluids although both normal and 'immune' fluids had some agglutinins for *E. coli* and erythrocytes. Wagner (1961) reviewed the question of acquired resistance to bacterial infection in insects and concluded that any increased resistance was nonspecific in character. Transplantation of tissue between different species or genera of insects is easily effected and is a standard technique for the study of insect physiology. Triplett *et al.* (1958) found no difference in the response of a sipunculid worm (*Dendrostomum*) to autologous or homologous tissues and Cushing (1962) is quoted by Good and Papermaster (1964) as finding no rejection of homografts in octopuses for at least 40 days. Several writers have mentioned the curious habit of certain nudibranch molluscs which feed on sea anemone tentacles of transplanting the functioning nematocysts of the anemone to their own tissues (Kepner, 1943).

In the body fluid of insects and other invertebrates there are wandering cells, haemocytes, which are actively phagocytic for bacteria of many types if these are introduced into the body cavities. These cells also aggregate rapidly around any foreign material such as a parasite. Salt (1960) found that when a foreign body is placed in a certain species of caterpillar it becomes coated with haemocytes. If it is now transferred to another insect of the same species there is no further attachment of cells. If, however, it is placed in a distant species of caterpillar an additional layer of cells is added. This can be taken as showing that there is some capacity to recognize foreignness as between the cells of unrelated species.

Cameron (1932) made detailed studies of the inflammatory response in the earthworm and differentiated several types of cell on staining qualities but did not relate these to any specific functions of the cells. The process of reaction to trauma was associated with an increase in the number of cells in the coelomic fluid and accumulation or proliferation of undifferentiated cells in the damaged region, with phagocytosis of necrotic tissue.

There are many other references available to work on other invertebrates but none seems to require any modification of the conclusions which can be drawn from those that have been mentioned. There is an adequate mechanism for dealing with invading micro-organisms by phagocytosis, and larger foreign bodies by coating with haemocytes, in most or all invertebrates, and repair of traumatic damage can be effected. Such reactions must imply a definite though relatively crude capacity to differentiate between 'self' tissues and foreign material. The limitations of this are shown by the readiness with which homografts (and even grafts from distinct but not too distant species) are accepted. There is no evidence at all for specific antibody responses nor for the presence of proteins at all resembling vertebrate immunoglobulins (Woods *et al.* 1958).

IMMUNE RESPONSES IN THE LOWER VERTEBRATES

A growing volume of experimental work reviewed by Good and Papermaster (1964) makes it clear that elasmobranchs and bony fishes as well as all higher vertebrates show all the standard aspects of an immune system. They possess a thymus or its equivalent, lymphocytes and plasma cells and immunoglobulins (usually of more than one type), and can manifest rejection of skin homografts and antibody responses to typical antigens.

If, as everything suggests, this machinery of adaptive immunity is an exclusively vertebrate endowment, the main interest at the evolutionary level is in regard to the immune capacity of the most primitive vertebrates and such protovertebrates as the sea-squirts (*Tunicata*) and Amphioxus. There can be no direct evidence from palaeontology of how vertebrates evolved from some invertebrate ancestor, since it is virtually certain that the essential stages involved small soft-bodied organisms that could leave no fossils. The most favoured suggestion seems to be that vertebrates developed from the larval forms of some primitive tunicates and passed through a form analogous to Amphioxus, eventually giving rise to early cyclostomes. The earliest known fossils of vertebrates are the Silurian ostracoderms. These were large heavily armoured forms, obviously highly specialized types, which had evolved from some soft-bodied basic stock. The ostracoderms are divided into two groups, the pteraspidomorphs and the cephalaspidomorphs which, according to Stensiö (1958), are related to the modern hagfishes and lampreys respectively. The hagfishes are the more primitive of the two; the species are wholly marine and the young have essentially the same morphology as the adult. The lampreys have both freshwater and marine species and all have a larval form limited to a freshwater habitat. The larval forms of Ammocoetes have many structural similarities to Amphioxus.

Immunological studies on the cyclostomes are almost confined to those of Good's group. Hagfish kept under laboratory conditions produced no antibody to any of the antigens used (Papermaster *et al.* 1962, 1964) and showed no evidence of immune elimination of a phage known to be fully antigenic in bony fish. Neither autografts nor homografts of skin were practicable. The lamprey has only a feeble immune capacity, failing to produce antibody against most of the antigens used. *Brucella* agglutinins, however, were produced and specific rejection of homografts with retention of autografts was recorded (Good and Papermaster, 1964).

In line with these results, Papermaster *et al.* (1964) found no evidence of immunoglobulin by immunoelectrophoresis in hagfish serum while there was a single line, approximately equivalent to

631

the Ig M line of higher forms, in lamprey serum. A primitive thymus in the form of epithelial cells associated with a few lymphocytes is recognizable in the lamprey but not in the hagfish, and the circulating mononuclear cells have a more lymphocytic character in the lamprey.

In fish and amphibians, antibody responses improve greatly if the temperature of the animal is raised from 8–10 ° to around 20 °, and in general the immune system is less active and less versatile than in warm-blooded animals. Precipitins to soluble antigens such as mammalian serum albumin are rarely produced. It appears that the characteristic complex structure of immunoglobulin developed early in vertebrate evolution. Marchalonis and Edelman (1965) studied the immunoglobulin of the dogfish, an elasmobranch, and demonstrated a single antigenic type of immunoglobulin present both as a 7 S monomer and a 17 S molecule, probably a simple pentamer, which was associated with all antibody activity. The monomer was of complex structure with chains apparently analogous to the light and heavy chains in mammalian Ig G antibodies.

## THE DEVELOPMENT OF ANTIGENIC HETEROGENEITY IN MAMMALIAN SPECIES

If our contention is correct that antigenic individuality is an important evolutionary necessity in vertebrates, one would expect that a high mutability of the genes concerned should be demonstrable. At first sight, adequate evidence for such variability is provided by the extreme multiplicity of histocompatibility types in the available strains of laboratory mice all derived from a single wild species. However, having regard to the complexity of the loci concerned, a large proportion of the observed genetic diversity could have arisen by the recombination of qualities present in wild stocks. The H2(a) allele is for instance associated with 15 distinguishable red cell antigens according to Pizarro *et al.* (1961) and although the H2 locus on chromosome 9 in mice is much the most important, there are certainly many other loci involved. According to Snell (1963) there are 18 alleles at the H2 locus with more than 26 associated antigens, while both H1 and H3 loci also show multiple alleles.

It is a well-recognized phenomenon that if a standard strain $X$ held by laboratory $A$ is transferred to a new laboratory $B$ and maintained separately in each case by the accepted technique for pure line maintenance, it will be found after a few years that strain $X/A$ will not accept a skin homograft from $X/B$ (Billingham *et al.* 1954; Lindner, 1963). This in itself is indicative of a high level of mutability in some locus concerned with histocompatibility. A direct study of such mutability has been reported by Bailey and Kohn (1965) in which they studied $C_{57}BL$, BALB/c $F_1$s, irradiated and non-irradiated, for skin-graft rejection using a technique which would indicate the existence of either 'gain' or 'loss' mutations. There was no significant effect of irradiation, and after making allowance for the occasional cluster of positive results pointing to a mutation in a parent or grandparent of the mice tested they obtained the high mutation rate of $6 \cdot 7 \times 10^{-3}$/gamete.

From the complexities of red blood cell antigens in man (Race and Sanger, 1962) and the growing evidence of an almost equal complexity of leucocytic antigens (Dausset and Rapaport, 1966) it is highly probable that a similar high mutation rate is characteristic of the human situation as well. In Bailey and Kohn's experiments, 30 out of 31 changes had the character of an additional antigen, the other instance showing both gain and loss. Studies on mutation by loss of antigen have been mainly concerned with the capacity of cancer cells to produce tumours in various hybrids. Here we are necessarily concerned with somatic mutation and the phenomena are discussed in another section.

The case of the golden hamster is of special interest since the whole stock of experimental animals throughout the world is derived from a single litter found in Palestine and given to Adler in 1919. By 1960, Billingham *et al.* had available three stocks which had been long separated but not rigidly inbred. These strains (MHA, CB and LSH) appeared to differ by one or two major histocompatibility genes, skin homografts being rejected in 10–13 days. Each stock was, however, sufficiently homogeneous to allow cross-transplantation of skin between individuals (Billingham and Silvers, 1964).

The evidence from both mice and hamsters points, therefore, to a high degree of lability in the genetic control of histocompatibility antigens. It seems, also, that many of the mutational changes are selectively neutral and become manifest simply as a random consequence of the process of strict brother–sister inbreeding. In general, when a mutation occurs within the breeding line by which a genotype changes from AA to AB, there is a 1:8 chance that the line will eventually become homozygous for BB. For every demonstrable change in histocompatibility there must therefore be about seven mutations in the same loci which do not reach expression. This may have implications for somatic mutation since the indications are strong that antigenic changes are co-dominant and would both be expressed in a cell in which one allele had mutated.

### THE EXTENT OF SOMATIC MUTATION INVOLVING HISTOCOMPATIBILITY ANTIGENS

In any approach to the general hypothesis of immunological surveillance the most difficult point to approach experimentally is the problem of whether, and if so to what extent, somatic mutation involves the appearance of new histocompatibility antigens. It is obvious that even with a high rate of mutation, equivalent to what must occur in the germ cell line in mice, not more than $10^{-4}$–$10^{-3}$ of the body cells would carry a 'new' histocompatibility antigen unless associated with that change there was a proliferative advantage of the type characteristic of malignant change. Such a small proportion could hardly be detected experimentally and I know of no attempt to do so. There is, however, one possibility that should be explored by the use of specific reagents for histocompatibility antigens modified for the use of the methods of immunofluorescence. The recognition in, say, the lymphoid tissues of a mouse of one strain of a *very small* proportion of cells fluorescing when treated with antibody specific for another strain, should not be beyond current experimental techniques.

Indirect evidence for the appearance of new antigens could perhaps be obtained by close survey of the changes in the graft-versus-host activity of lymphocytes from an individual at different ages. No specific studies of this sort seem to be on record but the impression from the work on the Simonsen–CAM reaction in my own laboratory was that no significant difference between individual birds of the same genetic make-up existed. If antigenic mutations are indirectly responsible for the appearance and proliferation of the pock-producing immunocytes they must occur with a considerable degree of regularity.

## THE REACTIVITY OF LYMPHOCYTES FROM NORMAL ANIMALS AGAINST ALLOGENEIC CELLS

There are now very large numbers of examples of normal cells or normal immunoglobulins which have a damaging effect on cells or tissues of allogeneic individuals. In general but not always, immunization of the donor animal with cells from the target animal (or a syngeneic individual) increases the intensity of the damage.

The reactivity is best shown in target animals which, because of immaturity or recent X-irradiation, lack capacity to resist and destroy cells being implanted. Attention will therefore be concentrated on two classical examples of graft-versus-host reaction, the normal lymphocyte transfer reaction of Brent and Medawar (1964, 1966) and the chorioallantoic pock form (Boyer, 1960) of the Simonsen reaction in chick embryos (Simonsen, 1957).

The normal lymphocyte transfer reaction as shown in guinea-pigs is aroused when lymphocytes from one guinea-pig are injected intradermally into a normal adult guinea-pig whose genetic status in relation to the donor is such as would allow an immunological reaction. When pure line strains H and 13 were used, lymphocytes from H or 13 produced reactions in $F_1$ (H × 13) animals but lymphocytes from $F_1$ were inert against both H and 13 strain recipients. This is the classical behaviour of a graft-versus-host reaction and categorizes the normal lymphocyte reaction as an immunological one (Brent and Medawar, 1966). The response is most clearly shown in recipients that have been X-irradiated (600 r whole-body irradiation) 24 hours previously. In these it shows three phases the most important of which, from our point of view, is the 'first inflammatory episode'. This may be visible within 12 hours and reaches a standard level by 24 hours. Thereafter it remains approximately unaltered till the fourth day, when there is a 'flare-up' lasting about two days, and then it rather suddenly fades out. If the donor has been immunized against the target strain by a skin graft, lymphocytes from the draining lymph node give an initial response which is much more active, but with the same time scale as occurs with cells from unimmunized donors. To produce the same score for response intensity at 24 hours, 8–10 times as many normal cells must be used. This 'recognition' response is not dependent on cell division, being uninfluenced by adequate doses of mitotic inhibitors.

Although Brent and Medawar do not firmly accept the interpretation, they have nothing to counter the view that I should adopt as best fitting the observations; namely, that in the population of lymphocytes injected into the target animal is a not insignificant proportion which have immune receptors which can react with specific ADs in the target tissues. In our present context, the question is whether the number of reactive cells in the lymphocyte population is or is not beyond the number which could be produced without specific prior immune proliferation. In view of the points made in regard to the potentialities of reaction of a single combining site this question may not be as crucial as it has seemed to be in the past.

In a very similar field of work the spleen response was used by Simonsen (1962) as a method for assaying the potency of cells from normal and immunized mice to produce graft-versus-host reactions in newborn $F_1$ hybrid mice. His most interesting finding was that when there was a major $H_2$ difference between the strains $C_3H(k)$ and $DBA(d)$ the number of cells needed to give a standard degree of splenic enlargement was only slightly reduced if the donor was immunized with target-type cells. On the other hand, when the difference was slight and both had the same $H_2$ antigen (k) as with $C_3H$ and $AKR$, the normal lymphocytes had a minimal effect which was greatly increased by immunization. The suggestion is clear that a much larger number of the immunocytes in a $C_3H(k)$ mouse had a natural capacity to react with $H_2$ antigen (d) than with the antigen corresponding to the minor difference between $C_3H$ and $AKR$.

This difference can have at least two distinct explanations. The first has been referred to in chapter 6 of Book 1, namely, that there may be segments of many patterns of combining site which are specially prone to react with ADs characteristic of histocompatibility antigens. A large proportion of the immunocytes concerned will be eliminated by reaction with the animal's own histocompatibility antigens. Of those that remain, the close relation of AKR to C3H will ensure that only a very small proportion of those which initially could react with AKR antigens will remain after censorship against the autologous C3H antigens. On a random basis, many more of the immunocytes will react with the more distant pattern of DBA antigens. The second possibility is that the relatively large number of 'anti-DBA' cells results from an antigenic stimulation of the corresponding clones as a result of emergence by somatic mutation of sufficient cells with the d antigen to give a low-grade stimulus leading to the proliferation of such clones.

A more quantitative approach is possible with chorioallantoic membrane titrations of active cells. Here the same rules hold as in mammalian graft-versus-host reactions (Burnet and Burnet, 1961). No foci appear on the chorioallantois if it is of the same pure line as the adult fowl supplying the blood leucocytes; $F_1$ leucocytes produce no foci on parental-type CAM but parental cells do produce foci on $F_1$ CAM. Schierman and Nordskog (1963) used inbred strains classified according to their b isohaemagglutinogens which correlate with histocompatibility. They tested the capacity of what can be called CC lymphocytes to produce foci on AA and on AB chick embryos. In general the number of foci produced with randomly bred material is about 1 per $10^4$ mononuclear cells in the inoculum. If each focus is the result of the activity of an immunocyte with a specific pattern corresponding to the antigen on the CAM, then CC lymphocytes should include, amongst others, lines a and b specifically reactive against A and B antigens respectively. If this is so, then CC cells placed on AA, BB and AB in standard amount should give x, y and x + y foci. To a close approximation this is what Schierman and Nordskog found to be the case. When cells from a B3 donor were used the mean number of foci were on B1/B1 CAM 65, 113, 74, 68 and on B1/B2 149, 175, 147, 100 with overall means for one foreign antigen of 80 and, for two, of 143. This points to a specific difference between the clones of cells in B3/B3 which react with B1 or B2 antigen.

The immunocytes concerned are present in relatively large numbers when we remember that there is no reason to believe that B1 or B2 antigens had ever been introduced into the donor B3/B3 fowl. The possibilities to be weighed will be precisely the same as for the normal lymphocyte transfer reaction, an inborn quality of the combining site or an antigenic stimulation by somatic mutant cells.

The orthodox graft-versus-host interpretation that has been applied to all these phenomena may need to be modified as a result of the experiments of Jones and Lafferty (1966) in sheep. Essentially the reaction appeared to be the same as the normal lymphocyte transfer of Brent and Medawar (1966) but since homozygous sheep were not available, the experiments were made with individual allogeneic donor and recipient animals. Lymphocytes from the donor gave an inflammatory reaction; similar cells from the recipient injected intradermally into its own skin did not. Crude phenol-extracted RNA from the lymphocytes behaved similarly to that from the allogeneic donor, producing inflammation which was absent when the extract from recipient's own cells was used. This may not be basically equivalent to the normal lymphocyte transfer reaction but the finding has sufficient resemblance to that of Mannick and Egdahl (1962) (see p. 465) to suggest that the same experiment should be carried out to establish in guinea-pigs whether parental RNA will provoke the reaction in $F_1$ or vice versa. The possibility that crude RNA-extracts contain other biologically significant components must be kept very much in mind. If the extracts do in fact function as m-RNA directing synthesis of a protein by the recipient's cells, it would obviously be of the greatest interest to know whether what is being synthesized is a different histocompatibility antigen allowing the recipient to mount a host-versus-graft reaction, or an immune receptor which would confer donor character on enough recipient lymphocytes to initiate a graft-versus-host reaction.

## THE CLONAL SELECTION THEORY OF CARCINOGENESIS

This was used as the title of a paper by Prehn (1964) but the general approach has been common coin of discussion amongst oncologists for many years and was firmly supported in a review made 10 years ago (Burnet, 1957*b*). The essence of this approach is that somatic mutation, whether 'spontaneous' or associated with the presence of chemical or viral carcinogens, will only become evident when a mutant cell can proliferate beyond its unaltered congeners and in some way overcome any local morphogenetic controls. The 'conversion to malignancy' is not a positive effect of the carcinogen, it is simply the only way in which cells of altered somatic-genetic quality can manifest their existence. Any one of a hundred different modifications of the genome might have such an effect and could have equally numerous relevant or irrelevant changes in surface antigens of the cell undergoing mutation. Where the local situation is conducive to local (non-malignant) proliferation of the vulnerable cells, correspondingly increased numbers of cells will be exposed to the mutagenic process with increased likelihood of the appearance of potentially neoplastic mutants. In addition the possibility is always present that there will be an increase of sequential somatic mutation in a clone with perhaps a minor proliferative advantage. Just as an antigen, according to clonal selection theory, does not convert a genetically neutral cell into one producing specific antibody but selects an appropriate 'mutant' for proliferation, a strictly analogous process is set going by the deposition of a chemical carcinogen in the tissues.

The role of virus in carcinogenesis is more controversial. Most workers consider that the effect is a direct one but at least an equally strong case can be made for the assumption that the only regular quality—the presence of a recognizable transplantation antigen—may be only an epiphenomenon with no relevance to neoplastic quality. The existence of the new antigen and its association with a neoplastic cell does, however, provide particularly suitable material to give a model of immunological surveillance which is here my primary interest.

Confining ourselves to neoplasms arising spontaneously or as a result of irradiation or the application of carcinogenic chemicals, evidence in favour of the view comes in the first instance from cytogenetics, and particularly from Ford and Clarke's (1963) study of primary neoplasms in heavily irradiated laboratory animals. During the stage of recovery and regeneration of their own lymphoid tissue in animals that survive or are saved by bone-marrow transplant, a stem cell with chromosomal damage may give rise to a clone which dominates bone-marrow and lymphoid tissue as shown by the frequency of its recognizable karyotype.

In a Chinese hamster a reticulosarcoma was shown to be made up of cells with two distinct karyotypes. The commonest had the normal diploid complement of 22 chromosomes but 3 of these were recognizably abnormal with chrommeres situated sub-terminally instead of almost centrally, as in the corresponding normal chromosomes. The second group had 23 chromosomes, including the same 3 abnormal markers and with an additional metacentric chromosome. The obvious deduction made by Ford and Clarke is that the two tumour cell types are members of the same clone and have arisen one from the other. In the same paper the karyotypes found in two cases of human leukaemia are similarly analysed. The situation in these is more complicated but a clonal process is again indicated. It is concluded that 'at some point a karyotype variant arises and proliferates differentially as compared with the remainder of the neoplastic population; among its progeny further karyotypic variants occur and a selective process is set up.'

The second relevant point is the multiplicity of the antigens (tumour-specific transplantation antigens) found in methyl-cholanthrene tumours from syngeneic hosts (Prehn and Main, 1957). These antigenic qualities persist on passage, which strongly suggests that each tumour usually represents a single clone derived from *one* emergent mutant cell (Klein and Klein, 1963). In addition to qualitative differences in tumour-specific transplantation antigens, such chemically induced tumours also show striking differences in their immunizing capacity (Old *et al.* 1962).

There is now a good deal of evidence to suggest that the effectiveness of methylcholanthrene in allowing the emergence of malignant cells of modified antigenic pattern depends on the local or general immunodepressive effect of the carcinogen (Prehn,

1963; Malmgren *et al.* 1952; Stjernswärd, 1965). This means in effect that the antigenically 'new' cell could expect to be recognized and destroyed in normal circumstance but, in the presence of a local concentration of carcinogen, no defence process is brought to bear on it and the tumour is initiated. Once a significant mass of tumour is in existence, presumably with a considerable steady leak of antigen into the lymph and blood, a state of unresponsiveness, possibly assisted by the immunosuppressive action of the persisting carcinogen, develops. Stjernswärd (1966) has shown that mice in which a methylcholanthrene sarcoma has developed and been surgically removed are immediately thereafter *more* susceptible to a small number of the autochthonous tumour cells than controls. This probably represents a state of specific immunological tolerance since the enhanced susceptibility is not shown to other strains of methylcholanthrene-induced sarcoma. Again, the implication is strong that we are dealing with a clone with a single common marker; in these cases a distinctive antigen takes the place of an abnormal chromosome.

Similar results have been obtained by Mikulska *et al.* (1966) with rats in which benzopyrene tumours had been produced. They found the same lack of any resistance immediately after removal of the autochthonous tumour but showed that immunity, both active and passively demonstrable by spleen cells, developed if testing was delayed for a few weeks after removal of the tumour.

### IMMUNOLOGICAL DEFENCE AGAINST IMPLANTATION OF TUMOUR CELLS

The basic concepts of histocompatibility differences between mouse strains were based on the results of tumour transplantation (Gorer, 1937; Little, 1941; Snell, 1953). In general, a spontaneous tumour arising in strain A will multiply on transplantation only in strain A mice. The differences so defined are equivalent to those in which skin grafting is used. A striking example of this equivalence is implicit in the method used by Snell (1958) for the production of co-isogenic lines. The objective was to produce a mouse AB in which the only genetic difference from A was the replacement of an allele at a histocompatibility locus. The technique was to hybridize A with the strain carrying the desired histocompatibility antigen and backcross repeatedly to A. At each alternate generation from the second onward, all offspring were tested with an A tumour and only a female resisting the tumour used in the next backcross to A. In this way, histocompatibility antigens allowing growth of the A tumour are eliminated while in due course A alleles come to occupy all other loci. In a series of such lines thirty out of thirty-eight were found to differ by H2 differences (Snell, 1963).

The essential point is that the H2, and to a lesser extent the other groups of histocompatibility genes in the mouse, are completely adequate for the purpose we are postulating—the need to prevent cancer cells being contagious to young animals. There is no need for further discussion of this point but there still remains an important area of controversy. This is the extent to which this resistance is a specifically immunological one.

The alternatives are (*a*) that postulated by the theory of immunological surveillance and (*b*) Hellström's (1966*a*, *b*) theory of allogeneic inhibition. On the particular form of (*a*) which would naturally arise from the general point of view adopted in this book, it is assumed that the control is mediated by immunocytes with combining sites (receptors) capable of specific union with the allogeneic histocompatibility AD and any increased population of such immunocytes which has arisen, either from previous stimulation by mutant cells or by rapid stimulation by antigen from the challenge tumour.

Alternative (*b*) holds that close contact of two cells with differing histocompatibility antigens is mutually damaging and does not necessarily require any factor of immune quality (receptor, sessile antibody, etc.) to cause this damage. This view is based primarily on the fact that when a tumour is adapted to grow well on transplantation in strain A mice, it can be shown by quantitative experiments that a higher dose of cells is needed to produce a tumour in the $F_1$ A × B than in pure line A hosts (Hellström, 1966*a*). Correlated with this is the finding that if a small dose of A tumour cells is mixed with a large excess of normal allogeneic cells plus PHA to induce close cell-to-cell contact and aggregation, the infectivity of the A tumour cells for A hosts is destroyed or diminished. This is not seen if syngeneic cells are added to the tumour cells (Hellström, 1966b). This type of phenomenon has also been demonstrated *in vitro* by adding allogeneic lymphoid cells from non-immunized donors to monolayers of tumour cells (E. Möller, 1965). Cytotoxic effects were produced, provided that aggregation and close contact were induced by adding PHA or a heterologous anti-mouse serum. No effect was obtained with syngeneic cells but it was observed with semi-syngeneic ($F_1$) cells. Klein (1966) considers that allogeneic inhibition may represent a primitive biological capacity present in invertebrates as well as vertebrates and suggests (*a*) that the adaptive immune system may have evolved from it and (*b*) that it could provide an adequate basis for a tissue surveillance system to eliminate mutant potentially malignant cells arising in the body.

Klein does not in any way doubt the importance of immune responses to tumour cells: he is concerned only with the resistance manifested without immunization. This aspect has already been discussed in relation to the normal lymphocytic transfer reaction and the Simonsen–CAM reaction (p. 635), and if the Schierman–Nordskog experiment cited there is valid it represents almost a formal proof of the importance of *immune* reactivity on the side of the lymphoid cell. Perhaps the most important point in favour of an immune mechanism is the fact that the specificity of histocompatibility antigens as determined by 'immediate' reactions is identical with the specificity that can be determined by humoral antibodies. Snell (1963) cites six different methods of demonstrating this in mice: red cell agglutination, red cell lysis, white cell agglutination, complement fixation, fluorescent labelling, cytotoxic action and enhancement. In addition, spleen cells from B mice immunized with normal thymus cells from strain A can be shown to be capable of destroying the tumour-producing capacity of an A tumour cell suspension (Winn, 1961) and of producing cytotoxic effects *in vitro* (Rosenau and Moon, 1961; Wilson, 1965).

The effect of allogeneic inhibition has been shown so far only with differing H2 antigens but any histocompatibility differences between a tumour and its autochthonous host are never of this intensity. Yet there is abundant evidence, largely from Klein and his colleagues (Klein *et al.* 1960; Klein, 1962), that specific anti-

643

body responses can be developed to magnify the difference to an effective level.

At the experimental level the strongest evidence for an immune-type surveillance mechanism can be derived from the behaviour of the oncogenic viruses (see chapter 13, Book I). The essential features are that tumours induced by polyoma virus show a common cellular antigen not demonstrable in normal cells or in tumours of other origin (Habel, 1961; Sjögren *et al.* 1961*a*, *b*). Immunity to challenge with polyoma tumour cells of syngeneic origin can be induced in adult mice either by infection with the virus or by inoculation of a polyoma tumour from an allogeneic animal (Sjögren, 1961). SV40 and adenovirus 12 behave in essentially the same fashion in hamsters. No cross-reactions are found between the cellular antigens of SV40 and polyoma tumours (Habel and Eddy, 1963; Defendi, 1963). Other evidence pointing to the importance of immune processes in preventing the develop-ment of tumours from virus-transformed cells comes from the effect of thymectomy carried out three days after birth (Miller *et al.* 1964*b*) on the result of inoculation of polyoma virus some weeks later. In the thymectomized mice there were eighteen out of twenty tumours, in the controls, one out of sixteen. Kirschstein *et al.* (1964) obtained eight out of forty-eight tumours from thy-mectomized mice inoculated with adenovirus 12, as against none out of eighty in sham-operated controls.

Early thymectomy has also a significant effect in enhancing the production of skin tumours by a polycyclic hydrocarbon (3,4-benzopyrene) (Miller *et al.* 1963*b*) and allows a mast cell tumour to produce lethal growth in a mouse strain differing from the tumour host by a H2 difference and normally quite insusceptible (McEntegart *et al.* 1963). Both results make it difficult to ascribe any major significance to allogeneic inhibition which should be unimpaired in a thymectomized animal.

This fairly elaborate argument against the significance of allo-geneic inhibition has recently become quite redundant as a result of the work of Beatrice Mintz and her collaborators. She has shown that it is possible to cause fusion *in vitro* of two early blastomeres of genetically dissimilar mice. When the composite is replaced in the uterus of a foster-mother it will in more than 50 per cent of cases develop into a healthy long-lived mouse, carrying cells of both origins. Mintz and Silvers (1967) have shown that healthy composites can be prepared from two blastomeres differing by major histocompatibility factors. Such mice are permanent chimeras with thousands of tolerated contacts between cells of different histocompatibility. There is no failure to develop the antigens. Appropriate skin transfers between the allophene and the two 'parent' strains make it clear that both antigens are developed in normal fashion.

The phenomena of allogeneic inhibition presumably represent a marginal effect of incomplete homozygosity or minor deviations from the standard assumption that in the $F_1$ there is a simple co-existence without interaction of the two parental antigens.

## IMMUNOLOGICAL ASPECTS OF HUMAN CANCER

*Spontaneous regression and chemotherapeutic cures*

The spontaneous cure of cancer is very rare but undoubtedly cases occur. Everson (1964), in a review of the literature, concluded that 130 cases conformed to all the requirements for acceptance as true regression. The three conditions showing the largest number of spontaneous and permanent regressions are: neuroblastoma, 28; hypernephroma, 21; chorioncarcinoma, 13. These are all relatively rare tumours and in at least the first and the third groups there are strong indications that immune processes play a significant part. Neuroblastoma is one of the commoner forms of malignant disease in childhood arising from the adrenal medulla or tissue of similar origin. It is highly relevant that, according to Beckwith and Perrin (1963), if the adrenals are examined histologically from an unselected series of autopsies on young infants, nodules of neuro-blastoma are relatively common, forty to fifty times commoner than the overall incidence of clinical neuroblastoma. When this is added to the finding of Koop *et al.* (1955) that recovery occurred in seven out of forty-four cases where no radical removal of the tumour was attempted, one can hardly avoid postulating an immunological process. The adrenal medulla will undoubtedly possess some tissue-specific antigens which are sufficiently inaccessible to be uninvolved in normal tolerance. With the development

645

of a tumour, antigenic material will be liberated—possibly a normal tissue component; possibly an aberrant component specific to the tumour—and a homograft reaction can be mounted against the tumour. Whether or not it is successful will depend on the briskness of the immune response which, in its turn, will probably depend on the age of the child when the tumour becomes immunogenic and the proliferative power (malignancy) of the individual tumour. Such an interpretation is admittedly speculative but it should provide some useful possibilities for immunological investigation of these cases. One would suggest, for instance, that serum from a case of retrogression should provide a reagent for fluorescent-antibody detection of the relevant tissue antigen and that appropriate histological study might demonstrate a cellular attack on a tumour nodule at some stage.

Chorioncarcinoma, which is the third most frequent retrogressing tumour, has two other features of special interest. It is highly susceptible to clinical cure by methotrexate and it is derived from embryonic tissue genetically and perhaps antigenically distinct from that of the host. Again, everything suggests strongly that there is the possibility of a strong immunological response to counter a high degree of intrinsic malignancy. Partial destruction by a cytotoxic agent would of course swing the situation to the advantage of the immune response (Li *et al.* 1958; Hertz *et al.* 1961, 1964).

There is now strong evidence that the Burkitt lymphoma of central Africa is quite unusually susceptible to chemotherapy (Burkitt, 1963, 1966). Here, one might also look with some reason for a specific antigen analogous to the cellular antigen induced by polyoma. It is known that an animal with incipient polyoma tumours developing after a neonatal injection of virus has a significant resistance to transplants of polyoma tumour cells (Sjögren *et al.* 1961 *b*) and there is much to suggest that the Burkitt lymphoma has a similar viral aetiology. Klein *et al.* (1966) obtained some evidence of an antibody in patients from Nigeria and in a moderately high proportion of unaffected Africans, which reacted by immunofluorescence with a surface antigen or Burkitt lymphoma cells.

In addition to neuroblastoma there are other instances where histological evidence of malignancy is more frequent than clinical evidence. The paper by Ashley (1965) is of special interest as it indicates the shape of the specific age-incidence of carcinoma of the prostate as judged (*a*) by the incidence of histologically identifiable carcinoma in autopsy material and (*b*) by reported deaths from cancer of the prostate. The incidence of the first was much higher and the rise with age was 1:3 on a log:log basis while deaths showed the characteristic steep 1:7 slope. Either the histologically diagnosed appearances are not carcinoma or there is some process, presumably immunological, which gradually becomes less effective with age.

Nodules are very common if searched for in routine thyroid material. In a series of 1,000 necropsies (Editorial, *Lancet*, 1955) about 50 per cent of thyroids showed one or more nodules. Of these, 21 per cent showed histological signs of malignancy. This is enormously higher than the standard death rate from thyroid cancer of around 6 per million per annum.

*Diminished resistance to tumour homografts in cancer patients*
The decreased immunological efficiency with age, at least as far as 'new' antigens are concerned, has been discussed in earlier chapters. This has raised the thought that clinical cancer is in some way the result of a lapse in immunological effectiveness. Southam *et al.* (1957) initiated studies of the inoculability of cultured cancer cells in normal volunteers and in cancer patients. The cell lines in the early investigations were not autologous and included both cancer cell lines and adapted lines derived from normal human cells. The normal individuals showed only a few examples of cell proliferation (four out of fifteen) which regressed by 14 days, while cancer patients showed twenty-one out of twenty-three takes from cancer cell lines and six out of twelve from 'normal' cell cultures.

Subsequent work by the same group showed that this difference was consistent and that acceptance of cancer cell homografts did not occur in debilitated elderly patients without cancer (Levin *et al.* 1964). Most cancer patients could be shown to develop isoantibodies to homografts of tumour (Itoh and Southam, 1964) and in recent work (Southam *et al.* 1966) tests were made using autologous cancer cell cultures to determine whether the patients' leucocytes were effective in inhibiting takes. About half the experiments showed an inhibitory effect but there was no evidence that those with effective leucocytes fared clinically any better than those with ineffective cells.

The overall result from these studies is to substantiate the hypothesis that immune factors may be important in preventing neoplastic disease but they also indicate, as might be expected, the complexity of the situation and give no indication as to the nature of the antigens concerned or the mechanism of control.

## AUTOIMMUNE PHENOMENA IN RELATION TO NEOPLASTIC DISEASE

Malignant disease is very common and occurs characteristically in older people so that purely coincidental association with a wide range of other disease conditions could be expected. There are on record, however, a considerable range of usually rather rare conditions associated with cancer and involving organs or systems not invaded by neoplastic cells. A number of these conditions appear to be autoimmune in character while others show systemic manifestations suggesting hormone overproduction. The following information on the latter is drawn mainly from the reviews by Greenberg *et al.* (1964), Lebovitz (1965) and Watson (1966).

Hypercalcaemia is a common finding in cancer, particularly when bone metastases are occurring, and can be expected at some time in the disease in 10–20 per cent of cancer patients. In addition to patients with bone metastases there are others with hypercalcaemia but without evidence of neoplastic invasion of bone. In several of these it is on record that removal of the tumour rectified the hypercalcaemia and, in one, that recurrence of the tumour was associated with a renewed rise in blood calcium (Plimpton and Gellhorn, 1956). Most of the tumours with these effects are of lung and kidney. There can be no doubt that a humoral agent of some sort is being liberated by the tumour and there is an increasing tendency to implicate the parathyroid glands (Tashjian *et al.* 1964). There seem to be two possibilities; either that parathyroid hormone itself is liberated or that in some way the parathyroid glands are stimulated to overaction. From our present point of view, one of the important aspects is that it is only a small minority of tumours which show this 'parathyroid' type of hypercalcaemia or any of the other hormonal anomalies to be mentioned.

In discussing the pathogenesis of the condition, Watson (1966) considers it is possible that more than one mechanism is at work. The possibilities are (a) that parathyroid hormone is actually produced by the tumour, (b) that an unrelated agent of similar pharmacological action is produced by the tumour and (c) that parathyroid hormone produces the immediate effect, the stimulus to the parathyroids coming from the tumour. The third is almost completely excluded in *most* cases by Rothschild *et al.*'s (1964) démonstration that complete parathyroidectomy in two typical cases had no effect in lowering the blood Ca level, and most writers have found no abnormality in the parathyroids *post mortem*. The possibility of some cases having this pathogenesis is based on the occasional recognition of enlarged hyperactive parathyroids and the fact that in some individuals corticosteroid treatment will lower the Ca blood level. To foreshadow later discussion this could represent a stimulation by a long-acting thyroid stimulator-like auto-antibody against a parathyroid tissue antigen being liberated by the tumour.

Bronchial carcinoma is a common cause of the hypercalcaemic syndrome and it is of great interest how often other anomalous systemic conditions are associated with these tumours. Without attempting a critical analysis of the reports, the following examples of hormonal anomaly directly associated with tumour (usually bronchial carcinoma) may be cited from Greenberg *et al.*'s review: Cushing's syndrome (presumably due to ACTH-like action), gynaecomastia; evidence for antidiuretic hormone (Amatruda *et al.* 1963); hypoglycaemia with insulin-like activity in tumour extracts (Silverstein *et al.* 1964).

So far we have been concerned only with direct or indirect hormonal manifestations but I am more closely concerned with the well known fact that distant effects of cancer may occur which are not hormonal in character and which have many of the signs of autoimmune processes. The best known examples are the auto-immune-type haemolytic anaemias and thrombocytopenias, to be seen in the later stages of leukaemia and lymphosarcomatous conditions with considerable regularity. In Dacie's experience from 1947 to 1961 there were 59 cases of secondary autoimmune haemolytic anaemia as against 108 idiopathic. Of these secondary cases, 24 were associated with neoplastic disease of the lymphoid system (Dacie, 1962). Lewis *et al.* (1966) have suggested that in fact most terminal cases of lympho-proliferative disorders will show some evidence of haemolytic anaemia or thrombocytopenia and they find that the onset of either condition tends to appear within a week or two of the initiation of radiation therapy or the use of radiomimetic drugs.

In trying to interpret this series of phenomena along the general lines we have adopted, a convenient starting-point is Metcalf's (1956*b*) observation that a factor analogous to the lymphocytosis-stimulating factor which he obtained from mouse thymus extracts was also present in the blood of patients with lymphatic leukaemia. We deduce from this that in a variety of neoplastic lymphoid tissues certain features of the internal environment of the thymus must be present without the full physiological and structural co-ordination of function characteristic of the normal thymus of the infant and child.

The most likely hypothesis is that the clones of autoimmune cells do not arise from the neoplastic clones of lymphoid cells but, as in other circumstances, from bone-marrow stem cells carrying genetic determinants for immune patterns which can react with body components. Such cells, when they lodge in neoplastic tissue of thymus-like quality, will, by hypothesis, undergo differentiation to immunocytes and, because of the unorganized quality of the neoplastic tissue will have an abnormally large chance of escaping to the general circulation. In terms of our general discussion, such cells will be relatively normal in other characteristics and hence susceptible to secondary controls to a sufficient extent to prevent pathological manifestations. When, however, cell lines carrying immune patterns reactive with the most active and accessible auto-antigens, namely red cells and platelets, are present in relatively large numbers the position must be a very unstable one. Any further sequential mutation along the path to full pathogenicity will be liable to initiate rapid proliferation with the corresponding symptoms. This therefore provides an appropriate basis for the effect of irradiation and radiomimetic drugs in triggering autoimmune disease.

Further study is needed as to why red cells and platelets are the preferred auto-antigens. It may in part be due simply to the fact that severe anaemia or a purpuric rash are striking signs calling at once for diagnosis. Minor, or even severe, autoimmune damage of other types might readily escape notice in an already severely ill patient.

The next immunologically relevant systemic effect of certain malignancies is the finding of Osserman and Takatsuki (1963) in regard to hypergammaglobulinaemia associated with prostatic and lower bowel malignancies. They were primarily interested in the fact that routine blood examination of hospital patients gave a number showing a high $\gamma$-globulin level with the characteristic sharp peak on electrophoresis which is conventionally regarded as indicating a monoclonal gammopathy. When they confined themselves to patients with such findings in the serum but with no signs, radiological or clinical, of multiple myelomatosis in its normal form, they found that among these patients there was a much higher incidence of lower bowel carcinoma and prostatic carcinoma than would be found in a similar group of hospital patients of similar age but with other reasons for primary entry into the hospital. It was quite striking that none of the commoner neoplasms such as bronchial carcinoma or gastric carcinoma were included in this series. Once again there is no accepted interpretation, but in view of the earlier discussion of a gut-associated hormone involved in plasma cell maturation it seems appropriate to develop that hypothesis in the present context.

Lymphocytes and plasma cells are extremely common in sections of prostate and lower bowel; they were recorded as numerous in the stroma of twelve out of fourteen tumours in this series, and one wonders whether the existence of a high concentration of bursal hormone in such a situation may have been responsible for the stimulation of a mutant cell to take on a rather innocent type of proliferation and $\gamma$-globulin production. From this, one could easily imagine a further sequential mutation giving rise to a clone capable of producing the full picture of multiple myelomatosis.

The final group of distant effects of cancer are best exemplified in the recent book and series of papers from London by Lord Brain and his associates (Brain and Norris, 1965). Bronchial carcinoma is not uncommonly associated with one or other of a considerable range of neuromyopathies. The relatively malignant form of bronchial carcinoma, histologically designated oat-celled, is particularly prone to cause either peripheral sensory- or encephalitic-type conditions (Davan *et al.* 1965). Cerebellar lesions may also be found with lung cancer but are rare and may also be associated with other types of cancer. The symptomatology and histopathology of these cerebellar degenerations is strikingly similar to the New Guinean disease kuru (Brain and Wilkinson, 1965). An indication that some (or all) of these conditions may be of autoimmune character is obtainable from the work of Wilkinson and Zeromski (1965). They found in a small group of patients with sensory-type neuropathy that serum reacted specifically (as judged by immunofluorescence) with cytoplasmic granules present in neurones from all parts of the central nervous system but absent in other parts of the body.

The association of dermatomyositis with cancer may have a similar significance. According to Williams (1959) about 15 per cent of cases are associated with cancer; often the tumour is at an early stage and quite small. The extent and distribution of the muscle and skin lesions are highly variable and have no relationship to the site or histological type of the associated tumour.

One final instance may be mentioned as a result of a personal communication from Kunkel (1966). He told me of a patient with severe rheumatoid arthritis with unusually pronounced serological reactions who, in the course of routine hospital examination, was shown to have an unsuspected bronchial carcinoma. On its removal the symptoms of arthritis and the serological reactivity rapidly disappeared. An abstract of this case has since been published (Litwin *et al.* 1966).

*Production of abnormal proteins by tumour cells*

This heterogeneous collection of systemic effects associated with neoplastic disease is, I believe, susceptible of interpretation in terms of a fairly uniform process. A tumour cell, like any other somatic cell, has potential (or genetic information) to produce any protein or antigen which can be synthesized in any tissue of the organism in which the tumour has developed. De-repression of a 'wrong' region of the genome resulting in the synthesis and liberation of an abnormal protein may occur in two different ways. It may result from a simple somatic mutation or may be a manifestation of increasing disturbance of the functioning of the genome as a cell line 'progresses' in malignancy and becomes aneuploid. As is always the case with somatic mutation, no demonstrable effect is produced unless the mutant cell can proliferate extensively. If the genetic change leading to production of an aberrant protein takes place at an early enough stage of the development of a tumour to allow extensive proliferation of descendants, significant amounts of the material will be produced. There is another alternative; that a metabolic change in a single cell is fortuitously magnified when for related or unrelated reasons the cell becomes neoplastic. It may be legitimate to summarize the situation by saying that somatic mutation giving rise to many sorts of metabolic anomalies is very common but to give evidence of an effect the mutant must be involved in a neoplastic process.

There are three major consequences of such liberation of aberrant antigens. The antigen, being specific for the tumour, may, if the tumour is still in a sufficiently vulnerable condition, provoke a proliferation of immunocytes adequate to eliminate the neoplastic tissue. If this fails to occur, the aberrant protein will go on being liberated and, if it is a pharmacologically potent substance like parathyroid hormone, will produce a clinical effect. Most proteins produced in excess by cancer cells will have no recognizable effect, though one might guess that with more knowledge of intertissue controls there will be an increase in the number recognizable. Apart from such direct effects, any protein carrying ADs toward which effective tolerance has not been developed will be liable to act as an antigen. This holds particularly for the 'inaccessible antigens' characteristic of all highly differentiated tissues such as the nervous system.

The commonest interpretation of carcinomatous neuromyopathy is that a virus, presumably not normally neuropathogenic, either gains access through the tumour or finds it possible to multiply within the CNS because of some remote metabolic effect of the tumour. Immunological processes have also been suggested, but only in general terms.

The following hypothesis is proposed to account more satisfactorily for the features (*a*) relation to a specific histological type, (*b*) limitation to a small proportion of those with the tumour, (*c*) wide variety of localization within the CNS and (*d*) presence of tissue-specific antibody in a proportion of cases.

It is proposed that the symptoms and lesions in the CNS represent an autoimmune attack by immunocytes (and/or antibody) directed toward ADs specific for CNS cells and 'inaccessible' in the sense of not being capable of producing normal tolerance. Different ADs are probably responsible for the anatomical distribution of lesions.

Proliferation of the pathogenic immunocytes is ascribed to the liberation of antigens with the same determinants by malignant cells. Proliferation (and antibody production if it occurs) presumably takes place in regional lymph nodes draining the tumour site. The circumstances which allow active symptomatic attack on the target tissues are unknown. As is the case in other forms of autoimmune disease the presence of antibody in serum is a useful index that autoimmune processes are active, but not that the antibody is itself the pathogenic agent.

Such an approach would be as applicable to dermatomyositis and to the case of rheumatoid disease mentioned earlier as to the group of neurological degenerations. It is essentially speculative but has both a solid background of observation and of analogies to experimental conditions and calls for specific investigational approaches. In the experimental field the most important example of this type of behaviour is the appearance of TL antigen in murine leukaemias not as yet associated with virus infection (Old *et al.* 1963). The TL antigen is found in thymus cells from a group of mouse strains and is absent in other strains which can therefore form antibody against the TL antigen. The most striking finding is that in leukaemias arising in TL– mice, the antigen appears in the circulating leukaemic cells. The evidence is that TL– mice carry the information for production of TL antigen but this is normally permanently repressed. In the development of leukaemia there is an abnormal de-repression leading to synthesis of TL antigen by all descendants of the affected clones. Basically we have the same situation in which a repressed cistron is de-repressed at some stage during the development of the leukaemic state and with proliferation of the neoplastic cells the normally hidden antigen is manifested.

At the level of clinical investigation there should be important opportunities for immunofluorescent work making use both of patient's serum and of sera from other sources specific for the type of antigen which, from the clinical picture, is being produced anomalously in the tumour. Study by standard methods for the presence of antigen in the tumour and of both antigen and antibody-producing cells in the draining lymph nodes could hardly fail to be illuminating.

# References

ADA, G. L. and LANG, P. G. (1966). Antigen in tissues. II. State of antigen in lymph node of rats given isotopically-labelled flagellin, haemocyanin or serum albumin. *Immunology*, **10**, 431.

ADA, G. L., NOSSAL, G. J. V. and AUSTIN, C. M. (1964a). Antigens in immunity. V. The ability of cells in lymphoid follicles to recognize foreignness. *Aust. J. exp. Biol. med. Sci.* **42**, 331.

ADA, G. L., NOSSAL, G. J. V. and PYE, J. (1964b). Antigens in immunity. III. Distribution of iodinated antigens following injection into rats *via* the hind footpads. *Aust. J. exp. Biol. med. Sci.* **42**, 295.

ADA, G. L., NOSSAL, G. J. V. and PYE, J. (1965). Antigens in immunity. XI. The uptake of antigen in animals previously rendered immunologically tolerant. *Aust. J. exp. Biol. med. Sci.* **43**, 337.

ADA, G. L., PARISH, C. R., NOSSAL, G. J .V. and ABBOT, A. (1968). The tissue localization, immunogenic, and tolerance-inducing properties of antigens and antigen-fragments. *Cold Spring Harb. Symp. quant. Biol.* **32**, 381.

ADINOLFI, M. (1965). Anti-I antibody in normal human newborn infants. *Immunology*, **9**, 43.

ADINOLFI, M., GLYNN, A. A., LINDSAY, M. and MILNE, C. M. (1966). Serological properties of γA antibodies to *Escherichia coli* present in human colostrum. *Immunology*, **10**, 517.

ADLER, F. L., FISHMAN, M. and DRAY, S. (1966). Antibody formation initiated *in vitro*. III. Antibody formation and allotypic specificity directed by ribonucleic acid from peritoneal exudate cells. *J. Immun.* **97**, 554.

AINBENDER, E., BERGER, R., HEVIZY, M. M., ZEPP, H. D. and HODES, H. L. (1966). Sex difference in serum immunoglobulin A (Ig A) polio-antibody. *Abstracts of the American Pediatric Society, 66th annual Meeting*, p. 15.

AISENBERG, A. C. and WILKES, B. (1967). Immunological tolerance induced by *cyclo*phosphamide assayed by plaque spleen cell method. *Nature, Lond.* **213**, 498.

ALBRIGHT, J. F. and MAKINODAN, T. (1965). Dynamics of expression of competence of antibody-producing cells, in *Molecular and Cellular Basis of Antibody Formation*, p. 427. Ed. J. Šterzl. Prague: Czecho-slovak Acad. Sci.

ALEXANDER, J. (1932). Some intracellular aspects of life and disease. *Protoplasm*, **14**, 296.

ALLEN, J. C., KUNKEL, H. G. and KABAT, E. A. (1964). Studies on human antibodies. II. Distribution of genetic factors. *J. exp. Med.* **119**, 453.

ALMEIDA, J., CINADER, B. and HOWATSON, A. (1963). The structure of antigen–antibody complexes: a study by electron microscopy. *J. exp. Med.* **118**, 327.

AMANO, S. and MARUYAMA, K. (1963). Electron microscopic studies on germinal centre cells of the lymph node. *Rep. Inst. Virus Res. Kyoto*, **6**, 157.

AMATRUDA, T. T., MULROW, P. J., GALLAGHER, J. C. and SAWYER, W. H. (1963). Carcinoma of the lung with inappropriate antidiuresis: demonstration of antidiuretic-hormone-like activity in tumor extract. *New Engl. J. Med.* **269**, 544.

AMBROSE, C. T. (1964). The requirements for hydrocortisone in antibody-forming tissue cultivated in serum-free medium. *J. exp. Med.* **119**, 1027.

AMBRUS, C. M. and AMBRUS, J. L. (1959). Regulation of the leucocyte level. *Ann. N.Y. Acad. Sci.* **77**, 445.

ANDERSEN, B. R., ABELE, D. C. and VANNIER, W. E. (1966). Effects of mild periodate oxidation on antibodies. *J. Immun.* **97**, 913.

ANFINSEN, C. B. and HABER, E. (1961). Studies on the reduction and reformation of protein disulphide bonds. *J. biol. Chem.* **236**, 1361.

ANFINSEN, C. B., HABER, E., SELA, M. and WHITE, F. H. (1961). The kinetics of formation of native ribonuclease during oxidation of the reduced polypeptide chain. *Proc. natn. Acad. Sci. U.S.A.* **47**, 1309.

ANGEVINE, C. D., ANDERSEN, B. R. and BARNETT, E. V. (1966). A cold agglutinin of Ig A class. *J. Immun.* **96**, 578.

ARCHER, G. T. (1963). Isolation of granules from eosinophil leucocytes and study of their enzyme content. *J. exp. Med.* **118**, 277.

ARCHER, G. T. (1966). The function of the eosinophil. *XIth Congr. Internat. Soc. Blood Transf., Sydney, Plenary Sessions*, p. 61.

ARCHER, G. T. and BLACKWOOD, A. (1965). Formation of Charcot–Leyden crystals in human eosinophils and study of the composition of isolated crystals. *J. exp. Med.* **122**, 173.

ARCHER, G. T. and HIRSCH, J. G. (1963). Motion picture studies on degranulation of horse eosinophils during phagocytosis. *J. exp. Med.* **118**, 287.

ARCHER, R. K. (1959). Eosinophil leucocytes and their reactions to histamine and 5-hydroxytryptamine. *J. Path. Bact.* **78**, 95.

ARGYRIS, B. J. (1963). Adoptive tolerance; transfer of the tolerant state. *J. Immun.* **90**, 29.

ARMITAGE, P. and DOLL, R. (1957). A two-stage theory of carcinogenesis in relation to the age distribution of human cancer. *Br. J. Cancer*, **11**, 161.

ASHERSON, G. L. (1964). Experimental production of autoantibody to gut antigens. *Proc. R. Soc. Med.* **57**, 813.

ASHERSON, G. L. and DUMONDE, D. C. (1963). Autoantibody production in rabbits. II. Organ-specific autoantibody in rabbits injected with rat tissues. *Immunology*, **6**, 19.

ASHLEY, D. J. B. (1965). On the incidence of carcinoma of the prostate. *J. Path. Bact.* **90**, 217.

ASKONAS, B. A. and HUMPHREY, J. H. (1958). Formation of antibody by isolated perfused lungs of immunized rabbits. The use of [$^{14}$C] amino acids to study the dynamics of antibody secretion. *Biochem. J.* **70**, 212.

## References

ASKONAS, B. A. and RHODES, J. M. (1965). Is antigen associated with macrophage RNA? In *Molecular and Cellular Basis of Antibody Formation*, p. 503. Ed. J. Šterzl. Prague: Czechoslovak Acad. Sci.

ASKONAS, B. A. and WILLIAMSON, A. R. (1966). Biosynthesis of immunoglobulins on polyribosomes and assembly of the Ig G molecule. *Proc. R. Soc.* B, **166**, 232.

ASPINALL, R. L. and MEYER, R. K. (1964). Effect of steroidal and surgical bursectomy and surgical thymectomy on the skin homograft reaction in chickens, in *The Thymus in Immunobiology*, p. 376. Ed. R. A. Good and A. E. Gabrielsen. New York: Hoeber.

ATTARDI, G., COHN, M., HORIBATA, K. and LENNOX, E. S. (1959). On the analysis of antibody synthesis at the cellular level. *Bact. Rev.* **23**, 213.

ATTARDI, G., COHN, M., HORIBATA, K. and LENNOX, E. S. (1964). Antibody formation by rabbit lymph node cells. I. Single cell responses to several antigens. *J. Immun.* **92**, 335.

ATWOOD, K. C. (1958). The presence of $A_2$ erythrocytes in $A_1$ blood. *Proc. natn. Acad. Sci. U.S.A.* **44**, 1054.

AUERBACH, R. (1960). Morphogenetic interactions in the development of the mouse thymus gland. *Devl. Biol.* **2**, 271.

AUSTEN, K. F. and BEER, F. (1964). The measurement of the second component of human ($C'2^{hu}$) by its interaction with $EAC'1a^{gp}4^{gp}$ cells. *J. Immun.* **92**, 946.

AZAR, H. A., NAUJOKS, G. and WILLIAMS, J. (1963). Role of the adult thymus in immune reactions. I. Observations on lymphoid organs, circulating lymphocytes and serum protein fractions of thymectomized or splenectomized adult mice. *Am. J. Path.* **43**, 213.

BACH, F. and HIRSCHHORN, K. (1964). Lymphocyte interactions: a potential histocompatibility test *in vitro*. *Science, N.Y.* **143**, 813.

BAGLIONI, C., ZONTA, L. A., CIOLI, D. and CARBONARA, A. (1966). Allelic antigenic factor Inv(a) of the light chains of human immunoglobulins: chemical basis. *Science, N.Y.* **152**, 1517.

BAILEY, D. W. and KOHN, H. I. (1965). Inherited histocompatibility changes in progeny of irradiated and unirradiated inbred mice. *Genet. Res.* **6**, 330.

BALFOUR, B. M., COOPER, E. H. and MEEK, E. S. (1967). Lymph node plasma cell production in secondary immune response, in *Germinal Centers in Immune Responses*, p. 126. Ed. H. Cottier, N. Odartchenko, R. Schindler and C. C. Congdon. Berlin: Springer.

BALL, W. D. and AUERBACH, R. (1960). *In vitro* formation of lymphocytes from embryonic thymus. *Expl Cell Res.* **20**, 245.

BANGHAM, D. R. (1961). The transmission of homologous serum proteins to the foetus and to the amniotic fluid in the rhesus monkey. *J. Physiol., Lond.* **153**, 265.

BARNES, D. H. W., FORD, C. E. and LOUTIT, J. W. (1964). Haemopoietic stem-cells. *Lancet*, i, 1395.

BARNETT, E. V., BAKEMEIER, R. F., LEDDY, J. P. and VAUGHAN, J. H.

(1965). Heterogeneity of antinuclear factors in lupus erythematosus and rheumatoid arthritis. *Proc. Soc. exp. Biol. Med.* **118**, 803.

BARNETT, E. V., DUMONDE, D. C. and GLYNN, L. E. (1963). Induction of autoimmunity to adrenal gland. *Immunology*, **6**, 382.

BARTH, W. F., MCLAUGHLIN, C. L. and FAHEY, J. L. (1965). The immunoglobulins of mice. VI. Response to immunization. *J. Immun.* **95**, 781.

BASSETT, E. W., TANENBAUM, S. W., PRYZWANSKY, K., BEISER, S. M. and KABAT, E. A. (1965). Studies on human antibodies. III. Amino acid composition of four antibodies from the same individual. *J. exp. Med.* **122**, 251.

BAUER, D. C. and STAVITSKY, A. B. (1961). On the different molecular forms of antibody synthesized by rabbits during the early response to a single injection of protein and cellular antigens. *Proc. natn. Acad. Sci. U.S.A.* **47**, 1667.

BEAUVIEUX, Y.-J. (1963). Données expérimentales sur l'un des aspects de la physiologie du thymus. Le thymocyte. Existence d'une thymocytose: signification fonctionelle. *C.r. hebd. Séanc. Acad. Sci., Paris*, **256**, 2914.

BECK, J. S. (1961). Variations in the morphological patterns of 'autoimmune' nuclear fluorescence. *Lancet*, i, 1203.

BECK, J. S., ANDERSON, J. R., MCELHINNEY, A. J. and ROWELL, N. R. (1962). Antinucleolar antibodies. *Lancet*, ii, 575.

BECK, J. S. and ROWELL, N. R. (1963). Transplacental passage of antinuclear antibody. *Lancet*, i, 134.

BECKER, A. J., MCCULLOCH, E. A. and TILL, J. E. (1963). Cytological demonstration of the clonal nature of spleen colonies derived from transplanted bone-marrow cells. *Nature, Lond.* **197**, 452.

BECKER, E. L. (1965). Small molecular weight inhibitors of complement action, in *CIBA Foundation Symposium on Complement*, p. 58. Ed. G. E. W. Wolstenholme and J. Knight. London: Churchill.

BECKWITH, J. B. and PERRIN, E. V. (1963). *In situ* neuroblastomas: a contribution to the natural history of neural crest tumors. *Am. J. Path.* **43**, 1089.

BENACERRAF, B. and LEVINE, B. B. (1962). Immunological specificity of delayed and immediate hypersensitivity reactions. *J. exp. Med.* **115**, 1023.

BENACERRAF, B. and MCCLUSKEY, R. T. (1963). Methods of immunologic injury to tissues. *A. Rev. Microbiol.* **17**, 263.

BENACERRAF, B., OJEDA, A. and MAURER, P. H. (1963a). Studies on artificial antigens. II. The antigenicity in guinea-pigs of arsanilic acid conjugates of copolymers of D or L amino acids. *J. exp. Med.* **118**, 945.

BENACERRAF, B., OVARY, Z., BLOCH, K. L. and FRANKLIN, E. C. (1963b). Properties of guinea-pig 7S antibodies. I. Electrophoretic separation of two types of guinea-pig 7S antibodies. *J. exp. Med.* **117**, 937.

BEN-EFRAIM, S., FUCHS, S. and SELA, M. (1963). Hypersensitivity to a synthetic polypeptide: induction of a delayed reaction. *Science, N.Y.* **139**, 1222.

BEN-EFRAIM, S. and MAURER, P. H. (1965). Antigenicity of synthetic polypeptides in inbred and random bred strains of guinea-pigs. *Fedn. Proc. Fedn. Am. Socs exp. Biol.* **24**, 181.

BENNETT, J. C., HOOD, L., DREYER, W. J. and POTTER, M. (1965). Evidence for amino acid sequence differences amongst proteins resembling the L chain subunits of immunoglobulin. *J. molec. Biol.* **12**, 81.

BERNHARD, W. and GRANBOULAN, N. (1960). Ultrastructure of immunologically competent cells, in *CIBA Foundation Symposium on the Cellular Aspects of Immunity*, p. 92. Ed. G. E. W. Wolstenholme and M. O'Connor. London: Churchill.

BERNHEIMER, A. W., CASPARI, E. and KAYSER, A. D. (1952). Studies on antibody formation in caterpillars. *J. exp. Zool.* **119**, 23.

BERNIER, G. M. and CEBRA, J. J. (1965). Frequency distribution of $\alpha$, $\gamma$, $\kappa$, and $\lambda$ polypeptide chains in human lymphoid tissues. *J. Immun.* **95**, 246.

BERNIER, G. M. and PUTNAM, F. W. (1964). Myeloma proteins and macroglobulins: hallmarks of disease and models of antibodies. *Prog. Hemat.* **4**, 160.

BILLINGHAM, R. E. (1964). Transplantation immunity and the maternal-fetal relation. *New Engl. J. Med.* **270**, 667.

BILLINGHAM, R. E., BRENT, L. and MEDAWAR, P. B. (1953). 'Actively acquired tolerance' of foreign cells. *Nature, Lond.* **172**, 603.

BILLINGHAM, R. E., BRENT, MEDAWAR, P. B. and SPARROW, E. M. (1954). Quantitative studies on tissue transplantation immunity. I. The survival times of skin homografts exchanged between members of different inbred strains of mice. *Proc. R. Soc.* B, **143**, 43.

BILLINGHAM, R. E., SAWCHUCK, G. H. and SILVERS, W. K. (1960). Studies on the histocompatibility genes of the Syrian hamster. *Proc. natn. Acad. Sci. U.S.A.* **46**, 1079.

BILLINGHAM, R. E. and SILVERS, W. K. (1962). Some factors that determine the ability of cellular inocula to induce tolerance to tissue homografts. *J. cell. comp. Physiol.* **60**, 183.

BILLINGHAM, R. E. and SILVERS, W. K. (1964). Syrian hamsters and transplantation immunity. *Plastic reconstr. Surg.* **34**, 329.

BILLINGHAM, R. E., SILVERS, W. K. and WILSON, D. B. (1965). A second study on the *H-Y* transplantation antigen in mice. *Proc. R. Soc.* B, **163**, 61.

BINAGHI, R. A. (1966). Production of 7S immunoglobulins in immunized guinea pigs. *J. Immun.* **97**, 159.

BIOZZI, G., STIFFEL, C., MOUTON, D., LIACOPOULOS, M., DECREUSE-FOND, C. and BOUTHILLIER, Y. (1966). Etude du phénomène de l'immuno-cyto-adhérence au cours de l'immunisation. *Annls Inst. Pasteur, Paris*, **110**, Suppl. 1.

BISHOP, D. W. and GUMP, D. (1961). Production of circulating antibody against defined antigen (BGG) by neonatal guinea pigs. *Proc. Soc. exp. Biol. Med.* **106**, 24.

BLACK, F. L. and ROSEN, L. (1962). Patterns of measles antibodies in residents of Tahiti and their stability in the absence of exposure. *J. Immun.* **88**, 725.

BLAKEMORE, F. and GARNER, R. J. (1956). The maternal transfer of antibodies in the bovine. *J. comp. Path.* 66, 287.

BLOOM, B. R. and BENNETT, B. (1966). Mechanism of a reaction *in vitro* associated with delayed hypersensitivity. *Science, N.Y.* 153, 81.

BLUMER, H., RICHTER, M., CUA-LIM, F. and ROSE, B. (1962). Precipitating and nonprecipitating antibodies in the primary and secondary immune responses: rate of decline, anaphylaxis-sensitizing capacity, and the effect of cortisone. *J. Immun.* 88, 669.

BORSOS, T., DOURMASHKIN, R. R. and HUMPHREY, J. H. (1964). Lesions in erythrocyte membranes caused by immune haemolysis. *Nature, Lond.* 202, 251.

BOUGHTON, B. and SPECTOR, W. G. (1963). 'Autoimmune' testicular lesions induced by injury to the contralateral testis and intradermal injection of adjuvant. *J. Path. Bact.* 86, 69.

BOYDEN, S. V. (1963). Cellular recognition of foreign matter. *Int. Rev. exp. Path.* 2, 311.

BOYDEN, S. V. (1964). Cytophilic antibody in guinea pigs with delayed-type hypersensitivity. *Immunology*, 7, 474.

BOYDEN, S. V. (1966). Natural antibodies and the immune response. *Adv. Immun.* 5, 1.

BOYDEN, S. V., NORTH, R. J. and FAULKNER, S. M. (1965). Complement and the activity of phagocytes, in *CIBA Foundation Symposium on Complement*, p. 190. Ed. G. E. W. Wolstenholme and J. Knight. London: Churchill.

BOYDEN, S. V., SORKIN, E. and SPÄRCK, J. V. (1960). Observations on the antibodies associated with spleen cells at different stages of immunization, in *Mechanisms of Antibody Formation*, p. 237. Ed. M. Holub and L. Jarošková. Prague: Czechoslovak Acad. Sci.

BOYER, G. (1960). Chorioallantoic membrane lesions produced by inoculation of adult fowl leucocytes. *Nature, Lond.* 185, 327.

BOYSE, E. A., OLD, L. J. and LUELL, S. (1963). Antigenic properties of experimental leukemias. II. Immunological studies *in vivo* with C57BL/6 radiation-induced leukemias. *J. natn. Cancer Inst.* 31, 987.

BRAIN, LORD and NORRIS, F. H. (eds) (1965). *The Remote Effects of Cancer on the Nervous System*. New York: Grune and Stratton.

BRAIN, W. R. and WILKINSON, M. (1965). Subacute cerebellar degeneration associated with neoplasms. *Brain*, 88, 465.

BRAMBELL, F. W. R. (1958). The passive immunity of the young mammal. *Biol. Rev.* 33, 488.

BRAMBELL, F. W. R. (1966). The transmission of immunity from mother to young and the catabolism of immunoglobulins. *Lancet*, ii, 1087.

BREINL, F. and HAUROWITZ, F. (1930). Chemische Untersuchung des Präzipitates aus Hämoglobin und Anti-hämoglobin-Serum und Bemerkungen über die Natur der Antikörper. *Hoppe-Seyler's Z. physiol. Chem.* 192, 45.

BRENNER, S. and MILSTEIN, C. (1966). Source of antibody variation. *Nature, Lond.* 211, 242.

661

BRENT, L. and GOWLAND, G. (1962). Induction of tolerance of skin homografts in immunologically competent mice. *Nature, Lond.* **196**, 1298.

BRENT, L. and GOWLAND, G. (1963). On the mechanism of immunological tolerance, in *Conceptual Advances in Immunology and Oncology*, p. 355. New York: Hoeber.

BRENT, L. and MEDAWAR, P. B. (1963). Tissue transplantation—a new approach to the 'typing' problem. *Br. med. J.* ii, 269.

BRENT, L. and MEDAWAR, P. B. (1964). Nature of the normal lymphocyte transfer reaction. *Nature, Lond.* **204**, 90.

BRENT, L. and MEDAWAR, P. B. (1966). Quantitative studies on tissue transplantation immunity. VII. The normal lymphocyte transfer reaction. *Proc. R. Soc.* B, **165**, 281.

BREYERE, E. J. and BARRETT, M. K. (1960). Prolonged survival of skin homografts in parous female mice. *J. natn. Cancer Inst.* **25**, 1405.

BREYERE, E. J. and BARRETT, M. K. (1963). Tolerance induced by parity in mice incompatible at the *H-2* locus. *J. natn. Cancer Inst.* **27**, 409.

BRIDGES, R. A., CONDIE, R. M., ZAK, S. J. and GOOD, R. A. (1959). The morphological basis of antibody formation during the neonatal period *J. Lab. clin. Med.* **53**, 331.

BROBERGER, O. and PERLMANN, P. (1959). Auto-antibodies in human ulcerative colitis. *J. exp. Med.* **110**, 657.

BROBERGER, O. and PERLMANN, P. (1963). *In vitro* studies of ulcerative colitis. I. Reactions of patients' serum with human fetal colon cells in tissue cultures. II. Cytotoxic action of white blood cells from patients on human fetal colon cells. *J. exp. Med.* **117**, 705, 717.

BROOKE, M. S. (1964). Breaking of immunological paralysis by injection of a specific depolymerase. *Nature, Lond.* **204**, 1319.

BROOKE, M. S. (1966). Studies on the induction, specificity, prevention and breaking of immunologic paralysis and immunity to pneumococcal polysaccharide. *J. Immun.* **96**, 364.

BROOKE, M. S. and KARNOVSKY, M. J. (1961). Immunological paralysis and adoptive immunity. *J. Immun.* **87**, 205.

BRYANT, B. J. and KELLY, L. S. (1958). Autoradiographic studies of leucocyte function. *Proc. Soc. exp. Biol. Med.* **99**, 681.

BRYCE, L. M. and BURNET, F. M. (1932). Natural immunity to staphylococcal toxin. *J. Path. Bact.* **35**, 183.

BRYSON, V. and VOGEL, H. J. (eds) (1965). *Evolving Genes and Proteins*. New York and London: Academic Press.

BUCKLEY, C. E., WHITNEY, P. L. and TANFORD, C. (1963). The unfolding and renaturation of a specific univalent antibody fragment. *Proc. natn. Acad. Sci. U.S.A.* **50**, 827.

BURCH, P. R. J. (1963). Autoimmunity: some aetiological aspects: inflammatory polyarthritis and rheumatoid arthritis. *Lancet*, i, 1253.

BURCH, P. R. J. (1966*a*). Spontaneous auto-immunity: Equations for age-specific prevalence and initiation-rates. *J. theor. Biol.* **12**, 397.

References

BURCH, P. R. J. (1966b). Age and sex distributions for some idiopathic non-malignant conditions in man. Some possible implications for growth-control and natural and radiation-induced ageing, in *Radiation and Ageing*, p. 117. Ed. P. J. Lindop and G. A. Sacher. London: Taylor and Francis.

BURCH, P. R. J. and BURWELL, R. G. (1965). Self and not-self: a clonal induction approach to immunology. *Q. Rev. Biol.* **40**, 252.

BURKHOLDER, P. M. (1961). Complement fixation in diseased tissues. I. Fixation of guinea-pig complement in sections of kidney from humans with membranous glomerulonephritis and rats injected with anti-rat kidney serum. *J. exp. Med.* **114**, 605.

BURKITT, D. (1963). A lymphoma syndrome in tropical Africa. *Int. Rev. exp. Path.* **2**, 67.

BURKITT, D. (1966). African lymphoma: observations on response to vincristine sulphate therapy. *Cancer, N.Y.* **19**, 1131.

BURNET, F. M. (1941). *The Production of Antibodies.* Melbourne: Macmillan.

BURNET, F. M. (1950). The natural history of surgical infection. *Ann. R. Coll. Surg.* **7**, 191.

BURNET, F. M. (1952). The pattern of disease in childhood. *Australas. Ann. Med.* **1**, 93.

BURNET, F. M. (1956). Concluding remarks. *Proc. R. Soc.* B, **146**, 90.

BURNET, F. M. (1957a). A modification of Jerne's theory of antibody production using the concept of clonal selection. *Aust. J. Sci.* **20**, 67.

BURNET, F. M. (1957b). Cancer—a biological approach. *Br. med. J.* i, 779, 782, 841.

BURNET, F. M. (1959). *The Clonal Selection Theory of Acquired Immunity.* Cambridge Univ. Press.

BURNET, F. M. (1961). Cellular aspects of immunology as manifested in the Simonsen reaction. *Yale J. Biol. Med.* **34**, 207.

BURNET, F. M. (1962). The immunological significance of the thymus: an extension of the clonal selection theory of immunity. *Australas. Ann. Med.* **11**, 79.

BURNET, F. M. (1963a). Theories of immunity, in *Conceptual Advances in Immunology and Oncology*, p. 7. New York: Hoeber.

BURNET, F. M. (1963b). Die Rolle des Thymus für die Immunität. *Natur. Rdsch., Stuttg.* **9**, 335.

BURNET, F. M. (1964). Pathology of the thymus with special reference to autoimmune disease. *NW. Med., Seattle*, **63**, 519, 599.

BURNET, F. M. (1965a). Somatic mutation and chronic disease. *Br. med. J.* i, 338.

BURNET, F. M. (1965b). Mast cells in the thymus of NZB mice. *J. Path. Bact.* **89**, 271.

BURNET, F. M. (1966). A possible genetic basis for specific pattern in antibody. *Nature, Lond.* **210**, 1308.

BURNET, F. M. The newer immunology: an evolutionary approach, in *Infectious Agents and Host Resistance* (in press). Ed. S. Mudd. New York: Saunders.

BURNET, F. M. and BURNET, D. (1961). Analysis of major histocompatibility factors in a stock of closely inbred white leghorn fowls using a graft-versus-host reaction on the chorioallantoic membrane. *Aust. J. exp. Biol. med. Sci.* **39**, 101.

BURNET, F. M. and FENNER, F. (1949). *The Production of Antibodies,* 2nd ed. Melbourne: Macmillan.

BURNET, F. M. and HOLMES, M. C. (1964). Thymic changes in the mouse strain NZB in relation to the autoimmune state. *J. Path. Bact.* **88**, 229.

BURNET, F. M. and HOLMES, M. C. (1965). The natural history of the NZB/NZW F1 hybrid mouse: a laboratory model of systemic lupus erythematosus. *Australas. Ann. Med.* **14**, 185.

BURNET, F. M. and LIND, P. E. (1961). Immunological function of the bursa, in *Rep. Walter and Eliza Hall Inst. Res. Path. Med.* 1960–61, p. 22.

BURNET, F. M. and MACKAY, I. R. (1965). Histology of a thymus removed surgically from a patient with severe untreated systemic lupus erythematosus. *J. Path. Bact.* **89**, 263.

BUSSARD, E. and HANNOUN, C. (1965). Examen des immunocytes en culture *in vitro*: individualization fonctionnelle et identification morphologique. *C. r. hebd. Séanc. Acad. Sci., Paris,* **260**, 6486.

BUTLER, W. T. and COONS, A. H. (1964). Studies in antibody production. XII. Inhibition of priming by drugs. *J. exp. Med.* **120**, 1051.

CAHN, R. D. and CAHN, M. B. (1966). Heritability of cellular differentiation: clonal growth and expression of differentiation in retinal pigment cells *in vitro*. *Proc. natn. Acad. Sci. U.S.A.* **55**, 106.

CAMERON, G. R. (1932). Inflammation in earthworms. *J. Path. Bact.* **35**, 933.

CAMERON, G. R. and SPECTOR, W. G. (1961). *The Chemistry of the Injured Cell.* Springfield: Thomas.

CAMPBELL, D. H. (1962). The physical properties of antigen-antibody complexes, in *Mechanism of Cell and Tissue Damage Produced by Immune Reactions,* Int. Symp. Immunopath., no. 2, p. 67. Ed. P. Grabar and P. Miescher. Basel: Schwabe.

CARBONARA, A. O., RODHAIN, J. A. and HEREMANS, J. F. (1963). Localization of $\gamma_{1A}$-globulin ($\beta_{2A}$-globulin) in tissue cells. *Nature, Lond.* **198**, 999.

CARSTAIRS, K. (1961). Transformation of the small lymphocyte in culture. *Lancet,* ii, 984.

CARSTAIRS, K. C., BRECKENRIDGE, A., DOLLERY, C. T. and WORLLEDGE, S. M. (1966). Incidence of a positive direct Coombs test in patients on α-methyldopa. *Lancet,* ii, 133.

CASTLEMAN, B. (1960). The pathology of the thymus gland in myasthenia gravis, in *Thymectomy for Myasthenia Gravis,* p. 70. Ed. H. R. Viets and R. S. Schwab. Springfield: Thomas.

CATSOULIS, E. A., FRANKLIN, E. C., ORATZ, M. and ROTHSCHILD, M. A. (1964). Gamma globulin metabolism in rabbits during the anamnestic response. *J. exp. Med.* **119**, 615.

CEBRA, J. J., COLBERG, J. E. and DRAY, S. (1966). Rabbit lymphoid cells differentiated with respect to $\alpha$-, $\gamma$- and $\mu$-heavy polypeptide chains and to allotypic markers AA1 and AA2. *J. exp. Med.* **123**, 547.

CELADA, F. (1966). Quantitative studies of the adoptive immunological memory in mice. I. An age-dependent barrier to syngeneic transplantation. *J. exp. Med.* **124**, 1.

CELADA, F. (1967). Quantitative studies of the adoptive immunological memory in mice. II. Linear transmission of cellular memory. *J. exp. Med.* **125**, 199.

CHASE, M. W. (1946). Inhibition of experimental drug allergy by prior feeding of the sensitizing agent. *Proc. Soc. exp. Biol. Med.* **61**, 257.

CHASE, M. W. (1966). Immunologic tolerance to defined antigens and haptens: introductory remarks. *Fedn. Proc. Fedn. Am. Socs exp. Biol.* **25**, 145.

CHASE, M. W., BATTISTO, J. R. and RITTS, R. E. (1963). The acquisition of immunologic tolerance *via* simple allergenic chemicals, in *Conceptual Advances in Immunology and Oncology*, p. 395. New York: Hoeber.

CHIN, P. H. and SILVERMAN, M. S. (1960). Studies on the transfer of antibody formation by iso- and heterotransplants. *J. Immun.* **85**, 120.

CHODIRKER, W. B. and TOMASI, T. B. (1963). Gamma globulins: quantitative relations in human serum and non-vascular fluids. *Science, N.Y.* **142**, 1080.

CLAMAN, H. N. and TALMAGE, D. W. (1963). Thymectomy: prolongation of immunological tolerance in the adult mouse. *Science, N.Y.* **141**, 1193.

CLARK, S. L., Jr. (1963). The thymus in mice of strain 129/J studied with the electron microscope. *Am. J. Anat.* **112**, 1.

CLEGHORN, T. E. (1961). The occurrence of certain rare blood group factors in Britain. *M.D. Thesis, University of Sheffield*, quoted R. R. Race and R. Sanger (1962), p. 109.

COCK, A. G. and SIMONSEN, M. (1958). Immunological attack on newborn chickens by injected adult cells. *Immunology*, **1**, 103.

COE, J. E. and SALVIN, S. B. (1963). The specificity of allergic reactions. VI. Unresponsiveness to simple chemicals. *J. exp. Med.* **117**, 401.

COHEN, E. P., NEWCOMB, R. W. and CROSBY, L. K. (1965). Conversion of nonimmune spleen cells to antibody-forming cells by RNA: strain specificity of the response. *J. Immun.* **95**, 583.

COHEN, M. W., JACOBSON, E. B. and THORBECKE, G. J. (1966). Gammaglobulin and antibody formation *in vitro*. V. The secondary response made by splenic white and red pulp with reference to the role of secondary nodules. *J. Immun.* **96**, 944.

COHEN, M. W., THORBECKE, G. J., HOCHWALD, G. M. and JACOBSON, E. G. (1963). Induction of graft-versus-host reaction in newborn mice by injection of newborn or adult homologous thymus cells. *Proc. Soc. exp. Biol. Med.* **114**, 242.

COHEN, S. (1966). General structure and heterogeneity of immunoglobulins. *Proc. R. Soc. B*, **166**, 114.

COHEN, S. and MCGREGOR, I. A. (1963). Gamma globulin and acquired immunity to malaria, in *Immunity to Protozoa*, p. 123. Ed. P. C. C.

Garnham, A. E. Pierce and I. Roitt. Oxford: Blackwell Scientific Publications.

COHEN, S. and PORTER, R. R. (1964*a*). Structure and biological activity of immunoglobulins. *Adv. Immun.* **4**, 287.

COHEN, S. and PORTER, R. R. (1964*b*). Heterogeneity of the peptide chains of γ-globulin. *Biochem. J.* **90**, 278.

COHN, Z. A. and MORSE, S. I. (1960). Functional and metabolic properties of polymorphonuclear leucocytes. I. Observations on the requirements and consequences of particle ingestion. II. The influence of a lipopolysaccharide endotoxin. *J. exp. Med.* **111**, 667, 689.

COLE, L. J. and DAVIS, W. E. (1961). Specific homograft tolerance in lymphoid cells of long-lived radiation chimeras. *Proc. natn. Acad. Sci. U.S.A.* **47**, 594.

COLE, L. J. and GARVER, R. M. (1961) Homograft-reactive large mononuclear leucocytes in peripheral blood and peritoneal exudates. *Am. J. Physiol.* **200**, 147.

COMBINED STUDY. (1966). Prevention of Rh haemolytic disease: results of the clinical trial. A combined study from centres in England and Baltimore. *Br. med. J.* ii, 907.

CONGDON, C. C. and DUBA, D. B. (1961). Bone marrow heterografting: use of isologous thymus in lethally irradiated mice. *Archs Path.* **71**, 311.

CONGDON, C. C. and GOODMAN, J. W. (1962). Changes in lymphatic tissue during foreign tissue transplantation, in *Proceedings of International Symposium on Tissue Transplantation*, p. 181. Ed. A. P. Cristoffanini and G. Hoecker. Santiago: University of Chile.

CONGDON, C. C. and HANNA, M. G. (1967). Comparison of existing theories on the function of germinal centers, in *Germinal Centers in Immune Responses*, p. 1. Ed. H. Cottier, N. Odartchenko, R. Schindler and C. C. Congdon. Berlin: Springer.

COON, H. G. (1966). The retention of differentiated cell function amongst clonal and subclonal progeny of precartilage and cartilage cells from chick embryos. *J. Cell Biol.* **23**, 20A.

COONS, A. H. (1958). The cytology of antibody formation. *J. cell. comp. Physiol.* **52** (Suppl.), 55.

COONS, A. H. (1965). The nature of the secondary response, in *Molecular and Cellular Basis of Antibody Formation*, p. 559. Ed. J. Šterzl. Prague: Czechoslovak Acad. Sci.

COONS, A. H., CREECH, H. J. and JONES, R. N. (1941). Immunological properties of an antibody containing a fluorescent group. *Proc. Soc. exp. Biol. Med.* **47**, 200.

COONS, A. H., CREECH, H. J. and JONES, R. N. (1942). The demonstration of pneumococcus antigens in tissues by the use of fluorescent antibody. *J. Immun.* **45**, 159.

COONS, A. H., LEDUC, E. H. and CONNOLLY, J. M. (1955). Studies on antibody production. I. A method for the histochemical demonstration of specific antibody and its application to the study of the hyperimmune rabbit. *J. exp. Med.* **102**, 49.

COOPER, E. H., BARKHAN, P. and HALE, A. J. (1961). Mitogenic activity of phytohemagglutinin. *Lancet*, ii, 210.

COOPER, G. N. and PILLOW, J. A. (1959). Experimental shigellosis in mice. II. Immunological responses to *Shigella dysenteriae* type 2 infections. *Aust. J. exp. Biol. med. Sci.* **37**, 201.

COOPER, H. L. and RUBIN, A. D. (1966). Synthesis of non-ribosomal RNA by lymphocytes: a response to PHA treatment. *Science, N.Y.* **152**, 516.

COOPER, M. D., GABRIELSEN, A. E., PETERSON, R. D. A. and GOOD, R. A. (1967). Ontogenetic development of the germinal centers and their function: relationship to the bursa of Fabricius, in *Germinal Centers in Immune Responses*, p. 28. Ed. H. Cottier, N. Odartchenko, R. Schindler and C. C. Congdon. Berlin: Springer.

COOPER, M. D., PETERSON, R. D. A. and GOOD, R. A. (1965). Delineation of the thymic and bursal lymphoid systems in the chicken. *Nature, Lond.* **205**, 143.

COOPER, M. D., PETERSON, R. D. A., SOUTH, M. A. and GOOD, R. A. (1966). The functions of the thymus system and the bursa system in the chicken. *J. exp. Med.* **123**, 75.

COSTEA, N., YAKULIS, V. and HELLER, P. (1966). Light-chain heterogeneity of cold agglutinins. *Science, N.Y.* **152**, 1520.

COTTIER, H., KEISER, G., ODARTCHENKO, N., HESS, M. and STONER, R. D. (1967 b). *De novo* formation and rapid growth of germinal centers during secondary responses to tetanus toxoid in mice, in *Germinal Centers in Immune Responses*, p. 270. Ed. H. Cottier, N. Odartchenko, R. Schindler and C. C. Congdon. Berlin: Springer.

COTTIER, H., ODARTCHENKO, N., SCHINDLER, R. and CONGDON, C. C. (eds) (1967 a). *Germinal Centers in Immune Responses*. Berlin: Springer.

CRABBÉ, P. (1967). *Signification du tissu lymphoïde des muqueuses digestives*. Brussels: Arscia; Paris: Maloine.

CRADDOCK, C. G., NAKAI, G. S., FUKUTA, H. and VANSLAGER, L. M. (1964). Proliferative activity of the lymphatic tissues of rats as studied with tritium-labeled thymidine. *J. exp. Med.* **120**, 389.

CRONKITE, E. P. (1964). Enigmas underlying the study of hemopoietic cell proliferation. *Fedn. Proc. Fedn. Am. Socs exp. Biol.* **23**, 649.

CRONKITE, E. P., JANSEN, C. R., MATHER, G., NIELSEN, N., USENIK, E. A., ADAMIK, E. and SIPE, C. R. (1962). Studies on lymphocytes. I. Lymphopenia by prolonged extra-corporeal irradiation of circulating blood. *Blood*, **20**, 203.

CROSS, A. M., LEUCHARS, E. and MILLER, J. F. A. P. (1964). Studies on the recovery of the immune response in irradiated mice thymectomized in adult life. *J. exp. Med.* **119**, 837.

CRUCHAUD, A. and COONS, A. H. (1964). Studies in antibody production. XIII. The effect of chloramphenicol on priming in mice. *J. exp. Med.* **120**, 1061.

CUDKOWICZ, G., BENNETT, M. and SHEARER, G. M. (1964 a). Pluripotent stem cell function of the mouse marrow 'lymphocyte'. *Science, N.Y.* **144**, 866.

CUDKOWICZ, G., UPTON, A. C., SHEARER, G. M. and HUGHES, W. L. (1964b). Lymphocyte content and proliferative capacity of serially transplanted bone marrow. *Nature, Lond.* **201**, 165.

CURTAIN, C. C. (1966). Hypergammaglobulinaemia in New Guinea. *Papua New Guin. med. J.* **9**, 145.

CURTAIN, C. C. and BAUMGARTEN, A. (1966). Immunocytochemical localization of the immunoglobulin factors Gm(a), Gm(b) and Inv(a) in human lymphoid tissue. *Immunology*, **10**, 499.

CURTIS, H. J. (1963). Biological mechanisms underlying the aging process. *Science, N.Y.* **141**, 686.

DACIE, J. V. (1962). *The Haemolytic Anaemias: Congenital and Acquired*, part 2, 2nd ed., p. 343. London: Churchill.

DACIE, J. V., CROOKSTON, J. H. and CHRISTENSON, W. N. (1957). 'Incomplete' cold antibodies: role of complement in sensitization to antiglobulin serum by potentially haemolytic antibodies. *Br. J. Haemat.* **3**, 77.

DAMESHEK, W. and SCHWARTZ, S. O. (1938). Hemolysins as the cause of clinical and experimental hemolytic anemias with particular reference to the nature of spherocytosis and increased fragility. *Am. J. med. Sci.* **196**, 769.

DARNBOROUGH, J. (1958). A unique case of auto-immune erythrocyte sensitization. *Br. med. J.* ii, 1451.

DAUSSET, J. and RAPAPORT, F. (1966). Le rôle des antigènes de groupes sanguins en histocompatibilité humaine. *XIth Congr. Internat. Soc. Blood Transf., Sydney, Plenary Sessions*, p. 296.

DAVAN, A. D., CROFT, P. B. and WILKINSON, M. (1965). Association of carcinomatous neuromyopathy with different histological types of carcinoma of the lung. *Brain*, **88**, 435.

DAVID, J. R., AL-ASKARI, S., LAWRENCE, H. S. and THOMAS, L. (1964a). Studies of delayed hypersensitivity *in vitro*. I. The specificity of delayed hypersensitivity *in vitro*. I. The specificity of cell migration. *J. Immun.* **93**, 264.

DAVID, J. R., LAWRENCE, H. S. and THOMAS, L. (1964b). The *in vitro* desensitization of sensitive cells by trypsin. *J. exp. Med.* **120**, 1189.

DAVIES, A. J. S., LEUCHARS, E., WALLIS, V. and KOLLER, P. (1966). The mitotic response of thymus-derived cells to antigenic stimulus. *Transplantation*, **4**, 438.

DEFENDI, V. (1963). Effect of $SV_{40}$ virus immunization on growth of transplantable $SV_{40}$ and polyoma virus tumors in hamsters. *Proc. Soc. exp. Biol. Med.* **113**, 12.

DEICHER, H. R. G., HOLMAN, H. R. and KUNKEL, H. G. (1959). The precipitin reaction between DNA and a serum factor in systemic lupus erythematosus. *J. exp. Med.* **109**, 97.

DEICHMILLER, M. P. and DIXON, F. J. (1960). The metabolism of serum proteins in neonatal rabbits. *J. gen. Physiol.* **43**, 1047.

DENT, C. E. and WALSHE, J. M. (1953). Primary carcinoma of the liver: description of a case with ethanolaminuria, a new and obscure metabolic effect. *Br. J. Cancer*, **7**, 166.

DENT, P. B. and GOOD, R. A. (1965). Absence of antibody production in the bursa of Fabricius. *Nature, Lond.* **207**, 491.

DE PETRIS, S., KARLSBAD, G. and PERNIS, B. (1963). Localization of antibodies in plasma cells by electron microscopy. *J. exp. Med.* **117**, 849.

DEUTSCH, H. F. and MORTON, J. I. (1957). Dissociation of human serum macroglobulin. *Science, N.Y.* **125**, 600.

DE VRIES, M. J. and HIJMANS, W. (1966). A deficient development of the thymic epithelium and auto-immune disease in NZB mice. *J. Path. Bact.* **91**, 487.

DIETRICH, F. M. and GREY, H. M. (1964). Quantity and quality of antibody produced following the termination of tolerance in C57BL/6 mice. *Nature, Lond.* **201**, 1236.

DIETRICH, F. M. and WEIGLE, W. O. (1963). Induction of tolerance to heterologous proteins and their catabolism in C57BL/6 mice. *J. exp. Med.* **117**, 621.

DIXON, F. J., FELDMAN, J. D. and VAQUEZ, J. J. (1961). Experimental glomerulonephritis: the pathogenesis of a laboratory model resembling the spectrum of human glomerulonephritis. *J. exp. Med.* **113**, 899.

DIXON, F. J., MAURER, P. H. and WEIGLE, W. O. (1955). Immunologic activity of pneumococcal polysaccharide fixed in the tissues of the mouse. *J. Immun.* **74**, 188.

DIXON, F. J. and WEIGLE, W. O. (1957). The nature of the immunologic inadequacy of neonatal rabbits as revealed by cell transfer studies. *J. exp. Med.* **105**, 75.

DIXON, F. J., WEIGLE, W. O. and ROBERTS, J. C. (1957). Comparison of antibody responses associated with the transfer of rabbit lymph-node, peritoneal exudate, and thymus cells. *J. Immun.* **78**, 56.

DOAK, S. M. A. and KOLLER, P. C. (1961). Homografts on isologous and homologous radiation mouse chimeras. *Transplantn Bull.* **27**, 444.

DODD, M. C., BIGLEY, N. J., GEYER, V. B., MCCOY, F. W. and WILSON, H. E. (1962). Autoimmune response in rabbits injected with rat and rabbit liver ribosomes. *Science, N.Y.* **137**, 688.

DONALDSON, V. H. and EVANS, R. R. (1963). A biochemical abnormality in hereditary angioneurotic edema: absence of serum inhibitor of C′1 esterase. *Am. J. Med.* **35**, 37.

DONATH, J. and LANDSTEINER, K. (1904). Ueber paroxysmale Hämoglobinurie. *Münch. med. Wschr.* **51**, 1590.

DONIACH, D. and ROITT, I. M. (1964). An evaluation of gastric and thyroid autoimmunity in relation to hematologic disorders. *Seminars Hemat.* **1**, 313.

DOOLITTLE, R. F. and BLOMBACK, B. (1964). Amino acid sequence investigations of fibrinopeptides from various mammals: evolutionary implications. *Nature, Lond.* **202**, 147.

DOOLITTLE, R. F. and SINGER, S. J. (1965). Tryptic peptides from the active sites of antibody molecules. *Proc. natn. Acad. Sci. U.S.A.* **54**, 1773.

DOUGLAS, G. W., THOMAS, L., CARR, M., CULLEN, N. M. and MORRIS, R.

(1959). Trophoblast in the circulating blood during pregnancy. *Am. J. Obstet. Gynec.* **78**, 960.

DRESSER, D. W. (1962). Specific inhibition of antibody formation. II. Paralysis induced in adult mice by small quantities of protein antigen. *Immunology*, **5**, 378.

DRESSER, D. W. (1963). Specific inhibition of antibody formation. III. Apparent changes in the half-life of bovine gamma globulin in paralysed mice. *Immunology*, **6**, 345.

DREYER, W. J. and BENNETT, J. C. (1965). The molecular basis of antibody formation: a paradox. *Proc. natn. Acad. Sci. U.S.A.* **54**, 864.

DUKOR, P., DIETRICH, F. M. and ROSENTHAL, M. (1966). Recovery of immunological responsiveness in thymectomized mice. *Clin. exp. Immun.* **1**, 391.

DUKOR, P., MILLER, J. F. A. P., HOUSE, W. and ALLMAN, V. (1965). Regeneration of thymic grafts. I. Histological and cytological aspects. *Transplantation*, **3**, 639.

DUMONDE, D. C. (1966). Tissue-specific antigens. *Adv. Immun.* **5**, 245.

DUTTON, R. W. and BULMAN, H. N. (1964). The significance of the protein carried in the stimulation of DNA synthesis by hapten-protein conjugates in the secondary response. *Immunology*, **7**, 54.

DUTTON, R. W. and EADY, J. (1962). Studies on the mechanism of antigenic stimulation. *Biochem. J.* **82**, 31 P.

DUTTON, R. W. and EADY, J. (1964). An *in vitro* system for the study of the mechanism of antigenic stimulation in the secondary response. *Immunology*, **7**, 40.

DUTTON, R. W. and HARRIS, G. (1963). The apparent transfer of antigen-specific stimulation of DNA synthesis in rabbit spleen cell suspensions. *Fedn Proc. Fedn Am. Socs exp. Biol.* **22**, 266.

EAST, J. and DE SOUSA, M. A. B. (1965). The thymus, autoimmunity and malignancy in New Zealand black mice, in *Conference on Murine Leukemia; Natn. Cancer Inst. Monogr.* no. 22, p. 605.

EAST, J. and PARROTT, D. M. (1965). The role of the thymus in autoimmune disease. *Acta allerg.* **20**, 227.

EDELMAN, G. M., BENACERRAF, B. and OVARY, Z. (1963a). Structure and specificity of guinea-pig 7S antibodies. *J. exp. Med.* **118**, 229.

EDELMAN, G. M., BENACERRAF, B., OVARY, Z. and POULIK, M. D. (1961). Structural differences among antibodies of different specificities. *Proc. natn. Acad. Sci. U.S.A.* **47**, 1751.

EDELMAN, G. M. and GALLY, J. A. (1964). A model for the 7S antibody molecule. *Proc. natn. Acad. Sci. U.S.A.* **51**, 846.

EDELMAN, G. M. and GALLY, J. A. (1967). Somatic recombination of duplicated genes: an hypothesis on the origin of antibody diversity. *Proc. natn. Acad. Sci. U.S.A.* **57**, 353.

EDELMAN, G. M., OLINS, D. E., GALLY, J. A. and ZINDER, N. T. (1963b). Reconstitution of immunologic reactivity by interaction of polypeptide chains of antibody. *Proc. natn. Acad. Sci. U.S.A.* **50**, 753.

EDITORIAL (1964). Thyroid nodules and thyroid cancer. *Lancet*, ii, 129.

EHRICH, W. E. and HARRIS, T. N. (1942). The formation of antibodies in the popliteal lymph node in rabbits. *J. exp. Med.* **76**, 335.

EHRLICH, P. (1900). On immunity with special reference to cell life. *Proc. R. Soc.* B, **66**, 424.

EISEN, H. N. (1959). Delayed-type hypersensitivity reactions, in *Mechanisms of Hypersensitivity*, p. 413. Ed. J. H. Shaffer, G. A. LoGrippo and M. W. Chase. London: Churchill.

EISEN, H. N. (1964). The immune response to a simple antigenic determinant. *Harv. Lect.* **60**, 1.

EISEN, H. N. and KARUSH, F. (1964). Immune tolerance and an extracellular regulatory role for bivalent antibody. *Nature, Lond.* **202**, 677.

EISEN, H. N., LITTLE, J. R., OSTERLAND, C. K. and SIMMS, E. S. (1968). A myeloma protein with antibody activity. *Cold Spring Harb. Symp. quant. Biol.* **32**, 75.

EITZMAN, D. V. and SMITH, R. T. (1959). Antibody response to heterologous protein in rabbits of varying maturity. *Proc. Soc. exp. Biol. Med.* **102**, 529.

ELBERG, S. S. (1960). Cellular immunity. *Bact. Rev.* **24**, 67.

ELKINS, W. L. (1966). The interaction of donor and host lymphoid cells in the pathogenesis of renal cortical destruction induced by a local graft-versus-host reaction. *J. exp. Med.* **123**, 103.

EPSTEIN, G. J., GOLDBERGER, R. F. and ANFINSEN, C. B. (1964). The genetic control of tertiary protein structure. *Cold Spring Harb. Symp. quant. Biol.* **28**, 437.

EPSTEIN, W. V. and TAN, M. (1966). Increase of L chain proteins in the sera of patients with systemic lupus erythematosus and the synovial fluids of patients with peripheral rheumatoid arthritis. *Arthritis Rheum.* **9**, 713.

ERNSTRÖM, U. (1965). Studies on growth and cytomorphosis in the thymo-lymphatic system. *Acta path. microbiol. scand.* Suppl. 178.

EVERETT, N. B. and TYLER (CAFFREY), R. W. (1967). Radioautographic studies of reticular and lymphoid cells in germinal centers of lymph nodes, in *Germinal Centers in Immune Responses*, p. 145. Ed. H. Cottier, N. Odartchenko, R. Schindler and C. C. Congdon. Berlin: Springer.

EVERSON, T. C. (1964). Spontaneous regression of cancer. *Ann. N. Y. Acad. Sci.* **114**, 721.

FAGRAEUS, A. (1948). Antibody production in relation to the development of plasma cells. *Acta med. scand.* **130**, Suppl. 204.

FAHEY, J. L. (1962). Heterogeneity of $\gamma$-globulins. *Adv. Immun.* **2**, 42.

FAHEY, J. L. and ROBINSON, A. G. (1963). Factors controlling serum $\gamma$-globulin concentration. *J. exp. Med.* **118**, 845.

FAHEY, J. L., WUNDERLICH, J. and MISHELL, R. (1964). The immunoglobulins of mice. I. Four major classes of immunoglobulins $7S\gamma_2$-, $7S\gamma_1$-, $\gamma_{1A}$ ($\beta_{2A}$)- and $18S\gamma_{1M}$-globulins. *J. exp. Med.* **120**, 223.

FAZEKAS de ST GROTH, S. and WEBSTER, R. G. (1963). The neutralization of animal viruses. III. Equilibrium conditions in the influenza virus-antibody system. IV. Parameters of the influenza virus-antibody system. *J. Immun.* **90**, 140, 151.

FAZEKAS de ST GROTH, S. and WEBSTER, R. G. (1964). The antibody response, in *CIBA Foundation Symposium on the Cellular Biology of Myxovirus Infections*, p. 246. Ed. G. E. W. Wolstenholme and J. Knight. London: Churchill.

FEINSTEIN, A., GELL, P. G. H. and KELUS, A. S. (1963). Immunochemical analysis of rabbit γ-globulin allotypes. *Nature, Lond.* **200**, 653.

FELDMAN, J. D. (1958). Electron microscopy of serum sickness nephritis. *J. exp. Med.* **108**, 957.

FELDMAN, J. D. (1964). Ultrastructure of immunologic processes. *Adv. Immun.* **4**, 175.

FELDMAN, M. (1963). Discussion of paper by Trentin and Fahlberg, in *Conceptual Advances in Immunology and Oncology*, p. 73. New York: Hoeber.

FELDMAN, M., GLOBERSON, A. and NACHTIGAL, D. (1963). The reactivation of the immune response in immunologically suppressed animals, in *Conceptual Advances in Immunology and Oncology*, p. 427. New York: Hoeber.

FELDMAN, M. and MEKORI, T. (1966a). Differentiation and immunological competence of cloned cell populations of lymphoid origin, in *CIBA Foundation Symposium on Experimental and Clinical Studies*, p. 86. Ed. G. E. W. Wolstenholme and R. Porter. London: Churchill.

FELDMAN, M. and MEKORI, T. (1966b). Antibody production by 'cloned' cell populations. *Immunology*, **10**, 149.

FELTON, L. D. (1932). Active immunization of white mice by a non-polysaccharide and probably non-protein derivative of the pneumococcus. *J. Immun.* **23**, 405.

FELTON, L. D., KAUFFMANN, G., PRESCOTT, B. and OTTINGER, B. (1955). Studies on the mechanism of the immunological paralysis induced in mice by pneumococcal polysaccharides. *J. Immun.* **74**, 17.

FENNESTAD, K. L. and BORG-PETERSEN, C. (1962). Antibody and plasma cells in bovine fetuses infected with *Leptospira saxkoebing*. *J. infect. Dis.* **110**, 63.

FERNANDO, N. V. P. and MOVAT, H. Z. (1963). Fine structural changes in lymphoid tissue after antigenic stimulation. *Fedn Proc. Fedn Am. Socs exp. Biol.* **22**, 373.

FICHTELIUS, K. E. (1953). On the fate of the lymphocyte. *Acta anat.* Suppl. **19**, 1.

FINN, J. and ARQUILA, E. R. (1965). Multiple gene control of insulin antibody production in guinea-pigs. *Fedn Proc. Fedn Am. Socs exp. Biol.* **24**, 181.

FINN, R., CLARKE, C. A., DONOHOE, W. T. A., MCCONNELL, R. B., SHEPPARD, P. M., LEHANE, D. and KULKE, W. (1961). Experimental studies on the prevention of Rh haemolytic disease. *Br. med. J.* i, 1486.

FISCHER, H. and HAUPT, I. (1961). Das cytolysierende Prinzip von Serumkomplement. *Z. Naturf.* **16B**, 321.

FISHMAN, M. (1961). Antibody formation *in vitro*. *J. exp. Med.* **114**, 837.

FISHMAN, M. and ADLER, F. L. (1963). Antibody formation initiated *in vitro.* II. Antibody synthesis in X-irradiated recipients of diffusion chambers containing nucleic acid from macrophages incubated with antigen. *J. exp. Med.* **117**, 595.

FISHMAN, M., HAMMERSTROM, R. D. and BOND, V. P. (1963). In vitro transfer of macrophage RNA to lymph node cells. *Nature, Lond.* **198**, 549.

FITCH, F. W., PIERCE, C., HUNTER, R. L., CANNON, D. and WISSLER, R. W. (1967). Recent observations on the origin and fate of antigen-stimulated cells in the rat spleen, in *Germinal Centers in Immune Responses*, p. 286. Ed. H. Cottier, N. Odartchenko, R. Schindler and C. C. Congdon. Berlin: Springer.

FLEISCHMAN, J. B., PORTER, R. R. and PRESS, E. M. (1963). The arrangement of the peptide chains in gamma globulin. *Biochem. J.* **88**, 220.

FLIEDNER, T. M. (1967). On the origin of tingible bodies in germinal centers, in *Germinal Centers in Immune Responses*, p. 218. Ed. H. Cottier, N. Odartchenko, R. Schindler and C. C. Congdon. Berlin: Springer.

FORD, C. E. and CLARKE, C. M. (1963). Cytogenetic evidence of clonal proliferation in primary reticular neoplasms, in *Proc. Can. Cancer Res. Conf.* **5**, 129.

FORD, C. E., HAMMERTON, J. L., BARNES, D. W. H. and LOUTIT, J. F. (1956). Cytological identification of radiation-chimaeras. *Nature, Lond.* **177**, 452.

FORD, C. E. and MICKLEM, H. S. (1963). The thymus and lymph nodes in radiation chimaeras. *Lancet*, i, 359.

FORD, C. E., MICKLEM, H. S. and GRAY, S. M. (1959). Evidence of selected proliferation of reticular cell clones in heavily irradiated mice. *Br. J. Radiol.* **32**, 280.

FORSEN, N. R. and CONDIE, R. M. (1963). Abolition of immunologic memory with 6-mercaptopurine. *Fedn Proc. Fedn Am. Socs exp. Biol.* **22**, 500.

FOWLER, R., SCHUBERT, W. K. and WEST, C. D. (1960). Acquired partial tolerance to homologous skin grafts in the human infant at birth. *Ann. N.Y. Acad. Sci.* **87**, 403.

FRANĚK, F. and NEZLIN, R. S. (1963). Recovery of antibody combining activity by interaction of different peptide chains isolated from purified horse antitoxins. *Folia microbiol., Praha*, **8**, 128.

FRANGIONE, B. and FRANKLIN, E. C. (1965). Structural studies of human immunoglobulins. Differences in the Fd fragments of the heavy chains in G myeloma proteins. *J. exp. Med.* **122**, 1.

FRANGIONE, B., PRELLI, F. and FRANKLIN, E. C. (1966). Further studies of the Fd fragments of Ig G myeloma proteins. *Fedn Proc. Fedn Am. Socs exp. Biol.* **25**, 374.

FRANKLIN, E. C. (1964a). The immune globulins: their structure and function and some techniques for their isolation. *Prog. Allergy*, **8**, 58.

FRANKLIN, E. C. (1964b). Structural studies of human 7S γ-globulin (G immunoglobulin). Further observations of a naturally occurring protein related to the crystallizable fast fragment. *J. exp. Med.* **120**, 691.

FRANKLIN, E. C. and FUDENBERG, H. H. (1964). Antigenic heterogeneity of human Rh antibodies, rheumatoid factors, and cold agglutinins. *Archs Biochem. Biophys.* **104**, 433.

FRANKLIN, E. C. and FUDENBERG, H. H. (1965). The heterogeneity of rheumatoid factor and the genetic control of polypeptide chains of gamma globulins. *Ann. N.Y. Acad. Sci.* **124**, 873.

FRANKLIN, E. C., FUDENBERG, H. H., MELTZER, M. and STANWORTH, D. R. (1962). The structural basis for genetic variations of normal human $\gamma$-globulins. *Proc. natn. Acad. Sci. U.S.A.* **49**, 914.

FRANKLIN, E. C., LOWENSTEIN, J., BIGELOW, B. and MELTZER, M. (1964). Heavy chain disease: a new disorder of serum $\gamma$-globulins. *Am. J. Med.* **37**, 332.

FRASER, A. S. and SHORT, B. F. (1958). Studies of sheep mosaic for fleece type. I. Patterns and origin of mosaicism. *Aust. J. biol. Sci.* **2**, 200.

FRASER, E. A. and HILL, J. P. (1915). The development of the thymus, epithelial bodies, and thyroid in the Marsupialia. I. *Trichosurus vulpecula. Phil. Trans. R. Soc.* B, **207**, 1.

FREEDMAN, M. H. and SELA, M. (1966). Recovery of specific activity upon reoxidation of completely reduced polyalanyl rabbit antibody. *J. biol. Chem.* **241**, 5225.

FREY, J. R., de WECK, A. L. and GELEICK, H. (1964). Immunological unresponsiveness in allergic contact dermatitis to dinitrochlorobenzene in guinea-pigs. *J. invest. Derm.* **42**, 41.

FRIEDMAN, H. (1965 a). Absence of antibody plaque forming cells in spleens of thymectomized mice immunized with sheep erythrocytes. *Proc. Soc. exp. Biol. Med.* **118**, 1176.

FRIEDMAN, H. (1965 b). Transfer of RNA extracts from immune donor spleen cells to *Shigella*-tolerant recipient mice. *Nature, Lond.* **207**, 1315.

FRIEDMAN, H. P., STAVITSKY, A. B. and SOLOMON, J. M. (1965). Induction *in vitro* of antibodies to phage T2: antigens in the RNA extract employed. *Science, N.Y.* **149**, 1106.

FUDENBERG, H. H., DREWS, G. and NISONOFF, A. (1964). Serologic demonstration of dual specificity of rabbit bivalent hybrid antibody. *J. exp. Med.* **119**, 151.

FUDENBERG, H. H., MANDY, W. J. and NISONOFF, A. (1962). Serologic studies of proteolytic fragments of rabbit agglutinating antibodies. *J. clin. Invest.* **41**, 2123.

GALLEY, R. and FELDMAN, M. (1966). The induction of antibody production in X-irradiated animals by macrophages that interacted with antigen. *Israel J. med. Sci.* **2**, 358.

GALTON, M., REED, P. B. and HOLT, S. F. (1964). The relation of thymic chimerism to actively acquired tolerance. *Ann. N.Y. Acad. Sci.* **120**, 191.

GARVER, R. M., SANTOS, G. W. and COLE, L. J. (1959). Specific hemagglutinins in X-irradiated, bone-marrow treated mice following differential immunization of host and donor. *J. Immun.* **83**, 57.

GELL, P. G. H. and BENACERRAF, B. (1961). Studies on hypersensitivity. IV. The relationship between contact and delayed hypersensitivity. A study of the specificity of cellular immune reactions. *J. exp. Med.* **113**, 571.

GELL, P. G. H. and KELUS, A. (1962). Deletions of allotypic γ-globulins in antibodies. *Nature, Lond.* **195**, 44.

GELZER, J. and KABAT, E. A. (1964*a*). Specific fractionation of human antidextran antibodies. II. Assay of human antidextran sera and specifically fractionated purified antibodies by microcomplement fixation and complement fixation inhibition techniques. *J. exp. Med.* **119**, 983.

GELZER, J. and KABAT, E. A. (1964*b*). Specific fractionation of human antidextran antibodies. III. Fractionation of anti-dextran by sequential extraction with oligosaccharides of increasing chain length. *Immunochemistry*, **1**, 303.

GENGHOF, D. S. and BATTISTO, J. R. (1961). Antibody production in guinea-pigs receiving 6-mercaptopurine. *Proc. Soc. exp. Biol. Med.* **107**, 933.

GERMUTH, F. G. (1953). A comparative histologic and immunologic study in rabbits of induced hypersensitivity of the serum sickness type. *J. exp. Med.* **97**, 257.

GESNER, B. M. and GOWANS, J. L. (1962). The fate of lethally irradiated mice given isologous and heterologous thoracic duct lymphocytes. *Br. J. exp. Path.* **43**, 431.

GITLIN, D. and CRAIG, J. M. (1963). The thymus and other lymphoid tissues in congenital agammaglobulinemia. I. Thymic alymphoplasia and lymphocytic hypoplasia and their relation to infection. *Pediatrics, Springfield*, **32**, 517.

GLICK, B. (1964). The bursa of Fabricius and the development of immunologic competence, in *The Thymus in Immunobiology*, p. 343. Ed. R. A. Good and A. E. Gabrielsen. New York: Hoeber.

GLICK, B., CHANG, T. S. and JAAP, R. C. (1956). The bursa of Fabricius and antibody formation. *Poult. Sci.* **35**, 224.

GOLD, E. R., MANDY, W. J. and FUDENBERG, H. H. (1965). Relation between Gm(f) and the structure of the γ-globulin molecule. *Nature, Lond.* **207**, 1099.

GOLDBERGER, R. F. and ANFINSEN, C. B. (1962). The reversible masking of amino groups in ribonuclease and its possible usefulness in the synthesis of the protein. *Biochemistry, Washington*, **1**, 401.

GOLDSTEIN, G. and MACKAY, I. R. (1965). Contrasting abnormalities in the thymus in systemic lupus erythematosus and myasthenia gravis: a quantitative histological study. *Aust. J. exp. Biol. med. Sci.* **43**, 381.

GOLDSTEIN, G. and WHITTINGHAM, S. (1966). Experimental autoimmune thymitis: an animal model of human myasthenia gravis. *Lancet*, ii, 315.

GOOD, R. A., BRIDGES, R. A. and CONDIE, R. M. (1960). Host-parasite relationships in patients with dysproteinemias. *Bact. Rev.* **24**, 115.

GOOD, R. A., COOPER, M. D., PETERSON, R. D. A., HOYER, J. R. and GABRIELSEN, A. E. (1967). Immunological deficiency diseases of man—

relationships to disturbances of germinal center formation, in *Germinal Centers in Immune Responses*, p. 386. Ed. H. Cottier, N. Odartchenko, R. Schindler and C. C. Congdon. Berlin: Springer.

GOOD, R. A., KELLY, W. D., ROTSTEIN, J. and VARCO, R. L. (1962). Immunological deficiency diseases: agammaglobulinemia, hypogammaglobulinemia, Hodgkin's disease and sarcoidosis. *Prog. Allergy*, **6**, 187.

GOOD, R. A., MARTINEZ, C. and GABRIELSEN, A. E. (1964). Clinical considerations of the thymus in immunobiology, in *The Thymus in Immunobiology*, p. 3. Ed. R. A. Good and A. E. Gabrielsen. New York: Hoeber.

GOOD, R. A. and PAPERMASTER, B. W. (1964). Ontogeny and phylogeny of adaptive immunity. *Adv. Immun.* **4**, 1.

GOODMAN, G. T. and KOPROWSKI, H. (1962). Macrophages as a cellular expression of inherited natural resistance. *Proc. natn. Acad. Sci. U.S.A.* **48**, 160.

GOODMAN, H. C., FAHEY, J. L. and MALMGREN, R. A. (1960). Serum factors in lupus erythematosus and other diseases reacting with cell nuclei and nucleoprotein extracts: electrophoretic, ultracentrifugal and chromatographic studies. *J. clin. Invest.* **39**, 1595.

GOODMAN, J. W. (1964). On the origin of peritoneal fluid cells. *Blood*, **23**, 18.

GORDON, J. and MACLEAN, L. D. (1965). A lymphocyte-stimulating factor produced *in vitro*. *Nature, Lond.* **208**, 795.

GORER, P. A. (1937). The genetic and antigenic basis of tumour transplantation. *J. Path. Bact.* **44**, 691.

GORMAN, J. G. and CHANDLER, J. G. (1964). Is there an immunologically incompetent lymphocyte? *Blood*, **23**, 117.

GOWANS, J. L. (1959). The recirculation of lymphocytes from blood to lymph in the rat. *J. Physiol., Lond.* **146**, 54.

GOWANS, J. L. (1962). The fate of parental strain small lymphocytes in $F_1$ hybrid rats. *Ann. N.Y. Acad. Sci.* **99**, 432.

GOWANS, J. L., GESNER, B. M. and MCGREGOR, D. D. (1961). The immunological activity of lymphocytes, in *Biological Activity of the Leucocyte*, CIBA Foundation Study Group no. 10, p. 32. Ed. G. E. W. Wolstenholme and M. O'Connor. London: Churchill.

GOWANS, J. L. and KNIGHT, E. J. (1964). The route of re-circulation of lymphocytes in the rat. *Proc. R. Soc. B*, **159**, 257.

GOWANS, J. L. and MCGREGOR, D. D. (1963). The origin of antibody-forming cells, in *Immunopathology*, Int. Symp. Immunopath., no. 3, p. 89. Ed. P. Grabar and P. Miescher. Basel: Schwabe.

GOWANS, J. L. and MCGREGOR, D. D. (1965). The immunological activities of lymphocytes. *Prog. Allergy*, **9**, 1.

GOWANS, J. L., MCGREGOR, D. D. and COWEN, D. M. (1963). The role of small lymphocytes in the rejection of homografts of skin, in *The Immunologically Competent Cell*, CIBA Foundation Study Group no. 16, p. 20. Ed. G. E. W. Wolstenholme and J. Knight. London: Churchill.

GRABAR, P., COURÇON, J., BARNES, D. W. H., FORD, C. E. and MICKLEM,

H.S. (1962). Study of the antigenic constituents of sera from mouse/rat chimaeras. *Immunology*, 5, 673.

GRABAR, P., COURÇON, J., ILBERG, P. L. T., LOUTIT, J. F. and MERRILL, J. P. (1957). Etude immuno-électrophorétique du sérum de souris irradiées par des doses létales de rayons X et protegées par des cellules de la moelle osseuse de rats. *C. r. hebd. Séanc. Acad. Sci., Paris*, 245, 950.

GRANICK, S. (1965). Evolution of heme and chlorophyll, in *Evolving Genes and Proteins*, p. 67. Ed. V. Bryson and H. J. Vogel. New York and London: Academic Press.

GRÄSBECK, R., NORDMAN, C., and de la CHAPELLE, A. (1963). Mitogenic action of antileucocyte immune serum on peripheral leucocytes *in vitro*. *Lancet*, ii, 385.

GRAY, W. R., DREYER, W. J. and HOOD, L. E. (1967). Mechanism of antibody synthesis: size differences between mouse Kappa chains. *Science, N.Y.* 155, 465.

GREEN, H., BARROW, P. and GOLDBERG, B. (1959). Effect of antibody and complement on permeability control in ascites, tumor cells and erythrocytes. *J. exp. Med.* 110, 699.

GREEN, I., INKELAS, M. and ALLEN, L. B. (1960). Hodgkin's disease; a maternal-to-foetal lymphocyte chimaera? *Lancet*, i, 30.

GREEN, I., VASSALLI, P. and BENACERRAF, B. (1967a). Cellular localization of anti-DNP-PLL and anti-conveyor albumin in genetic non-responder guinea-pigs immunized with DNP-PLL albumin complexes. *J. exp. Med.* 125, 527.

GREEN, I., VASSALLI, P., NUSSENZWEIG, V. and BENACERRAF, B. (1967b). Specificity of the antibodies produced by single cells following immunization with antigens bearing two antigenic determinants. *J. exp. Med.* 125, 511.

GREENBERG, E., DIVERTIE, M. G. and WOOLNER, L. (1964). A review of unusual systemic manifestations associated with carcinoma. *Am. J. Med.* 36, 106.

GRÉGOIRE, C. (1935). Recherches sur la symbiose lympho-épithéliale au niveau du thymus de mammifère. *Archs Biol. Paris*, 46, 717.

GRÉGOIRE, C. (1958). The cultivation in the living organism of the thymus epithelium of the guinea-pig and rat. *Q. Jl microsc. Sci.* 99, 511.

GREY, H. M. (1964). Studies on changes in the quality of rabbit BSA antibody following immunization. *Immunology*, 7, 82.

GREY, H. M. and KUNKEL, H. G. (1964). H-chain subgroups of myeloma proteins and normal 7S γ-globulin. *J. exp. Med.* 120, 253.

GREY, H. M., MANNIK, M. and KUNKEL, H. G. (1965). Individual antigenic specificity of myeloma proteins. *J. exp. Med.* 121, 561.

GROSSBERGER, A. L., RADZIMSKI, G. and PRESSMAN, D. (1962). Effect of iodination on the active site of several anti-hapten antibodies. *Biochemistry, Wash.* 1, 391.

GRUNDMANN, E. and HOBIK, H. P. (1967). Histological and autoradiographic studies of the follicles of neonatally thymectomized mice, in

## References

*Germinal Centers in Immune Responses*, p. 349. Ed. H. Cottier, N. Odart-chenko, R. Schindler and C. C. Congdon. Berlin: Springer.

GUMP, D. W. (1965). Proliferative effect of purified protein derivative (PPD) on human cells. *Fedn Proc. Fedn Am. Socs exp. Biol.* **24**, 182.

GUNN, A., MICHIE, W. and IRVINE, W. J. (1964). The thymus in thyroid disease. *Lancet*, ii, 776.

GUSTAFSSON, B. E. and LAURELL, C.-B. (1959). Gamma globulin production in germfree rats after bacterial contamination. *J. exp. Med.* **110**, 675.

HABEL, K. (1961). Resistance of polyoma virus-immune animals to transplanted polyoma tumors. *Proc. Soc. exp. Biol. Med.* **106**, 722.

HABEL, K. and EDDY, B. E. (1963). Specificity of resistance to tumor challenge of polyoma and SV40 virus-immune hamsters. *Proc. Soc. exp. Biol. Med.* **113**, 1.

HABER, E. (1964). Recovery of antigenic specificity after denaturation and complete reduction of disulphides in a papain fragment of antibody. *Proc. natn. Acad. Sci. U.S.A.* **52**, 1099.

HACKETT, E. and BEECH, M. (1961). Immunological treatment of case of choriocarcinoma. *Br. med. J.* ii, 1123.

HALBERT, S. P., BIRCHER, R. and DAHLE, E. (1961). The analysis of streptococcal infections. V. Cardiotoxicity of streptolysin O for rabbits *in vivo. J. exp. Med.* **113**, 759.

HALBERT, S. P., SWICK, L. and SONN, C. (1955). The use of precipitin analysis in agar for the study of human streptococcal infections. II. Ouchterlony and Oakley technics. *J. exp. Med.* **101**, 557.

HALL, J. G. and MORRIS, B. (1964). Effect of X-irradiation of the popliteal lymph node on its output of lymphocytes and immunological responsiveness. *Lancet*, i, 1077.

HALL, J. G. and MORRIS, B. (1965). The origin of the cells in the efferent lymph flow from a single lymph node. *J. exp. Med.* **121**, 901.

HALL, J. G., MORRIS, B., MORENO, G. D. and BESSIS, M. C. (1967). The ultrastructure and functions of the cells in lymph following antigenic stimulation. *J. exp. Med.* **125**, 91.

HALLANDER, H. and DANIELSSON, D. (1962). *In vitro* production of antibody by thoracic dust lymphocytes. *Acta path. microbiol. scand.* **56**, 75.

HAMMAR, J. A. (1936). *Die normalmorphologische Thymusforschung im letzten Vierteljahre.* Leipzig: Barth.

HANNA, M. G., MAKINODAN, T. and FISHER, W. D. (1967). Lymphatic tissue germinal center localization of [125]I-labeled heterologous and isologous macroglobulins, in *Germinal Centers in Immune Responses*, p. 86. Ed. H. Cottier, N. Odartchenko, R. Schindler and C. C. Congdon. Berlin: Springer.

HANNOUN, C. and BUSSARD, A. E. (1966). Antibody production by cells in tissue culture. I. Morphological evolution of lymph node and spleen cells in tissue culture. *J. exp. Med.* **123**, 1035.

HARBOE, M. (1965). Reaction of [131]I trace labeled cold agglutinin. *Ann. N.Y. Acad. Sci.* **124**, 491.

HARBOE, M. and DEVERILL, J. (1964). Immunological properties of cold haemagglutinins. *Scand. J. Haemat.* **1**, 223.

HARBOE, M., OSTERLAND, C. K., MANNIK, M. and KUNKEL, H. G. (1962). Genetic characters of human γ-globulins in human myeloma proteins. *J. exp. Med.* **116**, 719.

HARBOE, M., PANDE, H., BRANDTZAEG, P., TVETER, K. J. and HJORT, P. F. (1966). Synthesis of monoclonal γ-globulins following thymus transplantation in hypo-γ-globulinaemia with severe lymphocytopenia. *XIth Congr. Internat. Soc. Haematol., Sydney*, Abstr. p. 9.

HARGRAVES, M. M., RICHMOND, H. and MORTON, R. (1948). Presentation of two bone marrow elements: the 'tart' cell and the 'L.E.' cell. *Proc. Staff Meet. Mayo Clin.* **23**, 25.

HARRIS, H. (1954). Role of chemotaxis in inflammation. *Physiol. Rev.* **34**, 529.

HARRIS, H. (1966). Hybrid cells from mouse and man. *Proc. R. Soc. B*, **166**, 358.

HARRIS, J. E. and FORD, C. E. (1964). Cellular traffic of the thymus: experiments with chromosome markers. I. Evidence that the thymus plays an instructional part. *Nature, Lond.* **201**, 884.

HARRIS, J. E., FORD, C. E., BARNES, D. W. H. and EVANS, E. P. (1964). Cellular traffic of the thymus: experiments with chromosome markers. II. Evidence from parabiosis for an afferent stream of cells. *Nature, Lond.* **201**, 886.

HARRIS, T. N. and EHRICH, W. E. (1946). The fate of injected particulate antigens in relation to the formation of antibodies. *J. exp. Med.* **84**, 157.

HARRIS, T. N., GRIMM, E., MERTENS, E. and EHRICH, W. E. (1945). The role of the lymphocyte in antibody formation. *J. exp. Med.* **81**, 73.

HARRIS, T. N., HUMMELER, K. and HARRIS, S. (1966). Electron microscopic observations on antibody-producing lymph node cells. *J. exp. Med.* **123**, 161.

HARVEY, A. M. and JOHNS, R. J. (1962). Myasthenia gravis and the thymus. *Am. J. Med.* **32**, 1.

HAŠEK, M., HRABA, T. and MADAR, J. (1965). An attempt to characterize cellular background of partial tolerance. *Folia biol., Praha*, **11**, 318.

HAUROWITZ, F. (1965). Antibody formation. *Physiol. Rev.* **45**, 1.

HAUSCHKA, S. D. and KONIGSBERG, I. R. (1966). The influence of collagen on the development of muscle clones. *Proc. natn. Acad. Sci. U.S.A.* **55**, 119.

HEGE, J. S. and COLE, L. J. (1966). Antibody plaque forming cells in non-sensitized mice: effects of neo-natal thymectomy, phytohemagglutinin and X radiation. *Fedn Proc. Fedn Am. Socs exp. Biol.* **25**, 305.

HEILMAN, D. and MCFARLAND, W. (1965). Inhibition of tuberculin-induced mitogenesis in lymphocyte cultures by serum from tuberculous donors. *Fedn Proc. Fedn Am. Socs exp. Biol.* **24**, 182.

HELLSTRÖM, K. E. (1966a). Studies on allogeneic inhibition. I. Differential bahavior of mouse tumors transplanted to homozygous and F₁ hybrid hosts. *Int. J. Cancer*, **1**, 349.

679

HELLSTRÖM, K. E. (1966*b*). Studies on allogeneic inhibition. II. Inhibition of mouse tumor cell growth by *in vitro* contact with cells containing foreign H-2 antigens. *Int. J. Cancer*, **1**, 361.

HELYER, B. J. and HOWIE, J. B. (1963). Renal disease associated with positive lupus erythematosus tests in a cross-bred strain of mice. *Nature, Lond.* **197**, 197.

HEPTINSTALL, R. N. and GERMUTH, F. G., Jr (1957). Experimental studies on the immunologic and histologic effects of prolonged exposure to antigen. I. Distribution of allergic lesions following multiple injections of bovine albumin, bovine gamma globulin, and albumin and globulin together with special reference to the occurrence of granulomatous arteritis. *Bull. Johns Hopkins Hosp.* **100**, 71.

HERBERT, W. J. (1966). Antigenicity of soluble protein in the presence of high levels of antibody: a possible mode of action of the antigen adjuvants. *Nature, Lond.* **210**, 747.

HEREMANS, J. F., CRABBÉ, P. A. and MASSON, P. L. (1966). Biological significance of exocrine $\gamma$-A-immunoglobulin. *Acta med. scand.* **179**, Supp. 445, 84.

HEREMANS, J. F. and VAERMAN, J.-P. (1962). $\beta_{2A}$-globulin as a possible carrier of allergic reaginic activity. *Nature, Lond.* **193**, 1091.

HERTZ, R., LEWIS, J. and LIPSETT, M. B. (1961). Five years' experience with the chemotherapy of metastatic choriocarcinoma and related trophoblastic tumors in women. *Am. J. Obstet. Gynec.* **82**, 631.

HERTZ, R., ROSS, G. T. and LIPSETT, M. B. (1964). Chemotherapy in women with trophoblastic disease: choriocarcinoma, chorioadenoma destruens, and complicated hydatidiform mole. *Ann. N.Y. Acad. Sci.* **114**, 881.

HICKS, J. D. and BURNET, F. M. (1966). Renal lesions in the 'auto-immune' mouse strains NZB and F1 NZB × NZW. *J. Path. Bact.* **91**, 467.

HILDEMANN, W. H. (1964). Immunological properties of small lymphocytes in the graft-versus-host reaction in mice. *Transplantation*, **2**, 38.

HILDEMANN, W. H., LINSCOTT, W. D. and MORLING, M. J. (1962). Immunological competence of small lymphocytes in the graft-versus-host reaction in mice, in *CIBA Foundation Symposium on Transplantation*, p. 236. Ed. G. E. W. Wolstenholme and M. P. Cameron. London: Churchill.

HILGARD, H. R., YUNIS, E. and MARTINEZ, C. (1964). Treatment of wasting in thymectomized mice with splenic or thymic cells. *Fedn Proc. Fedn Am. Socs exp. Biol.* **23**, 287.

HILL, R. L., DELANEY, R., FELLOWS, R. E. and LEBOWITZ, H. E. (1966). The evolutionary origin of the immunoglobulins. *Proc. natn. Acad. Sci. U.S.A.* **56**, 1762.

HILSCHMANN, N. and CRAIG, L. C. (1965). Amino acid sequence studies of Bence Jones protein. *Proc. natn. Acad. Sci. U.S.A.* **53**, 1403.

HINRICHSEN, K. (1967). Thymidine-$^3$H in developing germinal centers, in *Germinal Centers in Immune Responses*, p. 152. Ed. H. Cottier, N. Odartchenko, R. Schindler and C. C. Congdon. Berlin: Springer.

HIRAMOTO, R. N. and HAMLIN, M. (1965). Detection of two antibodies in single plasma cells by the paired fluorescence technique. *J. Immun.* 95, 214.

HIRSCH, J. G. (1965). The eosinophil leucocyte, in *The Inflammatory Process*, p. 266. Ed. B. W. Zweifach, L. Grant and R. T. McCluskey. New York: Academic Press.

HIRSCHHORN, K., SCHREIBMAN, R. R., VERBO, S. and CRUSKIN, R. H. (1964). The action of streptolysin S on peripheral lymphocytes of normal subjects and patients with acute rheumatic fever. *Proc. natn. Acad. Sci. U.S.A.* 52, 1151.

HITZIG, W. H., BIRÓ, Z., BOSCH, H. and HUSER, H. J. (1958). Agammaglobulinämie und Alymphocytose mit Schwund des lymphatischen Gewebes. *Helv. paediat. Acta*, 13, 551.

HOLMES, M. C. and BURNET, F. M. (1964). Experimental studies of thymic function in NZB mice and the F1 hybrids with C3H. *Aust. J. exp. Biol. med. Sci.* 42, 589.

HOLUB, M. and ŘÍHA, I. (1960). Morphological changes in lymphocytes cultivated in diffusion chambers during the primary antibody response to a protein antigen, in *Mechanisms of Antibody Formation*, p. 30. Ed. M. Holub and L. Jarošková. Prague: Czechoslovak Acad. Sci.

HOLUB, M., ŘÍHA, I. and KAMARYTOVA, V. (1965). Immunological competence of different stages of the lymphoid cell, in *Molecular and Cellular Basis of Antibody Formation*, p. 477. Ed. J. Šterzl. Prague: Czechoslovak Acad. Sci.

HONG, R., PALMER, J. L. and NISONOFF, A. (1965). Univalence of half-molecules of rabbit antibody. *J. Immun.* 94, 603.

HOWARD, J. G., CHRISTIE, G. H., BOAK, J. L., EVANS-ANFOM, E. (1965). Evidence for the conversion of lymphocytes into liver macrophages during graft-versus-host reaction. *Colloques int. Cent. natn. Rech. scient.* 147, 95.

HOYER, L. W., CONDIE, R. M. and GOOD, R. A. (1960). Prevention of experimental allergic encephalomyelitis with 6-mercaptopurine. *Proc. Soc. exp. Biol. Med.* 103, 205.

HRABA, T. and MÁJSKÝ, A. (1963). The frequency of A group free erythrocytes during agglutination. *Folia biol., Praha*, 9, 271.

HULLIGER, L. and SORKIN, E. (1965). Formation of specific antibody by circulating cells. *Immunology*, 9, 391.

HUMBLE, J. G., JAYNE, W. H. W. and PULVERTAFT, H. J. V. (1956). Biological interaction between lymphocytes and other cells. *Br. J. Haemat.* 2, 285.

HUMPHREY, J. H. (1963). The non-specific globulin response to Freund's adjuvant, in *La Tolérance acquise et la tolérance naturelle à l'égard de substances antigéniques définies*. Colloques int. Cent. natn. Rech. scient. 116, 401.

HUMPHREY, J. H. (1964). Immunological unresponsiveness to protein antigens in rabbits. I. Duration of unresponsiveness following a single injection at birth. II. The nature of the subsequent antibody response. *Immunology*, 7, 449, 462.

*References*

HUMPHREY, J. H. (1967). Cell-mediated immunity: general perspectives. *Br. med. Bull.* **23**, 93.

HUMPHREY, J. H., AUSTEN, K. F. and RAPP, H. J. (1963). *In vitro* studies of reversed anaphylaxis with rat cells. *Immunology*, **6**, 226.

HUMPHREY, J. H. and DOURMASHKIN, R. R. (1965). Electron microscope studies of immune cell lysis, in *CIBA Foundation Symposium on Complement*, p. 175. Ed. G. E. W. Wolstenholme and J. Knight. London: Churchill.

HUMPHREY, J. H., PARROTT, D. M. V. and EAST, J. (1964). Studies on globulin and antibody production in mice thymectomized at birth. *Immunology*, **7**, 419.

HUMPHREY, J. H. and TURK, J. L. (1961). Immunological unresponsiveness in guinea pigs. I. Immunological unresponsiveness to heterologous serum proteins. *Immunology*, **4**, 301.

IKARI, N. S. (1964). Bactericidal antibody to *Escherichia coli* in germ-free mice. *Nature, Lond.* **202**, 379.

IMHOF, J., BALLIEUX, R., MUL, N. and POEN, H. (1966). Monoclonal and diclonal gammopathies. *Acta med. scand.* **179**, Suppl. 445, p. 102.

INGRAM, V. M. (1961). Gene evolution and the haemoglobins. *Nature, Lond.* **189**, 704.

ISAKOVIĆ, K. and JANKOVIĆ, B. D. (1964). Role of the thymus and the bursa of Fabricius in immune reactions in chickens. II. Cellular changes in lymphoid tissues of thymectomized, bursectomized and normal chickens in the course of first antibody response. *Int. Archs Allergy appl. Immun.* **24**, 296.

ISAKOVIĆ, K., SMITH, S. B. and WAKSMAN, B. H. (1965). Role of the thymus in tolerance. I. Transfer of specific unresponsiveness to BGG with thymus grafting. *J. exp. Med.* **122**, 1103.

ISHIZAKA, K. (1963). Gamma globulin and molecular mechanisms in hypersensitivity reactions. *Prog. Allergy*, **7**, 32.

ISHIZAKA, K. and ISHIZAKA, T. (1966). Physicochemical properties of reaginic antibody. III. Further studies on the reaginic antibody in $\gamma$A-globulin preparations. *J. Allergy*, **38**, 108.

ISHIZAKA, K., ISHIZAKA, T. and HORNBROOK, M. M. (1963). Blocking of Prausnitz-Küstner sensitization with reagin by normal human $\beta_{2A}$ globulin. *J. Allergy*, **34**, 395.

ISHIZAKA, K., ISHIZAKA, T. and HORNBROOK, M. M. (1966). Physicochemical properties of reaginic antibody. V. Correlation of reaginic activity with $\gamma$E-globulin antibody. *J. Immun.* **97**, 840.

ISHIZAKA, T. and ISHIZAKA, K. (1962). Biological activities of aggregated $\gamma$-globulin. V. Agglutination of erythrocytes and platelets. *J. Immun.* **89**, 709.

ITANO, H. A. (1957). The human hemoglobins, their properties and their genetic control. *Adv. Protein Chem.* **12**, 215.

ITOH, T. and SOUTHAM, C. M. (1964). Isoantibodies to human cancer cells in cancer patients following cancer homotransplants. *J. Immun.* **93**, 926.

JAFFE, W. P. and FECHHEIMER, N. S. (1966). Cell transport and the bursa of Fabricius. *Nature, Lond.* **212**, 92.

JANEWAY, C. A. and SELA, M. (1967). Synthetic antigens composed exclusively of L or D amino acids. I. Effect of optical configuration on the immunogenicity of synthetic polypeptides in mice. *Immunology,* **13**, 29.

JANKOVIĆ, B. D. and DVORAK, H. F. (1962). Enzymatic inactivation of immunologically competent lymph node cells in the 'transfer reaction'. *J. Immun.* **89**, 571.

JANKOVIĆ, B. D. and IŠVANESKI, M. (1963). Experimental allergic encephalomyelitis in thymectomized, bursectomized and normal chickens. *Int. Archs Allergy appl. Immun.* **23**, 188.

JANKOVIĆ, B. D. and LESKOWITZ, S. (1965). Restoration of antibody producing capacity in bursectomized chickens by bursal grafts in Millipore chambers. *Proc. Soc. exp. Biol. Med.* **118**, 1164.

JENNINGS, R. B. and EARLE, D. P. (1961). Post-streptococcal glomerulonephritis: histopathological and clinical studies of the acute, subsiding acute and early chronic latent stages. *J. clin. Invest.* **40**, 1525.

JENSEN, C. (1933). Antitoxin curve in children after active immunization with diphtheria anatoxin with special reference to duration of antitoxic immunity. *Acta path. microbiol. scand.* **10**, 137.

JENSEN, K. E., DAVENPORT, F. M., HENNESSY, A. V. and FRANCIS, T. (1956). Characterization of influenza antibodies by serum absorption. *J. exp. Med.* **104**, 199.

JERNE, N. K. (1955). The natural selection theory of antibody formation. *Proc. natn. Acad. Sci. U.S.A.* **41**, 849.

JERNE, N. K. and NORDIN, A. A. (1963). Plaque formation in agar by single antibody-producing cells. *Science, N.Y.* **140**, 405.

JOHNSON, F. R. and ROBERTS, K. B. (1964). The growth and division of human small lymphocytes in tissue culture: an electron microscopic study. *J. Anat.* **98**, 303.

JONAS, W. E., GURNER, B. W., NELSON, D. S. and COOMBS, R. R. A. (1965). Passive sensitization of tissue cells. I. Passive sensitization of macrophages by guinea-pig cytophilic antibody. *Int. Archs Allergy appl. Immun.* **28**, 86.

JONES, M. A. S. and LAFFERTY, K. J. (1966). The dermal reaction induced in sheep by homologous lymphocytes and an RNA fraction extracted from homologous lymphocytes. *XIth Congr. Internat. Soc. Blood Transf., Sydney, Abstracts,* p. 45.

JONJA, M. G. and STICH, H. F. (1963). The survival of abnormal cells in the thymus of mice. *Expl Cell Res.* **31**, 220.

KABAT, E. A. (1966). Structure and heterogeneity of antibodies. *Acta haemat.* **36**, 198.

KABAT, E. A. (1967). The paucity of species-specific amino acid residues in the variable regions of human and mouse Bence-Jones proteins and its evolutionary and genetic implications. *Proc. natn. Acad. Sci. U.S.A.* **57**, 1345.

KADOWAKI, J. I., THOMPSON, R. I., ZUELZER, W. W., WOOLEY, P. V., BROUGH, A. J. and GRUBER, D. (1965). XX/XY lymphoid chimaerism in congenital immunological deficiency syndrome with thymic alymphoplasia. *Lancet*, ii, 1152.

KAGEN, L. J. and BECKER, E. L. (1963). Inhibition of permeability globulins by C'-1 esterase inhibitor. *Fedn Proc. Fedn Am. Socs exp. Biol.* **22**, 613.

KANTOR, F. S. (1961). Renal glomerular lesions in mice following injection of streptococcal M protein in mice. *Yale J. Biol. Med.* **34**, 70.

KANTOR, F. S., OJEDA, A. and BENACERRAF, B. (1963). Studies on artificial antigens. I. Antigenicity of DNP polylysine and DNP copolymer of lysine and glutamic acid in guinea pigs. *J. exp. Med.* **117**, 55.

KAPLAN, M. H. (1964). Discussion, in *The Streptococcus, Rheumatic Fever and Glomerulonephritis*, p. 185. Ed. J. W. Uhr. Baltimore: Williams and Wilkins.

KAPLAN, M. H. (1965). Induction of autoimmunity to heart in rheumatic fever by streptococcal antigen(s) cross-reactive with heart. *Fedn Proc. Fedn Am. Socs exp. Biol.* **24**, 109.

KAPLAN, M. H. and DALLENBACH, F. D. (1961). Immunologic studies of heart tissue. III. Occurrence of bound gamma globulin in auricular appendages from rheumatic hearts. Relationship to certain histopathologic features of rheumatic heart disease. *J. exp. Med.* **113**, 1.

KARAKLIS, A., VALAES, T., PANTELAKIS, S. N. and DOXIADIS, S. A. (1964). Thymectomy in an infant with autoimmune haemolytic anaemia. *Lancet*, ii, 778.

KARUSH, F. (1956). The interaction of purified antibody with optically isomeric haptens. *J. Am. chem. Soc.* **78**, 5519.

KARUSH, F. (1957a). The interactions of purified anti-$\beta$-lactoside antibody with hapten. *J. Am. chem. Soc.* **79**, 3380.

KARUSH, F. (1957b). The role of disulphide bonds in antibody specificity. *J. Am. chem. Soc.* **79**, 5323.

KARUSH, F. (1958). Specificity of antibodies. *Trans. N.Y. Acad. Sci.* (Ser. II), **20**, 581.

KARUSH, F. (1962). The role of disulphide bonds in the acquisition of immunologic specificity. *J. Pediat.* **60**, 103.

KARUSH, F. and EISEN, H. N. (1962). A theory of delayed hypersensitivity. *Science, N.Y.* **136**, 1032.

KASAKURA, S. and LOWENSTEIN, L. (1965). A factor stimulating DNA synthesis derived from the medium of leucocyte cultures. *Nature, Lond.* **208**, 794.

KEKWICK, R. A. and MOLLISON, P. L. (1961). Blood group antibodies associated with the 19S and 7S components of human sera. *Vox Sang.* **6**, 398.

KENNEDY, J. C., TILL, J. E., SIMINOVITCH, L. and MCCULLOCH, E. A. (1966). The proliferative capacity of antigen-sensitive precursors of hemolytic plaque-forming cells. *J. Immun.* **96**, 973.

KEPNER, W. A. (1943). The manipulation of the nematocysts of *Pennaria tiarella* by *Aelis pilata*. *J. Morph.* **73**, 297.

KERMAN, R., SEGRE, D. and MYERS, W. L. (1967). Altered response to pneumococcal polysaccharide in offspring of immunologically paralysed mice. *Science, N.Y.* **156**, 1514.

KÉUNING, F. J. and BOS, W. H. (1967). Regeneration patterns of lymphoid follicles in the rabbit spleen after sublethal X-irradiation, in *Germinal Centers in Immune Responses*, p. 250. Ed. H. Cottier, N. Odartchenko, R. Schindler and C. C. Congdon. Berlin: Springer.

KIRSCHSTEIN, R. L., RABSON, A. S. and PETERS, E. A. (1964). Oncogenic activity of adenovirus 12 in thymectomized BALB/c and C3H/HeN mice. *Proc. Soc. exp. Biol. Med.* **117**, 198.

KLEIHAUER, E., BRAUN, H. and BETKE, K. (1957). Demonstration von fetalem Hämoglobin in den Erythrocyten eines Blutausstrichs. *Klin. Wschr.* **35**, 637.

KLEIN, G. (1962). Some features of tumor-specific antigens: a general discussion. *Ann. N.Y. Acad. Sci.* **101**, 170.

KLEIN, G. (1966). Tumor antigens. *A. Rev. Microbiol.* **20**, 223.

KLEIN, G., CLIFFORD, P., KLEIN, E. and STJERNSWÄRD, J. (1966). Search for tumor-specific immune reactions in Burkitt lymphoma patients by the membrane immunofluorescence reaction. *Proc. natn. Acad. Sci. U.S.A.* **55**, 1628.

KLEIN, G. and KLEIN, E. (1963). Antigenic properties of other experimental tumors. *Cold Spring Harb. Symp. quant. Biol.* **27**, 463.

KLEIN, G., SJÖGREN, H. O., KLEIN, E. and HELLSTRÖM, K. E. (1960). Demonstration of resistance against methylcholanthrene-induced sarcomas in the primary autochthonous host. *Cancer Res.* **20**, 1561.

KLEIN, J. J., GOTTLIEB, A. J., MONES, R. J., APPEL, S. H. and OSSERMAN, K. E. (1964). Thymoma in polymyosites. *Archs intern. Med.* **113**, 142.

KLINE, B. S., COHEN, M. B. and RUDOLPH, J. A. (1932). Histologic changes in allergic and non-allergic wheals. *J. Allergy*, **3**, 531.

KLINMAN, K. R., ROCKEY, J. H. and KARUSH, F. (1964). Valence and affinity of equine non-precipitating antibody to a haptenic group. *Science, N.Y.* **146**, 401.

KOBURG, E. (1967). Cell production and cell migration in the tonsil, in *Germinal Centers in Immune Responses*, p. 176. Ed. H. Cottier, N. Odartchenko, R. Schindler and C. C. Congdon. Berlin: Springer.

KOLLER, P. C., DAVIES, A. J. S. and DOAK, S. M. A. (1961). Radiation chimeras. *Adv. Cancer Res.* **6**, 181.

KOLLER, P. C. and DOAK, S. M. A. (1960). Quoted Koller *et al.* 1961. *Int. J. Radiat. Biol.* (Suppl.), p. 327.

KONIGSBERG, I. R. (1963). Clonal analysis of myogenesis. *Science, N.Y.* **140**, 1273.

KOOP, C. E., KIESEWETTER, W. B. and HORN, R. C. (1955). Neuroblastoma in childhood. *Surgery*, **38**, 272.

KOPROWSKI, H., NORTON, T. W., HUMMELER, K., STOKES, J., HUNT, A. D., FLACK, A. and JERVIS, J. A. (1956). Immunization of infants with living attenuated poliomyelitis virus: laboratory investigations of ali-

*References*

mentary infection and antibody response in infants under six months of age with congenitally acquired antibodies. *J. Am. med. Ass.* **162**, 1281.

KORNGOLD, L. (1961). Abnormal plasma components and their significance in disease. *Ann. N.Y. Acad. Sci.* **94**, 110.

KOSHLAND, M. E. (1966). Primary structure of immunoglobulins and its relationship to antibody specificity. *J. cell. comp. Physiol.* **67**, Supp. 1, 33.

KOSHLAND, M. E. and ENGLBERGER, F. M. (1963). Differences in the amino acid composition of two purified antibodies from the same rabbit. *Proc. natn. Acad. Sci. U.S.A.* **50**, 61.

KOSHLAND, M. E., ENGLBERGER, F. M. and SHAPANKA, R. (1964). Differences in amino acid composition of a third rabbit antibody. *Science, N.Y.* **143**, 1330.

KRITZMAN, J., KUNKEL, H. G., MCCARTHY, J. and MELLORS, R. C. (1961). Studies of a Waldenström-type macroglobulin with rheumatoid factor properties. *J. Lab. clin. Med.* **57**, 905.

KRUEGER, R. G. (1965). The effect of streptomycin on antibody synthesis *in vitro*. *Proc. natn. Acad. Sci. U.S.A.* **54**, 144.

KRUEGER, R. G. (1966). Properties of antibodies synthesized by cells *in vitro* in the presence and absence of streptomycin. *Proc. natn. Acad. Sci. U.S.A.* **25**, 1206.

KUHNS, W. J. and PAPPENHEIMER, A. M., Jr (1952). Immunochemical studies of antitoxin produced in normal and allergic individuals hyperimmunized with diphtheria toxoid. I. Relationship of skin sensitivity to purified diphtheria toxoid to the presence of circulating, non-precipitating antitoxin. *J. exp. Med.* **95**, 363.

KUNKEL, H. G. and TAN, E. M. (1964). Autoantibodies and disease. *Adv. Immun.* **4**, 351.

KUNKEL, H. G., YOUNT, W. J. and LITWIN, S. D. (1966). Genetically determined antigen of the Ne subgroup of gamma-globulin: detection by precipitin analysis. *Science, N.Y.* **154**, 1041.

LACHMANN, P. J. and COOMBS, R. R. A. (1965). Complement, conglutinin and immuno-conglutinins, in *CIBA Foundation Symposium on Complement*, p. 242. Ed. G. E. W. Wolstenholme and J. Knight. London: Churchill.

LACHMANN, P. J., MÜLLER-EBERHARD, H. J., KUNKEL, H. G. and PARONETTO, F. (1962). The localization of *in vivo* bound complement in tissue sections. *J. exp. Med.* **115**, 63.

LANDSTEINER, K. and CHASE, M. W. (1937). Studies on the sensitization of animals with simple chemical compounds. IV. Anaphylaxis induced by picryl chloride and 2:4-dinitrochlorobenzene. *J. exp. Med.* **66**, 337.

LANDSTEINER, K. and WIENER, A. S. (1940). An agglutinable factor in human blood recognized by immune sera for Rhesus blood. *Proc. Soc. exp. Biol. Med.* **43**, 223.

LANDY, M., SANDERSON, R. P., BERNSTEIN, M. T. and JACKSON, A. L. (1964). Antibody production by leucocytes in peripheral blood. *Nature, Lond.* **204**, 1321.

LANDY, M., SANDERSON, R. P., BERNSTEIN, M. T. and LERNER, E. M. (1965). Involvement of thymus in immune response of rabbits to somatic polysaccharides of gram-negative bacteria. *Science, N.Y.* **147,** 1591.

LANGE, K., WASSERMANN, E. and SLOBODY, L. B. (1960). The significance of serum complement levels for the diagnosis and prognosis of acute and subacute glomerulonephritis and lupus erythematosus disseminatus. *Ann. intern. Med.* **53,** 636.

LANNI, F. and LANNI, Y. T. (1954). Antigenic structure of bacteriophage. *Cold Spring Harb. Symp. quant. Biol.* **18,** 159.

LAPRESLE, C. and WEBB, T. (1960). Degradation of a protein antigen by intracellular enzymes, in *CIBA Foundation Symposium on the Cellular Aspects of Immunity,* p. 44. Ed. G. E. W. Wolstenholme and M. O'Connor. London: Churchill.

LARSON, B. L. and GILLESPIE, D. C. (1957). Origin of the major specific proteins in milk. *J. biol. Chem.* **227,** 565.

LAWRENCE, H. S. (1949). The cellular transfer of cutaneous hypersensitivity to tuberculin in man. *Proc. Soc. exp. Biol. Med.* **71,** 516.

LAWRENCE, H. S. (1955). The transfer in humans of delayed skin sensitivity of the tuberculin type with components of disrupted leucocytes. *J. clin. Invest.* **34,** 219.

LAWRENCE, H. S. (1959). The transfer of hypersensitivity of the delayed type in man, in *Cellular and Humoral Aspects of the Hypersensitive States,* p. 279. Ed. H. S. Lawrence. New York: Hoeber.

LAWRENCE, H. S. (1960). Some biological and immunological properties of transfer factor, in *CIBA Foundation Symposium on Cellular Aspects of Immunity,* p. 243. Ed. G. E. W. Wolstenholme and M. O'Connor. London: Churchill.

LEBOVITZ, H. E. (1965). Endocrine-metabolic syndromes associated with neoplasms, in *The Remote Effects of Cancer on the Nervous System,* p. 104. Ed. Lord Brain and F. H. Norris. New York: Grune and Stratton.

LEDDY, J. P. and BAKEMEIER, R. F. (1965). Structural aspects of human erythrocyte autoantibodies. I. L chain types and electrophoretic dispersion. *J. exp. Med.* **121,** 1.

LEDERBERG, J. (1959). Genes and antibodies. *Science, N.Y.* **129,** 1649.

LEONHARDT, T. (1957). Familial hypergammaglobulinaemia and systemic lupus erythematosus. *Lancet,* ii, 1200.

LESKOWITZ, S. (1967). Mechanism of delayed reactions. *Science, N.Y.* **155,** 350.

LESKOWITZ, S., JONES, V. E., and ZAK, S. J. (1966). Immunochemical study of antigenic specificity in delayed hypersensitivity. V. Immunization with monovalent low molecular weight conjugates. *J. exp. Med.* **123,** 229.

LEUCHARS, E., CROSS, A. M., DAVIES, A. J. S. and WALLIS, V. J. (1964). A cellular component of thymic function. *Nature, Lond.* **203,** 1189.

LEVEY, R. H. and MEDAWAR, P. B. (1966). Nature and mode of action of antilymphocytic antiserum. *Proc. natn. Acad. Sci. U.S.A.* **56,** 1130.

LEVEY, R. H., TRAININ, N. and LAW, L. W. (1963). Evidence for function of thymic tissue in diffusion chambers implanted in neonatally thymectomized mice. *J. natn. Cancer Inst.* **31**, 199.

LEVIN, A. G., CUSTODIO, D. B., MANDEL, E. E. and SOUTHAM, C. M. (1964). Rejection of cancer homotransplants by patients with debilitating non-neoplastic disease. *Ann. N.Y. Acad. Sci.* **120**, 410.

LEVINE, B. B., OJEDA, A. and BENACERRAF, B. (1963). Studies on artificial antigens. III. The genetic control of the immune response to hapten-poly-L-lysine conjugates in guinea-pigs. *J. exp. Med.* **118**, 953.

LEVINE, P. (1943). Serological factors as possible causes in spontaneous abortions. *J. Hered.* **34**, 71.

LEVINE, P., CELANO, M. J. and FALKOWSKI, F. (1965). The specificity of the antibody in paroxysmal cold hemoglobinuria. *Ann. N.Y. Acad. Sci.* **124**, 456.

LEVINE, P. and STETSON, R. E. (1939). An unusual case of intra-group agglutination. *J. Am. med. Ass.* **113**, 126.

LEWIS, F. B., SCHWARTZ, R. S. and DAMESHEK, W. (1966). X-irradiation and alkylating agents as possible 'trigger' mechanisms in the autoimmune complications of malignant lymphoproliferative disease. *Clin. exp. Immun.* **1**, 3.

LEWIS, J. P. and TROBAUGH, F. E., Jr (1964). Haematopoietic stem cells. *Nature, Lond.* **204**, 589.

LI, M. C., HERTZ, R. and BERGENSTAL, D. M. (1958). Therapy of choriocarcinoma and related trophoblastic tumors with folic acid and purine antagonists. *New Engl. J. Med.* **259**, 66.

LICHTENSTEIN, L. M. and OSLER, A. C. (1964). Studies on the mechanism of hypersensitivity phenomena. IX. Histamine release from human leucocytes by ragweed pollen antigen. *J. exp. Med.* **120**, 507.

LINDNER, O. E. A. (1963). Skin compatibility of different CBA sublines separated from each other in the course of varying numbers of generations. *Transplantation*, **1**, 58.

LINNA, J. and STILLSTRÖM, J. (1966). Migration of cells from the thymus to the spleen in young guinea pigs. *Acta path. microbiol. scand.* **68**, 465.

LINSCOTT, W. D. and WEIGLE, W. O. (1965). Induction of tolerance to BSA by means of whole-body X-radiation. *J. Immun.* **94**, 430.

LITT, M. (1963). Eosinophils in lymph nodes during the early phase of the primary immune response. *Fedn Proc. Fedn Am. Socs exp. Biol.* **22**, 380.

LITT, M. (1964). Eosinophils and antigen-antibody reactions. *Ann. N.Y. Acad. Sci.* **116**, 964.

LITTLE, C. C. (1941). The genetics of tumour transplantation, in *Biology of the Laboratory Mouse*, p. 279. Ed. G. D. Snell. Philadelphia: Blakiston.

LITWIN, S. D., ALLEN, J. C. and KUNKEL, H. G. (1966). Disappearance of the clinical and serologic manifestations of rheumatoid arthritis following a thoracotomy for a lung tumor. *Arthritis Rheum.* **9**, 865.

LITWIN, S. D. and KUNKEL, H. G. (1966). Studies on the major subgroup of human γG globulin heavy chains using two new genetic factors. *Fedn Proc. Fedn Am. Socs exp. Biol.* **25**, 371.

LITWIN, S. D. and KUNKEL, H. G. (1967). The genetic control of γ-globulin heavy chains: studies of the major heavy chain subgroup using multiple genetic markers. *J. exp. Med.* **125**, 847.

LONG, G., HOLMES, M. C. and BURNET, F. M. (1963). Auto-antibodies produced against mouse erythrocytes in NZB mice. *Aust. J. exp. Biol. med. Sci.* **41**, 315.

LOUTIT, J. F. (1962). Immunological and trophic functions of lymphocytes. *Lancet*, ii, 1106.

LOUTIT, J. F. (1963). Lymphocytes. *Br. J. Radiol.* **36**, 785.

LOVELL, R. and REES, T. A. (1961). Immunological aspects of colostrum, in *Milk: the Mammary Gland and its Secretion*, vol. 2, p. 363. Ed. S. K. Kon and A. T. Cowie. New York and London: Academic Press.

LURIE, M. (1942). Studies on the mechanism of immunity in tuberculosis: the fate of tubercle bacilli ingested by mononuclear phagocytes derived from normal and immunized animals. *J. exp. Med.* **75**, 247.

LYON, M. F. (1961). Gene action in the X-chromosome of the mouse (*Mus musculus* L.). *Nature, Lond.* **190**, 372.

MCCLUSKEY, R. T., BENACERRAF, B. and MCCLUSKEY, J. W. (1963). Studies on the specificity of the cellular infiltrate in delayed hypersensitivity reactions. *J. Immun.* **90**, 466.

MCCLUSKEY, R. T., BENACERRAF, B. and MILLER, F. (1962). Passive acute glomerulonephritis induced by antigen-antibody complexes solubilized in hapten excess. *Proc. Soc. exp. Biol. Med.* **111**, 764.

MCDEVITT, H. O. and SELA, M. (1965). Genetic control of the antibody response. I. Demonstration of determinant-specific differences in response to synthetic polypeptide antigens in two strains of inbred mice. *J. exp. Med.* **122**, 517.

MCENTEGART, M. G., ROSS, P. W. and BEST, P. V. (1963). Growth of the mast-cell tumour P.815 in thymectomised mice. *Lancet*, ii, 611.

MCFARLAND, W. and HEILMAN, D. H. (1965). Lymphocyte foot appendage: its role in lymphocyte function and in immunological reactions. *Nature, Lond.* **205**, 887.

MCGREGOR, D. D. and GOWANS, J. L. (1963). The antibody response of rats depleted of lymphocytes by chronic drainage from the thoracic duct. *J. exp. Med.* **117**, 303.

MCGREGOR, D. D. and GOWANS, J. L. (1964). Survival of homografts of skin in rats depleted of lymphocytes by chronic drainage from the thoracic duct. *Lancet*, i, 629.

MCINTIRE, K. R., SELL, S. and MILLER, J. F. A. P. (1964). Pathogenesis of the post-neonatal thymectomy wasting syndrome. *Nature, Lond.* **204**, 151.

MACKANESS, G. B. (1954). Growth of tubercle bacilli in monocytes from normal and vaccinated rabbits. *Am. Rev. Tuberc. pulm. Dis.* **69**, 495.

MACKANESS, G. B. (1962). Cellular resistance to infection. *J. exp. Med.* **116**, 381.

# References

MACKANESS, G. B. (1967). The relationship of delayed hypersensitivity to acquired cellular resistance. *Br. med. Bull.* **23**, 52.

MACKAY, I. R. (1965). The thymus in human disease, in *Rep. Walter and Eliza Hall Inst. Res. Path. Med.* 1964–65, p. 52.

MACKAY, I. R. and BURNET, F. M. (1963). *Autoimmune Diseases: Pathogenesis, Chemistry and Therapy*, p. 16. Springfield: Thomas.

MACKAY, I. R. and de GAIL, P. (1963). Thymic 'germinal centres' and plasma cells in systemic lupus erythematosus. *Lancet*, ii, 667.

MCLAREN, A. (1961). Induction of tolerance to skin homografts in adult mice treated with 6-mercaptopurine. *Transplantn Bull.* **28**, 479.

MÄKELÄ, O. (1964a). Evidence that different cells produce different kinds of antibody against the tail of T6. *Immunology*, **7**, 9.

MÄKELÄ, O. (1964b). Studies on the quality of neutralizing bacteriophage antibodies produced by single cells. II. Heterogeneity among anti-T2 cells as shown by different cross reactions. *Annls Med. exp. Biol. Fenn.* **42**, 152.

MÄKELÄ, O. and NOSSAL, G. J. V. (1961). Bacterial adherence: a method for detecting antibody production by single cells. *J. Immun.* **87**, 447.

MÄKELÄ, O. and NOSSAL, G. J. V. (1962). Autoradiographic studies on the immune response. II. DNA synthesis among single antibody-producing cells. *J. exp. Med.* **115**, 231.

MAKINODAN, T., GENGOZIAN, N. and CONGDON, C. C. (1956). Agglutinin production in normal, sublethally irradiated and lethally irradiated mice treated with mouse bone marrow. *J. Immun.* **77**, 250.

MAKINODAN, T., KASTENBAUM, M. A. and PETERSON, W. J. (1962). Radiosensitivity of spleen cells from normal and preimmunized mice and its significance to intact animals. *J. Immun.* **88**, 31.

MAKINODAN, T. and PETERSON, W. J. (1964). Growth and senescence of the primary antibody-forming potential of the spleen. *J. Immun.* **93**, 886.

MALMGREN, R. A., BENNISON, B. E. and MCKINLEY, T. W. (1952). Reduced antibody titers in mice treated with carcinogenic and cancer chemotherapeutic agents. *Proc. Soc. exp. Biol. Med.* **79**, 484.

MANNICK, J. A. and EGDAHL, R. H. (1962). Ribonucleic acid in 'transformation' of lymphoid cells. *Science, N.Y.* **137**, 976.

MARCHALONIS, J. and EDELMAN, G. M. (1965). Phylogenetic origins of antibody structure. I. Multichain structure of immunoglobulins in the smooth dogfish (*Mustelus canis*). *J. exp. Med.* **122**, 601.

MARCHESI, V. T. and GOWANS, J. L. (1964). The migration of lymphocytes through the endothelium of venules in lymph nodes: an electron microscope study. *Proc. R. Soc.* B, **159**, 283.

MARKOWITZ, A. S. and LANGE, C. F. (1964). Streptococcal related glomerulonephritis. I. Isolation immunochemistry and comparative chemistry of soluble fractions from type 12 nephritogenic streptococci and human glomeruli. *J. Immun.* **92**, 565.

MARSH, W. L. (1961). Anti-i: a cold antibody defining the Ii relationship in human red cells. *Br. J. Haemat.* **7**, 200.

MARSH, W. L. and JENKINS, W. J. (1961). Anti-i: a new cold antibody. *Nature, Lond.* **188**, 753.

MARSHALL, A. H. E. and WHITE, R. G. (1961). The immunological reactivity of the thymus. *Br. J. exp. Path.* **42**, 379.

MARSHALL, W. H. and ROBERTS, K. B. (1963). The growth and mitosis of human small lymphocytes after incubation with a phytohaemagglutinin. *Q. Jl exp. Physiol.* **48**, 146.

MÅRTENSSON, L. (1966a). Gm genes and γG molecules. *Acta Univ. lund.* Sec. 2, no. 3.

MÅRTENSSON, L. (1966b). Genes and immunoglobulins. *Vox Sang.* **11**, 521.

MARTINEZ, C. and GOOD, R. A. (1963). Transfer of tolerance to isologous mice, in *Conceptual Advances in Immunology and Oncology*, p. 417. New York: Hoeber.

MARTINEZ, C., SHAPIRO, F. and GOOD, R. A. (1961). Transfer of tolerance induced by parabiosis to isologous newborn mice. *Proc. Soc. exp. Biol. Med.* **107**, 553.

MARTINS, A. B., MOORE, W. D., DICKINSON, J. B. and RAFFEL, S. (1964). Cellular activities in hypersensitive reactions. III. Specifically reactive cells in delayed hypersensitivity: tuberculin hypersensitivity. *J. Immun.* **93**, 953.

MAURER, P. H., PINCHUK, P. and GERULAT, B. F. (1965). Antigenicity of polypeptides (poly alpha amino acids). XIV. Studies on immunological tolerance with structurally related synthetic polymers. *Proc. Soc. exp. Biol. Med.* **118**, 1113.

MAYER, M. M. (1965). Mechanism of haemolysis by complement, in *CIBA Foundation Symposium on Complement*, p. 4. Ed. G. E. W. Wolstenholme and J. Knight. London: Churchill.

MEHROTRA, T. N. (1960a). Individual specific nature of the cold autoantibodies of acquired haemolytic anaemia. *Nature, Lond.* **185**, 323.

MEHROTRA, T. N. (1960b). Immunological identification of the pathological cold auto-antibodies of acquired haemolytic anaemia as $\beta_{2M}$ globulin. *Immunology*, **3**, 265.

MEKORI, T., CHIECO-BIANCI, L. and FELDMAN, M. (1965). Production of clones of lymphoid cell populations. *Nature, Lond.* **206**, 367.

MELLORS, R. C. and BRZOSKO, W. J. (1962). Studies in molecular pathology. I. Localization and pathogenetic role of heterologous immune complexes. *J. exp. Med.* **115**, 891.

MELLORS, R. C., HEIMER, R., CORCOS, J. and KORNGOLD, L. (1959). Cellular origin of rheumatoid factor. *J. exp. Med.* **110**, 875.

MELLORS, R. C. and KORNGOLD, L. (1963). The cellular origin of human immunoglobulins ($\gamma_2$, $\gamma_{1M}$, $\gamma_{1A}$). *J. exp. Med.* **118**, 387.

MELTZER, M., FRANKLIN, E. C., FUDENBERG, H. and FRANGIONE, B. (1964). Single peptide differences between γ-globulins of different genetic (*Gm*) types. *Proc. natn. Acad. Sci. U.S.A.* **51**, 1007.

MEMORANDUM (1965). Nomenclature for human immunoglobulins. *Bull. Wld Hlth Org.* **30**, 447.

*References*

METCALF, D. (1955). The aetiological significance of differing patterns in the age incidence of cancer mortality. *Med. J. Aust.* i, 874.

METCALF, D. (1956a). A lymphocytosis stimulating factor in the plasma of chronic lymphatic leukaemic patients. *Br. J. Cancer,* 10, 169.

METCALF, D. (1956b). The thymic origin of the plasma lymphocytosis stimulating factor. *Br. J. Cancer,* 10, 442.

METCALF, D. (1963). The autonomous behaviour of normal thymus grafts. *Aust. J. exp. Biol. med. Sci.* 41, 437.

METCALF, D. (1964). Functional interactions between the thymus and other organs, in *The Thymus,* Wistar Inst. Symp. Monogr. no. 2, p. 53. Ed. V. Defendi and D. Metcalf. Philadelphia.

METCALF, D. (1966). *The Thymus: Its Role in Immune Responses, Leukaemia Development and Carcinogenesis,* Recent Results in Cancer Research, no. 5. Berlin: Springer Verlag.

METCALF, D., SPARROW, N. and WYLLIE, R. (1962). Alkaline phosphatase activity in mouse lymphoma tissue. *Aust. J. exp. Biol. med. Sci.* 40, 215.

METCALF, D. and WAKONIG-VAARTAJA, R. (1964). Stem cell replacement in normal thymus grafts. *Proc. Soc. exp. Biol. Med.* 115, 731.

METCHNIKOFF, E. (1892). *Leçons sur la pathologie comparée de l'inflammation.* Paris: Masson.

METZGER, H. (1967). Characterization of a human macroglobulin. V. A Waldenström macroglobulin with antibody activity. *Proc. natn. Acad. Sci. U.S.A.* 57, 1490.

METZGER, H. and MANNIK, M. (1964). Recombination of antibody polypeptide chains in the presence of antigen. *J. exp. Med.* 120, 765.

METZGER, H., WOFSY, L. and SINGER, S. J. (1964). The participation of A and B polypeptide chains in the active sites of antibody molecules. *Proc. natn. Acad. Sci. U.S.A.* 51, 612.

MEYER, R. K., RAO, M. A. and ASPINALL, R. L. (1959). Inhibition of the development of the bursa of Fabricius in embryos of the common fowl by 19-nortestosterone. *Endocrinology,* 64, 890.

MICHAEL, J. G. and ROSEN, F. S. (1963). Association of natural antibodies to Gram-negative bacteria with the $\gamma 1$ macroglobulins. *J. exp. Med.* 118, 619.

MICHAELIDES, M. C. and COONS, A. H. (1963). Studies on antibody production. V. The secondary response *in vitro. J. exp. Med.* 117, 1035.

MICKLEM, H. S. and BROWN, J. A. H. (1967). Germinal centers, allograft sensitivity and iso-antibody formation in skin allografted mice, in *Germinal Centers in Immune Responses,* p. 277. Ed. H. Cottier, N. Odartchenko, R. Schindler and C. C. Congdon. Berlin: Springer.

MICKLEM, H. S., FORD, C. E., EVANS, E. P. and GRAY, J. (1966). Interrelationships of myeloid and lymphoid cells: studies with chromosome-marked cells transferred into lethally irradiated mice. *Proc. R. Soc. B,* 165, 78.

MIESCHER, P. (1957). The antigenic constituents of the neutrophil leucocyte with special reference to the LE cell phenomenon. *Vox Sang.* 2, 145.

MIGITA, S. and PUTNAM, F. W. (1962). Antigenic relationships of Bence Jones proteins, myeloma globulins, and normal human γ-globulin. *J. exp. Med.* **117**, 81.

MIKULSKA, Z. B., SMITH, C. and ALEXANDER, P. (1966). Evidence for an immunological reaction of a host against its own actively growing primary tumor. *J. natn. Cancer Inst.* **36**, 29.

MILES, A. A. and WILHELM, D. L. (1955). Enzyme-like globulins from serum reproducing the vascular phenomena of inflammation. I. An activable permeability factor and its inhibitor in guinea-pig serum. *Br. J. exp. Path.* **36**, 71.

MILLER, J. F. A. P. (1962). Effect of neonatal thymectomy on the immunological responsiveness of the mouse. *Proc. R. Soc. B,* **156**, 415.

MILLER, J. F. A. P. (1963). Immunity and the thymus. *Lancet,* i, 43.

MILLER, J. F. A. P. (1964). The thymus and the development of immunologic responsiveness. *Science, N.Y.* **144**, 1544.

MILLER, J. F. A. P., DOAK, S. M. A. and CROSS, A. M. (1963a). Role of the thymus in recovery of immune mechanism in the irradiated adult mouse. *Proc. Soc. exp. Biol. Med.* **112**, 785.

MILLER, J. F. A. P. and DUKOR, P. (1964). *Die Biologie des Thymus.* Basel: Karger.

MILLER, J. F. A. P., GRANT, G. A. and ROE, F. J. C. (1963b). Effect of thymectomy on the induction of skin tumours by 3,4-benzopyrene. *Nature, Lond.* **199**, 920.

MILLER, J. F. A. P., LEUCHARS, E., CROSS, A. M. and DUKOR, P. (1964a). Immunologic role of the thymus in radiation chimeras. *Ann. N.Y. Acad. Sci.* **120**, 205.

MILLER, J. F. A. P., MARSHALL, A. H. E. and WHITE, R. G. (1962). The immunological significance of the thymus. *Adv. Immun.* **2**, 111.

MILLER, J. F. A. P., TING, R. C. and LAW, L. W. (1964b). Influence of thymectomy on tumor induction by polyoma virus in C57BL mice. *Proc. Soc. exp. Biol. Med.* **116**, 323.

MILLER, J. J. and NOSSAL, G. J. V. (1964). Antigens in immunity. VI. The phagocytic reticulum of lymph node follicles. *J. exp. Med.* **120**, 1075.

MILLS, J. and HARDEN, D. (1966). Immunological significance of antigen induced lymphocyte transformation *in vitro. Fedn Proc. Fedn Am. Socs exp. Biol.* **25**, 370.

MILSTEIN, C. (1965). Interchain disulphide bridge in Bence-Jones proteins and in γ-globulin *B* chains. *Nature, Lond.* **205**, 1171.

MILSTEIN, C. (1966). Variations in amino acid sequence near the disulphide bridges of Bence Jones proteins. *Nature, Lond.* **209**, 370.

MIMS, C. A. (1962). Experiments on the origin and fate of lymphocytes. *Br. J. exp. Path.* **43**, 639.

MIMS, C. A. (1964). Aspects of the pathogenesis of virus diseases. *Bact. Rev.* **28**, 30.

MINTZ, B. and SILVERS, W. K. (1967). Intrinsic immunologic tolerance. *Science, N.Y.* **158**, 1484.

*References*

MITCHELL, G. F. and MILLER, J. F. A. P. (1968). Immunological activity of thymus and thoracic-duct lymphocytes. *Proc. natn. Acad. Sci. U.S.A.* **59**, 296.

MITCHELL, J. and ABBOT, A. (1966). Ultrastructure of the antigen-retaining reticulum of lymph node follicles by electron microscopic autoradiography. *Nature, Lond.* **208**, 500.

MITCHELL, J. and NOSSAL, G. J. V. (1966). Mechanism of induction of immunological tolerance. I. Localization of tolerance-inducing antigen. *Aust. J. exp. Biol. med. Sci.* **44**, 211.

MITCHISON, N. A. (1964). Induction of immunological paralysis in two zones of dosage. *Proc. R. Soc. B,* **161**, 275.

MÖLLER, E. (1965). Contact-induced cytotoxicity by lymphoid cells containing foreign isoantigens. *Science, N.Y.* **147**, 873.

MÖLLER, G. (1964). Antibody-induced depression of the immune response: a study of the mechanism in various immunological systems. *Transplantation,* **2**, 405.

MÖLLER, G. (1965). 19S Antibody production against soluble lipopolysaccharide antigens by individual lymphoid cells *in vitro. Nature, Lond.* **207**, 1166.

MÖLLER, G. and WIGZELL, H. (1965). Antibody synthesis at the cellular level: antibody-induced suppression of 19S and 7S antibody response. *J. exp. Med.* **121**, 969.

MOORE, M. A. S. and OWEN, J. J. T. (1966). Experimental studies on the development of the bursa of Fabricius. *Devl Biol.* **14**, 40.

MORGAN, W. T. J. (1960). A contribution to human biochemical genetics: the chemical basis of blood-group specificity. (Croonian lecture.) *Proc. R. Soc. B,* **151**, 308.

MORRIS, B. (1966). Lymphoid cells—their role in the establishment of systemic immunity. *XIth Congr. Internat. Soc. Haematol., Sydney, Plenary sessions,* p. 25.

MORRIS, I. G. (1964). The effects of heterologous sera on the uptake of rabbit antibody from the gut of young mice. *Proc. R. Soc. B,* **148**, 84.

MOTA, I. (1964). The mechanism of anaphylaxis. I. Production and biological properties of 'mast cell sensitizing' antibody. *Immunology,* **7**, 681.

MOVAT, H. Z., LOVETT, C. A. and TAICHMAN, N. S. (1966). Demonstration of antigen on the surface of sensitized rat mast cells. *Nature, Lond.* **212**, 851.

MOWBRAY, J. F. (1963). Inhibition of immune responses by injection of large doses of a serum glycoprotein fraction. *Fedn Proc. Fedn Am. Socs exp. Biol.* **22**, 441.

MUDD, S. (1932). A hypothetical mechanism of antibody formation. *J. Immun.* **23**, 423.

MUELLER, A. P., WOLFE, H. R. and MEYER, R. K. (1960). Precipitin production in chickens. XXI. Antibody production in bursectomized chickens and in chickens injected with 19-nortestosterone on the fifth day of incubation. *J. Immun.* **85**, 172.

MUELLER, A. P., WOLFE, H. R., MEYER, R. K. and ASPINALL, R. L. (1962). Further studies on the role of the bursa of Fabricius in antibody production. *J. Immun.* **88**, 354.

MURAKAMI, W. T., van VUNAKIS, H., LEHRER, H. I. and LEVINE, L. (1962). Immunochemical studies on bacteriophage DNA. III. Specificity of the antibodies. *J. Immun.* **89**, 116.

MURPHY, G. E. and SWIFT, H. F. (1950). The induction of rheumatic-like cardiac lesions in rabbits by repeated focal infections with Group A streptococci: comparison with the cardiac lesions of serum disease. *J. exp. Med.* **91**, 485.

MURRAY, J. and CLARK, E. C. (1952). Production of anti-Rh in guinea-pigs from human erythrocytes. *Nature, Lond.* **169**, 886.

MUSCHEL, L. H. and JACKSON, J. E. (1963). Activity of the antibody-complement system and lysozyme against rough gram negative organisms. *Proc. Soc. exp. Biol. Med.* **113**, 881.

NAGAYA, H. and SIEKER, H. O. (1965). Allograft survival: effect of antiserums to thymus glands and lymphocytes. *Science, N.Y.* **150**, 1181.

NAIRN, R. C. (ed.) (1964). Immunological tracing: general considerations, in *Fluorescent Protein Tracing*, p. 103. 2nd ed. Edinburgh: Livingstone.

NAJARIAN, J. S. and FELDMAN, J. D. (1961). Passive transfer of tuberculin sensitivity by tritiated thymidine-labeled lymphoid cells. *J. exp. Med.* **114**, 779.

NASTUK, W. L., PLESCIA, O. J. and OSSERMAN, K. E. (1960). Changes in serum complement activity in patients with myasthenia gravis. *Proc. Soc. exp. Biol. Med.* **105**, 177.

NEEPER, C. A. and SEASTONE, C. V. (1963*a*). Mechanisms of immunologic paralysis by pneumococcal polysaccharide. I. Studies of adoptively acquired immunity to pneumococcal infection in immunologically paralyzed and normal mice. *J. Immun.* **91**, 374.

NEEPER, C. A. and SEASTONE, C. V. (1963*b*). Mechanisms of immunologic paralysis by pneumococcal polysaccharide. II. The influence of non-specific factors on the immunity of paralyzed mice to pneumococcal infection. *J. Immun.* **91**, 378.

NELSON, D. S. (1963*a*). Reaction to antigen *in vivo* of the peritoneal macrophages of guinea-pigs with delayed-type hypersensitivity: effects of anticoagulants and other drugs. *Lancet*, ii, 175.

NELSON, D. S. (1963*b*). Immune adherence. *Adv. Immun.* **3**, 131.

NELSON, D. S. (1965). Immune adherence, in *CIBA Foundation Symposium on Complement*, p. 222. Ed. G. E. W. Wolstenholme and J. Knight. London: Churchill.

NELSON, D. S. and BOYDEN, S. V. (1963). The loss of macrophages from peritoneal exudates following the injection of antigens into guinea-pigs with delayed-type hypersensitivity. *Immunology*, **6**, 264.

NELSON, D. S. and MILDENHALL, P. (1967). Studies in cytophilic antibodies. 1. The production by mice of macrophage cytophilic antibodies to sheep erythrocytes: relationship to the production of other antibodies and the development of delayed-type hypersensitivity. *Aust. J. exp. Biol. med. Sci.* **45**, 113.

NELSON, R. A. (1965). The role of complement in immune phenomena, in *The Inflammatory Process*, p. 819. Ed. B. W. Zweifach, L. Grant and R. T. McCluskey. New York and London: Academic Press.

NETER, E., WHANG, H. Y., LUDERITZ, O. and WESTPHAL, O. (1966). Immunological priming without production of circulating bacterial antibodies conditioned by endotoxin and its lipoid A component. *Nature, Lond.* **212**, 420.

NETTESHEIM, P. and MAKINODAN, T. (1965). Differentiation of lymphocytes undergoing an immune response in diffusion chambers. *J. Immun.* **94**, 868.

NISONOFF, A. (1963). Synthesis and properties of hybrid rabbit antibodies, in *Conceptual Advances in Immunology and Oncology*, p. 273. New York: Hoeber.

NJOKU-OBI, A. N. and OSEBOLD, J. W. (1962). Studies on mechanisms of immunity in listeriosis: I. Interaction of peritoneal exudate cells from sheep with *Listeria monocytogenes in vitro. J. Immun.* **89**, 187.

NOSSAL, G. J. V. (1958). Antibody production by single cells. *Br. J. exp. Path.* **39**, 544.

NOSSAL, G. J. V. (1962). Cellular genetics of immune responses, in *Adv. Immun.* **2**, 163.

NOSSAL, G. J. V. (1964). Studies on the rate of seeding of lymphocytes from the intact guinea-pig thymus. *Ann. N.Y. Acad. Sci.* **120**, 171.

NOSSAL, G. J. V. and ABBOT, A. (1966). Lymphoid cells and antibody formation, in *XIth Congr. Internat. Soc. Blood Transf., Sydney, Plenary sessions,* p. 5.

NOSSAL, G. J. V., ADA, G. L. and AUSTIN, C. M. (1964a). Antigens in immunity. II. Immunogenic properties of flagella, polymerized flagellin and flagellin in the primary response. *Aust. J. exp. Biol. med. Sci.* **42**, 283.

NOSSAL, G. J. V., ADA, G. L. and AUSTIN, C. M. (1964b). Antigens in immunity. IV. Cellular localization of $^{125}$I- and $^{131}$I-labelled flagella in lymph nodes. *Aust. J. exp. Biol. med. Sci.* **42**, 311.

NOSSAL, G. J. V. and LEDERBERG, J. (1958). Antibody production by single cells. *Nature, Lond.* **181**, 1419.

NOSSAL, G. J. V. and MÄKELÄ, O. (1962a). Autoradiographic studies of the immune response. I. The kinetics of plasma cell proliferation. *J. exp. Med.* **115**, 209.

NOSSAL, G. J. V. and MÄKELÄ, O. (1962b). Elaboration of antibodies by single cells. *A. Rev. Microbiol.* **16**, 53.

NOSSAL, G. J. V., SZENBERG, A., ADA, G. L. and AUSTIN, C. M. (1964c). Single cell studies on 19S antibody production. *J. exp. Med.* **119**, 485.

NOWELL, P. C. (1960). Phytohemagglutinin: an initiator of mitosis in cultures of normal human lymphocytes. *Cancer Res.* **20**, 462.

ODARTCHENKO, N., LEWERENZ, M., SORDAT, B., ROOS, B. and COTTIER, H. (1967). Kinetics of cellular death in germinal centers of mouse spleen, in *Germinal Centers in Immune Responses*, p. 212. Ed. H. Cottier, N. Odartchenko, R. Schindler and C. C. Congdon. Berlin: Springer.

OHNO, S. (1966). Cytologic and genetic evidence of somatic segregation in mammals, birds and fishes. *Proceedings of the Symposium of the Tissue Culture Association, San Francisco.*

OHNO, S., WEILER, C., POOLE, J., CHRISTIAN, L. and STENIUS, C. (1966). Autosomal polymorphism due to pericentric inversions in the deer mouse (*Petromyscus maniculatus*) and some evidence of somatic segregation. *Chromosoma*, **18**, 177.

OKABAYASHI, A. (1964). Induction of a disease resembling systemic lupus erythematosus in later stage of prolonged sensitization in rabbits. *Acta path. jap.* **14**, 345.

OLD, L. J., BOYSE, E. A., CLARKE, D. A. and CARSWELL, E. A. (1962). Antigenic properties of chemically induced tumors. *Ann. N. Y. Acad. Sci.* **101**, 80.

OLD, L. J., BOYSE, E. A., OETTGEN, H. F., de HARVEN, E., GEERING, G., WILLIAMSON, B. and CLIFFORD, P. (1966). Precipitating antibody in human serum to an antigen present in cultured Burkitt's lymphoma cells. *Proc. natn. Acad. Sci. U.S.A.* **56**, 1699.

OLD, L. J., BOYSE, E. A. and STOCKERT, E. (1963). Antigenic properties of experimental leukemias. I. Serological studies *in vitro* with spontaneous and radiation-induced leukemias. *J. natn. Cancer Inst.* **31**, 977.

OLD, L. J., BOYSE, E. A. and STOCKERT, E. (1964). Typing of mouse leukaemias by serological methods. *Nature, Lond.* **201**, 777.

OLSON, G. B. and WOSTMANN, B. S. (1966a). Lymphocytopoiesis, plasmacytopoiesis and cellular proliferation in nonantigenically stimulated germfree mice. *J. Immun.* **97**, 267.

OLSON, G. B. and WOSTMANN, B. S. (1966b). Cellular and humoral response of germfree mice stimulated with 7S HGG or *Salmonella typhimurium*. *J. Immun.* **97**, 275.

ONOUE, K., YAGI, Y., STELOS, P. and PRESSMAN, D. (1964). Antigen-binding activity of 6S subunits of $B_2$ macroglobulin antibody. *Science, N.Y.* **146**, 404.

OORT, J. and TURK, J. L. (1965). A histological and autoradiographic study of lymph nodes during the development of contact sensitivity in the guinea-pig. *Br. J. exp. Path.* **46**, 147.

ORGEL, L. E. (1963). The maintenance of the accuracy of protein synthesis and its relevance to ageing. *Proc. natn. Acad. Sci. U.S.A.* **49**, 517.

ORTEGA, L. G. and MELLORS, R. C. (1956). Analytical pathology. IV. The role of localized antibodies in the pathogenesis of nephrotoxic nephritis in the rat. *J. exp. Med.* **104**, 151.

ORTEGA, L. G. and MELLORS, R. C. (1957). Cellular sites of formation of gamma globulin. *J. exp. Med.* **106**, 627.

OSBORNE, J. J., DANCIS, J. and JULIA, J. E. (1952). Studies on diphtheria immunization in very young children. *Pediatrics, Springfield*, **9**, 736.

OSMOND, D. C. and EVERETT, N. B. (1964). Radioautographic studies of bone marrow lymphocytes *in vivo* and in diffusion chamber cultures. *Blood*, **23**, 1.

## References

OSOBA, D. and MILLER, J. F. A. P. (1964). The lymphoid tissues and immune responses of neonatally thymectomized mice bearing thymic tissues in millipore diffusion chambers. *J. exp. Med.* **119**, 177.

OSSERMAN, E. F. and TAKATSUKI, K. (1963). Plasma cell myeloma: gamma globulin synthesis and structure. *Medicine, Baltimore*, **42**, 357.

OSSERMAN, K. E. and WEINER, L. B. (1964). Studies in myasthenia gravis: immunofluorescent tagging of muscle striation with antibody from serums of 256 myasthenic patients. *Ann. N.Y. Acad. Sci.* **124**, 730.

OSTERLAND, C. K., MILLER, E. J., KARAKAWA, W. W. and KRAUS, R. M. (1966). Characteristics of streptococcal group-specific antibody isolated from hyperimmune rabbits. *J. exp. Med.* **123**, 599.

OUCHTERLONY, O. (1948). *In vitro* method for testing toxin-producing capacity of diphtheria bacilli. *Acta path. microbiol. scand.* **25**, 186.

OVARY, Z. and KARUSH, F. (1961). Studies on the immunologic mechanism of anaphylaxis. II. Sensitizing and combining capacity *in vivo* of fractions separated from papain digests of anti-hapten antibody. *J. Immun.* **86**, 146.

OWEN, R. D. (1945). Immunogenetic consequences of vascular anastomoses between bovine twins. *Science, N.Y.* **102**, 400.

PANUM, P. L. (1847). Beobachtungen über das Maserncontagium. *Virchows Arch. path. Anat. Physiol.* **1**, 492.

PAPERMASTER, B. W., CONDIE, R. M., FINSTAD, J. and GOOD, R. A. (1964). Evolution of the immune response. I. The phylogenetic development of adaptive immunologic responsiveness in vertebrates. *J. exp. Med.* **119**, 105.

PAPERMASTER, B. W., CONDIE, R. M. and GOOD, R. A. (1962). Immune response in the Californian hagfish. *Nature, Lond.* **196**, 355.

PAPPENHEIMER, A. M. (1956). Hypersensitivity of the delayed type. *Harvey Lect.* **52**, 100.

PARROTT, D. M. V. (1967). The integrity of the germinal center: an investigation of the differential localization of labeled cells in lymphoid organs, in *Germinal Centers in Immune Responses*, p. 168. Ed. H. Cottier, N. Odartchenko, R. Schindler and C. C. Congdon. Berlin: Springer.

PARROTT, D. M. V. and de SOUSA, M. A. B. (1966). Changes in the thymus-dependent areas of lymph nodes after immunological stimulation. *Nature, Lond.* **212**, 1316.

PARROTT, D. M. V., de SOUSA, M. A. B. and EAST, J. (1966). Thymus-dependent areas in the lymphoid areas of neonatally thymectomized mice. *J. exp. Med.* **123**, 191.

PARROTT, D. M. V. and EAST, J. (1962). Role of the thymus in neonatal life. *Nature, Lond.* **195**, 347.

PATERSON, P. Y. (1966). Experimental allergic encephalomyelitis and autoimmune disease. *Adv. Immun.* **5**, 131.

PATON, D. M. (1961). A theory of drug action based on the rate of drug-receptor combination. *Proc. R. Soc.* B, **154**, 21.

PAULING, L. (1940). A theory of the structure and process of formation of antibodies. *J. Am. chem. Soc.* **62**, 2643.

698

References

PEARMAIN, G. E., LYCETTE, R. R. and FITZGERALD, P. H. (1963). Tuberculin-induced mitosis in peripheral blood leucocytes. *Lancet*, i, 637.
PEARSON, C. M. and WOOD, F. D. (1959). Studies of polyarthritis and other lesions induced in rats by injection of mycobacterial adjuvant. I. General clinical and pathologic characteristics and some modifying factors. *Arthritis Rheumat.* **2**, 440.
PERKINS, E. H. and LEONARD, M. R. (1963). Specificity of phagocytosis as it may relate to antibody function. *J. Immun.* **90**, 228.
PERKINS, E. H. and MAKINODAN, T. (1964). Relative pool size of potentially competent antibody-forming cells of primed and nonprimed spleen cells grown in *in vivo* culture. *J. Immun.* **92**, 192.
PERLMANN, P., HAMMARSTRÖM, S., LAGERCRANTZ, R. and GUSTAFSSON, B. E. (1965). Antigen from colon of germfree rats and antibodies in human ulcerative colitis. *Ann. N.Y. Acad. Sci.* **124**, 377.
PERNIS, B. (1967). The immunoglobulins present in the germinal centers, in *Germinal Centers in Immune Responses*, p. 112. Ed. H. Cottier, N. Odartchenko, R. Schindler and C. C. Congdon. Berlin: Springer.
PERNIS, B. and CHIAPPINO, G. (1964). Identification in human lymphoid tissues of cells that produce group I or group II gamma-globulins. *Immunology*, **7**, 500.
PERNIS, B., CHIAPPINO, G., KELUS, A. S. and GELL, P. G. H. (1965). Cellular localization of immunoglobulins with different allotype specificities in rabbit lymphoid tissues. *J. exp. Med.* **122**, 853.
PERNIS, B., CHIAPPINO, G. and ROWE, D. S. (1966). Cells producing Ig D immunoglobulins in human spleen. *Nature, Lond.* **211**, 424.
PETERSON, R. D. A., COOPER, M. D. and GOOD, R. A. (1965). The pathogenesis of immunologic deficiency diseases. *Am. J. Med.* **38**, 579.
PETRAKIS, N. L., DAVIS, M. and LUCIA, S. P. (1961). The *in vivo* differentiation of human leukocytes into histiocytes, fibroblasts and fat cells in subcutaneous diffusion chambers. *Blood*, **17**, 109.
PIERCE, A. E. and FEINSTEIN, G. (1965). Biophysical and immunological studies on bovine immune globulins with evidence for selective transport within the mammary gland from maternal plasma to colostrum. *Immunology*, **8**, 106.
PINCHUCK, P. and MAURER, P. H. (1965). Antigenicity of polypeptides. XVI. Genetic control of immunogenicity of synthetic polypeptides in mice. *J. exp. Med.* **122**, 673.
PIZARRO, D., HOECKER, G., RUBINSTEIN, P. and RAMOS, A. (1961). The distribution in the tissues and the development of H2 antigens in the mouse. *Proc. natn. Acad. Sci. U.S.A.* **47**, 1900.
PLAYFAIR, J. H. L., PAPERMASTER, B. W. and COLE, L. J. (1965). Focal antibody production by transferred spleen cells in irradiated mice. *Science, N.Y.* **149**, 998.
PLESCIA, O. J., BRAUN, W. and PALCZUK, N. C. (1964). Production of antibodies to denatured DNA. *Proc. natn. Acad. Sci. U.S.A.* **52**, 279.
PLIMPTON, C. H. and GELLHORN, A. (1956). Hypercalcemia in malignant disease without evidence of bone destruction. *Am. J. Med.* **21**, 750.

References

POLLARA, B., COOPER, M. D. and GOOD, R. A. (1966). Studies of the mucous surface immunoglobulin system. *Fedn Proc. Fedn Am. Socs exp. Biol.* **25**, 653.

PORTER, R. R. (1963). Chemical structure of γ-globulin antibodies. *Br. med. Bull.* **19**, 197.

POTTER, M., APPELLA, E. and GEISSER, S. (1965). Variations in the heavy polypeptide chain structure of gamma myeloma immunoglobulins from an inbred strain of mice and a hypothesis as to their origin. *J. molec. Biol.* **14**, 361.

POTTER, M. and MACCARDLE, R. C. (1964). Histology of developing plasma cell neoplasia induced by mineral oil in BALB/c mice. *J. natn. Cancer Inst.* **33**, 497.

PREHN, R. T. (1963). Function of depressed immunological reactivity during carcinogenesis. *J. natn. Cancer Inst.* **31**, 791.

PREHN, R. T. (1964). A clonal selection theory of chemical carcinogenesis. *J. natn. Cancer Inst.* **32**, 1.

PREHN, R. T. and MAIN, J. M. (1957). Immunity to methylcholanthrene-induced sarcomas. *J. natn. Cancer Inst.* **20**, 207.

PRESSMAN, D. (1963). The chemical nature of the combining site of antibody molecules, in *Conceptual Advances in Immunology and Oncology*, p. 290. New York: Hoeber.

PUTNAM, F. W., TITANI, K. and WHITNEY, E. (1966a). Chemical structure of light chains: amino acid sequence of type K Bence Jones proteins. *Proc. R. Soc. B*, **166**, 124.

PUTNAM, F. W., TITANI, K., WHITNEY, E. and AVOGARDO, L. (1966b). Biological significance of sequence variations in Bence Jones proteins. *Fedn Proc. Fedn Am. Socs exp. Biol.* **25**, 373.

RACE, R. R. and SANGER, R. (1962). *Blood Groups in Man*, 4th ed. Oxford: Blackwell Scientific Publications.

RAFFEL, S. and NEWEL, J. M. (1958). The 'delayed hypersensitivity' induced by antigen-antibody complexes. *J. exp. Med.* **108**, 823.

RAMMELKAMP, C. H. (1964). Concepts of pathogenesis of glomerulonephritis derived from studies in man, in *The Streptococcus, Rheumatic Fever and Glomerulonephritis*, p. 289. Ed. J. W. Uhr. Baltimore: Williams and Wilkins.

RAPAPORT, F. T., LAWRENCE, H. S., MILLAR, J. W., PAPPAGIANIS, D. and SMITH, C. E. (1960a). The immunologic properties of coccidioidin as a skin test reagent in man. *J. Immun.* **84**, 368.

RAPAPORT, F. T., LAWRENCE, H. S., MILLAR, J. W., PAPPAGIANIS, D. and SMITH, C. E. (1960b). Transfer of delayed hypersensitivity to coccidioidin in man. *J. Immun.* **84**, 358.

RAUSCH, H. C. and RAFFEL, S. (1964). Antigenic uptake by specifically reactive cells in experimental allergic encephalomyelitis. *Ann. N.Y. Acad. Sci.* **122**, 297.

REIF, A. E. and ALLEN, J. M. V. (1964). The AKR thymic antigen and its distribution in leukemias and nervous tissues. *J. exp. Med.* **120**, 413.

REISS, E., MERTENS, E. and EHRICH, W. E. (1950). Agglutination of bacteria by lymphoid cells *in vitro*. *Proc. Soc. exp. Biol. Med.* **74**, 732.

RICHARDSON, M. and DUTTON, R. W. (1964). Antibody-synthesizing cells: appearance after secondary antigenic stimulation *in vitro*. *Science, N.Y.* **146**, 655.

RIDGES, A. P. and AUGUSTIN, R. (1964). An *in vitro* test for atopic reagins by double-layer leucocyte agglutination. *Nature, Lond.* **202**, 667.

RIEDER, R. S. and OUDIN, J. (1963). Studies on the relationship of allotypic specificities to antibody specificities in the rabbit. *J. exp. Med.* **118**, 627.

RITTENBERG, M. B. and NELSON, E. L. (1960). Macrophages, nucleic acid and the induction of antibody formation. *Am. Nat.* **94**, 321.

RIVERS, T. M. and SCHWENTKER, F. F. (1935). Encephalomyelitis accompanied by myelin destruction experimentally produced in monkeys. *J. exp. Med.* **61**, 689.

ROBBINS, J., EITZMAN, D. V. and SMITH, R. T. (1963). The catabolism of protein antigens in newborn and maturing rabbits. *J. exp. Med.* **118**, 959.

ROBBINS, W. H., HOLMAN, H. R., DEICHER, H. R. and KUNKEL, H. G. (1957). Complement fixation with cell nuclei and DNA in lupus erythematosus. *Proc. Soc. exp. Biol. Med.* **96**, 575.

ROBERTS, J. C. and DIXON, F. J. (1956). The morphology of antibody-producing lymph node and peritoneal cells transferred in immunologically inert recipients. *Am. J. Path.* **32**, 625.

ROBINSON, E., SHULMAN, J., BEN-HUR, N., ZUCKERMAN, H. and NEUMAN, Z. (1963). Immunological studies and behaviour of husband and foreign homografts in patients with chorionepithelioma. *Lancet*, i, 300.

ROCKEY, J. H., HANSON, L. A., HEREMANS, J. F. and KUNKEL, H. G. (1964*a*). $\beta_{2A}$ Aglobulinemia in two healthy men. *J. Lab. clin. Med.* **63**, 205.

ROCKEY, J. H., KLINMAN, N. R. and KARUSH, F. (1964*b*). Equine antihapten antibody. I. 7S $\beta_{2A}$- and 10S $\gamma_1$-globulin components of purified anti-$\beta$-lactoside antibody. *J. exp. Med.* **120**, 589.

ROGISTER, G. (1967). Development of lymphatic tissues in early thymectomized 'Swiss Albino' mice, in *Germinal Centers in Immune Responses*, p. 356. Ed. H. Cottier, N. Odartchenko, R. Schindler and C. C. Congdon. Berlin: Springer.

ROHOLT, O. A., RADZIMSKI, G. and PRESSMAN, D. (1963). Antibody combining site: the B polypeptide chain. *Science, N.Y.* **141**, 726.

ROHOLT, O. A., RADZIMSKI, G. and PRESSMAN, D. (1965*a*). Polypeptide chains of antibody: effective binding sites require specificity in combination. *Science, N.Y.* **147**, 613.

ROHOLT, O. A., RADZIMSKI, G. and PRESSMAN, D. (1965*b*). Preferential recombination of antibody chains to form effective binding sites. *J. exp. Med.* **122**, 785.

## References

ROSENAU, W. and MOON, H. D. (1961). Lysis of homologous cells by sensitized lymphocytes in tissue culture. *J. natn. Cancer Inst.* **27**, 471.

ROSEVEAR, J. W. and SMITH, E. L. (1961). Glycopeptides. I. Isolation and properties of glycopeptides from a fraction of human γ-globulin. *J. biol. Chem.* **236**, 425.

ROTH, N. (1962). Zur semiquantitativen Erfassung der beiden Serum-Immun-Globuline $\beta_{2A}$ und $\beta_{2M}$ im Neugeborenen- und Kindesalter. *Annls paediat.* **199**, 548.

ROTHMAN, W. and LIDEN, S. (1965). Isolation of a lymphoid cell protein with relation to delayed hypersensitivity. *Nature, Lond.* **208**, 389.

ROTHSCHILD, E. D., MYERS, W. P. L. and LAWRENCE, W. (1964). Cancer hypercalcemia and total parathyroidectomy. *Clin. Res.* **12**, 462.

ROWE, D. S. and FAHEY, J. L. (1965). A new class of human immunoglobulin. II. Normal serum Ig D. *J. exp. Med.* **121**, 185.

ROWE, D. S. and MCGREGOR, I. A. (1967). Pers. comm.

ROWLEY, D. (1962). Phagocytosis. *Adv. Immun.* **2**, 241.

ROWLEY, D. (1964). Endotoxin-induced changes in susceptibility to infections, in *Bacterial Endotoxins*, p. 359. Ed. M. Landy and W. Braun. New Brunswick: Institute of Microbiology, Rutgers University.

ROWLEY, D. and TURNER, K. J. (1966). Number of molecules required to promote phagocytosis of one bacterium. *Nature, Lond.* **210**, 496.

ROWLEY, D. A. and FITCH, F. W. (1965). The mechanism of tolerance produced in rats to sheep erythrocytes. II. The plaque-forming cell and antibody response to multiple injections of antigen begun at birth. *J. exp. Med.* **121**, 683.

ROWLEY, D. A. and FITCH, F. W. (1968). Clonal selection and inhibition of the primary response by antibody, in *Regulation of the Antibody Response*, p. 127. Ed. B. Cinader. Springfield: Thomas.

RUSSE, H. P. and CROWLE, A. J. (1965). A comparison of thymectomized and antithymocyte serum-treated mice in their development of hypersensitivity to protein antigens. *J. Immun.* **94**, 74.

RYTÖMAA, T. (1960). Organ distribution and histochemical properties of eosinophil granulocytes in the rat. *Acta path. microbiol. scand.* **50**, Supp. 140, 1.

SAGER, R. and RYAN, F. J. (1961). *Cell Heredity*, pp. 318, 319. New York and London: Wiley.

SAHIAR, K. and SCHWARTZ, R. S. (1965). The immunoglobulin sequence. I. Arrest by 6-mercaptopurine and restitution by antibody, antigen or splenectomy. *J. Immun.* **95**, 345.

ST PIERRE, R. L. and ACKERMAN, G. A. (1965). Bursa of Fabricius in chickens: possible humoral factor. *Science, N.Y.* **147**, 1307.

SAINTE-MARIE, G. and COONS, A. H. (1964). Studies on antibody production. X. Mode of formation of plasmacytes in cell transfer experiments. *J. exp. Med.* **119**, 743.

SAINTE-MARIE, G. and LEBLOND, C. P. (1958). Origin and fate of cells in the medulla of the rat thymus. *Proc. Soc. exp. Biol. Med.* **98**, 909.

SAINTE-MARIE, G. and LEBLOND, C. P. (1964). Thymus-cell population dynamics, in *The Thymus in Immunobiology*, p. 207. Ed. R. A. Good and A. E. Gabrielsen. New York: Hoeber.

SALT, G. (1960). Experimental studies in insect parasitism. XI. The haemocytic reaction of a caterpillar under varied conditions. *Proc. R. Soc. B*, **151**, 446.

SALVIN, S. B. and SMITH, R. F. (1960). The specificity of allergic reactions. I. Delayed versus Arthus hypersensitivity. *J. exp. Med.* **111**, 465.

SALVIN, S. B. and SMITH, R. F. (1961). The specificity of allergic reactions. III. Contact hypersensitivity. *J. exp. Med.* **114**, 185.

SALVIN, S. B. and SMITH, R. F. (1964). The specificity of allergic reactions. VII. Immunologic unresponsiveness, delayed hypersensitivity and circulating antibody to proteins and hapten–protein conjugates in adult guinea-pigs. *J. exp. Med.* **119**, 851.

SANTOS, G. W. (1967). Immunosuppressive drugs I. *Fedn Proc. Fedn Am. Socs exp. Biol.* **26**, 907.

SAWYER, W. A. (1931). Persistence of yellow fever immunity. *J. prev. Med., Baltimore*, **5**, 413.

SCHALLER, J., DAVIS, S. D., CHING, Y.-C., LAGUNOFF, D., WILLIAMS, C. P. S. and WEDGWOOD, R. J. (1966). Hypergammaglobulinaemia, antibody deficiency, autoimmune haemolytic anaemia and nephritis in an infant with familial lymphopenic immune defect. *Lancet*, ii, 825.

SCHAPIRA, F., DREYFUS, J. C. and SCHAPIRA, G. (1963). Anomaly of aldolase in primary liver cancer. *Nature, Lond.* **200**, 995.

SCHAYER, R. W. (1964). A unified theory of gluco-corticoid action. *Perspect. Biol. Med.* **8**, 71.

SCHEINBERG, S. L. and RECKEL, R. P. (1960). Induced somatic mutations affecting erythrocyte antigens. *Science, N.Y.* **131**, 1887.

SCHIERMAN, L. W. and NORDSKOG, A. W. (1963). Influence of the B blood group-histocompatibility locus in chickens on a graft-versus-host reaction. *Nature, Lond.* **197**, 511.

SCHINCKEL, P. G. and FERGUSON, K. A. (1953). Skin transplantation in the foetal lamb. *Aust. J. biol. Sci.* **6**, 533.

SCHLOSSMAN, S. and LEVINE, H. (1967). Immunochemical studies on delayed and Arthus-type hypersensitivity reactions. I. The relationship between antigenic determinant size and antibody combining site size. *J. Immun.* **98**, 211.

SCHOFIELD, F. S. (1957). The serum protein pattern of West Africans in Britain. *Trans. R. Soc. trop. Med. Hyg.* **51**, 332.

SCHROENLOHER, R. E., KUNKEL, H. G. and TOMASI, T. B. (1964). Activity of dissociated and reassociated 19S anti-γ-globulins. *J. exp. Med.* **120**, 1215.

SCHUBERT, W. K., FOWLER, R., MARTIN, L. W. and WEST, C. D. (1960). Homograft rejection in children with congenital immunological defects: agammaglobulinemia and Aldrich syndrome. *Transplantn Bull.* **26**, 125.

SCHUBOTHE, H. (1965). Current problems of chronic cold hemagglutinin disease. *Ann. N.Y. Acad. Sci.* **124**, 484.

SCHWAB, J. H. and CROMARTIE, W. J. (1957). Studies on a toxic cellular component of Group A streptococci. *J. Bact.* **74**, 673.

SCHWARTZ, J. H. and EDELMAN, G. M. (1963). Comparisons of Bence Jones proteins and polypeptide chains of myeloma globulins after hydrolysis with trypsin. *J. exp. Med.* **118**, 41.

SCHWARTZ, R. S. (1963). Alteration of immunity by antimetabolites, in *Conceptual Advances in Immunology and Oncology*, p. 137. New York: Hoeber.

SCHWARTZ, R. S. and ANDRÉ, J. (1960). Clearance of proteins from blood of normal and 6-mercaptopurine treated rabbits. *Proc. Soc. exp. Biol. Med.* **104**, 228.

SCHWARTZ, R. S. and DAMESHEK, W. (1959). Drug-induced immunological tolerance. *Nature, Lond.* **183**, 1682.

SELA, M. and MOZES, E. (1966). Dependence of the chemical nature of antibodies on the net electrical charge of antigens. *Proc. natn. Acad. Sci. U.SA.* **55**, 445.

SELL, S. (1967). Studies on rabbit lymphocytes *in vitro*. VI. The induction of blast transformation with sheep antisera to rabbit Ig A and Ig M. *J. exp. Med.* **125**, 393.

SELL, S. and GELL, P. (1965). Studies on rabbit lymphocytes *in vitro*. I. Stimulation of blast transformation with an anti-allotype serum. *J. exp. Med.* **122**, 423.

SERCARZ, E. and COONS, A. H. (1960). The exhaustion of specific antibody-producing capacity during a secondary antibody response. *Fedn Proc. Fedn Am. Socs exp. Biol.* **19**, 199.

SERCARZ, E. E. and COONS, A. H. (1963). The absence of antibody-producing cells during unresponsiveness to BSA in the mouse. *J. Immun.* **90**, 478.

SETLOW, R. B. (1964). Physical changes and mutagenesis. *J. cell. comp. Physiol.* **64**, Supp. 1, 51.

SHAPIRO, A. L., SCHARFF, M. D., MAIZEL, J. V. and UHR, J. W. (1966*a*). Synthesis of excess light chains of gamma globulin by rabbit lymph node cells. *Nature, Lond.* **211**, 243.

SHAPIRO, A. L., SCHARFF, M. D., MAIZEL, J. V. and UHR, J. W. (1966*b*). Polyribosomal synthesis and assembly of the H and L chains of gamma globulin. *Proc. natn. Acad. Sci. U.S.A.* **56**, 216.

SHELTON, E. and RICE, M. E. (1959). Growth of normal peritoneal cells in diffusion chambers: a study in cell modulation. *Am. J. Anat.* **105**, 281.

SHULMAN, A. (1964). Modern aspects of heterocyclic chemistry; the importance of structure/action studies to the chemist and biologist. *Proc. R. Aust. chem. Inst.* **31**, 41.

SHULMAN, N. R. (1964). A mechanism of cell destruction in individuals sensitized to foreign antigens and its implications in autoimmunity. *Ann. intern. Med.* **60**, 506.

SHULMAN, N. R., MARDER, V. J. and WEINRACH, R. S. (1965). Similarities between known antiplatelet antibodies and the factor responsible for thrombocytopenia in idiopathic purpura: physiologic, serologic and isotopic studies. *Ann. N.Y. Acad. Sci.* **124**, 499.

SILVERSTEIN, A. M. (1964). Ontogeny of the immune response. *Science*, *N.Y.* **144**, 1423.

SILVERSTEIN, A. M. and BOREK, F. (1966). Desensitization studies of delayed hypersensitivity with special reference to the possible role of high affinity antibodies. *J. Immun.* **96**, 953.

SILVERSTEIN, A. M. and LUKES, R. J. (1962). Foetal response to antigenic stimulus. I. Plasma-cellular and lymphoid reactions in the human foetus to intrauterine injection. *Lab. Invest.* **11**, 918.

SILVERSTEIN, A. M., UHR, J. W. and KRAMER, K. L. (1963). Foetal response to antigenic stimulus. II. Antibody production by the foetal lamb. *J. exp. Med.* **117**, 799.

SILVERSTEIN, M. N., WAKIM, K. G. and BAHN, R. C. (1964). Hypoglycemia associated with neoplasia: a review. *Am. J. Med.* **36**, 415.

SIMAR, L., BETZ, E. H. and LEJEUNE, G. (1967). Ultrastructural modifications of the lymph nodes after homologous skin grafting in the mouse, in *Germinal Centers in Immune Responses*, p. 60. Ed. H. Cottier, N. Odartchenko, R. Schindler and C. C. Congdon. Berlin: Springer.

SIMMONS, R. L. and RUSSELL, P. S. (1962). Antigenicity of mouse trophoblast. *Ann. N.Y. Acad. Sci.* **99**, 717.

SIMONSEN, M. (1957). The impact on the developing embryo and newborn animal of adult homologous cells. *Acta path. microbiol. scand.* **40**, 480.

SIMONSEN, M. (1960). Identification of immunologically competent cells, in *CIBA Foundation Symposium on Cellular Aspects of Immunity*, p. 122. Ed. G. E. W. Wolstenholme and M. O'Connor. London: Churchill.

SIMONSEN, M. (1962). The factor of immunization: clonal selection theory investigated by spleen assays of graft-versus-host reaction, in *CIBA Foundation Symposium on Transplantation*, p. 185. Ed. G. E. W. Wolstenholme and M. P. Cameron. London: Churchill.

SIMPSON, J. A. (1960). Myasthenia gravis—a new hypothesis. *Scott. med. J.* **5**, 419.

SINGER, S. J. and DOOLITTLE, R. F. (1966). Antibody active sites and immunoglobulin molecules. *Science*, *N.Y.* **153**, 13.

SISKIND, G. W. and HOWARD, J. G. (1966). Studies on the induction of immunological unresponsiveness to pneumococcal polysaccharide in mice. *J. exp. Med.* **124**, 417.

SJÖGREN, H. O. (1961). Further studies on the induced resistance against isotransplantation of polyoma tumors. *Virology*, **15**, 214.

SJÖGREN, H. O., HELLSTRÖM, I. and KLEIN, G. (1961*a*). Transplantation of polyoma virus-induced tumors in mice. *Cancer Res.* **21**, 329.

SJÖGREN, H. O., HELLSTRÖM, I. and KLEIN, G. (1961*b*). Resistance of polyoma virus immunized mice to transplantation of established polyoma tumors. *Exp. Cell Res.* **23**, 204.

SLOAN, H. E. (1943). The thymus in myasthenia gravis. *Surgery, St Louis*, **13**, 154.

SMITH, E. L. (1946). The immune proteins of bovine colostrum and plasma. *J. biol. Chem.* **164**, 345.

SMITH, R. T. (1960). Response to active immunization of human infants during the neonatal period, in *CIBA Foundation Symposium on Cellular Aspects of Immunity*, p. 348. Ed. G. E. W. Wolstenholme and M. O'Connor. London: Churchill.

SMITHIES, O. (1964). Chromosomal rearrangements and protein structure. *Cold Spring Harb. Symp. quant. Biol.* **29**, 309.

SMITHIES, O. (1967). Antibody variability. *Science, N.Y.* **157**, 267.

SNELL, G. D. (1953). The genetics of transplantation. *J. natn. Cancer Inst.* **14**, 691.

SNELL, G. D. (1958). Histocompatibility genes of the mouse. II. Production and analysis of isogenic resistant lines. *J. natn. Cancer Inst.* **21**, 843.

SNELL, G. D. (1963). The immunology of tissue transplantation, in *Conceptual Advances in Immunology and Oncology*, p. 323. New York: Hoeber.

SOUTH, M. A., GOOD, R. A. and WOLHEIM, F. A. (1966). Salivary Ig A levels in normal children and in children with various pathological conditions. *Fedn Proc. Fedn Am. Socs exp. Biol.* **25**, 490.

SOUTHAM, C. M., BRUNSCHWIG, A., LEVIN, A. G. and DIZON, Q. S. (1966). Effect of leucocytes on transplantability of human cancer. *Cancer, N.Y.* **19**, 1743.

SOUTHAM, C. M., MOORE, A. E. and RHOADS, C. P. (1957). Homotransplantation of human cell lines. *Science, N.Y.* **125**, 158.

SPECTOR, W. G. (1967). Histology of allergic inflammation. *Br. med. Bull.* **23**, 35.

SPECTOR, W. G. and LYKKE, A. W. J. (1966). The cellular evolution of inflammatory granulomata. *J. Path. Bact.* **92**, 163.

SPECTOR, W. G. and WILLOUGHBY, D. A. (1964). Endogenous mechanisms of injury in relation to inflammation, in *CIBA Foundation Symposium on Cellular Injury*, p. 74. Ed. A. V. S. de Reuck and J. Knight. London: Churchill.

SPIEGELBERG, H. L. and WEIGLE, W. O. (1965). The catabolism of homologous and heterologous 7S gamma globulin fragments. *J. exp. Med.* **121**, 323.

SPRINGER, G. F., HORTON, R. E. and FORBES, M. (1959). Origin of anti-human blood group B agglutinins in germfree chicks. *Ann. N.Y. Acad. Sci.* **78**, 272.

STAPLES, P. J., GERY, I. and WAKSMAN, B. H. (1966). Role of the thymus in tolerance. III. Tolerance to bovine γ-globulin after direct injection into the shielded thymus of irradiated rats. *J. exp. Med.* **124**, 127.

STEINBERG, A. G. and WILSON, J. A. (1963). Hereditary globulin factors and immune tolerance in man. *Science, N.Y.* **140**, 303.

STENSIÖ, E. (1958). Classe des cyclostomes. Les cyclostomes fossiles ou ostracodermes, in *Traité de zoologie*, vol. 13, part 1, pp. 173–425. Ed. P. P. Grassé. Paris: Masson.

STENT, G. S. (1963). *Molecular Biology of Bacterial Viruses*, p. 61. San Francisco: Freeman.

ŠTERZL, J. (ed.) (1965). *Molecular and Cellular Basis of Antibody Formation*. Prague: Czechoslovak Acad Sci.

ŠTERZL, J. (1966). Immunological tolerance as the result of terminal differentiation of immunologically competent cells. *Nature, Lond.* **209**, 416.

ŠTERZL, J., FRANĚK, F., ŘÍHA, I., KOSTKA, J. and LANC, A. (1961). Synthesis of gamma globulin in colostrum-free and germ-free newborns: significance of intestinal microorganisms for antibody formation and bactericidal activity of sera, in *Plasma Proteins and Gastrointestinal Tract in Health and Disease*, p. 199. Ed. M. Schwartz and P. Vesin. Copenhagen: Munksgaard.

STEVENS, K. M. (1964). The aetiology of systemic lupus erythematosus. *Lancet*, ii, 506.

STJERNSWÄRD, J. (1965). Immunodepressive effect of 3-methylcholanthrene: antibody formation at the cellular level and reaction against weak antigenic homografts. *J. natn. Cancer Inst.* **35**, 885.

STJERNSWÄRD, J. (1966). Effect of noncarcinogenic and carcinogenic hydrocarbons on antibody-forming cells measured at the cellular level *in vitro*. *J. natn. Cancer Inst.* **36**, 1189.

STOLLAR, D., LEVINE, L., LEHRER, H. I. and van VUNAKIS, H. (1962). The antigenic determinants of denatured DNA reactive with lupus erythematosus serum. *Proc. natn. Acad. Sci. U.S.A.* **48**, 874.

STOLLERMAN, G. H. (1964). The epidemiology of primary and secondary rheumatic fever, in *The Streptococcus, Rheumatic Fever and Glomerulonephritis*. Ed. J. W. Uhr. Baltimore: Williams and Wilkins.

STONER, R. D. and BOND, V. P. (1963). Antibody formation by transplanted bone marrow, spleen, lymph nodes and thymus cells in irradiated recipients. *J. Immun.* **91**, 185.

STRAUSS, A. J. L., van der GELD, H. W. R., KEMP, P. G., EXUM, E. D. and GOODMAN, H. C. (1965). Immunological concomitants of myasthenia gravis. *Ann. N.Y. Acad. Sci.* **124**, 744.

STUTMAN, O. and ZINGALE, S. B. (1964). Immunological reactivity of thymic autograft in rat. *Proc. Soc. exp. Biol. Med.* **117**, 389.

SULZBERGER, M. B. (1929). Hypersensitiveness to arsphenamine in guinea-pigs. I. Experiments in prevention and in desensitization. *Archs Derm. Syph.* **20**, 669.

SÜSSDORF, D. H. (1960). Repopulation of the spleen of X-irradiated rabbits by tritium-labeled lymphoid cells of the shielded appendix. *J. infect. Dis.* **107**, 108.

SÜSSDORF, D. H. and DRAPER, L. R. (1956). The primary hemolysin response in rabbits following shielding from X rays or X irradiation of the spleen, appendix, liver or hind legs. *J. infect. Dis.* **99**, 129.

SUTER, E. and RAMSEIER, H. (1964). Cellular reactions in infection. *Adv. Immun.* **4**, 117.

SUTHERLAND, D. E. R., ARCHER, O. K. and GOOD, R. A. (1964). Role of the appendix in development of immunologic capacity. *Proc. Soc. exp. Biol. Med.* **115**, 673.

SVEHAG, S. E. and MANDEL, B. (1964 a). The formation and properties of poliovirus-neutralizing antibody. I. 19S and 7S antibody formation: differences in kinetics and antigen dose requirement for induction. *J. exp. Med.* **119**, 1.

SVEHAG, S. E. and MANDEL, B. (1964 b). The formation and properties of poliovirus-neutralizing antibody. II. 19S and 7S antibody formation: differences in antigen dose requirement for sustained synthesis, anamnesis, and sensitivity to X-irradiation. *J. exp. Med.* **119**, 21.

SZENBERG, A. and WARNER, N. L. (1961). Large lymphocytes and the Simonsen phenomenon. *Nature, Lond.* **191**, 920.

SZENBERG, A. and WARNER, N. L. (1962). Dissociation of immunological responsiveness in fowls with a hormonally arrested development of lymphoid tissue. *Nature, Lond.* **194**, 146.

SZENBERG, A. and WARNER, N. L. (1964). Immunological reactions of bursaless fowls to homograft antigens. *Ann. N.Y. Acad. Sci.* **120**, 150.

SZENBERG, A., WARNER, N. L., BURNET, F. M. and LIND, P. E. (1962). Quantitative aspects of the Simonsen phenomenon. II. Circumstances influencing the focal counts obtained on the chorioallantoic membrane. *Br. J. exp. Path.* **43**, 129.

TALIAFERRO, W. H. and JAROSLOW, B. N. (1960). The restoration of hemolysin formation in X-rayed rabbits by nucleic acid derivatives and antagonists of nucleic acid synthesis. *J. infect. Dis.* **107**, 341.

TALMAGE, D. W. (1957). Allergy and immunology. *A. Rev. Med.* **8**, 239.

TALMAGE, D. W. (1959 a). Mechanism of the antibody response, in *A Symposium on Molecular Biology*, p. 91. Ed. R. E. Zirkle. University of Chicago Press.

TALMAGE, D. W. (1959 b). Immunological specificity: an alternative to the classical concept. *Science, N.Y.* **129**, 1463.

TAN, E. M. and KAPLAN, M. H. (1962). Renal tubular lesions in mice produced by streptococci in intraperitoneal diffusion chambers: role of streptolysin S. *J. infect. Dis.* **110**, 55.

TAN, E. M. and KUNKEL, H. G. (1966). Characteristics of a soluble nuclear antigen precipitating with sera of patients with systemic lupus erythematosus. *J. Immun.* **96**, 464.

TANIGAKI, N., YAGI, Y., MOORE, C. E. and PRESSMAN, D. (1966). Immunoglobulin production in human leukemia cell lines. *J. Immun.* **97**, 634.

TASHJIAN, A. H., LEVINE, L. and MUNSON, P. L. (1964). Immunochemical identification of parathyroid hormone in non-parathyroid neoplasms associated with hypercalcemia. *J. exp. Med.* **119**, 467.

TAYLOR, R. B. (1963). Immunological competence of thymus cells after transfer to thymectomized recipients. *Nature, Lond.* **199**, 873.

TAYLOR, R. B. (1964). An effect of thymectomy on recovery from immunological paralysis. *Immunology*, **7**, 595.

TERRY, W. D. (1966). Skin-sensitizing activity related to γ-polypeptide chain characteristics of human Ig G. *J. Immun.* **95**, 1041.

TERRY, W. D. and FAHEY, J. L. (1964). Subclasses of human $\gamma_2$-globulin based on differences in the heavy polypeptide chains. *Science, N.Y.* **146**, 400.

THEILER, M. and CASALS, J. (1958). The serological reactions in yellow fever. *Am. J. trop. Med. Hyg.* **7**, 585.

THOMSEN, O. and KETTEL, K. (1929). Die Stärke der menschlichen Isoagglutinine und entsprechenden Blutkörperchenrezeptoren in verschiedenen Lebensaltern. *Z. ImmunForsch. exp. Ther.* **63**, 67.

THORBECKE, G. J. (1960). Gamma globulin and antibody formation *in vitro*. I. Gamma-globulin formation in tissues from immature and normal adult rabbits. *J. exp. Med.* **112**, 279.

THORBECKE, G. J., ASOFSKY, R. M., HOCHWALD, G. M. and SISKIND, G. W. (1962). Gamma globulin and antibody formation *in vitro*. III. Induction of secondary response at different intervals after the primary: the role of secondary nodules in the preparation for the secondary response. *J. exp. Med.* **116**, 295.

THORBECKE, G. J. and COHEN, M. W. (1964). Immunological competence and responsiveness of the thymus, in *The Thymus*; Wistar Inst. Symp. Monogr., no. 2, p. 33. Ed. V. Defendi and D. Metcalf. Philadelphia.

THORBECKE, G. J., COHEN, M. W., JACOBSON, E. B. and WAKEFIELD, J. D. (1967). The production of memory cells by the white pulp of the spleen in rabbits, in *Germinal Centers in Immune Responses*, p. 259. Ed. H. Cottier, N. Odartchenko, R. Schindler and C. C. Congdon. Berlin: Springer.

THORN, G. W., FORSHAM, P. H., PRUNTY, F. T. and HILLS, A. G. (1948). A test for adrenal cortical insufficiency. *J. Am. med. Ass.* **137**, 1005.

TILL, J. E. and MCCULLOCH, E. A. (1961). A direct measurement of the radiation sensitivity of normal mouse bone marrow cells. *Radiat. Res.* **14**, 213.

TITANI, K., WHITELY, E., AVOGARDO, L. and PUTNAM, F. W. (1965). Immunoglobulin structure: partial amino acid sequence of a Bence Jones protein. *Science, N.Y.* **149**, 1090.

TITANI, K., WIKLER, M. and PUTNAM, F. W. (1967). Evolution of immunoglobulins: structural homology of Kappa and Lambda Bence Jones proteins. *Science, N.Y.* **155**, 828.

TOBLER, R. and COTTIER, H. (1958). Familiäre Lymphopenie mit Agammaglobulinämie und schwerer Moniliasis: die 'essentielle Lymphocytophthise' als besondere Form der Frühkindlichen Agammaglobulinämie. *Helv. paediat. Acta*, **13**, 313.

TOMASI, T. B., TAN, E. M., SOLOMON, A. and PRENDERGAST, R. A. (1965). Characteristics of an immune system common to certain external secretions. *J. exp. Med.* **121**, 101.

TRENTIN, J. J. and FAHLBERG, W. J. (1963). An experimental model for studies of immunologic competence in irradiated mice repopulated with 'clones' of spleen cells, in *Conceptual Advances in Immunology and Oncology*, p. 66. New York: Hoeber.

TRENTIN, J. J., WOLF, N., CHENG, V., FAHLBERG, W., WEISS, D. and BONHAG, R. (1967). Antibody production by mice repopulated with limited numbers of clones of lymphoid cell precursors. *J. Immun.* **98**, 1326.

TRIPLETT, E. L., CUSHING, J. E. and DURALL, G. L. (1958). Observations on some immune reactions of the sipunculid worm *Dendrostomum zostericolum*. *Am. Nat.* **92**, 287.

TROWELL, O. A. (1958). The lymphocyte. *Int. Rev. Cytol.* **7**, 236.

TURK, J. L. (1962). The passive transfer of delayed hypersensitivity in guinea pigs by the transfusion of isotopically-labelled lymphoid cells. *Immunology*, **5**, 478.

TURK, J. L. and OORT, J. (1967). Germinal center activity in relation to delayed hypersensitivity, in *Germinal Centers in Immune Responses*, p. 311. Ed. H. Cottier, N. Odartchenko, R. Schindler and C. C. Congdon. Berlin: Springer.

TURK, J. L. and WILLOUGHBY, D. A. (1967). Central and peripheral effects of anti-lymphocytic serum. *Lancet*, i, 249.

TYAN, M. L., COLE, L. J. and NOWELL, P. C. (1966). Fetal liver and thymus roles in the ontogenesis of the mouse immune system. *Transplantation*, **4**, 79.

UHR, J. W. (1966). Delayed hypersensitivity. *Physiol. Rev.* **46**, 359.

UHR, J. W. and BAUMANN, J. B. (1961). Antibody formation. I. The suppression of antibody formation by passively administered antibody. *J. exp. Med.* **113**, 935.

UHR, J. W., DANCIS, J., FRANKLIN, E. C., FINKELSTEIN, M. and LEWIS, E. W. (1962*a*). The antibody response to bacteriophage φX 174 in newborn premature infants. *J. clin. Invest.* **41**, 150.

UHR, J. W. and FINKELSTEIN, M. S. (1963). Antibody formation. IV. Formation of rapidly and slowly sedimenting antibodies and immunological memory to bacteriophage φX 174. *J. exp. Med.* **117**, 457.

UHR, J. W., FINKELSTEIN, M. S. and BAUMANN, J. B. (1962*b*). Antibody formation. III. The primary and secondary antibody response to bacteriophage φX 174 in guinea pigs. *J. exp. Med.* **115**, 655.

UHR, J. W. and PAPPENHEIMER, A. M., Jr. (1958). Delayed hypersensitivity. III. Specific desensitization of guinea pigs sensitized to protein antigens. *J. exp. Med.* **108**, 891.

UHR, J. W., SALVIN, S. B. and PAPPENHEIMER, A. M., Jr. (1957). Delayed hypersensitivity. II. Induction of delayed hypersensitivity in guinea pigs by means of antigen–antibody complexes. *J. exp. Med.* **105**, 11.

UHR, J. W. and WEISSMANN, G. (1965). Intracellular distribution and degradation of bacteriophage in mammalian tissues. *J. Immun.* **94**, 544.

VALETTE-ROBIN, Y. (1964). Un essai de contrôle expérimental des théories sélectives de la synthèse des anticorps. *Archs int. Physiol. Biochim.* **72**, 700.

van der GELD, H. (1964). Anti-heart antibodies in the post-pericardiotomy and post-myocardial infarction syndromes. *Lancet*, ii, 617.

van der GELD, H., TEWFELTKAMP, G., van LOGHEM, J. J., OSTERHUIS, H. J. C. H. and BIEMOND, A. (1963). Multiple antibody production in myasthenia gravis. *Lancet*, ii, 373.

van FURTH, R. (1964). The formation of immunoglobulins by human tissues *in vitro*. Doctoral thesis, Leiden.

van FURTH, R., SCHUIT, H. R. E. and HIJMANS, W. (1966a). The formation of immunoglobulins by human tissues *in vitro*. III. Spleen, lymph nodes, bone marrow and thymus. *Immunology*, 11, 19.

van FURTH, R., SCHUIT, H. R. E. and HIJMANS, W. (1966b). The formation of immunoglobulins by human tissues *in vitro*. IV. Circulating lymphocytes in normal and pathological conditions. *Immunology*, 11, 29.

VASSALLI, P. and MCCLUSKEY, R. T. (1964). The pathogenic role of fibrin deposition in immunologically induced glomerulonephritis. *Ann. N.Y. Acad. Sci.* 116, 1052.

VAUGHAN, R. B. (1965a). The discriminative behaviour of rabbit phago-cytes. *Br. J. exp. Path.* 46, 71.

VAUGHAN, R. B. (1965b). The comparative *in vitro* phagocytic activity of rabbit polymorphonuclear leucocytes and macrophages. *Br. J. exp. Path.* 46, 82.

VOLKMAN, A. and GOWANS, J. L. (1965a). The production of macrophages in the rat. *Br. J. exp. Path.* 46, 50.

VOLKMAN, A. and GOWANS, J. L. (1965b). The origin of macrophages from bone marrow in the rat. *Br. J. exp. Path.* 46, 62.

von PIRQUET, C. (1908). Das Verhalten der kutanen Tuberkulinreaktion wahrend der Masern. *Dt. med. Wschr.* 34, 1297.

VOSS, E. W. and BAUER, D. C. (1966). Rabbit antibody protein associated with polyribosomes. *Fedn Proc. Fedn Am. Socs exp. Biol.* 25, 369.

WAGNER, R. R. (1961). Acquired resistance to bacterial infection in insects. *Bact. Rev.* 25, 100.

WAKEFIELD, J. D., COHEN, M. W., MCCLUSKEY, J. and THORBECKE, G. J. (1967). The fate of lymphoid cells from the white pulp at the peak of germinal center formation, in *Germinal Centers in Immune Responses*, p. 183. Ed. H. Cottier, N. Odartchenko, R. Schindler and C. C. Congdon. Berlin: Springer.

WAKEFIELD, J. D., THORBECKE, G. J., OLD, L. J. and BOYSE, E. A. (1966). Production of immunoglobulins by established cell lines from leukemia and Burkitt lymphoma. *Fedn Proc. Fedn Am. Socs exp. Biol.* 25, 660.

WAKSMAN, B. H., PEARSON, C. M. and SHARP, J. T. (1960). Studies of arthritis and other lesions induced in rats by injection of mycobacterial adjuvant. II. Evidence that the disease is a disseminated immunologic response to exogenous antigen. *J. Immun.* 85, 403.

WALDENSTRöM, J. (1960). Studies on conditions with disturbed gamma globulin formation (gammopathies). *Harvey Lect.* 56, 211.

WALDENSTRöM, J. (1962). Monoclonal and polyclonal gammopathies and the biological system of gamma globulin. *Prog. Allergy*, 6, 320.

WALTERS, M. N. I. and WILLOUGHBY, D. A. (1965). Indomethacin: a new anti-inflammatory drug: its potential use as a laboratory tool. *J. Path. Bact.* **90**, 641.

WALZER, R. A. and EINBINDER, J. (1963). Immunoleukopenia as an aspect of hypersensitivity to propylthiouracil. *J. Am. med. Ass.* **184**, 743.

WARNER, N. L. (1964). Quantitative aspects of the Simonsen phenomenon. V. Studies on the apparent loss of specificity of transplanted cells on passage. *Br. J. exp. Path.* **45**, 459.

WARNER, N. L. and BURNET, F. M. (1961). The influence of testosterone treatment on the development of the bursa of Fabricius in the chick embryo. *Aust. J. biol. Sci.* **14**, 580.  •

WARNER, N. L., HERZENBERG, L. A. and GOLDSTEIN, G. (1966). Immunoglobulin isoantigens (allotypes) in the mouse. II. Allotypic analysis of three $\gamma G_2$-myeloma proteins from (NZB × BALB/c) $F_1$ hybrids and of normal $\gamma G_2$-globulins. *J. exp. Med.* **123**, 707.

WARNER, N. L., SZENBERG, A. and BURNET, F. M. (1962). The immunological role of different lymphoid organs in the chicken. I. Dissociation of immunological responsiveness. *Aust. J. exp. Biol. med. Sci.* **40**, 373.

WATSON, L. (1966). Calcium metabolism and cancer. *Australas. Ann. Med.* **15**, 359.

WEIDANZ, W. P. and LANDY, M. (1963). A simplified method for bactericidal assay of natural antibodies against Gram-negative bacteria. *Proc. Soc. exp. Biol. Med.* **113**, 861.

WEIGLE, W. O. (1961). The immune response of rabbits tolerant of BSA to the injection of other serum albumens. *J. exp. Med.* **114**, 111.

WEIGLE, W. O. (1962). Termination of acquired immunological tolerance to protein antigens following immunization with altered protein antigens. *J. exp. Med.* **116**, 913.

WEIGLE, W. O. (1964). The immune response of BSA tolerant rabbits to injections of BSA following the termination of the tolerant state. *J. Immun.* **92**, 791.

WEIGLE, W. O. (1965). The induction of autoimmunity in rabbits following injection of heterologous or altered homologous thyroglobulin. *J. exp. Med.* **121**, 289.

WEILER, I. J. (1965). Antibody production by determined cells in diffusion chambers. *J. Immun.* **94**, 91.

WEILER, I. J. and WEILER, E. (1965). Association of immunologic determination with the lymphocyte fraction of the peritoneal fluid cell population. *J. Immun.* **95**, 288.

WEINER, W. and VOS, G. H. (1963). Serology of acquired hemolytic anemias. *Blood*, **22**, 606.

WEISBLUM, B., GONANO, F., von EHRENSTEIN, G. and BENZER, S. (1965). A demonstration of coding degeneracy for leucine in the synthesis of protein. *Proc. nat. Acad. Sci. U.S.A.* **53**, 328.

WEISER, R. S. and LAXSON, C. (1962). The fate of fluorescein-labeled soluble antigen-antibody complex in the mouse. *J. infect. Dis.* **111**, 55.

WEISSMAN, I. L. (1967). Thymus cell migration. *J. exp. Med.* **126**, 291.

WESSLÉN, T. (1952). Studies on the role of lymphocytes in antibody production. *Acta derm.-vener., Stockh.* **32**, 265.

WEST, C. D., HONG, R. and HOLLAND, N. H. (1962). Immunoglobulin levels from the newborn period to adulthood and in immunoglobulin deficiency states. *J. clin. Invest.* **41**, 2054.

WHANG, J., FREI, E., TJIO, J. H., CARBONE, P. P. and BRECHER, G. (1963). The distribution of the Philadelphia chromosome in patients with chronic myelogenous leukemia. *Blood,* **22**, 664.

WHITE, R. G. (1958). Antibody production by single cells. *Nature, Lond.* **182**, 1383.

WHITE, R. G. (1963). Functional recognition of immunologically competent cells by means of the fluorescent antibody technique, in *The Immunologically Competent Cell,* CIBA Foundation Study Group no. 16, p. 6. Ed. G. E. W. Wolstenholme and J. Knight. London: Churchill.

WHITE, R. G. and MARSHALL, A. H. E. (1962). The autoimmune response in myasthenia gravis. *Lancet,* ii, 120.

WIENER, A. S., UNGER, L. J., COHEN, L. and FELDMAN, J. (1956). Type-specific cold auto-antibodies as a cause of acquired hemolytic anemia and hemolytic transfusion reactions: biologic test with bovine red cells. *Ann. intern. Med.* **44**, 221.

WIGZELL, H. (1966). Antibody synthesis at the cellular level: antibody-induced suppression of 7S antibody synthesis. *J. exp. Med.* **124**, 953.

WIGZELL, H., MÖLLER, G. and ANDERSSON, B. (1965). Studies at the cellular level of the 19S immune response. *Acta path. microbiol. scand.* **66**, 530.

WIKLER, M., TITANI, K., SHINODA, T. and PUTNAM, F. W. (1967). The complete amino acid sequence of a λ type Bence-Jones protein. *J. biol. Chem.* **242**, 1668.

WILKINSON, P. C. and ZEROMSKI, J. (1965). Immunofluorescent detection of antibodies against neurones in sensory carcinomatous neuropathy. *Brain,* **88**, 528.

WILLIAMS, C. A. and GRABAR, P. (1955). Immunoelectrophoretic studies on serum proteins. II. Immune sera; antibody distribution. *J. Immun.* **74**, 397.

WILLIAMS, G. M. (1966a). Antigen localization in lymphopenic states. I. Localization pattern following chronic thoracic duct drainage. *Immunology,* **11**, 467.

WILLIAMS, G. M. (1966b). Antigen localization in lymphopenic states. II. Further studies on whole body X-irradiation. *Immunology,* **11**, 475.

WILLIAMS, R. C. (1959). Dermatomyositis and malignancy: a review of the literature. *Ann. intern. Med.* **50**, 1174.

WILLOUGHBY, D. A. (1966). The lymph node permeability factor: a possible mediator of delayed hypersensitivity reactions. *XIth Congr. Internat. Soc. Blood Transf., Sydney, Abstracts,* p. 43.

WILLOUGHBY, D. A., SPECTOR, W. G. and BOUGHTON, B. (1964). A lymph-node permeability factor in the tuberculin reaction. *J. Path. Bact.* **87**, 353.

## References

WILLOUGHBY, D. A., WALTERS, M. N. I. and SPECTOR, W. G. (1965). Lymph node permeability factor in the dinitrochlorobenzene skin hypersensitivity reaction in guinea-pigs. *Immunology*, **8**, 578.

WILMERS, M. J. and RUSSELL, P. A. (1963). Autoimmune haemolytic anaemia in an infant treated by thymectomy. *Lancet*, ii, 915.

WILSON, D. B. (1965). Quantitative studies on the behavior of sensitized lymphocytes *in vitro*. I. Relationship of the degree of destruction of homologous target cells to the number of lymphocytes and to the time of contact in culture and consideration of the effects of isoimmune serum. *J. exp. Med.* **122**, 143.

WILSON, R., SJODIN, K. and BEALMEAR, M. (1964). The absence of wasting in thymectomized germfree (axenic) mice. *Proc. Soc. exp. Biol. Med.* **117**, 237.

WINN, H. J. (1961). Immune mechanisms in homo-transplantation. II. Quantitative assay of the immunologic activity of lymphoid cells stimulated by tumor homografts. *J. Immun.* **86**, 228.

WISSLER, R. W., ROBSON, M. J., FITCH, F., NELSON, W. and JACOBSON, L. O. (1953). The effects of spleen shielding and subsequent splenectomy upon antibody formation in rats receiving total-body X-irradiation. *J. Immun.* **70**, 379.

WITTEN, T. A., WANG, W. L. and KILLIAN, M. (1963). Reaction of lymphocytes with purified protein derivative conjugated with fluorescein. *Science, N.Y.* **142**, 596.

WOLFE, H. R., MUELLER, A., NEESS, J. and TEMPELIS, C. (1957). Precipitin production in chickens. XVI. The relationship of age to antibody production. *J. Immun.* **79**, 142.

WOLSTENHOLME, G. E. W. and KNIGHT, J. (eds) (1963). *The Immunologically Competent Cell*. Ciba Foundation Study Group no. 16. London: Churchill.

WOLSTENHOLME, G. E. W. and KNIGHT, J. (eds) (1965). *CIBA Foundation Symposium on Complement*. London: Churchill.

WOODS, K. R., ENGLE, R. L., PERT, J. H. and PAULSEN, E. C. (1958). Starch-gel electrophoresis of some invertebrate sera. *Science, N.Y.* **127**, 519.

WORLLEDGE, S. M., CARSTAIRS, K. C. and DACIE, J. V. (1966). Autoimmune haemolytic anaemia associated with α-methyldopa therapy. *Lancet*, ii, 135.

YOFFEY, J. M. and COURTICE, F. C. (1956). *Lymphatics, Lymph and Lymphoid Tissue*. London: Arnold.

YOFFEY, J. M., HANKS, G. A. and KELLY, L. (1958). Some problems of lymphocyte production. *Ann. N.Y. Acad. Sci.* **73**, 47.

YOUNG, I. and FRIEDMAN, H. (1967). The morphologic demonstration of antibody formation in follicles of lymphoid tissue, in *Germinal Centers in Immune Responses*, p. 102. Ed. H. Cottier, N. Odartchenko, R. Schindler and C. C. Congdon. Berlin: Springer.

YUNIS, E. J., HILGARD, H. R., MARTINEZ, C. and GOOD, R. A. (1965). Studies on immunologic reconstitution of thymectomized mice. *J. exp. Med.* **121**, 607.

YUNIS, E. J., MARTINEZ, C. and GOOD, R. A. (1964). Failure to reconstitute neonatally thymectomized mice by 'successful' rat thymus transplantation. *Nature, Lond.* **204**, 664.

ZETTERVALL, O., SJÖQUIST, J., WALDENSTRÖM, J. and WINBLAD, S. (1966). Serological activity of myeloma type globulins. *Clin. exp. Immun.* **1**, 213.

ZUCKER-FRANKLIN, D. (1963). The ultrastructure of cells in human thoracic duct lymph. *J. Ultrastruct. Res.* **9**, 325.

ZUCKERKANDL, E. (1965). The evolution of hemoglobin. *Scient. Am.* **212**, no. 5, 110.

ZUCKERKANDL, E. and PAULING, L. (1965). Evolutionary divergence and convergence in proteins, in *Evolving Genes and Proteins*, p. 97. Ed. V. Bryson and H. J. Vogel. New York and London: Academic Press.

# Index

Adrenal, autoimmune disease, 278–80, 594

Agammaglobulinaemia
absence of germinal centres, 403
acquired, as autoimmune, 410, 506
congenital, 20, 92, 506, 517; DH present, 581
measles normal in, 28
relation to GALT, 410

Allogeneic inhibition, 221, 297–8, 642–3
as tissue surveillance system, 643

Anfinsen, C. B., 325

Antibody
amino acid analyses, 418, 437–41
bacteriophage, 79
cytophilic, 251, 385, 490, 509
definition, 31
feedback control, 53, 121, 477, 523
flagellar, 79, 533
foetal, 196, 222; plasma cells in foetus, 222
heterogeneity, 7, 35, 36, 102, 126, 330, 424, 436
immunoglobulin is antibody, 203, 508
incomplete, 510
monoclonal or oligoclonal, 437, 438–41
natural, 511, 516
sheep red cell, 165; see also Myeloma proteins

Antibody plaques (Jerne), 19, 43, 79–80, 165–6, 383–4
antigen-coated RBCs, 166
plaque-forming cells, 166, 215; entry into thymus, 351; low in germ-free animals, 511; in unimmunized animals, 165, 484

Antibody production, 189–212
to bacterial polysaccharides, 197, 546
in blood, 388
changes, with age, 396

double producers, 199, 323, 331, 457–8, 533–6
foetal production, 196, 222
genetics of capacity, 143
Ig M–Ig G change, 196–8, 216
inhibition by antibody, 120–1, 198, 207, 214, 550
in organ culture, 485
primary response, 194–9, 513, 524, 527
secondary response, 191–3, 526–7
by single cells, 78–80

Antigen–antibody complex, 492–4
adsorption of complement, 161
cell-surface damage by, 492
effect on eosinophils, 478–9
fibrin deposition, 494
immune-adherence, 161, 493
kidney lesions (in SLE) 264, (in experimental animals) 493
passive cutaneous anaphylaxis (PCA), 493
in rheumatic fever, 281

Antigenic determinant, 117–19

Antigens, 32
dextran, 34, 118
'good' antigens, 189, 211, 524–6, 533–6
haptens, 118, 184, 208
localization *in vivo*, 148, 474–7
presentation as stimulus, 205
staphylococcal toxoid, 526
synthetic polypeptides, 34, 118, 208, 442–3, 496–7; with D-amino acids, 554

Arbovirus infection, 233–6
cross-reactive antibodies, 331, 462
multiplication in macrophages, 473
neutralizing antibody, 233–4
pathogenesis, 233
persistence of antibody, 234
yellow fever immunity, 28, 235

Aschoff, L., 15, 148, 467

Autoimmune disease, 255–85
acquired agammaglobulinaemia, 410

Lymphocytes (*cont.*)
mobility, 225, 341, 343, 398, 513, 515
motility, 87, 90, 340–1
origin, 48, 83, 354
peritoneal cavity, 389, 473
relation to DPC, 156
source of nucleotides, etc., 225
thoracic duct, 46, 369
thymus, 46, 49, 83, 354
transfer reaction, 635
uropod, 341
vulnerability, 95, 225, 367
Lymphocytes, blast transformation
by allotypic antiserum, 167–8, 343
by antigen, 167, 343
by PHA, 28, 47, 76, 81, 167–9, 342, 381

Macrophages, 18, 148–55, 467–77
action in preparing antigen, 464, 467–70
bactericidal power, 471
evolution, 147
growth of virus in, 473
Kupffer's cells, 151, 152
metabolic activation, 472
mobility influenced by adjacent re-action, 472, 504
monocytes, 149
origin, 152–3, 344
peritoneal, 150, 152, 389, 464, 472, 503; origin from bone marrow, 153, 344, 361
recognition of foreignness, 470
removal of excess antigen, 470, 548
uptake of pneumococcal antigen, 547
Mast cells, 147, 159–60
in activated lymph nodes, 476
anaphylaxis, 160, 491, 501
from donor bone-marrow, 361
heteroplastic origin, 159, 501
histamine liberation, 159, 176, 491
Maternal-foetal relationship, 240–4
Burkitt lymphoma, 570
chorioncarcinoma, 569
foetus as homograft, 567
graft-versus-host reaction, 241, 570
placental passage of Ig G, 154, 241
Rh disease, 241–3, 568
selective effect, 244

trophoblast function, 567–8
Y antigen, 569
Measles and rubella
foetal antibody in rubella, 517
persistence of immunity, 511
Medawar, P. B., 220
Metchnikoff, E., 3, 15, 300
Miller, J. F. A. P., 46, 355, 503
Monoclonal immunoglobulins
haemolytic anaemia, 259, 609
myeloma proteins, 101–2, 528
rabbit antibody, 126, 184, 509
Waldenström's gammopathy, 20, 145, 528; association with cancer, 651
Myasthenia gravis, 271–4
antibodies, 272, 621
germinal centres in thymus, 622
myo-epithelial cells of thymus, 272, 621
thymectomy in, 271
thymic tumours, 273, 622
Myeloma proteins, 528–30
of antibody character, 103, 130, 439–40; specific for dinitro-phenyl hapten, 440, 529
Bence-Jones, protein, 101
comparison with antibody, 439
failure of cell maturation, 529
heterogeneity as class, 528
in man, 20, 32, 99, 101, 438–41
in mice, 45, 439

Nossal, G. J. V., 89

Opsonization, 29, 154
allowing uptake of flagellin, 477
for absorption to macrophages, 470, 472
needed for immunogenicity, 53, 154

Pasteur, L., 3
Pauling, L., 5, 6, 34, 128, 320, 324, 445
Phenotypic restriction, 14, 136
in immunocytes, 323, 447, 531–6
in immunoglobulins, 114, 459, 509
Lyon phenomenon, 136, 460
one cell, one antibody rule, 5; in guinea pigs, 535; in rats, 533
in purified antibodies, 530–1
somatic segregation, 461